Lecture Notes in Artificial Intelligence 11323

Subseries of Lecture Notes in Computer Science

LNAI Series Editors

Randy Goebel
University of Alberta, Edmonton, Canada
Yuzuru Tanaka
Hokkaido University, Sapporo, Japan
Wolfgang Wahlster
DFKI and Saarland University, Saarbrücken, Germany

LNAI Founding Series Editor

Joerg Siekmann
DFKI and Saarland University, Saarbrücken, Germany

More information about this series at http://www.springer.com/series/1244

Guojun Gan · Bohan Li
Xue Li · Shuliang Wang (Eds.)

Advanced Data Mining and Applications

14th International Conference, ADMA 2018
Nanjing, China, November 16–18, 2018
Proceedings

 Springer

Editors
Guojun Gan
University of Connecticut
Storrs, CT, USA

Xue Li
The University of Queensland
Brisbane, QLD, Australia

Bohan Li
Nanjing University of Aeronautics
and Astronautics
Nanjing, China

Shuliang Wang
Beijing Institute of Technology
Beijing, China

ISSN 0302-9743   ISSN 1611-3349 (electronic)
Lecture Notes in Artificial Intelligence
ISBN 978-3-030-05089-4   ISBN 978-3-030-05090-0 (eBook)
https://doi.org/10.1007/978-3-030-05090-0

Library of Congress Control Number: 2018962542

LNCS Sublibrary: SL7 – Artificial Intelligence

This Springer imprint is published by the registered company Springer Nature Switzerland AG
The registered company address is: Gewerbestrasse 11, 6330 Cham, Switzerland

Preface

The 14th International Conference on Advanced Data Mining and Applications (ADMA) was held in Nanjing, one of the most ancient cities in China. Over the years, ADMA has grown to become a flagship conference in the field of data mining and applications. One of the major goals of ADMA is to bring together data mining researchers from around the world to share their original data mining findings and practical data mining experiences. In the ear of big data and artificial intelligence (AI), data mining is becoming an important option for developing data-driven applications.

For ADMA 2018, we received 104 papers from 30 different countries including American, European, Middle-East, and Pacific-Asian countries. Each paper was assigned to at least three Program Committee members to review. All papers were rigorously reviewed and had at least three reviews. At the end, 23 papers were accepted as spotlight research papers with long presentation and 22 were accepted as regular research papers with short presentation. The conference program of ADMA 2018 was also complemented by several outstanding keynotes and tutorials given by world-renowned experts Xuemin Lin, Ekram Hossain, Guoren Wang, Yang Yu, Jie Tang as well as an invited industry keynote talk session, delivered by several invited industry speakers. We would like to particularly thank those speakers for contributing their insights and visions of the future of data mining technology in this dynamic research field, where many puzzling terms are emerging, such as blockchain, strong AI, and common-sense mining on networks of multi-modality data.

We greatly appreciate the Program Committee members' tremendous efforts to complete the review reports before the deadline. We would like to thank the external reviewers for their time and comprehensive reviews and recommendations. Their professional work was crucial to the final paper selection and production of the high-quality technical program for ADMA 2018.

This high-quality program would not have been possible without the expertise and dedication of our Program Committee members. We would like to express our gratitude to all individuals, institutions, and sponsors that supported ADMA 2018. We are grateful to all the chairs who are actively involved in the organization of this conference including attracting submissions, compiling all accepted papers, and working with the Springer team to produce the proceedings, managing the website. Our special thanks to the publicity chair, Guojun Gan, for editing, the local organization chair, Bohan Li, for the local arrangements ensuring the conference ran smoothly, and the registration chair, Donghai Guan, for handling the registration process. We would like to express our sincere thanks to Weitong (Tony) Chen for helping with paper submission process and the smooth running of the conference program based on his rich experiences. We would also like to thank Michael Sheng, Aixin Sun, Gao Cong, and Wei Luo for their contribution to the conference. Furthermore, we would like to acknowledge the support of the members of the conference Steering Committee.

Finally, we would like to thank all researchers, practitioners, and volunteer students who contributed with their work and participated in the conference.

With the new challenges in data mining research, we hope the participants in the conference and the readers of the proceedings will enjoy the research outcome of ADMA 2018.

October 2018
                                                                Xue Li
                                                          Joao Gama
                                                          Bing Chen
                                                       Songcan Chen
                                                      Shuliang Wang
                                                  Xingquan (Hill) Zhu

Organization

Main Organizing Committee

Honorary Chair

Zhiqiu Huang   Nanjing University of Aeronautics and Astronautics, China

General Chairs

Xue Li   University of Queensland, Australia
Joao Gama   University of Porto, Portugal

Acting Chair

Bing Chen   Nanjing University of Aeronautics and Astronautics, China

Program Chairs

Songcan Chen   Nanjing University of Aeronautics and Astronautics, China
Shuliang Wang   Beijing Institute of Technology, China
Xingquan (Hill) Zhu   Florida Atlantic University, USA

Demo Chairs

Zhifeng Bao   RMIT, Australia
Jianqiu Xu   Nanjing University of Aeronautics and Astronautics, China

Proceedings Chair

Yunlong Zhao   Nanjing University of Aeronautics and Astronautics, China

Awards Committee Chair

Aixin Sun   Nanyang Technological University, Singapore

Publicity Chair

Guojun Gan   University of Connecticut, USA

Data Mining Competition Chair/Pacific Chair

Luo Wei   Deakin University, Australia

Special Issue Chair

Daoqiang Zhang   Nanjing University of Aeronautics and Astronautics, China

Sponsorship Chairs

Donghai Guan	Nanjing University of Aeronautics and Astronautics, China
Xiangping Zhai	Nanjing University of Aeronautics and Astronautics, China

Local Chair

Bohan Li	Nanjing University of Aeronautics and Astronautics, China

Web Chair

Xin Li	Nanjing University of Aeronautics and Astronautics, China

Program Committee

Bin Guo	Northwestern Polytechnical University, USA
Bin Yao	Shanghai Jiao Tong University, China
Bin Zhao	Nanjing Normal University, China
Bin Zhou	National University of Defense Technology, China
Chandra Prasetyo Utomo	University of Queensland, Australia
Changdong Wang	Sun Yat-Sen University, China
Chuan Shi	Beijing University of Posts and Telecommunications, China
Dechang Pi	Nanjing University of Aeronautics and Astronautics, China
Guandong Xu	University of Technology Sydney, Australia
Hongxu Chen	University of Queensland, Australia
Hongzhi Wang	Harbin Institute of Technology, China
Hongzhi Yin	University of Queensland, Australia
Jianxin Li	University of Western Australia
Jingfeng Guo	Yanshan University, China
Lina Yao	University of New South Wales, Australia
Luyao Liu	University of Queensland, Australia
Meng Wang	Xi'an Jiaotong University, China
Michael Sheng	Macquarie University, Australia
Min Yao	Zhejiang University, China
Moscato Pablo	University of Newcastle, Australia
Nguyen Hung	Griffith University, Australia
Peiquan Jin	University of Science and Technology of China
Qilong Han	Harbin Engineering University, China
Rui Mao	Shenzhen University, China
Sen Wang	Griffith University, Australia
Shuai Ma	BeiHang University, China
Tong Chen	University of Queensland, Australia
Unankard Sayan	Maejo University, Thailand
Wei Zhang	Macquarie University, Australia
Weitong Chen	University of Queensland, Australia
Wenjie Ruan	University of Oxford, UK
Xiaolin Qin	Nanjing University of Aeronautics and Astronautics, China

Xiaoyang Tan	Nanjing University of Aeronautics and Astronautics, China
Xin Wang	Tianjin University, China
Xin Zhao	University of Queensland, Australia
Xiu Fang	Macquarie University, Australia
Yan Jia	National University of Defense Technology, China
Yanhui Gu	Nanjing Normal University, China
Yongxin Tong	BeiHang University, China
Yue Lin	Northeast Normal University, China
Yunjun Gao	Zhejiang University, China
Yuwei Peng	Wuhan University, China
Zongmin Ma	Nanjing University of Aeronautics and Astronautics, China

Steering Committee

Jie Cao	Nanjing University of Finance and Economics, China
Xue Li (Chair)	University of Queensland, Australia
Shuliang Wang	Beijing Institute of Technology, China
Michael Sheng	University of Adelaide, Australia
Jie Tang	Tsinghua University, China
Kyu-Young Whang	Advanced Institute of Science and Technology, South Korea
Min Yao	Zhejiang University, China
Osmar Zaiane	University of Alberta, Canada
Chengqi Zhang	University of Technology Sydney, Australia
Shichao Zhang	Guangxi Normal University, China

Contents

Data Mining Foundations

Big Data

Text and Multimedia Mining

Miscellaneous Topics

Data Mining Foundations

Efficiently Mining Constrained Subsequence Patterns

Abdullah Albarrak[1]([✉]), Sanad Al-Maskari[2], Ibrahim A. Ibrahim[3,4],
and Abdulqader M. Almars[3]

[1] Al Imam Mohammad Ibn Saud Islamic University, Riyadh, Saudi Arabia
amsbarrak@imamu.edu.sa
[2] Sohar University, Sohar, Oman
smaskari@soharuni.edu.om
[3] University of Queensland, Brisbane, Australia
{i.ibrahim,a.almars}@uq.edu.au
[4] Minia University, Minya, Egypt
i.ibrahim@minia.edu.org

Abstract. Big time series data are generated daily by various application domains such as environment monitoring, internet of things, health care, industry and science. Mining this massive data is a very challenging task because conventional data mining algorithms are unable to scale effectively with massive time series data. Moreover, applying a global classification approach to a highly similar and noisy data will hinder the classification performance. Therefore, utilizing constrained subsequence patterns in data mining applications increases the efficiency, accuracy, and could provide useful insight into the data.

To address the above mentioned limitations, we propose an efficient subsequence processing technique with preferences constraints. Then, we introduce a sub-patterns analysis for time series data. The sub-pattern analysis objective is to maximize the interclass separability using a localization approach. Furthermore, we make use of the deviation from a correlation constraint as an objective to minimize in our problem, and we include users preferences as an objective to maximize in proportion to users' preferred time intervals. We experimentally validate the efficiency and effectiveness of our proposed algorithm using real data to demonstrate its superiority and efficiency when compared to recently proposed correlation-based subsequence search algorithms.

1 Introduction

Time series data nowadays is in continuous increase in terms of size and complexity. It is being generated, gathered and stored in unprecedented rate, whether for the purpose of financial analysis (e.g., exchange rates, stock market), environment monitoring [1–4], health care [13,16], social networks [17]. This increase easily overwhelms data mining users when applying data mining applications on these ever-growing time series data.

© Springer Nature Switzerland AG 2018
G. Gan et al. (Eds.): ADMA 2018, LNAI 11323, pp. 3–16, 2018.
https://doi.org/10.1007/978-3-030-05090-0_1

One fundamental, preprocessing step of mining time series data is extracting representative features from the raw time series data [5–7]. In this paper, we focus on extracting correlated subsequences patterns [8,10,11,15,18] as features, where correlation is the Pearson Correlation Coefficient (ρ). Our choice of Pearson correlation is based on the fact that it is the most suitable measure for meaningful comparisons of time series, as stated in the literature [6,9,14].

In reality, it is not realistic to assume perfect separation between different classes. In fact, in time series data, a high overlapping could exist; due to noise, highly similar objects, unknown factors, and limited understanding. Therefore, instead of using the full time series sequences, we propose to use the most interesting and discriminant subsequences, i.e., the correlated subsequence patterns. We refer to such subsequences as sub-patterns (SPs). By identifying the most discriminant SP, highly overlapped classes can be discriminated without the need to store, process and extract features for all data. Additionally, in cases of highly noisy and similar data set, using global features can impact the classification performance. Finally, using SPs in data mining applications is much faster than current classification approaches, making it a better candidate for mining big data.

Interestingly, identifying SPs is not trivial, it is actually CPU and I/O intensive. Hence, several works have focused on optimizing it using complex indexing methods and pruning techniques. For instance, SKIP [10] was proposed to find the longest subsequence of two time series having a correlation above a threshold value θ, without any prior knowledge of the subsequence length m. Jocor [11] focus on finding the subsequence with the highest correlation value such that it has a length above a threshold value ml. While the above works are successful in extracting subsequences satisfying correlation thresholds, they fall short in addressing correlation thresholds as *ranges* (i.e., correlation between θ_1 and θ_2) and *users preferences*.

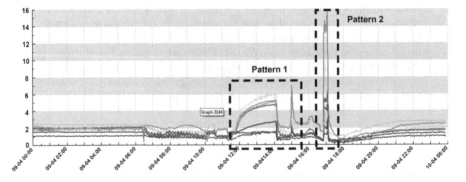

Fig. 1. A snippet of air quality sensor data. Large redundant, meaningless and irreverent data exist which does not contribute positively in the classification process. The two most significant sections are labeled Pattern 1 and Pattern 2.

Example 1. Figure 1 shows a snippet of a large air quality time series data from multiple sensors. Because most of these time series are redundant, they do not contribute positively in the classification process. However, it can be seen that pattern 1 and pattern 2 are the most significant sections since they provide highly distinguishable features.

Subsequence patterns 1 and 2 in Example 1 are identified by two constraints:

1. The subsequences must exhibit a correlation within a target range, and
2. The subsequences must be as close as possible to specific time interval.

Employing users' preferences to extract correlated subsequences promotes optimization opportunities with guarantees on the result's accuracy. Plainly, our proposed efficient algorithm strives to reduce CPU cost through harnessing users' preferences to limit the search space, and incrementally estimate subsequences correlation by using precomputed data in an efficient manner.

We summarize our contributions as follows:

– Incorporating users preferences and correlation ranges as objectives to extract constrained subsequences patterns from raw time series data.
– Proposing an efficient *CPU-centric algorithm* to find optimal constrained subsequences patterns by utilizing: (a) *users preferences* to limit, navigate and prune the search space, and (b) *multiple copies of cumulative arrays* which summarize time series values.
– Conducting experiments on real data to evaluate our algorithm and show the efficiency it provides when compared to recent algorithm for subsequence time series search.

2 Preliminaries and Definitions

We assume there are k data series $x_1, x_2, ..., x_k$ such that all series are of equal length. A data series x_i is a series of n consecutive values $x_i = \{v_1, v_2, ..., v_n\}$ that have an implicit ordering. For instance, x_i is a *time series* if its values ordering is based on a `timestamp` domain (e.g., date and/or time). While our work is based on time series, it can be generalized to data series too.

A subsequence of x_i is constructed by a *time interval* $[s, e]$, $x_i[s, e] = \{v_j, v_{j+1}, ..., v_e\}$ where $j = s$, $s < e$ and $e \leq n$, as shown in Fig. 2.

Our focus in this paper is on mining correlated subsequence patterns. The correlation is computed between a pair of synchronized subsequences $x_i[s, e]$ and $x_j[s, e]$ constructed from time series x_i and x_j, respectively. Henceforth, we refer to the pair $x_i[s, e]$ and $x_j[s, e]$ as a candidate subsequence $S_c(x_i, x_j, s, e)$ or briefly S_c when there is no need to specify which time series and what time interval.

As discussed in Example 1, the most interesting subsequence patterns must be within a certain correlation range. In other words, subsequences with the minimum deviation from a target correlation range are more preferable. We explain next how to support this objective in details (Table 1).

Table 1. Symbols and definitions

Symbol	Definition
x_i	A time series
n	Length of time series
m	Length of subsequence time series
$x_i[s, e]$	A subsequence of x_i
λ	Weight of preference
tc	Target correlation range $[c_l - c_u]$
S_I	Input(initial) subsequence
S_c	Candidate subsequence
$\Delta^\rho_{S_c}$	Correlation deviation of S_c
$\Delta^E_{S_c}$	Preference deviation of S_c
Δ_{S_c}	Total deviation of S_c

2.1 Correlation Deviation

The deviation in correlation Δ^ρ for a candidate subsequence S_c from a target correlation range $tc = [c_l - c_u]$ is a measure of how far the correlation value of S_c is to the target range. Formally:

$$\Delta^\rho_{S_c} = \begin{cases} 0 & \text{if } \rho(S_c) \in [c_l - c_u] \\ \rho(S_c) - c_l & \text{if } \rho(S_c) < c_l \\ \rho(S_c) - c_u & \text{if } \rho(S_c) > c_u \end{cases} \tag{1}$$

where $\rho(S_c)$ is a function that returns the *Pearson correlation coefficient* of the subsequences in S_c. The more $\Delta^\rho_{S_c}$ approaches zero, the more S_c is preferable.

Pearson correlation coefficient of two subsequences x and y (in S_c) of length m is computed as follows [10]:

$$\rho(S_c) = \rho(x, y) = \frac{\sum_{i=1}^{m} x_i y_i - \sum_{i=1}^{m} x_i \sum_{i=1}^{m} y_i}{\sqrt{\sum_{i=1}^{m} x_i^2 - (\sum_{i=1}^{m} x_i)^2}\sqrt{\sum_{i=1}^{m} y_i^2 - (\sum_{i=1}^{m} y_i)^2}} \tag{2}$$

But *how* to obtain any candidate subsequence in the first place? We explain the *how* next.

2.2 Enumerating Subsequences

Our algorithm enumerates candidate subsequences by recursively expanding and contracting the initial subsequence $S_I(x, y, s, e)$. Specifically, there are two operations applied on the time interval $[s, e]$, as shown in Fig. 2.

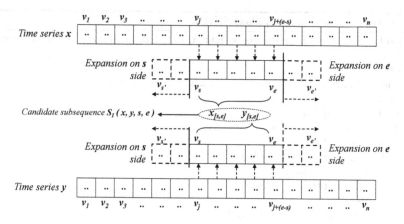

Fig. 2. Enumerating subsequences from $S_I(x, y, s, e)$ by expanding either sides s or e, i.e., $\mathcal{LE}, \mathcal{RE}$

1. **Expansion:** to expand $[s, e]$ from either sides s or e by δ.
   For instance, $[\hat{s}, e]$ is expanded from s side by δ such that $\hat{s} = s - \delta$ while $[s, \hat{e}]$ is expanded from e side by δ such that $\hat{e} = e + \delta$.
   We encode those two operations as $\mathcal{LE}$ (left expansion) and $\mathcal{RE}$ (right expansion).
2. **Contraction:** to contract $[s, e]$ from either sides s or e by δ.
   For instance, $[\hat{s}, e]$ is contracted from s side by δ such that $\hat{s} = s + \delta$ while $[s, \hat{e}]$ is contracted from e side by δ such that $\hat{e} = e - \delta$.
   Similarly, we encode those two operations as $\mathcal{LC}$ (left contraction) and $\mathcal{RC}$ (right contraction).

With those two operations, our algorithm is able to recursively generate all possible $\frac{n(n+1)}{2}$ combinations of candidate subsequences. To remove any approximation and to ensure no possible candidate subsequence is missed, we set $\delta = 1$.

Now, the question is, how to judge whether a subsequence S_c is more beneficial than another one? We answer this question by quantifying the benefit of a subsequence as a preference deviation, which is explained next.

2.3 Preference Deviation

Since users prefer similar subsequences to their first impression (i.e., S_I), we include users preferences as an objective. It is a normalized value that indicates how far S_c is from the input subsequence S_I. A subsequence that is far from a user's preference will exhibit a high deviation, and vise versa:

$$\Delta_{S_c}^E = \frac{|S_I.s - S_c.s|}{n} + \frac{|S_I.e - S_c.e|}{n} \tag{3}$$

where n is the length of the two time series. Note that the lower the value of $\Delta_{S_c}^E$, the more beneficial S_c is.

Algorithm 1. Baseline (simplified SKIP)

1: **Input:** $S_I(x,y,s,e)$, preference weight λ, target correlation tc, minimum length
ml, α
2: **Return:** S_p, best;
3: calculate S_x^α, $S_{x^2}^\alpha$, S_y^α, $S_{y^2}^\alpha$ and S_{xy}^α for whole time series x, y
4: best $= 1$; $S_p \leftarrow S_I$
5: **for** ($l = n$ to ml) **do**
6: **for** ($t = 0$ to $n - l$) **do**
7:   $S_i \leftarrow (x, y, t, (t + l - 1))$;
8:   **for** ($z \in \{x, y, x^2, y^2, xy\}$) **do**
9:    **if** ($t = 0$) **then**
10:     Compute $\sum z$ by S_z^α
11:    **else**
12:     Update $\sum z$ incrementally
13:   Compute $\rho(S_i)$;
14:   $\Delta_{S_i} = \lambda \Delta_{S_i}^E + (1 - \lambda)\Delta_{S_i}^\rho$;
15:   **if** ($\Delta_{S_i} <$ best) **then**
16:    best $= \Delta_{S_i}$; $S_p \leftarrow S_i$;
17: return S_p, best;

2.4   Problem Definition

Now, we are in position to formally define the problem of extracting constrained subsequences from raw time series data.

Definition 1. *Given $S_I(x, y, s, e)$, and a target correlation range tc. Find the optimal subsequence $S_p(x, y, \hat{s}, \hat{e})$ such that S_p minimizes the deviation in correlation $\Delta_{S_p}^\rho$ while considering users preferences by minimizing the preference deviation $\Delta_{S_p}^E$* ∎

As stated in Definition 1, our objective is to look for the subsequence S_p that minimizes the overall deviation defined as:

$$\Delta_{S_p} = \lambda \Delta_{S_p}^E + (1 - \lambda)\Delta_{S_p}^\rho \tag{4}$$

The parameter λ is used to control the trade off between satisfying the correlation and preference deviation. Setting $\lambda = 0$ means users preferences are neglected, which is a special case of our general objective.

3   Methodology

Mining for the optimal subsequence pattern is essentially a search problem. That is, to find S_p, an algorithm has to iterative over all possible combinations of subsequences, compute the objective Eq. (4), while keeping the one with the minimum deviation.

Iterating over all subsequences is obviously an intensive task and incurs high computational overhead. Hence, we tackle this problem from this angle, and propose a computational-centric (Sect. 3.3) algorithm with efficient optimization techniques.

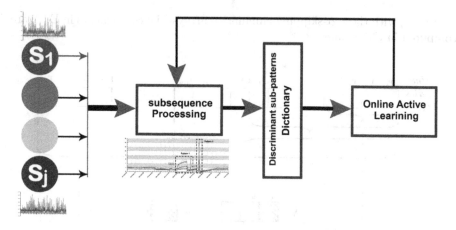

Fig. 3. Our proposed method for subsequence pattern mining.

3.1 Proposed Methods

Figure 3 illustrates our proposed method. Instead of using the full time series data for training which consist of irrelevant, noisy and weakly-labeled data, we focus on extracting most relevant, discriminant and meaningful subsets, which can be used to accurately represent the target class. For instance, Fig. 1 shows a small section of a training data for air quality sensors. From this data, we extract only the most interesting subsequences to be used in the training set (pattern 1 and pattern 2). The process of identifying most interesting, and relevant subsequence is automated and does not require expensive human manual labeling. Finally, all identified SPs will be added to the SP Dictionary. The SP Dictionary is a smaller subset of the full training data set which contain discriminant, relevant and representative sub-patterns of the original data.

3.2 Computations of Correlation

Computations cost of correlation in Eq. 2 increases linearly with n (i.e., $\mathcal{O}(n)$). Specifically, each of the summation components in Eq. 2 will perform n summation operations. With the observation that Eq. 2 can be computed incrementally [15], [10] proposed the α-*skipping cumulative array* to compute it in $\mathcal{O}(\alpha)$ time, where $\alpha \lll n$.

In the α-skipping cumulative array [10], each time series x of length n has two cumulative arrays; the sum of values and the sum of square values, S_x^α and $S_{x^2}^\alpha$, resp. Those arrays are of length $\frac{n}{\alpha}$. An element in those arrays is computed as follows (see Fig. 4):

$$S_w^\alpha[j] = \sum_{i=1}^{j*\alpha} w, \text{ if } (j * \alpha) \bmod \alpha = 0$$

where $j = (1, 2, ..., \frac{n}{\alpha})$, and $w \in \{x[i], (x[i])^2, x[i]y[i]\}$

Hence, with those α-skipping cumulative arrays, the components in Eq. 2 are computed in $\mathcal{O}(\alpha)$ time.

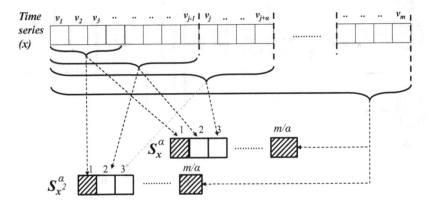

Fig. 4. Constructing α-cumulative sum arrays S_x^α and $S_{x^2}^\alpha$ for time series x of length m. $\alpha = 4$

Next, we describe two algorithms which utilize the α-skipping cumulative arrays to find S_p. Then, we show our algorithm which incorporates the preference objective to optimize the search for S_p without sacrificing the solution quality.

3.3   Computational-Centric Algorithms

In this section we discuss three algorithms to find the optimal solution S_p. We start by showing a brute force algorithm that we consider as a baseline to compare against the other algorithms.

Baseline: This is a simplified version of SKIP [10] which was proposed to find the longest correlated subsequence of multiple time series. We simplified and adapted the algorithm for two time series, i.e., $O = \{x, y\}$.

Essentially, Baseline utilizes two nested for-loops to generate the combinations of subsequences. The outer-loop specifies the right hand side of the time interval, while the inner-loop specifics the left hand side. In each iteration, the interval represents a candidate subsequence S_i. Then, Baseline computes correlation using the cumulative arrays that are generated for the whole time series in advance. This result in a complexity of $O(\alpha)$ only.

Then, the values computed from the previous step are updated incrementally to compute correlation for the next candidate subsequences, until the inner-loop finishes. Specifically, the next candidate subsequence is generated by shifting the interval to the right by one step, which means adding one value to the right hand side, and subtracting one value from the left.

Baseline++ (limited search space): Differently than the previous algorithm, Baseline++ limits the search space before navigating. Hence, it searches a small

---

Algorithm 2. Baseline++ (limited search space)

1: **Input:** $S_I(x, y, s, e)$, preference weight λ, target correlation tc, minimum length ml, α
2: **Return:** S_p, best;
3: $best = (1\text{-}\lambda)\Delta^\rho_{S_I}$; $S_p \leftarrow S_I$;
4: $i = S_I.s$; $j = S_I.e$;
5: **while** ($i \geq 0 \text{ ——— } j \leq n$) **do**
6:     $S_i \leftarrow (i,j)$;
7:     **if** ($\lambda\Delta^E_{S_i} > best$) **then**
8:        $maxs = i$; $maxe = j$; break;
9:     $i \leftarrow i - 1$; $j \leftarrow j + 1$;
10: calculate S^α_x, $S^\alpha_{x^2}$, S^α_y, $S^\alpha_{y^2}$ and S^α_{xy} for time interval $[maxs, maxe]$ of x, y
11: $n = maxe - maxs + 1$;
12: **for** ($l = n$ to ml) **do**
13:     **for** ($t = maxs$ to $n - l$) **do**
14:        $S_i \leftarrow (x, y, t, (t + l - 1))$;
15:        **for** ($z \in \{x, y, x^2, y^2, xy\}$) **do**
16:           **if** ($t = maxs$) **then**
17:              Compute $\sum z$ by S^α_z
18:           **else**
19:              Update $\sum z$ incrementally
20:        Compute $\rho(S_i)$;
21:        $\Delta_{S_i} = \lambda\Delta^E_{S_i} + (1 - \lambda)\Delta^\rho_{S_i}$;
22:        **if** ($\Delta_{S_i} < best$) **then**
23:           $best = \Delta_{S_i}$; $S_p \leftarrow S_i$;
24: **return** S_p, best;

---

part of the time series instead of the whole time series. To achieve that, it uses the preference objective to limit the search space (lines 5–10).

Specifically, it defines two variables $maxs$ and $maxe$ as new boundaries for the search space. Then, starting from S_I's boundaries (line 4), it enters a while-loop (line 5) until: (1) both sides of the time series have been reached, or (2) the preference deviation reached its maximum possible value (lines 7,8).

There are two prominent differences between Baseline and Baseline++. Firstly, the generation of the cumulative arrays. The cumulative arrays in Baseline++ are generated for a sub time interval of x and y, which are essentially the new found boundaries $[maxs, maxe]$ (line 10), not the whole interval as in Baseline.

Secondly, the new boundaries $[maxs, maxe]$ in Baseline++ will decrease the number of candidate subsequences (when $\lambda > 0$). As a result, the cost of search will decrease too. Though, Baseline++'s technique in optimizing the search cost is limited by the subsequence/time series length ratio.

Next, we introduce our algorithm *Incremental* (INC) which uses the preference deviation as an optimization technique to prune unpromising subsequences. INC further optimize the computations of correlation by a simple technique:

Algorithm 3. Incremental (INC)

1: **Input:** $S_I(x, y, s, e)$, preference weight λ
2: **Return:** S_p, best;
3: Define and compute $S_x^{\mathcal{A}}$, $S_{x^2}^{\mathcal{A}}$, $S_y^{\mathcal{A}}$, $S_{y^2}^{\mathcal{A}}$, $S_{xy}^{\mathcal{A}}$ for $[s, e]$
4: where $\mathcal{A} \in \{\mathcal{RE}, \mathcal{RC}, \mathcal{LE}, \mathcal{LC}\}$
5: $best = (1\text{-}\lambda)\Delta_{S_I}^{\rho}$; $S_p \leftarrow S_I$;
6: $\mathcal{Q}$.push(S_I);
7: **while** ($\mathcal{Q} \neq \phi$ & $\Delta_{TH} \leq best$) **do**
8:    $S_c \leftarrow$ Enumerate($\mathcal{Q}$.pop());
9:    **for** (each $S_i \in S_c$ s.t. S_i not visited) **do**
10:       Lookup $\mathcal{A}$ of S_i , i.e., $\{\mathcal{RE}, \mathcal{RC}, \mathcal{LE}, \mathcal{LC}\}$
11:       Update $S_x^{\mathcal{A}}$, $S_{x^2}^{\mathcal{A}}$, $S_y^{\mathcal{A}}$, $S_{y^2}^{\mathcal{A}}$, $S_{xy}^{\mathcal{A}}$ incrementally
12:       Compute $\rho(S_i)$;
13:       $\Delta_{S_i} = \lambda\Delta_{S_i}^{E} + (1 - \lambda)\Delta_{S_i}^{\rho}$;
14:       **if** ($\Delta_{S_i} < best$) **then**
15:          $best = \Delta_{S_i}$; $S_p \leftarrow S_i$;
16:       $\Delta_{TH} = \lambda\Delta_{S_i}^{E}$;
17:       $\mathcal{Q}$.push(S_i);
18: return S_p, best;

creating and maintaining four instances of the cumulative arrays, one for each enumeration operation: $\mathcal{LE}, \mathcal{RE}, \mathcal{LC}, \mathcal{RC}$.

Incremental (INC): Essentially, INC starts from S_I to generate all possible subsequences with the help of an auxiliary function Enumerate() and a priority queue $\mathcal{Q}$. This auxiliary function takes a subsequence as an input, and performs the four enumeration operations defined earlier in Sect. 2.2 to produce the set S_c. Then, for each $S_i \in S_c$, the corresponding cumulative arrays instance are updated incrementally. After that, the correlation is computed from this instance, and S_i is pushed into $\mathcal{Q}$. Once all candidate subsequences in S_c have been processed, INC pops an un-enumerated subsequence S_i from $\mathcal{Q}$ and calls Enumerate(), and so on. At any time, the candidate subsequences in $\mathcal{Q}$ are ascendingly sorted from the closest to S_I to the furthest based on Eq. 3.

To avoid an exhaustive search, INC abandons the search when reaching a point where any candidate subsequence to be generated has a preference deviation higher than the best deviation found so far, or when the queue becomes empty, as shown in Algorithm 3 line 7.

4   Experiments Setup and Results

We conducted a set of experiments on real-world dataset to evaluate our algorithms on a PC with Intel Core i7 @ 3.40 GHz, 16 GB of RAM. Algorithms were coded in Java. We report the results next.

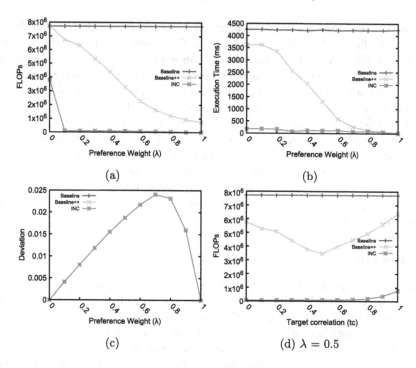

Fig. 5. Set of experiments on real dataset extracted from *Yahoo! Finance*.

4.1 Setup

Table 2 summarizes all parameters used throughout the experiments and the dataset settings.

Table 2. Parameters of the experiments

Parameter	Default	Range
Preference weight (λ)	0.5	0.0–1.0
Target correlation (tc)	-	0.0–1.0
Query/time series ratio	20%	10%–100%

Dataset: We have experimented with a real dataset which contains 61 time series. Each time series represents the historical daily closing price of a company listed in the US stocks market from 2/1/2009 to 31/12/2013, $n = 1,258 \approx 10^3$. This dataset was manually extracted from *Yahoo! Finance* and had to be merged and processed to be suitable for experimenting. Also, we have experimented with a synthetically generated time series dataset using the Random Walk model [12]. There are a total of 10,000 time series in this dataset, with length $n = 2,000$.

Performance Measures: We compare the computational-centric algorithms against Baseline and Baseline++ using the CPU cost as an indicator for efficiency. The CPU cost of an algorithm is the number of float point operations (FLOPs) the algorithm had to make to find the optimal solution. Specifically, those operations are the addition, subtraction, multiplication and division operations of numbers made to obtain the Pearson correlation.

As for effectiveness, we use the total deviation Δ defined in Eq. 4.

We averaged FLOPs and Δ for a workload of 10^3 input sequences. Specifically, each item in the workload consists of 10 pairs (chosen randomly out of k) time series, 10 uniformly distributed time intervals $[s, e]$ and 10 uniformly distributed target correlation ranges $[c_l - c_u]$ where: $0 \leq c_l, c_u \leq 1$ and $c_l \leq c_u$. Note that a pair of time series and a time interval construct one instance of input sequence S_I.

4.2   Results

Preference Weight (λ): To show the impact of λ, we evaluated the algorithms with our workload while varying λ and averaged the result. Figure 5a and c show the average FLOPs and average deviation respectively.

All three algorithms achieve the same accuracy in terms of deviation. However, they exhibit different costs. When compared to the Baseline and Baseline++, INC manages to find the same solution as them but with almost half the cost when $\lambda = 0$. The reason is, INC keeps four sets of the cumulative arrays. One for each move. This enables INC to update the cumulative arrays by only adding or subtracting one value (one float point operation) from the previously computed cumulative arrays, while Baseline and Baseline++ have to add and subtract two values (two float point operations) from the previously computed cumulative arrays.

By increasing the preference weight, both Baseline++ and INC costs decrease by a considerable amount while Baseline stays constant. The reason is: increasing the weight of preference for Baseline++ and INC is orthogonal to increasing the tightness of the search space in Baseline++ and to early abandoning the search by the threshold Δ_{TA} in INC. Also, Fig. 5c shows a tradeoff between the conflicting two objectives, which follows a semi-bell shape pattern.

Target Correlation (tc): To show the relationship between tc and FLOPs, we set $\lambda = 0.5$ and varied tc such that $c_l = c_u$. As Fig. 5d shows, Baseline exhibit a constant performance regardless of the value tc, while Baseline++ is the most affected by tc. The closer tc is to 0.5 (which happens to be the average correlation among all time series in the dataset), the less work Baseline++ will have to do to achieve the optimal solution. In other words, Baseline++ will achieve tighter search space if tc is close from the input's correlation $\rho(S_I)$ since *best* in Algorithm 2 line 3 will be very close from zero.

5 Conclusion

We have addressed the challenging problem of finding constrained subsequence patterns in time series data, to increase the efficiency and accuracy of data mining applications. Then, we proposed our efficient subsequence processing techniques and illustrated the reasons behind their design choices. Finally, we empirically demonstrated the efficiency of our techniques using real and synthetic dataset.

Acknowledgments. We would like to thank Lemma solutions (www.lemma.com.au) for their help during the production of this paper.

References

1. Al-Maskari, S., Bélisle, E., Li, X., Le Digabel, S., Nawahda, A., Zhong, J.: Classification with quantification for air quality monitoring. In: Bailey, J., Khan, L., Washio, T., Dobbie, G., Huang, J.Z., Wang, R. (eds.) PAKDD 2016. LNCS (LNAI), vol. 9651, pp. 578–590. Springer, Cham (2016). https://doi.org/10.1007/978-3-319-31753-3_46
2. Al-Maskari, S., Guo, W., Zhao, X.: Biologically inspired pattern recognition for e-nose sensors. In: Li, J., Li, X., Wang, S., Li, J., Sheng, Q.Z. (eds.) ADMA 2016. LNCS, vol. 10086, pp. 142–155. Springer International Publishing, Cham (2016). https://doi.org/10.1007/978-3-319-49586-6_10
3. Al-Maskari, S., Ibrahim, I.A., Li, X., Abusham, E., Almars, A.: Feature extraction for smart sensing using multi-perspectives transformation. In: Wang, J., Cong, G., Chen, J., Qi, J. (eds.) ADC 2018. LNCS, vol. 10837, pp. 236–248. Springer, Cham (2018). https://doi.org/10.1007/978-3-319-92013-9_19
4. Al-Maskari, S., Li, X., Liu, Q.: An effective approach to handling noise and drift in electronic noses. In: Wang, H., Sharaf, M.A. (eds.) ADC 2014. LNCS, vol. 8506, pp. 223–230. Springer, Cham (2014). https://doi.org/10.1007/978-3-319-08608-8_21
5. Fu, T.C.: A review on time series data mining. Eng. Appl. Artif. Intell. **24**(1), 164–181 (2011)
6. Gavrilov, M., Anguelov, D., Indyk, P., Motwani, R.: Mining the stock market (extended abstract): which measure is best? In: Proceedings of the Sixth ACM SIGKDD International Conference on Knowledge Discovery and Data Mining, 20–23 August 2000, Boston, MA, USA, pp. 487–496 (2000)
7. Ghazavi, S.N., Liao, T.W.: Medical data mining by fuzzy modeling with selected features. Artif. Intell. Med. **43**(3), 195–206 (2008)
8. Ibrahim, I.A., Albarrak, A.M., Li, X.: Constrained recommendations for query visualizations. Knowl. Inf. Syst. **51**(2), 499–529 (2017)
9. Keogh, E.J., Kasetty, S.: On the need for time series data mining benchmarks: a survey and empirical demonstration. Data Min. Knowl. Discov. **7**(4), 349–371 (2003)
10. Li, Y., U, L.H., Yiu, M.L., Gong, Z.: Discovering longest-lasting correlation in sequence databases. PVLDB **6**(14), 1666–1677 (2013)
11. Mueen, A., Hamooni, H., Estrada, T.: Time series join on subsequence correlation. In: 2014 IEEE International Conference on Data Mining, ICDM 2014, 14–17 December 2014, Shenzhen, China, pp. 450–459 (2014)

12. Mueen, A., Nath, S., Liu, J.: Fast approximate correlation for massive time-series data. In: Proceedings of the ACM SIGMOD International Conference on Management of Data, SIGMOD 2010, 6–10 June 2010, Indianapolis, Indiana, USA, pp. 171–182 (2010)
13. Raghupathi, W., Raghupathi, V.: Big data analytics in healthcare: promise and potential. Health Inf. Sci. Syst. $2(1)$, 1 (2014)
14. Rakthanmanon, T., et al.: Searching and mining trillions of time series subsequences under dynamic time warping. In: The 18th ACM SIGKDD International Conference on Knowledge Discovery and Data Mining, KDD 2012, 12–16 August 2012, Beijing, China, pp. 262–270 (2012)
15. Sakurai, Y., Papadimitriou, S., Faloutsos, C.: BRAID: stream mining through group lag correlations. In: Proceedings of the ACM SIGMOD International Conference on Management of Data, 14–16 June 2005, Baltimore, Maryland, USA, pp. 599–610 (2005)
16. Utomo, C., Li, X., Wang, S.: Classification based on compressive multivariate time series. In: Cheema, M.A., Zhang, W., Chang, L. (eds.) ADC 2016. LNCS, vol. 9877, pp. 204–214. Springer, Cham (2016). https://doi.org/10.1007/978-3-319-46922-5_16
17. Nahar, V., Al-Maskari, S., Li, X., Pang, C.: Semi-supervised learning for cyberbullying detection in social networks. In: Wang, H., Sharaf, M.A. (eds.) ADC 2014. LNCS, vol. 8506, pp. 160–171. Springer, Cham (2014). https://doi.org/10.1007/978-3-319-08608-8_14
18. Zhu, Y., Shasha, D.: Statstream: statistical monitoring of thousands of data streams in real time. In: Proceedings of 28th International Conference on Very Large Data Bases, VLDB 2002, 20–23 August 2002, Hong Kong, China, pp. 358–369 (2002)

Slice_OP: Selecting Initial Cluster Centers Using Observation Points

Md Abdul Masud$^{(\boxtimes)}$, Joshua Zhexue Huang, Ming Zhong, Xianghua Fu, and Mohammad Sultan Mahmud

Big Data Institute, College of Computer Science and Software Engineering, Shenzhen University, Shenzhen 518060, China
{masud,zx.huang,mingz,fuxh,sultan}@szu.edu.cn

Abstract. This paper proposes a new algorithm, Slice_OP, which selects the initial cluster centers on high-dimensional data. A set of observation points is allocated to transform the high-dimensional data into one-dimensional distance data. Multiple Gamma models are built on distance data, which are fitted with the expectation-maximization algorithm. The best-fitted model is selected with the second-order Akaike information criterion. We estimate the candidate initial centers from the objects in each component of the best-fitted model. A cluster tree is built based on the distance matrix of candidate initial centers and the cluster tree is divided into K branches. Objects in each branch are analyzed with k-nearest neighbor algorithm to select initial cluster centers. The experimental results show that the Slice_OP algorithm outperformed the state-of-the-art Kmeans++ algorithm and random center initialization in the k-means algorithm on synthetic and real-world datasets.

Keywords: Initial cluster center · Clustering algorithm
Center initialization · Observation point

1 Introduction

The Lloyd's classic clustering algorithm k-means [15] is a centroid-based clustering method, which is efficient for partitioning the large data. The k-means algorithm selects random initial cluster centers for the given K number of clusters. The centers are recomputed and objects are reassigned until the convergence of algorithm. The random initial cluster centers are used as input parameter to the k-means algorithm for performing the clustering solution. Therefore, the k-means algorithm yields inconsistent clustering results and provides unsatisfactory partitioning solution. The quality of the clustering result highly depends on the selection of initial cluster centers.

The algorithm, namely, Kmeans++ [2] is one of the most widely used methods, which provides a better alternative selection process of initial cluster centers for the k-means algorithm. The core advantages of this algorithm are easy to use and theoretical guarantees on partition quality without any assumption of data.

© Springer Nature Switzerland AG 2018
G. Gan et al. (Eds.): ADMA 2018, LNAI 11323, pp. 17–30, 2018.
https://doi.org/10.1007/978-3-030-05090-0_2

In Kmeans++ algorithm, the initial cluster centers are obtained using an adaptive sampling technique called D^2 weighting. First, it chooses an initial center uniformly at random from input data. Then, the remaining $K - 1$ initial centers are estimated sequentially to the previously selected initial centers and objects in data with D^2 weighting. In the clustering step, the partition is refined using Lloyd's algorithm, guaranteed the quality, and converged to a locally optimal solution in finite time. The disadvantage of Kmeans++ is that both the center initialization and the clustering steps compute the sum-of-squares distances between objects in clusters and cluster centers. As a result, this algorithm does not scale to massive datasets and the initialization step cannot select the objects properly as initial centers from data with many clusters.

Our recently proposed I-nice method [17] effectively identifies the number of clusters and initial cluster centers from large data and data with many clusters. However, this method is not proper choice on imbalanced datasets that contain majority and minority clusters.

There are many other studies on methods for selecting the initial cluster centers [4,5,12,13,18,22]. But, these algorithms select improper objects as initial centers on real-world datasets and many cases algorithms are not able to estimate initial centers from complex data. However, random selection is still commonly used in practice for its simplicity and efficiency.

The aim of this research is to select the initial cluster centers on high-dimensional data. We formulate an innovative method Slice_OP which is an abbreviation of **S**electing the **i**nitial cluster **c**entres with **O**bservation **P**oints. The Slice_OP algorithm performs an observation process for selecting initial cluster centers in data including majority and minority clusters. In Slice_OP, a set of observation points is allocated to bring the clustering information from data. The selection process of initial cluster centers is presented with Fig. 1. Five observation points are assigned to compute distances between observation points and objects in a two-dimensional data in Fig. 1(a). The distance data is used to build Gamma mixture models (GMMs) which are fitted with the expectation-maximization (EM) algorithm. The best-fitted model is selected with the second-order Akaike information criterion (AICc) and objects in each component of the best-fitted model are analyzed to estimate candidate initial centers for each observation point. Similarly, several candidate initial centers are estimated for all observation points in Fig. 1(b). We compute distance matrix among candidate initial centers and build a cluster tree Fig. 1(c). The cluster tree is divided into K branches and the candidate initial centers in each branch are analyzed with k-nearest neighbor (k-NN) algorithm to select the initial cluster center in Fig. 1(d).

The proposed method describes a straightforward and effective algorithmic solution for selecting the initial cluster centers. The Slice_OP method transforms high-dimensional data into one-dimensional data which speed the iterative clustering steps up.

The remainder of this paper is organized as follows. First, we present a brief overview of related works in Sect. 2. We introduce an innovative Slice_OP algo-

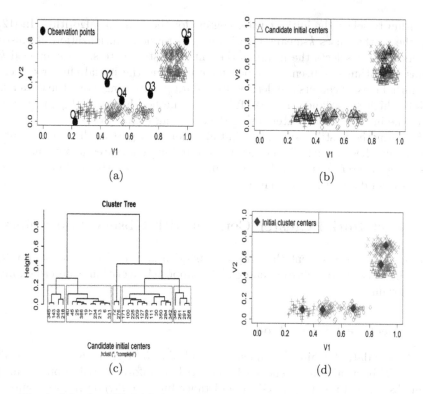

Fig. 1. Selection of initial cluster centers on data with observation points.

rithm and present the steps of algorithm in Sect. 3. We present the experimental results of the proposed and other methods in Sect. 4. Finally, we draw conclusions and discuss future work in Sect. 5.

2   Related Works

Several methods [12,13,22] have been proposed to select the initial cluster centers as input parameter for the clustering algorithms. A refinement algorithm refines the initial points from a set of random sub-samples on data for the k-means clustering process [4]. An optimized k-means approach has been developed in [8] where the maximum variation and the minimum correlation axes are estimated for selecting the initial cluster centers. In Min-Max algorithm, weights are assigned to the clusters in proportion to the variances to deal with initialization problem [20].

In model-based clustering methods, multiple models are built and a model selection criterion is used to select the best-fitted model. Therefore, a final model is obtained with a number of components [3,9–11].

The centers of each sub-sample are considered as the initial cluster centers of data. The authors in [22] have developed a new neighborhood-based clustering algorithm to select the initial cluster centers for the k-means algorithm. In [12], first the method uses k-means algorithm with random initialization of cluster centers, then it selects the mean as the initial cluster centers. An enhanced k-means algorithm has been developed for estimating the initial cluster centers. The initial cluster centers are derived from data partitioning along the data axis with the highest variance. The careful seeding method Kmeans++ [2] is used to select the initial cluster centers using the sampling technique.

Some methods still use random initialization as a part of the selection process. Using these initial cluster centers, there is no guarantee about quality of clustering. Therefore, it is still a remaining problem to estimate initial cluster centers from data with many clusters.

3   Select Initial Cluster Centers with Observation Points

In this section, we present the proposed algorithm, Slice_OP, which can automatically select initial cluster centers. The proposed algorithm is described into the following steps:

3.1   Select Best-Fitted Models

Let R^d be a data domain of d dimensions and $\mathcal{X} \subset R^d$ a set of N objects in R^d. Let $p \in R^d$ be a randomly generated point with a uniform distribution. Define p as an observation point to $\mathcal{X}$. Given a distance function $d(.)$ on R^d, we compute all distances between observation point p and N objects of $\mathcal{X}$ and transform into a set of distances $X_p = \{x_1, x_2, \ldots, x_N\}$. Given a different observation point, we can compute a different distance distribution from $\mathcal{X}$.

We consider that a distance distribution contains multiple peaks, i.e., the data contain more than one cluster. Therefore, the distance distribution can be modeled as GMMs. The method is well developed to solve the parameters [21]. Let $X_p = \{x_1, x_2, \ldots, x_N\}$ be a set of normalized distance values. The GMM of X_p is defined as:

$$p(x|\theta) = \sum_{j=1}^{M} \pi_j g(x|\theta_j), \quad x \geq 0 \tag{1}$$

where M is the number of the Gamma components, π_j is the mixing proportion of component j, and θ_j are the parameters of Gamma component j, including shape parameter α_j and scale parameter β_j.

We use the expectation-maximization (EM) algorithm to solve the GMMs. In the expectation step of the EM algorithm, given the parameters θ^n, which were estimated in the nth previous iteration; the membership weight $p(Z_i = j|x_i, \theta^n)$ of each element x_i in X_p in each component j is computed. Then, the expected values of log-likelihood function with respect to the latent random variables

$Z = \{z_i\}$ are computed as follows:

$$Q(\theta|\theta^n, X_p) = \sum_{i=1}^{N} \sum_{j=1}^{M} \{p(Z_i = j|x_i, \theta^n) \log \pi_j + p(Z_i = j|x_i, \theta^n) \log(g(x_i|\theta_j))\}$$

(2)

The first part of Eq. (2) can be optimized using Lagrange multiplier. Then, taking the first derivatives with respect to each π_j and make them equal to zero. After simplification of the derivative equations, we can calculate $\hat{\pi}_j$ as

$$\hat{\pi}_j = \frac{1}{N} \sum_{i=1}^{N} p(Z_i = j|x_i, \theta^n)$$

(3)

For estimating the parameters, we optimize the second term of Eq. (2). We take the first derivatives with respect to α_j and make them equal to zero. After doing some mathematical manipulations, we get:

$$\log(\hat{\alpha}_j) - \psi(\hat{\alpha}_j) = \log \left(\frac{\sum_{i=1}^{N} x_j p(Z_i = j|x_i, \theta^n)}{\sum_{i=1}^{N} p(Z_i = j|x_i, \theta^n)} \right) - \left(\frac{\sum_{i=1}^{N} p(Z_i = j|x_i, \theta^n) \log x_i}{\sum_{i=1}^{N} p(Z_i = j|x_i, \theta^n)} \right)$$

(4)

where $\psi(x) = \frac{\partial \log(\Gamma(x))}{\partial x} = \frac{\Gamma'(x)}{\Gamma(x)}$ is called the Digamma function. We can use a Newton-type algorithm [6] to calculate $\hat{\alpha}_j$.

Again, we take the first derivatives of the second term of Eq. (2) with respect to β_j and make them equal to zero as follows, we get:

$$\hat{\beta}_j = \frac{1}{\alpha_j} \frac{\sum_{i=1}^{N} x_i p(Z_i = j|x_i, \theta^n)}{\sum_{i=1}^{N} p(Z_i = j|x_i, \theta^n)}$$

(5)

With Eq. (5) and the estimated value of $\hat{\alpha}_j$ from Eq. (4), we can estimate $\hat{\beta}_j$.

Thus, $(Mmax - 1)$ Gamma models are fitted with EM algorithm for each observation point. We use minimum value of AICc [19] to select the best-fitted model. The AICc is calculated as follows.

$$AICc = -2 \log(L(\theta^*)) + 2q \left(\frac{N}{N - q - 1} \right)$$

(6)

where $L(\theta^*)$ is the maximum log-likelihood, N is the number of objects, and q is the number of parameters in the GMM model.

For each observation point computed distance distribution, one best-fitted model is selected by using Eq. (6). In this way, a set of best-fitted models are obtained on the distance distributions which are computed with a set of observation points.

3.2   Estimate the Candidate Initial Centers

In this section, we perform the data abstraction by choosing several informative objects which bring the clustering information from one-dimensional data as well as original data. For each observation point, we obtain a best-fitted model along with the number of components, where each component contains a set of objects. These objects in a component are analyzed to estimate the candidate initial centers.

Generally, objects in a dataset are distributed into different clusters where several clusters may contain a large number of objects and some clusters may contain a small number of objects. A best-fitted model is described with different components and each component is considered as a cluster [3]. In model-based clustering, it is assumed that the extracted objects in a component are derived from a specific cluster. But, in real scenario, majority objects in a component may be belonged to a specific cluster in some cases and other objects may be belonged to different clusters. The majority objects are likely to be derived from the clusters with large number of objects. However, it is hard to ensure that the majority objects in a component will be extracted from a cluster with a small number of objects. Therefore, the cluster with a small number of objects fails to provide the clustering information to the data abstraction step.

To address this issue, we select two objects as two candidate initial centers in each component of best-fitted model so that all representative clustering information are retrieved from the input data. We choose one object which has the smallest distance to other objects in the component as the first candidate initial center k_1, and another object which has the largest distance to other objects in the component as the second candidate initial center k_2. The first candidate initial center is likely to be a representative object from the cluster with a large number of objects and the second initial center is likely to be a representative object from the cluster with a small number of objects. Similarly, we select the candidate initial cluster centers $k^p = \{k_1^c, k_2^c\}_{c=1}^{c-\max}$ from all components of the best-fitted model, where $c - max$ is the number of components. One best-fitted model is related with one observation point, all candidate initial centers $k = \{k^p\}_{p=1}^{P}$ are selected from the best-fitted models of all observation points, where P is the number of observation points.

3.3   Select Initial Cluster Centers

We compute Euclidean distance matrix among candidate initial centers and build a cluster tree based on the distances. We divide the cluster tree into K branches where the candidate initial centers in each branch are close to each other. The candidate initial centers in each branch are analyzed with the k-NN algorithm to estimate a candidate initial center, which deserves the most compact k neighboring objects. This selected candidate initial center has the smallest distance from others in a branch, which is chosen as an initial cluster center K_1 in the branch. The K initial cluster centers are selected from the K branches of the cluster tree.

Algorithm 1: Slice_OP: Select intial cluster centers with observation points

---

1 **Input:** Input data $\mathcal{X}$
2 **Output:** Initial cluster centers
3 **Initialization:** P is the number of observation points and $Mmax$ is the maximum number of GMMs
4 **Select best-fitted models:**
5 **for** $p := 1$ to P **do**
6   Compute the distance vector X_p between $\mathcal{X}$ and p ;
7   **for** $M := 2$ to $Mmax$ **do**
8   CALL EM algorithm to solve $GMM(p, M)$;
9   Select best-fitted model $GMM(p)$ using minimum AICc ;
10 Keep all selected models $GMMs(p)$ with different number of components c for all p ;
11 **Estimate candidate initial centers:**
12 **for** $p := 1$ to P **do**
13   **for** $c := 1$ to c-max **do**
14   Select the first candidate initial center k_1 with the smallest distance to other objects in c ;
15   Select the second candidate initial center k_2 with the largest distance to other objects in c ;
16   Keep candidate initial centers $k^p = \{k_1^c, k_2^c\}_{c=1}^{c-max}$ for model $GMM(p)$;
17 Keep all candidate initial centers $k = \{k^p\}_{p=1}^P$ for P observation points ;
18 **Select initial cluster centers:**
19 Compute distance matrix $d(k)$ among candidate initial centers k ;
20 Build a cluster tree T from $d(k)$ and T is divided into K branches ;
21 **for** $b := 1$ to K **do**
22   Select the most densest candidate initial center as initial cluster centers K_1 from branch b ;
23 K initial cluster centers are obtained on input data $\mathcal{X}$;

The estimation steps of identifying the initial cluster centers are presented in Algorithm 1.

Clustering: The initial cluster centers are used to cluster the input data. We use k-means algorithm for its efficiency and simplicity.

4 Experiments

In this section, we present the experimental results of Slice_OP algorithm on both synthetic and real-world datasets to demonstrate the performance of proposed algorithm in comparing to other methods.

4.1 Datasets

We have used several synthetic and real-world datasets to evaluate the clustering results. The synthetic datasets are generated as below.

We have generated 10 synthetic datasets with different configurations. The characteristics of synthetic datasets are given in Table 1. Each dataset is generated as follows. Given the number of instances N, the number of cluster K, the number of dimensions d, and overlap threshold v, K vectors with d dimensions are randomly generated. The elements of each vector are randomly selected integers between 1 and K inclusive and no vector is equal to any other vector. From each vector, a cluster center is calculated as the values of elements minus 0.5. In this way, K cluster centers are obtained. After that, the distances between cluster centers are computed. The half of the shortest distance between cluster centers is taken as radius of clusters. The variances in the diagonal of the covariance matrix are computed by squaring the radius and multiplying it with overlap v. Then, the cluster center vector and covariance matrix are used as the parameters to the multivariate Gaussian distribution to generate a set of points in the d dimensional space that forms a cluster in Gaussian distribution. K Gaussian clusters are generated independently and merged in one dataset.

Table 1. Characteristics of the ten synthetic datasets.

Number	Datasets	Instances (N)	Features (d)	Classes (K)	Overlap (v)
1	Syn_Data1	500	2	5	0.010
2	Syn_Data2	1000	2	8	0.010
3	Syn_Data3	1000	2	10	0.008
4	Syn_Data4	900	2	12	0.004
5	Syn_Data5	2000	10	15	0.005
6	Syn_Data6	2500	20	18	0.006
7	Syn_Data7	6000	12	22	0.004
8	Syn_Data8	8000	15	24	0.003
9	Syn_Data9	1000	50	25	0.002
10	Syn_Data10	10000	50	30	0.001

Next, we chose 10 real-world datasets from the UCI machine learning repository [7] and the KEEL repository [1]. The characteristics of real-world datasets are given in Table 2.

Table 2. Characteristics of the ten real-world datasets.

Number	Datasets	Instances	Features	Classes
1	Waveform	5000	21	3
2	Vehicle	846	18	4
3	Breast tissue	106	10	6
4	Led7Digit	500	7	10
5	Yeast	1484	8	10
6	Page-blocks	5472	10	5
7	Optdigits	5620	64	10
8	Ecoli	336	8	8
9	Segment	3210	19	7
10	Abalone	4174	8	28

4.2   Experimental Settings and Evaluation Methods

The proposed algorithm is used to select the initial cluster centers from dataset. The initial cluster centers are applied as input parameter in the k-means clustering algorithm to the appropriate partition. Similarly, the Kmeans++ algorithm is used to select the initial cluster centers which are applied to the k-means algorithm for clustering results.

In clustering results with random initial cluster centers, we run the k-means algorithm 10 times where a set of random initial cluster centers are estimated each time for performing the clustering solution. We compute the mean of clustering results with random initial cluster centers as average clustering accuracy.

In these experiments, P is the number of observation points in the data space, M is the number of components in each GMM, and $Mmax$ is the maximum number of GMMs in each dataset. The value of P is set up to 10. For each dataset, the value of $Mmax$ is set as the true number of clusters plus 5, and the value of M is selected from 2 to $Mmax$. The value of K is the number of clusters in dataset.

To measure the clustering results, two evaluation criteria are used in the experiments. They are defined as follows:

1. **Purity:** Purity is the percentage of the objects classified correctly [16]. Purity is computed as follows:

$$Purity = \frac{1}{N} \sum_{i=1}^{K} \max_j |C_i \cap Y_j| \tag{7}$$

where N is the number of objects in the dataset, K is the number of clusters, C_i is the set of objects in cluster i and Y_j is the set of objects in class j that has the maximum intersection with cluster i among all sets of classes. The range of purity is between 0 and 1.

2. **Normalized Mutual Information (NMI):** NMI measures the normalized mutual information between class label and cluster assignment which are considered as random variables [14]. Let C be the random variable representing the cluster assignment of objects, and Y be the random variable representing the class label of objects. The NMI is computed as follows:

$$NMI = \frac{2I(C;Y)}{H(C) + H(Y)} \tag{8}$$

where $I(C;Y)$ is the mutual information of C and Y; and $H(C)$ and $H(Y)$ are the entropies of C and Y respectively. The range of NMI is also between 0 and 1.

4.3   Experimental Results and Analysis

In this section, we present and discuss the experimental results with the proposed and other methods. We have used both synthetic and real-world data to show the improvement of clustering performance by using the Slice_OP algorithm. The results show that the use of the automatically selected initial cluster centers can increase the clustering accuracy and efficiency of the k-means clustering process.

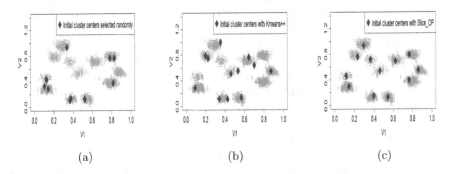

(a)   (b)   (c)

Fig. 2. Performance comparison of three methods in terms of selecting the initial cluster centers on the two dimensional synthetic Syn_Data4 dataset.

We demonstrate the performance comparison of three methods in selecting initial cluster centers on the synthetic Syn_Data4 dataset in Fig. 2. The two-dimensional Syn_Data4 dataset contains 12 clusters. The Fig. 2(a) shows the random initialization on data, where 7 true clusters are assigned with 12 random initial centers, 5 true clusters are empty, and few clusters are assigned with two or more initial centers. In this case, several dense regions of objects are not allocated with initial centers. This scenario is definitely not a good initialization for clustering process, which leads to produce the unsatisfactory clustering result. The selection of initial centers with the Kmeans++ algorithm is shown in Fig. 2(b), where 8 true clusters are assigned with 12 initial centers, 4 true

Table 3. Comparison of clustering results with Slice_OP and other algorithms on synthetic datasets.

Datasets	Purity			NMI		
	Random	Kmeans++	Slice_OP	Random	Kmeans++	Slice_OP
Syn_Data1	0.812	0.860	**0.985**	0.777	0.806	**0.952**
Syn_Data2	0.925	0.897	**0.990**	0.885	0.842	**0.973**
Syn_Data3	0.849	0.903	**0.991**	0.877	0.893	**0.980**
Syn_Data4	0.812	0.883	**0.997**	0.877	0.928	**0.995**
Syn_Data5	0.921	0.951	**0.966**	0.936	0.944	**0.966**
Syn_Data6	**0.935**	0.878	0.902	**0.945**	0.919	0.931
Syn_Data7	0.906	**0.948**	0.901	0.934	**0.955**	0.935
Syn_Data8	0.917	0.881	**0.965**	0.945	0.922	**0.961**
Syn_Data9	0.835	0.945	**0.964**	0.916	0.961	**0.974**
Syn_Data10	0.848	0.928	**1.000**	0.918	0.965	**1.000**

clusters are empty, and 4 true clusters are assigned with double initial centers. The Kmeans++ algorithm can not allocate the initial centers on all true clusters. The initialization with Kmeans++ is also not a reasonable allocation for clustering process but it is better than random initialization.

On the other hand, the Slice_Op algorithm successfully selects the initial centers from all true clusters. Each initial center is estimated from each dense region on data in Fig. 2(c). The automatically identified initial cluster centers are able to improve the clustering accuracy significantly in the k-means clustering process.

Table 3 shows the clustering results of the 10 synthetic datasets. The clustering results are measured in terms of purity and NMI. Column Random shows the average clustering results of the k-means algorithm with random initial cluster centers and the Kmeans++ column indicates that the clustering results are obtained by using the initial cluster centers identified with the Kmeans++ algorithm. The Slice_OP column indicates that the clustering results are obtained by using the initial cluster centers selected with Slice_OP. The results show that the Slice_OP method outperformed Kmeans++ and random initial cluster centers significantly in terms of both purity and NMI in most cases.

The clustering results in terms of purity and NMI on real-world datasets are presented in Table 4. The results demonstrate that the clustering performance with Slice_OP selected initial centers is better than the Kmeans++ selected initial centers, and random initial centers in most cases. The average clustering results generated with random initial cluster centers are worser than Kmeans++ selected initial centers.

Table 4. Comparison of clustering results with Slice_OP and other algorithms on real-world datasets.

Datasets	Purity			NMI		
	Random	Kmeans++	Slice_OP	Random	Kmeans++	Slice_OP
Waveform	0.530	0.530	0.530	0.362	0.362	0.362
Vehicle	0.443	**0.451**	0.434	0.183	0.185	**0.189**
Breast tissue	0.433	0.433	**0.452**	0.362	0.362	0.358
Led7Digit	0.514	0.502	**0.592**	0.491	0.507	**0.543**
Yeast	0.511	0.517	**0.523**	0.255	0.256	**0.262**
Page-blocks	0.900	0.900	0.900	0.049	0.049	**0.058**
Optdigits	0.771	0.800	**0.803**	0.734	0.747	**0.757**
Ecoli	0.815	0.821	**0.836**	0.588	0.590	**0.591**
Segment	0.536	**0.557**	0.471	0.504	**0.518**	0.426
Abalone	0.275	0.275	**0.276**	0.175	0.176	**0.180**

The fact of the success of proposed method is that the proposed algorithm, Slice_OP, is able to select initial centers from the exact distribution of data not from class labeled of data. The clustering information is retrieved by estimating the candidate initial centers from different informative regions of data, which are investigated to select the final set of initial centers.

The Slice_OP algorithm can select the initial cluster centers effectively on data with many clusters such as synthetic Syn_Data10 dataset along with 30 clusters and real-world Abalone dataset along with 28 clusters.

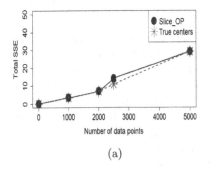
(a)

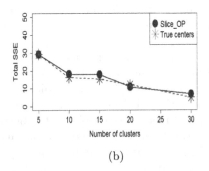
(b)

Fig. 3. Comparison of total sum-of-squares error in clustering results by using Slice_OP generated initial cluster centers and true centers.

The Fig. 3 shows a comparison test of total sum-of-squares error (SSE) in clustering results by using initial cluster centers selected with Slice_OP algorithm and true centers. We compute the means of objects in each class and assign them as true centers in the dataset. The Fig. 3(a) presents the compactness of clustering solutions between Slice_OP generated initial centers and true centers in the terms of number of data points and the total SSE of clustering results. To analyze the performance of clustering results, the number of data points is increased from 1000 to 5000. Two curves using initial centers with Slice_OP and true centers are maintaining similar total SSE for increasing the data points. We see that the compactness of clustering results with Slice_OP is similar to the compactness of clustering results with true centers. The Fig. 3(b) demonstrates the compactness of clustering solutions between Slice_OP generated initial centers and true centers in the terms of number of clusters and the total SSE of clustering results. To analyze the performance of clustering results, the number of clusters is increased from 5 to 30. The illustration indicates that both Slice_OP generated initial centers and true centers effectively improve the total SSE simultaneously by increasing the number of clusters. Therefore, the quality of clustering results with Slice_OP is equivalent to the quality of clustering results with true centers from data.

5   Conclusions and Future Work

We introduce a new approach to identifying the initial cluster centers for the k-means algorithm on complex data with many clusters. In the data abstraction step, the candidate initial centers are estimated from input data. These candidate centers represent the clustering information from different angles of data distributions. The proposed algorithm builds a cluster tree from candidate initial centers and analyzes the objects of K branches for selecting the initial cluster centers. The selected initial cluster centers are used in the k-means algorithm to cluster on the input high-dimensional data.

We have shown that the initial cluster centers generated with Slice_OP improve the clustering accuracy in terms of purity and NMI on both synthetic and real-world datasets. In addition, the proposed method is able to identify the initial cluster centers on data with many clusters. The current research can be applied in the semi-supervised clustering method for selecting the pairwise constraints. Furthermore, we can investigate this method on big data analytics.

Acknowledgment. This paper was supported by National Natural Science Foundations of China (under Grant No. 61473194 and 61472258) and Shenzhen-Hong Kong Technology Cooperation Foundation (under Grant No. SGLH20161209101100926).

References

1. Alcalafdez, J., Fernandez, A., Luengo, J., Derrac, J., Garcia, S.: KEEL data-mining software tool: data set repository, integration of algorithms and experimental analysis framework. Soft Comput. **17**, 255–287 (2011)
2. Arthur, D., Vassilvitskii, S.: k-means++: the advantages of careful seeding. In: Symposium on Discrete Algorithms (SODA), pp. 1027–1035. Society for Industrial and Applied Mathematics (2007)

3. Banfield, J.D., Raftery, A.E.: Model-based Gaussian and non-Gaussian clustering. Biometrics **49**(3), 803–821 (1993)
4. Bradley, P.S., Fayyad, U.M.: Refining initial points for k-means clustering. In: Proceedings of 15th International Conference on Machine Learning, pp. 91–99. Morgan Kaufmann, San Francisco (1998)
5. Deelers, S., Auwatanamongkol, S.: Enhancing k-means algorithm with initial cluster centers derived from data partitioning along the data axis with the highest variance. Int. J. Phys. Math. Sci. **1**(11), 518–523 (2007)
6. Dennis, J.E., Schnabel, R.B.: Numerical Methods for Unconstrained Optimization and Nonlinear Equations. Prentice-Hall, Englewood Cliffs (1983)
7. Dheeru, D., Karra Taniskidou, E.: UCI machine learning repository (2017). http://archive.ics.uci.edu/ml
8. Erisoglu, M., Calis, N., Sakallioglu, S.: A new algorithm for initial cluster centers in k-means algorithm. Pattern Recogn. Lett. **32**, 1701–1705 (2011)
9. Figueiredo, M.A.T., Jain, A.K.: Unsupervised learning of finite mixture models. IEEE Trans. Pattern Anal. Mach. Intell. **24**(3), 381–396 (2002)
10. Fraley, C., Raftery, A.E.: Model-based clustering, discriminant analysis, and density estimation. J. Am. Stat. Assoc. **97**(458), 611–631 (2002)
11. Fraley, C., Raftery, A.E.: Bayesian regularization for normal mixture estimation and model-based clustering. J. Classif. **24**(2), 155–181 (2007)
12. Jain, A.K., Dubes, R.C.: Algorithms for Clustering Data. Prentice Hall, Englewood Cliffs (1998)
13. Khan, S.S., Ahmad, A.: Cluster center initialization algorithm for k-means clustering. Pattern Recogn. Lett. **25**(11), 1293–1302 (2004)
14. Kuncheva, L.I., Hadjitodorov, S.: Using diversity in cluster ensembles. In: International Conference on System, Man and Cybernetics, vol. 2, pp. 1214–1219 (2004)
15. Lloyd, S.P.: Least squares quantization in PCM. IEEE Trans. Inf. Theory **28**(2), 129–137 (1982)
16. Manning, C.D., Raghavan, P., Schutze, H.: Introduction to Information Retrieval. Cambridge University Press, Cambridge, New York (2008). http://opac.inria.fr/record=b1127339
17. Masud, M.A., Huang, J.Z., Wei, C., Wang, J., Khan, I., Zhong, M.: I-nice: a new approach for identifying the number of clusters and initial cluster centres. Inf. Sci. **466**, 129–151 (2018)
18. Sbhatia, M.P., Khurana, D.: Analysis of initial centers for k-means clustering algorithm. Int. J. Comput. Appl. **71**(5), 9–12 (2013)
19. Sugiura, N.: Further analysis of the data by Akaike' s information criterion and the finite corrections. Commun. Stat.-Theory Methods **7**(1), 13–26 (1978)
20. Tzortzis, G., Likas, A.: The minmax k-means clustering algorithm. Pattern Recogn. **47**(7), 2505–2516 (2014)
21. Vegas-Sáchez-Ferrero, G., et al.: Gamma mixture classifier for plaque detection in intravascular ultrasonic images. IEEE Trans. Ultrason. Ferroelectr. Freq. Control **61**(1), 44–61 (2014)
22. Ye, Y., Huang, J.Z., Chen, X., Zhou, S., Williams, G., Xu, X.: Neighborhood density method for selecting initial cluster centers in k-means clustering. In: Ng, W.-K., Kitsuregawa, M., Li, J., Chang, K. (eds.) PAKDD 2006. LNCS (LNAI), vol. 3918, pp. 189–198. Springer, Heidelberg (2006). https://doi.org/10.1007/11731139_23

HierArchical-Grid CluStering Based on DaTA Field in Time-Series and the Influence of the First-Order Partial Derivative Potential Value for the ARIMA-Model

Krid Jinklub and Jing Geng[(⊠)]

School of Computer Science and Technology, Beijing Institute of Technology, Beijing, China
Janegeng@bit.edu.cn

Abstract. Extend the function of static-time dataframe clustering algorithm (HASTA: HierArchical-grid cluStering based on daTA field) to be able to cluster the time-series dataframe. The algorithm purposed to use a set of "first-partial derivative potential value" given from HASTA in the multiple dataframes as the input to the autoregressive integrated moving average (ARIMA) under preliminary parameters. The ARIMA model could perform the pre-labeling task for the cluster(s) in the connected dataframe on the same time-series data. Calculating the structural similarity as a distance measure between timeframe, ARIMA would mark the high potential grid(s) as the cluster tracker. As the result, the ARIMA model could interpreting and reasoning cluster movement phenomena in the systematic approach. This integration is the attempt to show the influent power of data field in the term of knowledge representation.

Keywords: Knowledge discovery · Pattern recognition · Data field
Time-series · Time-series clustering · ARIMA

1 Introduction

Data clustering is the technique in order to obtain the set of knowledge about how the single entity behaves or correlated with others entity on the same plane or space. The clustering could divide data into the groups in an unsupervised manner (not rely on prior knowledge) or a supervised manner (rely on the engineering process or systematic process) or even combination of them (semi-supervise) in order to achieve the clustering goal. A number of clustering algorithm has been published and utilized in the history. The clustering algorithms could be aligned or differentiated, depends on the sub-goal or the bias behind the procedures (or even behind the data itself) [1]. Literally, this is the main problem concern the clustering algorithm. In order to find "similarity" matrices which could reasonably form the group in the way that mirror the reality and nature of data. And, find "distance" which perfectly divide the group of similar objects into the well-defined and separated group. The clustering is one of (or part of) the knowledge discovery tool. This information could form or initiate the beginning of the process in order to acquire knowledge or the knowledge even could be embedded in the

© Springer Nature Switzerland AG 2018
G. Gan et al. (Eds.): ADMA 2018, LNAI 11323, pp. 31–41, 2018.
https://doi.org/10.1007/978-3-030-05090-0_3

information. As the application point of view, the clustering (or clustering analysis) could play a major role in data compression, improving the other algorithms on the efficient aspect. Whether for understanding or utility, cluster analysis has long been used in a wide variety of fields: psychology and other social sciences, biology, statistics, pattern recognition, information retrieval, machine learning, and data mining. Moreover, clustering is used for exploratory data analysis for summarization and generalization.

2 Literature Review and Related Work

To cluster the data, there is a number of algorithms corresponding to the strategy or technique. All of them would invent in order to deal with numerous natures of data whether about the volume, the distribution, the complexity, the time-constraint, or even the generalization [2]. In this literature, both of the backing theory and model for the purposed algorithm and experimental task would be reviewed into 3 main parts.

2.1 Data Field

The data field mentioned here is not the data field in the database design. In the other hands, this data field is the novelty concept to understand how data entry interact with each other by relying on the physical field as inspiration. In physics, a field is one of the basic forms of particle existence [3]. A particle, that diffuses its energy into a physical space, generates a field which also acts on other particles simultaneously. Similar to physics, each data object is treated as a particle with certain mass and diffuses its data energy to the whole data field in order to demonstrate its existence and action in the tasks of spatial data [4]. In the context, all the fields from different local particles are superposed in the global physical space. And the superposition enables the field to characterize the interaction among different particles. the mutual interaction among data objects. In the case of clustering, a strong function of short-range field can enlarge the interactional strength among data objects nearby [5], whereas diminishing the strength of objects far away. The potential function is the mathematically descriptive form of the distribution low of a data field. The data space $\Omega \subseteq R^P$ contains an arbitrary point $x = (x_1, x_2, x_3, \ldots, x_P)^T$ exists in Ω. The scalar potential function of the specific point in the space could be rewritten as in the Eq. 1.

$$\varphi(x_i) = m_i \times K\left(\frac{\|x - x_i\|}{\sigma}\right) \tag{1}$$

When the m_i is the mass of x_i and $K(x)$ is the selective unit potential function to express the low that x_i always radiates its data energy in the same way in its data field. In the other hand, the unit potential function cloud be the representation (or reflect) of the density of the data distribution. The unit potential function could hold the parameters and properties up to the type of the function. For example, the distance $\|x - x_i\|$ and the impact factor σ could be changed according to the situation in the data space and prior condition.

As the field, the data field could be virtually created with a set of data objects $A = (x_1, x_2, x_3, \ldots, x_n)$ in the same space $\Omega \subseteq R^P$. The potential at any point in the field could be written as the Eq. 2.

$$\varphi(x) = \varphi_A(x) = \sum_{i=1}^{n} \varphi_i(x) = \sum_{i=1}^{n} \left(m_i \times K\left(\frac{\|x - x_i\|}{\sigma}\right)\right) \qquad (2)$$

With the scalar potential form, the gradient operator could be insisted to construct the vector intensity as written in Eq. 3.

$$F(x) = \nabla\varphi(x) = C \times \sum_{i=1}^{n} \left((x_i - x) \cdot m_i \times K\left(\frac{\|x - x_i\|}{\sigma}\right)\right)$$
$$= \sum_{i=1}^{n} \left((x_i - x) \cdot m_i \times K\left(\frac{\|x - x_i\|}{\sigma}\right)\right) \qquad (3)$$

The result of the gradient could produce the vector pointing to the center of the sub-field. The constant C could be ignored for mathematical convenient. Both of the scalar form and vector form model, could use in the different contexts and conditions [6].

With this interpretation, the data field can enjoy many of properties and hypothesis exists in the physical field. For example, using mass to represent the strength of connection (potential), or form the topology of equipotential on computed data radiations. To form the virtual field, in which derived from the corresponding physical field, a few of behaviors should be concerned (i.e., short-range with source and temporal behavior). Each data point can radiate the distribution force, which allows a potential function (the distribution law of a data field) be formulated and calculated from every point in the data space.

2.2　HASTA Algorithm

HASTA (HierArchical-grid cluStering based on daTA field) [3], is proposed to quantize the dataset as the matrices. The quantized field could be shown as the grid alignment. Each of the grid could be interpreted as one of the entities in the data field which holding all the quantized data objects. To expand the focal point of the clustering algorithm, the larger quantized matrices could be counted in the process. The 2nd order-quantized field (hyper-grid) is the other grid to form the hierarchical grid. The interaction between grid and hyper-grid could form the potential value based on Eq. 3. Each of grid in a quantized field would accumulate the potential values interacted with each of grid in a 2nd order-quantized field. The potential could be specified to all of the feature vector space in the data object. Clustering centers of HASTA are defined to locate where the maximum value of each feature potential crossover. Edges of the cluster in HASTA are identified by analyzing the first-order partial derivative of potential value, thus the full size of arbitrarily shaped clusters can be detected. After using the flood-filled algorithm to label the cluster in the quantized field, the data objects in the raw field could be recognized the cluster they belonged to. The process could simply show as Fig. 1.

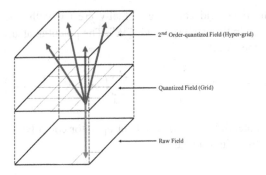

Fig. 1. HASTA overview

2.3   Auto-Regression Integrated Moving Average (ARIMA)

An ARIMA model is a class of statistical models for analyzing and forecasting time series data [7]. ARIMA is an acronym that stands for Autoregressive Integrated Moving Average. It is a generalization of the simpler Autoregressive Moving Average and adds the notion of integration. Each of these components is explicitly specified in the model as a parameter. A standard notation is used of $ARIMA(p, d, q)$ where the parameters are substituted with integer values to quickly indicate the specific ARIMA model being used. Adopting an ARIMA model for a time series assumes that the underlying process that generated the observations is an ARIMA process. This may seem obvious but helps to motivate the need to confirm the assumptions of the model in the raw observations and in the residual errors of forecasts from the model.

$$ARIMA(p, d, q) : \left(1 - \sum_{i=1}^{p} (\phi_i B^i)\right) \left((1 - B)^d\right) x_t = \epsilon_t (1 + \sum_{j=1}^{q} (\theta_j B^j)) \qquad (4)$$

The idea of combining 3 of the literature for extent the time-series clustering appears, when the HASTA [8] using the first-order partial derivative of potential value to determine the potential cluster center. In which, the set partial derivative value could consider as the distance measure within the single timeframe. In order to connect internal timeframe for track the cluster, the algorithm should be able to understand or to map the cluster from previous time frame with the current timeframe. In meantime, the ARIMA model behaves like distance matrices for time-series algorithm and the first-order partial derivative of potential value in the single timeframe.

3   Research Methodology

As mentioned in the literature review and related work, the idea of customizing the distance measure to extend the single timeframe clustering algorithm to time-series clustering algorithm is quite complicated. This methodology would declare the step to achieve the goal using both of the experimental approach and semantic approach [9].

3.1 Clustering Validity and Evaluation Criteria

In this section evaluation method for clustering [10], a number of view and axis have been purposed from the various interesting research. On the articles in time-series mining, the evaluation of time-series mining should follow some disciplines which are recommended as the various ranges like the used dataset should be published and freely available for implementation, the bias must be avoided by careful design of the experiments. Moreover, the definition of clusters also depends on the user, the domain, and it is subjective [11]. However, owing to the classified data labeled by a human judge or by their generator (in synthetic datasets), the result can be evaluated by using some measures; Rand Index, Adjusted Rand Index, Entropy, Purity, Jaccard, F-measure, FM, CSM, and MNI are used for the evaluation process (as shown in Fig. 2). All of these clustering evaluation criteria have values ranging from 0 to 1, where 1 corresponds to the case when ground truth and finding clusters are identical (except Entropy which is conversed and called cEntropy).

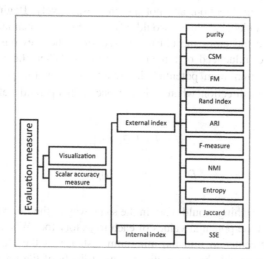

Fig. 2. Evaluation measure hierarchy widely used

The visualization and scalar measurements are the major technique for the evaluation of clustering quality which also is known as clustering validity in some articles [12]. And with the implication of visualization, the experiment would use this measure for the verification process.

3.2 Clustering Algorithm

With the extended version for HASTA, the program should be able to cluster data in each of the time points in the series separately. At this step, the experiment would like to adopt the calculation of simple distance measurement (Manhattan distance on two dimensions grid-space) as written in Eq. 5 to simply track the cluster movement on connected-dataframe. In the equation, i is the cluster that need to be labeled. C_{t+1}^i is the detected center of the cluster in current dataframe. Finding the manhattan distance with all the cluster in the previous dataframe C_t^i and select the minimum value of to rewrite the label of C_{t+1}^k to from the C_{t+1}^i. The experiment should show the limitation of the tracking method using direct way of measurement either the time complexity or accuracy.

$$C_{t+1}^i = \min_{k \in C_t} \left(Distance \left(C_{t+1}^k, C_t^i \right) \right) \tag{5}$$

Combining the ARIMA to solve the problem with the usage of the key element from HASTA is the crucial point for this experiment. The tracking could contain 2 parts; the forward prediction and the backward matching. The movement of the cluster could be defined in the discrete direction in Eq. 6 when $0, 1, 2, 3, 4$ are moving to the right, top, left, bottom direction, and not moving respectively. Beginning with backward matching, the unlabeled cluster would rely on the set of potential in the quantized field to observe the trajectory the cluster moved from the previous timeframe. The trajectory would be influenced or hinted by the Eq. 7. When the D is the direction which produce the minimum potential difference of C_t^i in the previous timeframe $\left(C_{t-1}^i \right)$ with selected direction when B is the backshift operand defined in ARIMA integration operand.

$$d \in \{0, 1, 2, 3, 4\} \tag{6}$$

$$D_d^{C_i} = \min_d (1 - B) p_{C_{t-1}^i}^d \tag{7}$$

The backward matching would work in the same way as the Eq. 5 but the factors of determination count the potential values of each trajectory too. When the Eq. 5 highly rely on the cluster center to find the minimum distance as the cluster tracker [13]. The ARIMA model would able to predict the most-likely of the trajectory the cluster will move to, based on the historical data. The forward prediction would predict the cluster trajectory from series defined as Eq. 8 using the ARIMA model as shown in Eq. 9. Then the matching would be used for inherit the cluster ID and labeling as shown in Fig. 3.

$$potential^d \in \left\{ p_{C_0^i}^{d'}, p_{C_1^i}^{d'}, \cdots p_{C_{t-1}^i}^{d'} \right\} \tag{8}$$

$$p_{C_t^i}^d = \underset{potential^d}{ARIMA}(p, d, q) \tag{9}$$

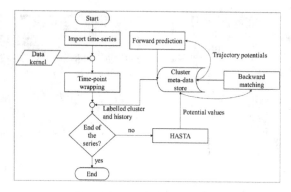

Fig. 3. Algorithm flowchart

4 Experiment and Discussion

In this experiment, the dataset would be clustered and continuously track along the timeline in the time-series dataframe [14]. After resampling the time-series, the aggregation function could be defined to deal with the quantized dataset. Using aggregate function to observe the feature space. The movement of features could be plotted and tracked in Fig. 4. For the preliminary tracking method, the mean of X and Y have been calculated as the center of the cluster. The nearest Manhattan distance (Eq. 5) of the cluster is the tracking caliber for this preliminary mapping.

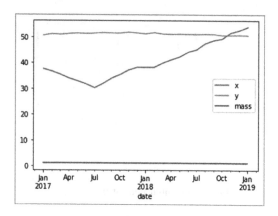

Fig. 4. Time wrapped features summary

The potential grid mechanism could detect the cluster peak (Fig. 5), according to the previous experiment. In the first interval frame (first-second frame), the Manhattan distance still is used in order to initialize the clustering bond (to solve the frozen-start problem). The forward prediction started at the second to the third frame using ARIMA for each trajectory with (p, d, q) equal to $(1, 1, 1)$. The reason is to match the short-range field property of data field [6].

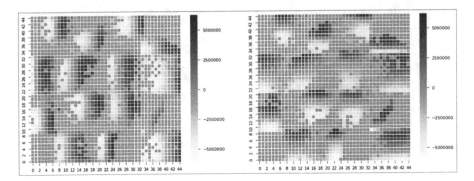

Fig. 5. Potential value of each features

Each of grid in the quantized field could contains the potential value from interaction with hyper-grid. And the result of HASTA could embed the set of potential values belong to the same HASTA step labeling. With a set of potential set, the time-series could be extracted as the Fig. 6. After the cluster has been initialized, the meta-data would be stored in structured object for backward matching and forward prediction process.

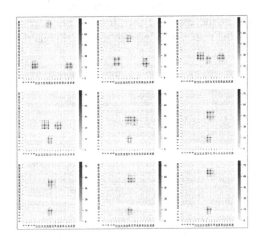

Fig. 6. Potential field of time-series

The model could predict the direction in the next frame (the third frame) using the Eq. 7 to select the direction mask. In this case, the cluster#1 is moving to the top of the viewpoint scope. The backward matching produces the result similar to the Fig. 7 (left), matching the trajectory and inherit the cluster id to the current frame after the result of forwarding prediction as Fig. 7 (right) could combine and label the cluster frame after frame along the time-series. With preliminary parameter for $ARIMA(1, 1, 1)$, the result of tracking algorithm could be compared with using K-mean

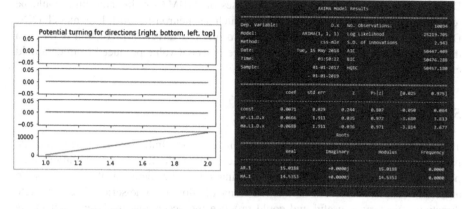

Fig. 7. (left) Backward matching, (right) Forward prediction

on the Manhattan distance for each of cluster as shown in Fig. 8. The result yields the consistent of the tracker in the movement between frame 7–12. The cluster could correctly inherit the previous frame label using combination of backward matching and forward prediction.

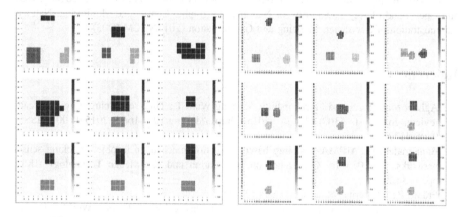

Fig. 8. Comparison between ARIMA-TS (left) and K-mean tracking (right)

The referral HASTA and ARIMA related tracking algorithm is the preliminary cluster and tracker tool extend from hierarchical and grid-based clustering tool. It could notice the some of the complex cluster movement like merge, coliid. In this kind of data explanation is useful in the real-world situation e.g. tell the group of related origin of current tracked cluster. The time-related operation could be extended in case of the complexity of the time-event involved; like seasonal, variance-deceased data. In the sense that this algorithm taking the time-series index as the pin-point, the further experiment could be designed like adopting more advance time-measurement like

DTW to manage to data and so on. The uses of ARIMA in this algorithm could be replaced with another prediction model, such as neuron-network-like model (NN, DNN, RNN, LSTM) to produce the same result. The direction prediction is the area that could be able to improve when using the potential value produced from data field calculation.

5   Conclusion

This clustering belongs to multi-step clustering category. Initialized with k-mean at the first internal time point, with the time point clustering, the temporal proximity of time points has been calculated using the combination of the hierarchical and grid-based algorithm. The cluster could be tracked along the flow of time-series. The 4-discrete direction trajectory is useful and could produce the acceptable tracking result in the grid-based data between the multiple frames. Moreover, the first-order partial derivative potential value could be used as the distance measure to calculate in the time-series clustering. With ARIMA integration, the cluster tracker could soundly reason the cluster movement.

Acknowledgment. This work was supported by the National Key Research and Development Program of China (No.2016YFB0502600), the National Natural Science Fund of China (61472039), Beijing Institute of Technology International Cooperation Project (GZ2016085103), and Open Fund of Key Laboratory for National Geographic Census and Monitoring, National Administration of Surveying, Mapping and Geoformation (2017NGCMZD03).

References

1. Aghabozorgi, S., Seyed Shirkhorshidi, A., Ying Wah, T.: Time-series clustering - a decade review. Inf. Syst. **53**(C), 16–38 (2015). https://doi.org/10.1016/j.is.2015.04.007. ISSN: 0306-4379
2. Banaezadeh, F.: ARIMA-modeling based prediction mechanism in object tracking sensor networks. In: 2015 7th Conference on Information and Knowledge Technology (IKT), pp. 1–5 (2015)
3. Fränti, P.: Clustering basic benchmark (2015)
4. Giachetta, G., Mangiarotti, L., et al.: Advanced Classical Field Theory. World Scientific, Singapore (2009)
5. Li, D., Wang, S., Li, D.: Spatial Data Mining. Springer, Heidelberg (2015). https://doi.org/10.1007/978-3-662-48538-5
6. Li, D., Wang, S., Yuan, H., Li, D.: Software and applications of spatial data mining. Wiley Interdiscip. Rev.: Data Min. Knowl. Discov. (Wiley Online Libr.) **6**, 84–114 (2016)
7. Ng, R.T., Han, J.: CLARANS: a method for clustering objects for spatial data mining. IEEE Trans. Knowl. Data Eng. **14**(5), 1003–1016 (2002)
8. Sardá-Espinosa, A.: Comparing time-series clustering algorithms in R using the dtwclust Package. Accessed May 2017
9. Steinbach, M., Ertöz, L., Kumar, V.: The challenges of clustering high dimensional data. In: Wille, L.T. (ed.) New Directions in Statistical Physics, pp. 273–309. Springer, Heidelberg (2004). https://doi.org/10.1007/978-3-662-08968-2_16

10. Vo, V., Luo, J., Vo, B.: Time series trend analysis based on k-means and support vector machine. Comput. Inform. **35**, 111–127 (2016)

11. Press, W.H., Teukolsky, S.A., Vetterling, W.T., Flannery, B.P.: Gaussian mixture models and k-means clustering. In: Numerical Recipes: The Art of Scientific Computing. Cambridge University Press, New York (2007)

12. Wang, S., Chen, Y.: HASTA: a hierarchical-grid clustering algorithm with data field. Int. J. Data Warehous. Min. (IJDWM) (IGI Global) **10**, 39–54 (2014)

13. Yan, Z.: Traj-ARIMA: a spatial-time series model for network-constrained trajectory. In: Proceedings of the Third International Workshop on Computational Transportation Science, pp. 11–16 (2010)

14. Zhang, T., Ramakrishnan, R., Livny, M.: Data Min. Knowl. Discov. **1**, 141–182 (1997)

Anomaly Detection with Changing Cluster Centers

Zhang Peng[✉] and Zhou Liang[✉]

Nanjing University of Aeronautics and Astronautics, Nanjing 210016, China
gegzhang@yeah.net, 1012010845@qq.com

Abstract. In view of the complexity of information system environment and the diversity of security requirements, many scholars proposed intrusion detection methods based on outlier mining. In order to meet the security requirement of condition guarantee information system, this paper proposes an anomaly detection with changing cluster centers (ADCCC). The rough set algorithm is used to reduce the sample set, and the number of sample repeats is determined on the basis of the duplicated degrees. The algorithm determines whether the sample is an outlier sample, mainly by changing the cluster center before and after adding a sample. Based on the overall deviation degree of the sample set, we can determine whether the sample set is an anomaly sample. Experimental results show that ADCCC algorithm has higher detection rate for anomaly detection.

Keywords: Anomaly detection · Rough set · Sample reduction
Outliers

1 Introduction

With the rapid development of the Internet, information system has been used in every aspect of daily life and commercial industry. The condition guarantee system is an integrated platform for the whole life cycle management of enterprise project construction, and has high security requirements. Meanwhile, attacks on network systems are increased diversely, so researchers are prompted to develop a variety of technologies to cope such circumstance. In the application of system security technology, on the one hand, the security architecture and strategy can be used to design the system. Besides, the used cryptographic system information transmission, access control and intrusion detection technologies can ensure operation safety, reliability and stability. On the other hand, the requirement of real-time analysis and processing must be considered, which requires reducing the complexity and detection time of intrusion detection model. In order to improve the processing precision and efficiency of intrusion detection model, artificial intelligence and data mining technology are widely used in the field of intrusion detection [1], such as naive Bayesian classification [2], KNN [3], decision tree [4], artificial neural network [5] and so on.

At present, the commonly used intrusion detection methods are generally divided into two categories: misuse detection and anomaly detection. The misuse detection must be defined by experts in advance, and with low false alarm rate. However,

© Springer Nature Switzerland AG 2018
G. Gan et al. (Eds.): ADMA 2018, LNAI 11323, pp. 42–54, 2018.
https://doi.org/10.1007/978-3-030-05090-0_4

adaptability and maintainability of misuse detection are bad, which can not detect new or unknown attacks, so does anomaly detection can. The anomaly detection achieves intrusion detection through comparing the monitored system behavior or the user's and resource usage pattern with the normal mode. Besides, the adaptability of anomaly detection is good, but the false detection rate is high. However, because of the advantages, anomaly detection has become the focus of study in the field of intrusion detection.

Breunig et al. [6] proposed a local outlier factor (LOF) for each object in the dataset. This was the first definition of outliers, and it also quantified outliers (quantifying which kind of objects are outliers). The position of outliers is relative, and each object considers only a limited neighbor. The LOF value of a data object was obtained by density comparison between itself and its neighbors, but it was only based on the density of the object of itself. It is important to note that the approach based on density does not explicitly divide the object into outliers. (If necessary, users can select a threshold to separate the LOF of two kinds.)

Zhang et al. [7] proposed outlier detection method based on local distances to find outliers from data sets. A local distance outlier factor (LDOF) determines the degree of deviation of the object from its neighborhoods (which cluster is closer to it). The overall complexity of calculated LDOF of all points in the dataset is $O(n^2)$, where the number of points in the data set is n.

Rough set theory [8] is a mathematical theory put forward by Pawlak in Poland. Rough set theory can be used to study uncertain data and fuzzy knowledge. In recent years, rough set theory has been widely applied in many fields such as data analysis, knowledge description, and so on. In the field of data analysis and data mining, the technology of attribute reduction based on rough sets is widely used [9, 10]. Through this technology, it can maintain better classification ability while attribute reduction and feature extraction to unlabeled attributes used for classification.

In this paper, an anomaly detection method based on changing cluster centers (ADCCC) is proposed. We analyze the intrusion behavior in the network by analyzing the influence of repeated samples added to the sample center. In the modeling, we preprocess the data and delete the redundant information irrelevant to classification, and then model all the existing normal behavior data and the anomaly behavior data. The contour of the abnormal behavior is established while the normal behavior contour established, and the double-contour hybrid detection model is formed.

2 Algorithm Design

The first step of ADCCC algorithm is to determine the value of k in the k-means algorithm by genetic algorithm. Next, the calculation of cluster centers and rough set theory are applied to the sample selection and sample reduction. We use outlier degree, anomaly judgment and threshold calculation to determine whether the sample is abnormal. In this process, the reduction of data sets, the Euclidean distance value and the over sampling method are mainly used [11]. After that, we test the three indicators of the sample: detection rate, false alarm rate and classification accuracy rate (Fig. 1).

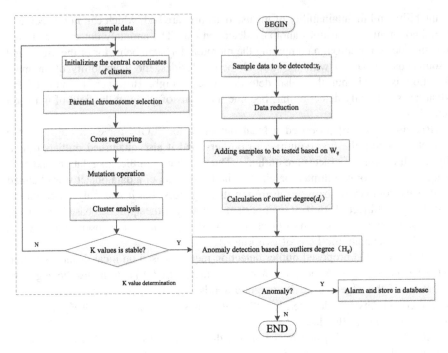

Fig. 1. ADCCC algorithm flow

2.1 K Value Determination

The specific steps and procedures of the genetic k-means algorithm are as follows:

- Step1 Sample data preprocessing: In order to solve the scale of continuity feature varies, this paper standardizes the initial data based on formula (1) to (3):

$$V = \frac{(|X_1 - AVG| + |X_2 - AVG| + \ldots \ldots + |X_n - AVG|)}{n} \tag{1}$$

$$AVG = \frac{(X_1 + X_2 + \ldots \ldots + X_n)}{n} \tag{2}$$

$$X_i' = \frac{X_i - AVG}{V} \tag{3}$$

X_i is the sample variable. n represents the number of variables.

- Step2 Initialization: Assuming that all chromosomes are the central coordinates of the cluster and n-dimensional, the length of each chromosome is n * k for k clusters.

- Step3 Parental Chromosome Selection: Two chromosomes with the highest fitness are selected as parental chromosomes, so it is necessary to specify the fitness function that makes the clustering as close as possible internally and the clustering time as loose as possible externally. Therefore, this paper defines fitness function as formula (4):

$$f = \frac{\sqrt{\sum_{i,j=1}^{k} \left\| c_i - c_j \right\|^2}}{\sqrt{\sum_{i=1}^{k} \sum_{j=1}^{N} \left\| x_j - c_i \right\|^2}} \tag{4}$$

c represents the cluster center, and x represents the data points except cluster centers.

- Step4 Cross Regrouping: A new chromosome is obtained by crossing part of the parent chromosome selected in step 2 with a certain probability.
- Step5 Mutation Operation: Some genes of the new chromosome obtained in step 3 mutate based on a certain probability. As the k value tends to stabilize, the mutation probability will be reduced in time.
- Step6 Cluster analysis and K-value determination: When the genetic k-means anomaly detection sub-module receives the data that the misuse detection module can not detect, it can judge whether the data is a new intrusion attack according to whether the data is outlier.

2.2 Selection and Preprocessing of Reference Samples

Samples Selection. When we use the cluster center as a reference sample to determine whether the newly added sample is an outlier, the center of the cluster is affected by the outliers [12], even though the computational complexity is low, and the results may be deviant. Therefore, we determine the outlier degree of the sample by the distance from the cluster center, and the outliers may not be correctly identified under the anomaly threshold of set. In particular, compared with adding a normal sample, adding an outlier sample has a greater impact on change of the center [13]. Even if the cluster center is selected as a reference sample, it is not appropriate to directly use the distance between the sample and the cluster center of its cluster to measure the degree of the outlier.

Using the cluster center of the normal sample set as the reference sample, we compare the change of the cluster center before and after the sample addition. The degree of the outlier of the newly added sample is judged, and the separation of the sample is calculated. Then the relationship between the outlier degrees and abnormal thresholds of the detected samples is used as a basis for judging whether the samples to be detected are abnormal samples. ADCCC is an anomaly detection method based on outlier mining. It takes the cluster center of the sample as the reference, and measures the outliers of each sample with the change of the cluster center.

For anomaly detection based on outliers, the vital problem is how to identify and select reference samples, and there are two kinds. First, we can select samples from the network access behavior of the information system directly, or we can get samples through proper classification and statistics of network access behavior of the system [14]. In general, samples in the same cluster tend to have a lot of similar properties, they can not only meet the needs of the sample selection, but also greatly reduce the number of samples.

Sample Reduction. In the face of complex application of network intrusion detection, it is difficult to obtain a good outlier mining result using only cluster centers as reference samples. The reason for this is that in network intrusion detection applications, each class of flow sample (such as normal flow, U2R flow) contains a number of specific types. Moreover, some of behaviors of the sample from different categories have similarities [12]. Therefore, after selecting the appropriate samples, it is necessary to preprocess the samples. Namely, we need to delete the redundant attributes and retain the necessary attributes.

The sample preprocessing in this paper is based on the rough set theory and dependency of attributes to reduce the redundancy attribute in the sample [15]. Then we get the reduction set of the flow sample. In short, after the sample enters the detection system, we label the classification attributes of the sample. We select a subset from the unlabeled attribute set (empty set at first), and calculate the change of attribute dependency of all unlabeled attributes in the sample in turn. If the attribute dependency is larger, the attribute will put into the subset. When judging remains once, we make a judgement whether the attribute dependency of the subset after adding attributes is equal to the attribute dependency of the original unlabeled attribute set. If the dependencies of the two attributes are equal, then the subset is the set of attribute reduction. Otherwise, we regard the attribute subset as the basis, and carry out the above operation until the attribute reduction set conformed to the condition is obtained.

Initial Cluster Center Calculation. In order to accurately identify some attacks, we extract a number of reference samples from the normal samples set. The calculation of cluster center (extracted by k-means algorithm) is similar to the calculation of cluster center method under the average number form, and it represents the average vector of all the samples within the range of its belonged. We can choose a number of cluster centers in the normal sample set as a reference sample.

The k-means algorithm is the most commonly used clustering algorithm at present. But the k-means algorithm is sensitive to the presupposed number k of clusters, and the choice of k has a great influence on the clustering. Therefore, we use an improved k-means algorithm to partition the sample set:

1. A cluster center is selected randomly to calculate the distance r_i between the rest samples and the center.
2. Get the total distance R for all r_i summation and calculate r_i/R, then select the cluster center corresponding to the maximum ratio as the second cluster.
3. Calculate the distance from the rest points to the nearest cluster center and repeat steps (2) until k cluster centers are obtained.

4. Use the k-means clustering algorithm to divide the network access behavior set X into k clusters C_1, C_2, ..., C_k, and calculate the cluster centers of each cluster c_1, c_2, ..., c_k.
5. Select $m(m \leq k)$cluster centers c_1, c_2, ..., c_m as a reference sample of the normal sample set from c_1, c_2, ..., c_k.

Generally, the number of chosen samples is more than 1% of the total number of samples, which may prevent the selection of noisy sample clusters or mislabeled samples containing too few samples.

2.3   Calculation of Outlier Degree

In this paper, before analyzing the sample, we need to do a reduction of the sample. After the reduction of attributes, we determine whether the sample is an outlier sample through the distance operation and sample model construction and analysis.

For attribute set $R \subseteq A$, if R satisfies $R = \{ X:X \subseteq C, \gamma_C(D) = \gamma_X(D)\}$, then R is an attribute reduction of attribute set A. Among them, C represents an unlabeled attribute set for classification. D represents a clustering attribute set. From the definition, we can see that the attribute reduction set is not the only one, so we can choose the attribute reduction set with the smallest cardinality as the final attribute reduction set.

After getting the reduction set, we represent it as clustering result data set $Q = \{c_q,$ $n_q, d_q, w_q\}$. Among them, Q is a cluster set after the AP classification of the normal sample data set. c_q represents the cluster center corresponding to the q-th cluster. n_q represents the number of samples associated with the cluster. d_q represents the maximum distortion of the cluster center c_q (the maximum distance between c_q and the samples contained in the cluster which c_q corresponded). w_q is the repeatability of the new samples in the cluster (the range of w_q is 0–1). The decision of whether the new added sample set X belongs to the intrusion event is transformed into computing the outlier degree of the sample in the set X relative to the cluster Q_i. Assuming that x_i is a newly added sample in X, we calculate the Euclidean distance between x_i and cluster centers. Then we find the cluster center c_q and the corresponding cluster C_q from Q while c_q is the nearest cluster center to x_i. We select C_q as the reference sample of the x_i decision. After that, the sample x_i is added to the cluster C_q and the new cluster center is calculated and defined as c_q'. The cluster center c_q' can be calculated through the following formula:

$$c_q' = \frac{1}{n_q + 1} * \sum_{x_j \in (C_q \cup \{x_i\})} x_j \qquad (5)$$

Obtaining the new cluster center c_q', we use the relative Euclidean distance to measure the relative distance d_i between two cluster centers of cluster C_q before and after adding x_i, that is:

$$d_i = dis(c_q, c'_q)/d_q \tag{6}$$

d_i indicates the outlier degree of all samples in C_q which is the nearest cluster from x_i. When d_i is larger than 1, x_i is an outlier sample and the event may be an anomaly. However, in the experiment, we find that when the number of normal samples contained in the sample set of Q is huge, the effect is not obvious when an outlier sample is added. It is often difficult to observe its obvious change. Therefore, in order to adapt the method to the large-scale training set and consider the difference of the number of training samples from different clusters, we use the method of sample differential replication. According to the number of normal samples of the reference sample set, we copy x_i times on the basis of a certain weight [16] (proportion). When each sample is added, we magnify the influence of the nearest cluster center (actual reference sample) to increase the division of d_i calculation. In general, when we add a regular sample, the influence of multiple conventional samples after the copy is still not very significant to d_i. When we replicate multiple outliers and add outliers sample to the nearest cluster, the sample has a significant influence on the calculation of d_i.

So, we assume that the number of replicating multiple samples x_i in X is $w_q * n_q$. c_q is the cluster center of cluster C_q which is nearest to x_i. n_q is the number of samples contained in C_q. When $w_q * n_q x_i$ are added to C_q, sample size of C_q is $n_q + w_q * n_q$, and the new cluster center c'_q is:

$$
\begin{aligned}
c'_q &= \frac{1}{n_q + w_q * n_q} * \sum_{x_j \in C_q \cup \{x_i, x_i, \dots x_i\}} x_j \\
&= \frac{1}{n_q(1 + w_q)} * \left(\sum_{x \in C_q} x_q + w_q * n_q * x_i \right) \\
&= \frac{1}{(1 + w_q)} * \frac{1}{n_q} \sum_{x \in C_q} x_q + \frac{w_q}{(1 + w_q)} * x_i) \\
&= \frac{1}{(1 + w_q)} * C_q + \frac{w_q}{(1 + w_q)} * x_i
\end{aligned}
\tag{7}
$$

To obtain the cluster center c'_q, we use formula (6) to calculate the corresponding outlier degree d_i. No matter how much w'_q is, the change in the position of the cluster center of adding an outlier into a cluster is still larger than the change of adding a normal one. Therefore, the greater the value of d_i is, the more likely an outlier sample is x_i.

But with the sample added, if the set contains a small number of samples and the time of repetitions is too much, we can see that the position of the cluster center closer to the newly added sample. Therefore, when increasing the number of duplicate samples, the number of samples added should be determined by the total impact of each point in the sample set.

The total impact value of the sample is calculated as follows:

$$H_q = \frac{\sum_{i=0, i \neq q}^{n} |(d_i \log d_i)|}{n_q} \tag{8}$$

2.4   Establishment of Normal Behavior Contour and Anomaly Detection

In order to establish normal behavior contours, after calculating the outlier degrees corresponding to the normal samples in X, we need an abnormal threshold τ to determine which normal samples can be used to establish the normal behavior contour. The specific method of judgment is as follows: the outlier degree of each sample in X compared with the value of the abnormal threshold τ, and the normal training samples are less than the abnormal threshold as the samples that can be used to establish the normal behavior contour. Moreover, in the experiment, an exorbitant outlier degree cannot be used to establish the normal behavior contour, so the anomaly detection method proposed in this paper can adapt to the normal sample training set mixed with a small amount of noise data (which may correspond to the intrusion behavior sample).

For a network anomaly detection method, different abnormal thresholds often have a greater impact on its detection performance (such as detection rate and false positive rate). Generally speaking, if the abnormal threshold is exorbitant, the detection rate of the anomaly detection method will be very low; if the abnormal threshold is too low, the false alarm rate will be exorbitant. Therefore, we need to get the threshold set through some calculations. We select a percentage β of a sample from the training set (the normal sample set X), and set the outlier degree of the sample as the abnormal threshold τ. In detail, we define a rank number α. The outlier degree of the α-th sample (as a threshold sample) is regarded as abnormal threshold τ. In order to get the abnormal threshold τ quickly, we can sort all the samples in X (assuming n samples in X) according to the outlier degrees, and then use the percentage $\beta(\beta = \alpha/n)$ to directly obtain the outlier degree which rank is α. For example, it is assumed that the normal training set has 100 samples. If β is set to 10%, we select the outlier degree of the ranking tenth directly in the ordered outlier degree sequence (total 100 elements) and set it as an abnormal threshold τ.

After getting the abnormal threshold τ, we can judge whether the new flow sample x_i is an abnormal flow sample by combining adding a sample strategy with over sampling. The specific way of judgment is to calculate the outlier degree d_i of x_i relative to the actual reference sample in the normal sample set X (the closest cluster center to x_i among m cluster center $c_1, c_2, ..., c_m$). If d_i is larger than the abnormal threshold τ, x_i is judged to be abnormal flow. Conversely, it is determined as normal flow.

The concrete construction and the detection process of ADCCC algorithm is shown as follow.

Algorithm: anomaly detection method based on changing cluster centers

Input: normal sample set X, specified percentage β set for abnormal thresholdτ , the number of repeated sampling w_q, the flow sample x_t to be detected, all Sample Number n in X

(1) Extract m cluster centers from the normal sample set X by k-means algorithm

(2) Calculate the dependency of all non-labeled attributes in the sample and find the attribute reduction set

(3) Calculate the Euclidean distances between x_t and c_q. c_q is the cluster center nearest to x_t and C_q is the cluster corresponding to c_q.

(4) Add x_t to the cluster c_q with w_q and calculate the new cluster center c_q' after the new sample x_t added by formula (7)

(5) Calculate the relative distance d_t between the cluster centers before and after x_t increase based on formula (6)

(6) Calculate the total impact value H_q of the sample set according to formula (8)

(7) Select the $n*\beta$-th score of the samples and set it as the abnormal thresholdτ .

(8) if $H_q < \tau$ then

(9) the test sample is normal

(10) else the test sample is abnormal

(11) end if

Output: the attribute of the flow sample x_t to be detected (normal or abnormal)

2.5 Complexity Analysis

In order to get the abnormal threshold τ, after optimizing the sample attributes, we only need to calculate outlier degrees of all samples related attributes in X, and sort them. The time complexity of optimizing attribute is $O(n)$. The time complexity required for calculation of the outlier degree of each sample in X is $O(nmp)$. The minimum time complexity required for sorting the outlier degrees of these samples is $O(nlogn)$. After obtaining an abnormal threshold τ, the time complexity and memory requirement for calculating the degree of outliers of a sample to be detected and judging whether it is an abnormal sample are $O(mp)$. Therefore, if the sample set containing S samples is detected, the time complexity of ADCCC is $O(smp)$.

From the above analysis, when s and p are fixed, the complexity of using ADCCC to complete anomaly sample set detection depends on the value of m. Because the number of clusters k divided by k-means algorithm is often far less than n, and $m \leq k$, so the time complexity and memory requirement of ADCCC in anomaly detection are low.

3 Experiment and Analysis

3.1 Experimental Data Set and Evaluation Index

The experimental data of this article are based on KDD CUP99 data set [17–19]. In the normal behavior contour construction, training sets are all normal category samples selected from the KDD CUP99 10% training set. Table 1 shows the sample distribution of these two sets.

Table 1. Sample distribution of experimental data sets

Class	Training set sample distribution	Test set sample distribution
Normal	95342	57413
U2R	0	264
R2L	0	21475
DoS	0	26547
Prb	0	3245
Total	95342	108944

Each original sample in KDD CUP99 dataset is characterized by 41 features, 38 of which are numerical features and the other 3 are character type features. The experiment consists of two steps in the data preprocessing module. The first step is to map some character types in the 41 features of each connection record into continuous features. Namely, we remove some features that have less impact on the classification, and replace some character features with numeric features. The second step is the classification and judgment of numerical characteristics.

Furthermore, we use several commonly evaluation indexes to evaluate the detection performance of ADCCC: detection rate, false alarm rate and classification accuracy.

3.2 Experimental Results and Analysis

This section gives ADCCC performance of binary detection (normal or anomaly) for the KDD CUP99 independent test set (corrected files) and various samples (identified as normal or anomaly) detection. Because the abnormal threshold usually has a great influence on the detection results, in formula (7), we can see that when the value w_q is different and there is a certain influence on the cluster center. In this subject, we determinate the influence on detection rate, the false positive and the classification accuracy when the value w_q is different by the experiment.

Table 2. The performance of binary detection of ADCCC with different β

β	Detection rate(%)	False positive(%)	Classification accuracy(%)
1%	91.03	1.56	92.47
3%	91.48	2.65	92.62
5%	91.79	3.50	92.71
7%	91.85	3.98	92.66
10%	91.96	4.83	92.58

As seen from Table 2, the binary detection rate of the ADCCC increases with the increasing of β (the abnormal threshold reduced constantly), but the false positive also increases. As the abnormal threshold continues to decrease, a part of samples recognized as normal when the abnormal threshold is high will be judged as abnormal when the abnormal threshold is low. And this part of samples contain not only some real attributes for the attack samples, but also some real attributes for the normal samples, they are judged to be abnormal samples. Some of them are abnormal, and some of them are normal. So the method improves the detection rate while the false alarm rate also increases. This is an inherent balance property of anomaly detection methods. From the current situation, it is very difficult for this method to break through this constraint relationship.

Then we study the detection performance of ADCCC under different repeated sampling time w_q. The number of clusters divided into 6 training sets, and β is 5%. Table 3 shows the binary detection performance of ADCCC.

Table 3. The binary detection performance of ADCCC under different values w_q

w_q	Detection rate (%)	False positive (%)	Classification accuracy (%)
0.2%	91.97	3.15	92.91
0.5%	91.98	3.09	92.94
0.7%	91.92	3.24	92.86
0.9%	91.85	3.32	92.80
1.0%	91.79	3.45	92.72

As can be seen from Table 3, the value w_q has a parabolic effect on the whole experiment. In the whole experiment, when w_q is 0.5, the classification accuracy is the highest; the false positive is the lowest. When w_q closer to 1, the detection rate, the lower the classification accuracy, the higher the false positive.

4 Conclusion

In view of the research and application of system intrusion detection, scholars have been studying and discussing all kinds of possible problems. Generally, network intrusion behavior can be regarded as outlier deviating from normal flow behavior. Outliers are regarded as mining technology for network intrusion detection. In this paper, the clustering center is used as a reference sample. By combining the rough set optimization algorithm and the over sampling "adding a sample" strategy, the outlier degrees of each sample are calculated and an abnormal threshold is defined to determine whether the normal behavior contour can be established and the anomaly detection can be carried out. This anomaly detection method has the characteristics of high detection efficiency and can adapt to a small amount of noise data. Experiments show that this method has the ability to detect known and unknown anomaly with relatively low detection rate and false alarm rate.

References

1. Guo, C., Zhou, Y.J., Ping, Y., et al.: Efficient intrusion detection using representative instances. Comput. Secur. **39**(4), 255–267 (2013)
2. Wang, Y., et al.: A novel intrusion detection system based on advanced Naive Bayesian classification. In: Long, K., Leung, V.C.M., Zhang, H., Feng, Z., Li, Y., Zhang, Z. (eds.) 5GWN 2017. LNICST, vol. 211, pp. 581–588. Springer, Cham (2018). https://doi.org/10.1007/978-3-319-72823-0_53
3. Feng, Y., Shigan, Y.U., Liu, H.: Network intrusion detection based on KNN-IPSO selecting features. Comput. Eng. Appl. (2014)
4. Jain, Y.K.: An efficient intrusion detection based on decision tree classifier using feature reduction. In: The First International Symposium on Data, Privacy, and E-Commerce, ISDPE 2007 (2007). 2012:c1
5. Hu, L., Zhang, Z., Tang, H., et al.: An improved intrusion detection framework based on artificial neural networks. In: International Conference on Natural Computation, pp. 1115–1120. IEEE (2016)
6. Breunig, M.M., Kriegel, H.P., Ng, R.T., et al.: LOF: identifying density-based local outliers. ACM Sigmod Rec. **29**(2), 93–104 (2000)
7. Zhang, K., Hutter, M., Jin, H.: A new local distance-based outlier detection approach for scattered real-world data. In: Theeramunkong, T., Kijsirikul, B., Cercone, N., Ho, T.-B. (eds.) PAKDD 2009. LNCS (LNAI), vol. 5476, pp. 813–822. Springer, Heidelberg (2009). https://doi.org/10.1007/978-3-642-01307-2_84
8. Pawlak, Z.: Rough set theory and its applications to data analysis. Cybern. Syst. **29**(29), 661–688 (2010)
9. Chen, Q., Liangxiao, J., Chaoqun, L.: Not always simple classification: learning super parent for class probability estimation. Expert Syst. Appl. **42**(13), 5433–5440 (2015)
10. Chen, H., Li, T., Luo, C., et al.: A decision-theoretic rough set approach for dynamic data mining. IEEE Trans. Fuzzy Syst. **23**(6), 1958–1970 (2015)
11. Lee, Y.J., Yeh, Y.R., Wang, Y.C.F.: Anomaly detection via online oversampling principal component analysis. IEEE Trans. Knowl. Data Eng. **25**(7), 1460–1470 (2013)
12. Bhuyan, M.H., Bhattacharyya, D.K., Kalita, J.K.: NADO: network anomaly detection using outlier approach. In: International Conference on Communication, Computing & Security, ICCCS 2011, February, Odisha, India, DBLP, pp. 531–536 (2011)
13. Xie, S., Nie, H.: Retinal vascular image segmentation using genetic algorithm plus FCM clustering. In: Proceedings of the 3rd International Conference on Intelligent System Design and Engineering Applications, Piscataway, NJ, pp. 1225–1228. IEEE (2013)
14. Shang, W., Zeng, P., Wan, M., et al.: Intrusion detection algorithm based oil OCSVM in industrial control system. Secur. Commun. Netw. **9**(10), 1040–1049 (2016)
15. Jaddi, N.S., Abdullah, S.: An interactive rough set attribute reduction using great deluge algorithm. In: Zaman, H.B., Robinson, P., Olivier, P., Shih, T.K., Velastin, S. (eds.) IVIC 2013. LNCS, vol. 8237, pp. 285–299. Springer, Cham (2013). https://doi.org/10.1007/978-3-319-02958-0_27
16. Jiang, L., Zhang, H.: Weightily averaged one-dependence estimators. In: Yang, Q., Webb, G. (eds.) PRICAI 2006. LNCS (LNAI), vol. 4099, pp. 970–974. Springer, Heidelberg (2006). https://doi.org/10.1007/978-3-540-36668-3_116

17. KDD-CUP-99 task description [EB/OL]. http://kdd.ics.uci.edu/databases/kddcup99/task. html. Accessed 29 June 2016
18. KDD cup 1999 data [EB/OL]. http://kdd.ics.uci.edu/databases/kddcup99/kddcup99.html. Accessed 29 June 2016
19. Singh, P., Garg, S., Kumar, V., et al.: A testbed for SCADA cyber security and intrusion detection. In: Proceedings of International Conference on Cyber Security of Smart Cities, Industrial Control System and Communications (2015). 1.6

A Novel Feature Selection-Based Sequential Ensemble Learning Method for Class Noise Detection in High-Dimensional Data

Kai Chen[1], Donghai Guan[1,2(✉)], Weiwei Yuan[1,2], Bohan Li[1,2], Asad Masood Khattak[3], and Omar Alfandi[3]

[1] College of Computer Science and Technology,
Nanjing University of Aeronautics and Astronautics, Nanjing, China
kaichen35315@foxmail.com, {dhguan,yuanweiwei,bhli}@nuaa.edu.cn
[2] Collaborative Innovation Center of Novel Software Technology
and Industrialization, Pittsburgh, USA
[3] College of Technological Innovation, Zayed University, Abu Dhabi, UAE
{Asad.Khattak,Omar.Alfandi}@zu.ac.ae

Abstract. Most of the irrelevant or noise features in high-dimensional data present significant challenges to high-dimensional mislabeled instances detection methods based on feature selection. Traditional methods often perform the two dependent step: The first step, searching for the relevant subspace, and the second step, using the feature subspace which obtained in the previous step training model. However, Feature subspace that are not related to noise scores and influence detection performance. In this paper, we propose a novel sequential ensemble method SENF that aggregate the above two phases, our method learns the sequential ensembles to obtain refine feature subspace and improve detection accuracy by iterative sparse modeling with noise scores as the regression target attribute. Through extensive experiments on 8 real-world high-dimensional datasets from the UCI machine learning repository [3], we show that SENF performs significantly better or at least similar to the individual baselines as well as the existing state-of-the-art label noise detection method.

Keywords: Noise Filtering · Sequential ensemble · Feature selection

1 Introduction

Noise detection is an important issue for machine learning [5,14]. Some research work has analyzed in detail the impact of class noise on the performance of the classifier. For example, in [5,15,19], the author introduced that the classification performance of the k nearest neighbor classifier is affected by the class noise. Research shows that the optimal value of k depends on the number of training

© Springer Nature Switzerland AG 2018
G. Gan et al. (Eds.): ADMA 2018, LNAI 11323, pp. 55–65, 2018.
https://doi.org/10.1007/978-3-030-05090-0_5

instances and the number of class noise. In general, smaller values of k can achieve higher performance with a small amount of class noise. However, as the number of class noise increases, the optimal k value of the nearest neighbor algorithm also monotonically increases. Therefore, it is necessary to propose techniques for eliminating noise or reducing its adverse effects.

High-dimensional data is ubiquitous in a wide range of practical applications, such as the millions of transactions in stock market surveillance, and the thousands of gene expression features in bioinformatics [13]. However, noise detection in high-dimensional data is a challenging task. High-dimensional data contains a large number of irrelevant attributes that can interfere with the performance of the classifier. In label noise detection tasks, classifier performance directly affects the accuracy of noise detection. Irrelevant features can also cause curse of dimensionality. Traditional methods often perform the two dependent step: the first step, searching for the relevant subspace, and the second step, using the feature subspace which obtained in the previous step training model. It has an obvious separate phase of dimensionality reduction which is undertaken before training the model [13]. However, feature subspace that are not related to noise scores and influence detection performance. In the literature [12], a sparse regression method was used to detect outlier in high-dimensional data. Inspired by it, we propose a sequential integration method to solve the high-dimensional problem of noise detection.

In this paper, we introduce a novel feature selection based sequential ensemble learning method (SENF) for class noise detection in high-dimensional data. The specific details of our method are as follows: first, SENF uses a given noise detection method to calculate the probability of an instance being mislabeled., and define a threshold function to select a set of noise candidate (A set of instances that have a high probability of being defined as noise). Second, the SENF performs sparse regression on the noise candidate set by using the noise score as the target attribute and the original feature space as the predictor to select the feature most relevant to the noise score. Note that we perform a regression on the candidate noise set rather than the entire original data set. Finally, a new detection model is trained using the feature subspace obtained in the second step instead of all feature spaces, and apply this new detector to generate a refine noise score. The above steps are performed iteratively, and a set of noise scores are generated until the sparse regression loss function does not decrease. In our method, the two tasks of noise detection and feature selection are combined and the two tasks interact to achieve the best result. In our approach, we use majority voting methods [2] to obtain the noise score, our method uses a set of learning algorithms to create a set of classifier to score the class label of data objects. In detail, we use n classifiers to score a positive instance (the label is positive).

In general, our contributions are following:

1. We introduce a novel Sequential Ensemble Noise Filtering (SENF) method for detect noise in high-dimensional feature space. SENF defines a sequential ensemble regression that combines the two tasks of noise detection and feature

selection, and mutually achieve optimal results at the termination of the iteration.

2. We evaluate our method on 8 real-world high-dimensional datasets from the UCI machine learning repository, we show that SENF performs significantly better or at least similar to the individual baselines as well as the existing state-of-the-art label noise detection method.

2 Related Work

2.1 Noise Filter Methods for Class Noise

In the existing literature, there are two commonly used methods for noise processing: algorithm level approaches [4,16] and data level approaches [?]

Some noise detection algorithms rely on classifier predictions: classification, voting, and k-fold cross-validation [5]. The predictions of classifier can be used to identify mislabeld data objects, therefore it is called classification filtering [6,9]. For example, in the literature [8,17], a classifier was trained using a data set with label noise and all instances that are misclassified by the classifier were removed. The classification filter performs k-fold cross-validation on the training data, training data is divided into k subsets. Then a set of classifiers is trained from the union of any $k - 1$ subsets. Afterwards, use these classifiers to detect noise instance in the excluded subset, and eliminating the label noise. However, since this method uses only a single classifier, a large number of instances are removed. In order to solve this problem, there are also some extended methods [11] that uses four different learning algorithms to learn a set of classifier and are combined by voting to detect mislabeled instances. There are two voting strategies used to identify misclassified instances: consensus vote and majority vote. (a) consensus voting removes instances which are identified misclassified by all the classifiers; (b) majority voting removes instances which are identified misclassified by more than a half of the classifiers. The above method can be applied to any classifier, however, these methods also has some inherent flaws. As discussed in [1], classifier with superior performance are necessary for classification filtering and learning in the presence of label noise may precisely produce low performance classifiers. Classification filtering suffered a paradox.

Voting filters are also removes noisy samples in multiple iterations until no more instances are removed. The IPF [10] describe that elimination of noise iteratively results in a more accurate noise filtering by using several classifiers built with the same learning. However, IPF only considers an algorithm to build a filter. The sensitivity of the algorithm to noise determines the performance of the filter. Also cannot benefit from the total information collected from multiple models which build from different algorithm. In the paper [10], using sample bagging to produce m classifiers to construct voting filters. Produce m training sample sets by resampling, then use this set of sample sets to build m classifiers. Finally, these classifiers are used to classify all instances in the original training data.

2.2 Feature Selection Techniques

Principal component analysis (PCA) is one of the most commonly used dimension reduction technologies in high-dimensional data. The PCA looks for a linear combination of original variables so that the derived variable captures the maximum variance. The success of PCA is due to the following two important factors: (a) The PCA captures the maximum variability in the feature space, thus ensuring minimal information loss. (b) The PCA are uncorrelated, so we can talk about one PCA without involving other components. However, PCA also has an obvious drawback, that is, each principal component is a linear combination of all features. This makes it often difficult to interpret the derived principal components.

Feature bagging is a feature extraction framework to handle high-dimensional data by using feature bagging. Feature bagging randomly selects a set of random feature subsets of size between $\lfloor \frac{d}{2} \rfloor$ and d from the original feature space, where d is the size of the dimension. Perform ensemble learning on this set of feature subspaces.

In the previous studies on noise detection in high-dimensional data, the usual practice was to divide feature detection and mislabeled instances detection into two separate tasks. However, this strategy may lead to a consequence that the resulting feature subset may not be related to the noise score. For methods that require class labels to obtain feature weights, one dilemma may be encountered in that the original data contains class noise. Therefore, obtaining the feature correlation coefficient may also be unreliable. In this paper we propose a new method to solve the above problem. The details will be introduced in the next section.

3 Proposed Approach

3.1 Sequential Ensemble for Noise Filter

The SENF approach builds a set of sequential ensembles to mutually refine noise score and feature extraction. As shown in Fig. 1, the detailed flow of our algorithm is as follows. Give a data set containing N instances $X = (X_1, X_2, ..., X_N)$, where $X_i = (X_{i1}, X_{i2}, ..., X_{id})$, each instance is described by a d-dimensional feature space. y is a vector containing all instance labels (Which contains the wrong labels). At the $t-th$ iteration, the noise score for all instances was obtained from the previous iteration. First of all, SENF define a Cantelli's inequality-based as noise score thresholding function which is to identify a set of most likely label noise $S^t \in R^{L^t \times (d+1)}$. L^t is the size of the noise candidate set, and $L^t \ll N$. S^t uses the original d-dimensional space, class labels, constitutes a dimensional space. Afterwards, we use lasso to define a sparse regression model ψ, SENF takes NS^{t-1} as a target attribute, and aggregates the original d-dimensional space and class labels as predictors. Then, applies sparse regression function

$\psi(S^t, NS^{t-1})$ to produce a new data set X'_t with a set of optimal features m^t, i.e. $X'_t = R^{N \times m^t}$. SENF uses noise scoring function H on X'_t to re-compute noise score of all instances, and obtain a new vector NS^t. Repeat the above process until $mse^t \geq mse^{t-1}$, and the entire iteration process constitutes a sequential ensemble model. Finally, we use sample bagging to build a set of sequential ensemble to produce the final noise score of all instances.

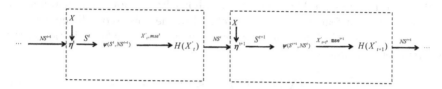

Fig. 1. This is an example of sequential ensemble. X is the original data set, NS is a vector that contains the noise score of all instances, η is a threshold function used to obtain noise candidate set, S is a noise candidate set, ψ is a regression function, X' is the new sample set after dimension reduction. mse is an empirical error, H is a scoring function.

3.2   Threshold Function

In order to select a set of noise candidate set, we use Cantelli's Inequality as a threshold function, which provides an upper bound for false positives. The specific definition of the threshold function is as follows:

Given the score of all instances, the greater the noise score of instance, the more likely noise. Let μ and δ^2 represent the expected value and variance respectively. Then noise candidate set can be defined as follows:

$$S = \{(X_i, y_i | \eta(NS_i, a) > 0)\}, \forall X_i \in X, y_i \in y, NS_i \in NS. \tag{1}$$

where $\eta(NS_i, a) = NS_i - \mu - a\delta$, a is a user-specified parameter. We use Cantelli's Inequality as a threshold function, which provides an upper bound for false positives $\frac{1}{1+a^2}$. Cantelli's inequality as a threshold function can well guaranteed the refinement of feature selection and noise scores in the iterative process.

3.3   Sparse Regression Function

SENF uses lasso to define a sparse regression model, which takes NS^{t-1} as a target attribute, and aggregates the original d-dimensional space and class labels as predictors. Then, applies sparse regression function to produce a new data set with a set of optimal features. The regression function uses three parameters: noise candidate set $S \in R^{L \times (d+1)}$, the noise score from the previous iteration, and regularization parameter λ. The specific definition of the regression function is as follows:

$$\phi(S, NS, \lambda) = argmin_w \left(\frac{1}{2L} \sum_{i=1}^{L} (NS_i - S_i^T W) + \lambda \|W\|_1 \right), \forall NS_i \in NS, S_i \in S. \tag{2}$$

Where W is the correlation coefficient vector $W = (w_1, w_2, ..., w_d)$. When the regularization parameter λ value is given, solving Eq. 2 above can yield a set of non-zero coefficient features, which are related to the noise score. Then we get a new dataset $X' \in R^{N \times m}$ with m relevant features. The specific definition of the new data set is as follows:

$$X' = \{X_i | w_i \neq 0, 0 \leq i \leq d\}, \forall w_i \in W. \tag{3}$$

The value of the regularization parameter λ is crucial to the performance of the lasso regression model. In order to minimizes the mean square error mse, we use 10-fold cross-validation to get the optimal λ.

3.4   Detailed Implementation

Algorithm 1 gives a detailed process of SENF generating noise scores. First of all, the algorithm needs to input three parameters: data set X, the parameter of threshold function η, and bagging size n. Step 2–3, use the original feature space to calculate the noise score NS^0 of all instances. Here we use a majority vote to score each instance [21]. Let the initial value of the mean squared error mse be 1. Step 4–10 is a complete iterative process and is the core step of sequential ensemble. In detail, step 6, using Cantelli's Inequality as a threshold function to obtain a set of candidate noises set S^t. In step 7, we use lasso to define a sparse regression model, which takes NS^{t-1} as a target attribute, and aggregates the original d-dimensional space and class labels as predictors. This step gets a correlation coefficient vector W and mean squared error mse. Step 8, according to the non-zero element in the correlation coefficient vector, we can obtain a new dataset with a set of relevant features. According to this new data set, we can get more accurate noise scores of all instance. Repeat steps 4–10 until $mse^t \geq mse^{t-1}$, according to our experiments, it takes only a few iterations to converge to optimality. We aggregate the scores generated during the iteration to produce one sequential ensemble score. As follows:

$$score(X) = \frac{1}{T} \sum_{t=1}^{T} NS^t(X). \tag{4}$$

where NS^t is the noise score generated during each iteration.

Algorithm 1. SENF

Require: DataSet: X, Threshold Funtion Parameter: a, Bagging Size: n
Ensure: The Noise Scores: $final\_scores$
1: **for** $j = 0 \rightarrow n$ **do**
2:     $NS^0 \leftarrow H(X)$
3:     $mse^0 \leftarrow 1, t \leftarrow 0$
4:     **repeat**
5:         $t \leftarrow t + 1$
6:         $S^t \leftarrow \eta(NS^{t-1}, a)$
7:         $W^t, mse^t \leftarrow \psi(S^t, NS^{t-1}, \lambda^t)$
8:         $X' \leftarrow \{X_{\cdot i} | w_i^t \neq 0, 0 \leq i \geq d\}$
9:         $y^t \leftarrow H(X')$
10:     **until** $mse^t \geq mse^{t-1}$
11:     $score_j(X) \leftarrow \frac{1}{T} \sum_{t=1}^{T} NS^t$
12: **end for**
13: $final\_scores \leftarrow \frac{1}{n} \sum_{j=1}^{n} score_j(X)$
14: **return** $final\_scores$

3.5   Obtain the Noise Set

After obtaining the noise score of all instances, we compare the score of each instance with 0.5. When the noise score of the instance is greater than 0.5, it means that more than half of the classifiers consider the instance to be noise. Specific detail as show in Algorithm 2:

Algorithm 2. Noise Filtering

Require: Noise Score of All Instance: $final\_scores$
Ensure: The Noise Set: NE
1: $NE \leftarrow \{\}$
2: **for** $i = 0 \rightarrow N$ **do**
3:     **if** $final\_scores[i] > 0.5$ **then**
4:         $NE \leftarrow NE \cup \{i\}$
5:     **end if**
6: **end for**
7: **return** NE

---

4   Experiments

Using standard evaluation metrics, we empirically compared SENF with the eight baseline methods on 10 real-world benchmark datasets from the UCI machine learning repository [3]. The experiment method is to train the classifier using a dataset that has been removed the noise (noise set is detected by each noise filter), and then evaluates the accuracy of the classifier to classify the test set.

4.1   DataSet

Table 1 summarizes the details of 8 real-world datasets used in our experiments. The procedures of each dataset preparation are presented below. In order to make the data set meet our task, we randomly add noise to the data set according to the noise ratio (0.1, 0.2, 0.3, 0.4). There is a common feature of these datasets is that they have high dimension.

Table 1. The information of DataSets

Name	N	d
Optdigits	1120	64
Musk	2034	166
Ad	773	1558
Isolet	480	617
Covtype	37877	54
MlProve	3059	56
Biodeg	1055	41
Spambase	4601	57

4.2   Test Accuracy

We compared the performance of SENF with several baseline methods, namely 1-ENN [18], 3-ENN [18], CF, MF [2]. We have also used two feature selection methods: featuring Bagging (FB) and REFCV [7]. The k-ENN method is based on the neighbor's label to identify whether the instance is a noise. Consensus filtering (CF) thinks that only when all classifiers consider an instance to be mislabled, it is marked as noise. Majority Filtering define the instance as noise if and only if more than half of the classifiers think the instance is mislabeled.

Table 2 presents the test accuracy of all compared algorithms on each benchmark dataset which contain noise with different noise ratio. Experimental results show that SENF statistically is superior to other algorithms on most of the datasets. Due to limited space, more experimental results are shown in the appendix. The k-ENN algorithm performs better when the noise is relatively small, but the performance degrades rapidly with the increase of the noise ratio. The conditions of the CF algorithm are too strict, resulting in much noise remaining in the data set that has not been removed. MF algorithm is the best performance of all baseline algorithms. The MF is the most stable algorithm among all baseline algorithms. For the original data set we performed the following processing, we selected 0.8 of the data set as the training sample and the rest as the test set.

Table 2. Perform accuracy on DataSet optdigits

Method	Accuracy				
	0.1	0.2	0.3	0.4	AVG
1-ENN	0.9777	0.9330	0.7813	0.6339	0.8315
3-ENN	**1.0000**	0.9598	0.9107	0.7054	0.8940
CF	0.8929	0.8080	0.7098	0.6384	0.7623
MF	0.9911	0.8080	0.9688	0.7679	0.8839
1-ENN-FB	0.9464	0.9152	0.7991	0.5893	0.8125
3-ENN-FB	0.9955	0.9554	0.8125	0.6920	0.8638
CF-FB	0.8973	0.8036	0.7366	0.5982	0.7589
MF-FB	0.9911	0.9911	0.9777	0.9018	0.9654
1-ENN-REFCV	0.9777	0.9286	0.8527	0.6830	0.8605
3-ENN-REFCV	0.9955	0.9554	0.8750	0.7188	0.8862
CF-REFCV	0.8616	0.8080	0.6652	0.6830	0.7545
MF-REFCV	0.9955	0.9955	0.9866	0.9241	0.9754
SENF	**1.0000**	**1.0000**	**0.9911**	**0.9866**	**0.9944**

5   Conclusions

This paper proposes a new noise filtering method SENF in high dimentional data. SENF incorporates noise monitoring and dimensionality reduction, which learns the sequential ensembles to obtain refine feature subspace and Improve detection accuracy by iterative sparse modeling with noise scores as the regression target attribute. We have compared the proposed method with other well-known filters over a large number of real-world datasets with different ratios of label noise. Experiments show that our method performs better than other baseline methods, the statistical analysis supports our conclusions [20].

Acknowledgements. This research was supported by Nature Science Foundation of China (Grant No. 61672284), Natural Science Foundation of Jiangsu Province (Grant No. BK20171418), China Postdoctoral Science Foundation (Grant No. 2016M591841), Jiangsu Planned Projects for Postdoctoral Research Funds (No. 1601225C). This research was also supported by the Fundamental Research Funds for the Central Universities (No. NS2016089). Meanwhile, this research work was supported by Zayed University Research Cluster Award # R18038.

References

1. Angelova, A., Abu-Mostafam, Y., Perona, P.: Pruning training sets for learning of object categories. In: 2005 IEEE Computer Society Conference on Computer Vision and Pattern Recognition, CVPR 2005, vol. 1, pp. 494–501. IEEE (2005)
2. Brodley, C.E., Friedl, M.A.: Identifying mislabeled training data. J. Artif. Intell. Res. 11, 131–167 (1999)
3. Dheeru, D., Karra Taniskidou, E.: UCI machine learning repository (2017)
4. Folleco, A., Khoshgoftaar, T.M., Van Hulse, J., Bullard, L.: Identifying learners robust to low quality data. In: 2008 IEEE International Conference on Information Reuse and Integration, IRI 2008, pp. 190–195. IEEE (2008)
5. Frénay, B., Verleysen, M.: Classification in the presence of label noise: a survey. IEEE Trans. Neural Netw. Learn. Syst. 25(5), 845–869 (2014)
6. Gamberger, D., Lavrac, N., Groselj, C.: Experiments with noise filtering in a medical domain. In: ICML, pp. 143–151 (1999)
7. Guyon, I., Weston, J., Barnhill, S., Vapnik, V.: Gene selection for cancer classification using support vector machines. Mach. Learn. 46, 389–422 (2002)
8. Jeatrakul, P., Wong, K.W., Fung, C.C.: Data cleaning for classification using misclassification analysis. J. Adv. Comput. Intell. Intell. Inf. 14(3), 297–302 (2010)
9. Khoshgoftaar, T.M., Rebours, P.: Generating multiple noise elimination filters with the ensemble-partitioning filter. In: 2004 Proceedings of the 2004 IEEE International Conference on Information Reuse and Integration, IRI 2004, pp. 369–375. IEEE (2004)
10. Khoshgoftaar, T.M., Rebours, P.: Improving software quality prediction by noise filtering techniques. J. Comput. Sci. Technol. 22(3), 387–396 (2007)
11. Miranda, A.L.B., Garcia, L.P.F., Carvalho, A.C.P.L.F., Lorena, A.C.: Use of classification algorithms in noise detection and elimination. In: Corchado, E., Wu, X., Oja, E., Herrero, Á., Baruque, B. (eds.) HAIS 2009. LNCS (LNAI), vol. 5572, pp. 417–424. Springer, Heidelberg (2009). https://doi.org/10.1007/978-3-642-02319-4_50
12. Pang, G., Cao, L., Chen, L., Lian, D., Liu, H.: Sparse modeling-based sequential ensemble learning for effective outlier detection in high-dimensional numeric data (2018)
13. Pechenizkiy, M., Tsymbal, A., Puuronen, S., Pechenizkiy, O.: Class noise and supervised learning in medical domains: the effect of feature extraction. In: 2006 19th IEEE International Symposium on CBMS 2006 Computer-Based Medical Systems, pp. 708–713. IEEE (2006)
14. Sáez, J.A., Galar, M., Luengo, J., Herrera, F.: INFFC: an iterative class noise filter based on the fusion of classifiers with noise sensitivity control. Inf. Fusion 27, 19–32 (2016)
15. Sánchez, J.S., Pla, F., Ferri, F.J.: Prototype selection for the nearest neighbour rule through proximity graphs. Pattern Recogn. Lett. 18(6), 507–513 (1997)
16. Teng, C.M.: Dealing with data corruption in remote sensing. In: Famili, A.F., Kok, J.N., Peña, J.M., Siebes, A., Feelders, A. (eds.) IDA 2005. LNCS, vol. 3646, pp. 452–463. Springer, Heidelberg (2005). https://doi.org/10.1007/11552253_41
17. Thongkam, J., Xu, G., Zhang, Y., Huang, F.: Support vector machine for outlier detection in breast cancer survivability prediction. In: Ishikawa, Y., et al. (eds.) APWeb 2008. LNCS, vol. 4977, pp. 99–109. Springer, Heidelberg (2008). https://doi.org/10.1007/978-3-540-89376-9_10

18. Wilson, D.L.: Asymptotic properties of nearest neighbor rules using edited data. IEEE Trans. Syst. Man Cybern. **3**, 408–421 (1972)
19. Wilson, D.R., Martinez, T.R.: Reduction techniques for instance-based learning algorithms. Mach. Learn. **38**(3), 257–286 (2000)
20. Yue, L., Chen, W., Li, X., Zuo, W., Yin, M.: A survey of sentiment analysis in social media. Knowl. Inf. Syst. 1–47 (2018)

Possibilistic Information Retrieval Model Based on a Multi-terminology

Wiem Chebil[1]($\boxtimes$), Lina F. Soualmia[2], and Mohamed Nazih Omri[1]

[1] MARS Research Laboratory, University of Sousse, Sousse, Tunisia
chebil.wiem@yahoo.com, mohamednazih.omri@fsm.rnu.tn
[2] LITIS-TIBS EA 4108, Normandie University, Rouen University, Rouen, France
lina.soualmia@litislab.fr

Abstract. We proposed in this paper a new approach for information retrieval intitled Conceptual Information Retrieval Model (CIRM). Our contribution is to exploit possibilistic networks (PN) and a multi-terminology in order to extract and disambiguate terms and then to retrieve documents. The two measures of possibility and necessity were used to select the relevant concept of an ambiguous term. Thus, the user query and unstructured documents are described throught a conceptual representation. Concepts were then filtered and ranked. Finally, a possibilistic network was exploited to match documents and queries. Two biomedical terminologies were exploited which are the MeSH thesaurus (Medical Subject Headings) and the SNOMED-CT ontology (Systematized Nomenclature of Medicine of Clinical Terms). The experimentations performed with CIRM on the OHSUMED corpus showed encouraging results: the improvement rates are +43.18% and +43.75% in terms of Main Average Precision and Normalized Discounted Cumulative Gain when compared to the baseline.

Keywords: Information retrieval · Possibilistic networks
Terms disambiguation · Concepts · Biomedical terminologies

1 Introduction

The aim of an Information Retrieval System (IRS) is to acquire, organise, store and find the most relevant information on the Web for user. Documents retrieved by an IRS is a result of a process triggered by a user query. The process of an IRS begins with indexing resources retrieved on the web and indexing the query. Then the index of user query is matched to the index of each document of the corpus. We can distinguish two types of indexing. The first is the free indexing that allows the extraction of free terms without using a controlled resource which lead to extract terms that not conform to the topic of the document. The second uses controlled vocabularies, for example the biomedical thesaurus MeSH [1]. The controlled matching allows a conceptual representation, thus, relevant terms that didn't occur in documents can be extracted, which not the case of free

© Springer Nature Switzerland AG 2018
G. Gan et al. (Eds.): ADMA 2018, LNAI 11323, pp. 66–79, 2018.
https://doi.org/10.1007/978-3-030-05090-0_6

term indexing. The indexing process usually starts with the pre-treatment [2] followed by the extraction of concepts. These later are then ranked according a score. Many approaches have been proposed to improve the IR process (Sect. 2). However, some of these approaches used one measure to evaluate a document or a term [4–15]. A unique measure reflects the incompleteness of the relevance information [3]. Others approaches didn't explore external and semantic resources [3–9, 12–14].

To further increase the performance of the IRS, it's necessary to deal with the ambiguity of terms. A term is ambiguous if it may be matched to more than one concept so to more than one sense. For example the term "cold" in a document may be matched to a MeSH concept and to three SNOMED-CT [16] concepts which are respectively "cold temperature", "Chronic obstructive lung disease (disorder)", "Common cold (disorder)" and "Cold sensation quality". The aim of the Word Sense Disambiguation (WSD) is to find the right sense for an ambiguous word. Several approaches have been proposed to disambiguate the sense of a term (Sect. 2). However, some of these approaches didn't perform a conceptual representation [22, 23]. Others, exploited one measure to select the right sense of a term and evaluate the relevance of a concept [17–23].

To improve the performance of the IRS, we proposed a new approach denoted Conceptual Information Retrieval Model (CIRM). This later exploits possibilistic networks (PN) and a multi-terminology to extract and disambiguate terms in order to have a conceptual representation of documents and queries and match them. The PN is a graph model founded on a robust inference mechanism for reasoning under condition of uncertainty and has been exploited in different fields [3]. The PN exploits two measures of possibility and necessity to improve the evaluation of relevance.

This paper is organized as follows: the section Related Work presents the related work in information retrieval and WSD. This is followed by detailing of the steps of the proposed approach. The steps are: term extraction, concepts assignment, terms disambiguation, filtering, final ranking and matching queries and documents. In Experimental Evaluations, Results and Discussion, we describe the experimentations and the results generated, which is followed by the Analysis of Results and Discussion. Finally, in the Conclusion section, we conclude and present some future work.

2   Related Work

Several approaches for information retrieval have been proposed. The Vector Space Model (VSM) [4] computed a similarity between the query terms and the documents terms. Several similarities could be used by the VSM such as cosines similarity. Another IR model is the probabilistic model. This model proposed to compute a probability of relevance of a document for a query. It is based essentially on a mathematical model based on the theory of probability [5,6]. The result of the matching between the document and the query is a set of documents ranked according their probabilities. The Okapi model BM25 [7] is based on the probabilistic model and exploits probabilities to match a query and documents. The BM25 model takes into account the term frequency in document and the length of documents compared to the average length of all documents in the collection. The language model [8] associates a language model to each document. The probability that a query can be generated by the language model of a document was computed. Turtle [9] proposed a Bayesian inference network of IR which is composed of the document network and the query network. The VSM was used to extract MeSH terms [10]. These later were ranked with a statistic and semantic weight. Two measures of possibility and necessity was exploited [3] to evaluate the similarity between the documents and the query. Dinh and Tamine [11] exploited a multi-terminology to find the most relevant documents. Prasath et al. [12] presented a cross language information retrieval approach to retrieve information written using a language other than the language of the user query. Wang et al. [13] proposed a generalized ensemble model (gEnM) for documents ranking that linearly combines multiple rankers. Sneiders [14] exploited co-occurrence and proximity of query terms in the document and Vector Space Model to retrieve documents. Ensan and Bagheri [15] proposed a language model for semantic document retrieval. This approach is based on a probabilistic reasoning model for calculating the conditional probability of query concepts given values assigned to document concepts.

Word Sens Disambiguation (WSD) approaches can be classified in knowledge based approaches, supervised approaches and unsupervised free-knowledge approaches. Knowledge approaches are based on external resources as MeSH. Among these approaches we can cite the work of Jimeno Yepes and Berlanga [17] that exploited the Naive Bayes to compute a probability of a context of an ambiguous word knowing the concept (the sense). The concept having the highest probability was matched to the polycemy word. Lesk [18] computed a similarity between the context of the ambigious word and the description of the candidates concepts. Voorhees [19] matched document words to wordnet sunsets. If a word had more than one match the authors considered that the right synset (sense) is the one having in his neighbors more words in commun with the local context of the word in the document. Tulkens and et al. [20] proposed to combine word representation created on large corpora with definitions from UMLS (Unified Medical Language System) [21] to create a concept representation. This later was compared to the representation of the context of ambiguous terms. The supervised approaches for WSD use a training set and features to learn a classifier how to select the right sense of a word. However, these approaches need a large training set which not always available. These approaches can exploit methods such as decision tree [22] and neural networks [23]. We can cite also Panchenko et al. [24] as unsupervised free knowledge approaches. Panchenko et al. performed a meta−combination of dependency features with a language model to disambiguate a word.

3 The Proposed Approach

A document is denoted D_i and a concept is denoted C_f. $D_i \in \{D_1....D_N\}$, N is the number of documents in the corpus. $C_f \in \{C_1....C_M\}$, M is the number of concepts in UMLS that correspond to MeSH descriptors and SNOMED-CT concepts. Each concept is composed of a set of terms. A term is denoted T_j. $T_j \in \{T_1....T_R\}$ with R the number of terms in MeSH and SNOMED-CT. A query is denoted Q_h. $Q_h \in \{Q_1....Q_A\}$, A is the number of queries. The process of CIRM is composed of three tasks. The first one is computing a similarity between D_i and C_f ($sim(D_i, C_f)$). The second is to compute a similarity between C_f and a query Q_h ($sim(Q_h, C_f)$). The same method was applied to compute $sim(D_i, C_f)$ and $sim(Q_h, C_f)$ with replacing a query by a document. The third task is to match Q_h and D_i (algorithm 1). The method that computes $sim(D_i, C_f)$ and $sim(Q_h, C_f)$ is composed of six steps: pre-treatment, term extraction, concepts assignment, terms disambiguation, filtering and final ranking. The steps of pre-treatment and filtering are the same in the work of Chebil et al. [2,25–27].

3.1 Terms Extraction

The step of terms extraction is performed using the possibilistic network [2]. The $Sim(D_i, T_j)$ is the sum of the possibility of D_i knowing T_j ($\Pi(D_i/T_j)$) and the necessity of D_i knowing T_j ($N(D_i/T_j)$)(Eq. 1). This step and computing $\Pi(D_i/T_j)$ and $N(D_i/T_j)$ are detailed in Chebil et al. [2].

$$Sim(D_i, T_j) = \Pi(D_i/T_j) + N(D_i/T_j) \tag{1}$$

3.2 Concepts Assignment

The score of a concept C_f is the maximum score of its terms (Eq. 4). The term having the maximum score is the Representative Term (RT). The $\Pi(D_i/C_f)$ is equal to $\Pi(D_i/RT_e)$ (Eq. 1). The $N(D_i/C_f)$ is equal to $N(D_i/RT_e)$ (Eq. 2). RT_e is the RT of C_f. $RT_e \in \{RT_1....RT_K\}$ with K the number of RT of concepts in MeSH and SNOMED-CT.

$$\Pi(D_i/C_f) = \Pi(D_i/RT_e) \tag{2}$$

$$N(D_i/C_f) = N(D_i/RT_e) \tag{3}$$

$$Sim(D_i, C_f) = Sim(D_i, RT_e) = max_{T_j \in T(C_f)}(Sim(D_i, T_j)) \tag{4}$$

With $T(C_f)$ a set of terms of a concept C_f.

3.3   Terms Disambiguation

An extracted term can belong to more than one concept therefore to more than one sense. To deal with this ambiguity, we propose to keep the concept having the maximum value of the $Sim(D_i, C_f)$. Thus, if we have a term T_j having more than one concept (more than one sense), the appropriate sense of T_j is computed as follows (Eq. 5):

$$C_f^* = argmax_{C_s \in C(T_j)}(Sim(D_i, C_s)) \tag{5}$$

With $C(T_j)$ a set of concepts of a term T_j.

In the next step, the concepts are ranked according $sim(D_i, C_f)$. A step of filtering is then performed using UMLS [2,25–27]. The concepts are finally re-ranked using $sim(D_i, C_f)$.

The $sim(Q_h, C_f)$ was computed in the same way as $sim(D_i, C_f)$ with replacing D_i with Q_h. We kept only the first six concepts (experimentally tuned) to index the query after final ranking of concepts according to $sim(Q_h, C_f)$. We kept the first fifteen concepts [2] to represent each document after final ranking of concepts according to $sim(D_i, C_f)$. The next step of CIRM is matching Q_h and D_i.

3.4   Matching Queries and Documents: Computing $sim(D_i, Q_h)$

We matched quey and documents using one of the following three methods: VSM, BM25 or a possibilistic network. In fact, documents were ranked using similarities computed using one of the three methods. Then we tested them to choose the most performant method (Sect. 4).

Matching Query and Documents Using VSM. The Index of the Query (IQ) was matched with the Index of a Document (ID) using Eq. 6 ($sim(D_i, Q_h) = sim(ID, IQ)$).

$$sim(D_i, Q_h) = \frac{\sum_{q=1}^{NC} sim(Q_h, C_q) \times sim(D_i, C_q)}{\sqrt{\sum_{q=1}^{NC} sim(Q_h, C_q)^2 \times sim(D_i, C_q)^2}} \tag{6}$$

IQ: Index of the Query; NC: Number of concepts in IQ; ID: Index Document; $sim(Q_h, C_q)$: computed using Eq. 4; $sim(D_i, C_q)$: computed using Eq. 4; $C_q \in$ IQ; If $C_q \notin$ ID then $sim(D_i, C_q) = 0$.

Matching Query and Documents Using BM25. We exploited the BM25 weight measure to match query with documents. In fact, we computed the Eq. 7 that exploits the $sim(D_i, C_q)$ (Eq. 4) with $C_q \in$ IQ.

$$sim(D_i, Q_h) = \sum_{q=1}^{NC} log \frac{N - n_q + 0.5}{n_q + 0.5}$$

$$* \frac{sim(D_i, C_q)(k_1 + 1)}{sim(D_i, C_q) + k_1(1 - b + b\frac{len}{avgdl})} \tag{7}$$

N: is the number of documents in collection; **NC**: is the number of concepts in IQ (the Index of Query); n_q: is the number of documents represented by the concept q; k_1 **and** b: are tuned parameters; **len**: is the length of documents (number of distinct words in the corpus); **avgdl**: is the average length of documents; $sim(D_i, C_q)$: is the similarity between a document D_i and a concept C_q of the query (computed using Eq. 4); If $C_q \notin$ ID then $sim(D_i, C_q) = 0$

Matching Query and Documents Using a PN. To match the query to a document we exploited a PN [3]. Thus we computed $\Pi(D_i/Q_h)$ and $N(D_i/Q_h)$. The $Sim(D_i, Q_h)$ is the sum of the possibility and the necessity (Eq. 8).

$$Sim(D_i, Q_h) = \Pi(D_i/Q_h) + N(D_i/Q_h) \tag{8}$$

To compute $\Pi(Q_h \wedge D_i)$ we used $\Pi(D_i/C_q)$ (Eqs. 2 and 3) instead of $TF \times IDF$ [3] to compute $\Pi(C_q/D_i)$ (Eqs. 9 and 10 and Table 1). Thus comparing to the work of Boughanem et al. [3] we exploited a conceptual representation of the queries and documents instead of terms and the classical measure $TF \times IDF$. The PN is based on the Bayesian Network, and as the Bayes rule we have [2, 28, 29]:
Table 1 summaries the conditional possibility $\Pi(C_q/D_i)$. $C_q \in$ IQ.

Table 1. Conditional possibility $\Pi(C_q/D_i)$

Π	Document D	
Concept	d_i	$\neg d_i$
C_q	Eq. 9	Eq. 10
$\neg C_q$	1	1

$$\Pi(C_q/D_i) = \frac{\Pi(D_i/C_q)\Pi(C_q)}{\Pi(D_i)} \tag{9}$$

$$\Pi(C_q/\overline{D_i}) = \frac{\Pi(\overline{D_i}/C_q)\Pi(C_q)}{\Pi(\overline{D_i})} \tag{10}$$

Algorithm 1. Computing $sim(D_i, Q_h)$

Require: Sim, Sim1, Poss: vectors
Ensure: Simlast, Simlast1, Simlast2 : vectors
  {computing $sim(D_i, Q_h)$ using cosinus similarity}
  for <i=1 to N> do
    for <q=1 to NC> do
      for <h=1 to A> do
        $Simlast\ [i][h] \rightarrow Cosinus\ (Sim\ [h][q],\ Sim1\ [i][q])$
      end for
    end for
  end for
  {computing $sim(D_i, Q_h)$ using BM25 similarity}
  for <i=1 to N> do
    for <q=1 to NC> do
      for <h=1 to A> do
        $Simlast1[i][h] \rightarrow Sim\ BM25\ (Sim[i][q])$
      end for
    end for
  end for
  {computing $sim(D_i, Q_h)$ using PN}
  for <i=1 to N> do
    for <q=1 to NC> do
      $Poss1[q][i] \rightarrow FunctionP\ (Poss[i][q], Poss[q]\ )$
    end for
  end for
  for <i=1 to N> do
    for <q=1 to NC> do
      for <h=1 to A> do
        $Simlast2[i][h] \rightarrow FunctionP1\ (Poss1[q][i])$
      end for
    end for
  end for

Sim: is a vector that contains the similarities computed in the Eq. 1;
Sim1: is a vector that contains the similarities computed in the Eq. 1 with replacing a document by a query; **CosinusSim:** is a function that computes the Eq. 6; **SimBM25:** is a function that computes the Eq. 7; **Poss:** is a vector that contains the $\Pi(D_i/C_q)$; **FunctionP:** is a function that computes the $\Pi(C_q/D_i)$; **Poss1:** is a vector that contains the $\Pi(C_q/D_i)$; **FunctionP1:** is a function that computes $sim(D_i, Q_h)$ using the vector Poss; **Simlast:** is a vector that contains the similarity $sim(D_i, Q_h)$ computed using the VSM; **Simlast1:** is a vector that contains the similarity $sim(D_i, Q_h)$ computed using the BM25; **Simlast2:** is a vector that contains the similarity $sim(D_i, Q_h)$ computed using the PN; **N:** is the number of documents that will be indexed; **NC:** is the number of concepts in IQ(the Index of Query); **A:** is the number of queries in the collection test.

4 Experimental Evaluations, Results and Discussion

The corpus used to carry out all the experimentations was OHSUMED[1] [30] document collection that was used for the TREC-9 Filtering Track. Details on this corpus are presented in Table 2 and Fig. 1. Each query was replicated by four searchers, two physicians experienced in searching and two medical librarians. The results were assessed for relevance by a different group of physicians, using three point scale: definitely, possibly, or not relevant. We consider the possibly relevant and relevant documents as relevant for the binary relevance. To compare the manual index of OHSUMED documents to the automatic index we replace the MeSH terms by their corresponding UMLS concepts.

To assess whether CIRM improves the IR process, we computed the precision of CIRM using the three methods (VSM, BM25 and PN) when the first 5, 20 and 50 documents are retrieved. We also computed the MAP (Mean Average Precision) and the improvement rate to compare the performance of our approach with the baseline. We measure also the performance of CIRM using the NDCG (Normalized Discounted Cumulative Gain) for each query. Compared to the other measures that are based on a binary relevance (relevant or non-relevant), NDCG (Eqs. 11 and 12) integrates ratings for degree of relevance into evaluation of effectiveness. The NDCG measure evaluates if the IRS allows to rank the most relevant documents at the first ranks. In addition, to show the statistically significant improvements of CIRM, we computed the paired-sample t-tests between means of each ranking obtained by each experimented approach and the baseline. This later was the approach of Duy and Tamine [11] (Table 3). This approach is among the latest approaches that exploited a multi-terminology for IR which justifies our choice for this baseline. This later exploited MaxMatcher+ to index documents and then applied a voting method to keep relevant concepts among candidate concepts selected using four terminologies. In our experimentations we used two terminologies instead of four to compare the baseline to our approach.

$$DCG = \sum_{i=1}^{p} \frac{2^{rating(res(i))} - 1}{log(1 + i)} \qquad (11)$$

Rating(res(i)) = 0 if res(i) is irrelevant, res(i)) = 1 if res(i) is possibly relevant and res(i)) = 2 if res(i) is relevant. p is the number of documents retrieved in response to a query.

$$NDCG = \frac{DCG}{oDCG} \qquad (12)$$

with oDCG is the optimal DCG (computed using optimal ranking).

To choose the number of concepts that will index the quey, we evaluated the performance of CIRM-PN with different number of concepts (Table 4).

To evaluate the added value of the step of disambiguation we tested CIRM with the step of disambiguation and without this later (Table 5). To analyse

---

[1] https://trec.nist.gov/.

```
<top>

<num> Number: OHSU12

<title> 49 yo B male with hypotension, hypokalemia, and low
aldosterone.

<desc> Description:

isolated hypoaldosteronism, syndromes where
hypoaldosteronism and hypokalemia occur concurrently

</top>
```

Fig. 1. Example of a query

Table 2. Statistics about test corpus

The number of documents (words)	384,566
The average length of documents (words)	100
The number of queries	63
The average length of a query	12
The number of MeSH terms	4904
The average number of relevant documents/queries	50

Table 3. Evaluation of CIRM

Approach	MAP	P@5	P@20	P@50
[11](baseline)	0.44	0.54	0.49	0.43
CIRM-VSM	0.52(+18.82%)	0.64(+17.39%)	0.55(+12.21%)	0.47(+10.20%)
CIRM-BM25	0.45(+2.26%)	0.56(+3.29%)	0.50(+2.03%)	0.45(+4.87%)
CIRM-PN	0.63(+43.18%)*	0.71(+30.02%)*	0.63(+29.12%)*	0.57(+33.64%)*

CIRM-VSM: CIRM was tested using VSM (Eq. 6) to match queries and documents.
CIRM-BM25: CIRM was tested using BM25 (Eq. 7) to match queries and documents.
CIRM-PN: CIRM was tested using PN (Eq. 8) to match queries and documents.
*: a significant change at $p < 0.05$.

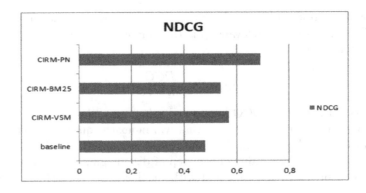

Fig. 2. Comparing the NDCG of CIRM to the baseline

Table 4. The performance of CIRM-PN using different values of NC

Measure	NC = 3	NC = 4	NC = 5	NC = 6	NC = 7
MAP	0.24	0.41	0.60	0.63	0.61
NDCG	0.31	0.48	0.65	0.67	0.69
P@5	0.23	0.45	0.69	0.71	0.69
P@20	0.21	0.42	0.61	0.63	0.59
P@50	0.25	0.35	0.54	0.57	0.56

NC: Number of Concepts that index the query Q_h.

even more the step of disambiguation we tested CIRM using only MeSH and then only SNOMED-CT with and without the step of disambiguation (Table 6, Fig. 3). The aim of this experimentation is to show which terminology has more ambiguous terms.

Table 5. Comparing the performance of CIRM-PN with and without disambiguation

Approach	MAP	NDCG	P@5	P@20	P@50
A	0.55	0.58	0.62	0.56	0.49
B	0.63(+14.51%)	0.70(+20.68%)	0.71(+14.49%)	0.63(+13.01%)	0.57(+17.31%)

A: CIRM-PN without disambiguation (baseline)
B: CIRM-PN with disambiguation

Table 6. Comparing the performance of CIRM-PN using MeSH or SNOMED-CT with and without disambiguation

Approach	MAP	NDCG	P@5	P@20	P@50
(1)	0.51	0.59	0.61	0.51	0.45
(2)	0.53(+3.89%)	0.62(+5.08%)	0.62(+1.79%)	0.52(+1.75%)	0.47(+4.43%)
(3)	0.45	0.51	0.48	0.46	0.41
(4)	0.51(+13.30%)	0.61(+19.60%)	0.55(+14.55%)	0.56(+21.38%)	0.52(+26.45%)

(1): CIRM using MeSH without a step of disambiguation (baseline)
(2): CIRM using MeSH with a step of disambiguation
(3): CIRM using SNOMED-CT without a step of disambiguation (baseline)
(4): CIRM using SNOMED-CT with a step of disambiguation

As expected, results in Table 3 showed the effectiveness of CIRM. In fact CIRM-VSM, CIRM-BM25 and CIRM-PN outperform the baseline. According to Table 3 the improvement rates of CIRM-PN compared to the baseline are +43.18%, +43.75% and +29.12% respectively in terms of MAP, NDCG (Fig. 2) and precision at 20. The good performance of CIRM-PN in terms of NDCG (Fig. 2)

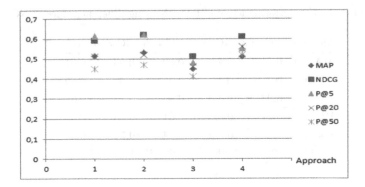

Fig. 3. The performance of CIRM using MeSH or SNOMED-CT with and without disambiguation

indicates that our approach allows a better ranking of the documents comparing to the manual one. In addition, CIRM is statistically significant compared to the baseline and the other approaches. The results in Chebil et al. [2] show that possibilistic networks and filtering improve extraction of relevant concepts and their ranking. This improvement contributed according to Table 3 and Fig. 2 to improve also the ranking of documents comparing to the baseline. This later is a multi-terminology based approach that exploited only one measure to rank documents which is BM25. It's clear also that CIRM-PN is more effective than CIRM-VSM and CIRM-BM25. This result highlights that exploiting the possibility and necessity degrees combined with a conceptual representation improves the performance of documents retrieval. Based on the fact that CIRM-PN outperformed CIRM-VSM and CIRM-BM25, we exploited CIRM-PN in the next experimentations (Tables 4, 5 and 6). Table 4 shows that CIRM-PN achieved better performance when NC=6. In fact precision at different ranks and MAP decrease when NC > 6 and NC < 6. This result allows as to keep NC = 6 when indexing queries. We can notice in Table 5 an improvement of the performance of CIRM-PN when a step of disambiguation was exploited. In fact the improvement rates of CIRM-PN with disambiguation are +14.51%, +20.68%, +14,49%, +13.01%, and +17.31% respectively in terms of MAP, NDCG, P@5, P@20 and P@50. These results emphasize the importance of the step of disambiguation of terms in the process of IR. In addition, it' s clear in Table 6 and Fig. 3 that the method of disambiguation improves CIRM when using SNOMED-CT more than using MeSH (the improvement rates are +13.30%, +19.60% +14.55%, +21.38%, and +26.45% respectively in terms of MAP, NDCG, P@5, P@20 and P@50 when using SNOMED-CT compared to the baseline). We can explain these results by the fact that SNOMED-CT may has more ambiguous terms than MeSH. Furthermore, results in Tables 5 and 6 confirm the effectiveness of using a multi-terminology specially MeSH and SNOMED-CT instead of using one terminology. In fact, precisions in different ranks, MAP, and NDCG of CIRM-PN when using MeSH and SNOMED-CT are higher than those of CIRM-PN when using only one terminology (MeSH or SNOMED-CT). This result reflects the positif impact of exploiting a multi-terminology in the process of IR.

5 Conclusion

Our contribution in this paper is to exploit possibilistic networks and terminologies to improve retrieving unstructured documents and their ranking in an IRS. The two measures of possibility and necessity and terminologies were exploited to extract and disambiguate terms. Thus, documents and the query are represented with ranked concepts. We tested then three methods to match queries and documents that exploited the similarities computed to describe documents and queries. The methods are the VSM, the BM25 and the PN. The exploitation of this later showed better results than the two other methods. In fact, the PN used the two measures of possibility and necessity to represent the documents and the queries with concepts and retrieve documents instead of the classical measures($TF \times IDF$ or BM25). In addition, the experimentations of CIRM carried out on OHSUMED corpus highlighted the added value of our approach since this later outperforms the baseline. Furthermore, the results emphasize the importance of the step of terms disambiguation in the process of IRS. We aim in the future to add a step of query expansion to further improve the performance of an IRS. It's interesting also in the future to exploit more terminologies. The good results of CIRM exploited on a biomedical database motivated us to plan to exploit CIRM on other fields.

References

1. Nelson, S.J., Johnson, W.D., Humphreys, B.L.: Relationships in medical subject heading. In: Relationships in the Organization of Knowledge, pp. 171–184 (2001)
2. Chebil, W., Soualmia, L.F., Omri, M.N., Darmoni, S.J.: Indexing biomedical documents with a possibilistic network. J. Assoc. Inf. Sci. Technol. **67**(4), 928–941 (2016)
3. Boughanem, M., Brini, A., Dubois, D.: Possibilistic networks for information retrieval. Int. J. Approx. Reason. **50**(7), 957–968 (2009)
4. Salton, G., Wong, A., Yang, C.S.: A vector space model for automatic indexing. Commun. ACM **18**(11), 613–620 (1975)
5. Salton, G., Fox, E., Wu., H.: Extended Boolean information retrieval. Commun. ACM **26**(11), 1022–1036 (1983)
6. Robertson, S.E., Jones, K.S.: Relevance weighting of search terms. J. Am. Soc. Inf. Sci. JASIS **27**(3), 129–146(1976)
7. Robertson, S., Walker, S.: Some simple effective approximations to the 2-Poisson model for probabilistic weighted retrieval. In: Proceedings of the International ACM-SIGIR Conference, pp. 232–241. ACM (1994)
8. Ponte, J.M., Croft, W.B.: A language modeling approach to information retrieval. In: Proceedings of the 21st Annual International ACM SIGIR, pp. 275–281. ACM, Melbourne (1998)

9. Turtle, H., Croft, W.B.: Inference networks for document retrieval. In: ACM SIGIR Conference on Research and Development in Information Retrieval, pp. 1–24. ACM, Brussels (1990)

10. Majdoubi, J., Tmar, M., Gargouri, F.: Using the MeSH thesaurus to index a medical article: combination of content, structure and semantics. In: Velásquez, J.D., Ríos, S.A., Howlett, R.J., Jain, L.C. (eds.) KES 2009. LNCS, vol. 5711, pp. 277–84. Springer, Heidelberg (2009). https://doi.org/10.1007/978-3-642-04595-0_34

11. Dinh, D., Tamine, L.: Towards a context sensitive approach to searching information based on domain specific knowledge sources. J. Web Sem. **12**, 41–52 (2012)

12. Prasath, R., Sarkar, S., O'Reilly, P.: Improving cross language information retrieval using corpus based query suggestion approach. In: Gelbukh, A. (ed.) CICLing 2015. LNCS, vol. 9042, pp. 448–457. Springer, Cham (2015). https://doi.org/10.1007/978-3-319-18117-2_33

13. Wang, Y., Choi, I., Liu, H.: Generalized ensemble model for document ranking in information retrieval. Comput. Sci. Inf. Syst. **14**(1), 123–151 (2016)

14. Sneiders, E.: Text retrieval by term co-occurrences in a query based vector space. In: Proceedings of COLING the 26th International Conference on Computational Linguistics, Osaka, Japan, pp. 2356–2365 (2016)

15. Ensan, F., Bagheri, E.: Document retrieval model rough semantic linking. In: Proceedings of WSDM, pp. 181–190. ACM, Cambridge (2017)

16. SNOMED-CT - SNOMED International. http://www.snomed.org/snomed-ct

17. Yepes, A.J., Berlanga, R.: Knowledge based word-concept model estimation and refinement for biomedical text mining. J. Biomed. Inform. **53**, 300–307 (2015)

18. Lesk, M.: Automatic sense disambiguation using machine readable dictionaries: how to tell a pine cone from an ice cream cone. In: Proceedings of the Annual International Conference on Systems Documentation, pp. 24–26. ACM, Toronto (1986)

19. Voorhees, E.M.: Using WordNet to disambiguate word senses for text retrieval. In: ACM SIGIR Conference, pp. 171–180. ACM (1993)

20. Tulkens, S., Šuster, S., Daelemans, W.: Using distributed representations to disambiguate biomedical and clinical concepts. arXiv preprint arXiv: 1608.05605 (2016)

21. Bodenreider, O.: The unified medical language system (UMLS): integrating biomedical terminology. Nucleic Acids Res. **32**(4), 267–270 (2004)

22. Parameswarappa, S., Narayana, V.N.: Kannada word sense disambiguation using decision list. IJETTCS **2**(3), 272–278 (2013)

23. Dayu, Y., Richardson, J., Doherty, R., Evans, C., Altendorf, E.: Semi-supervised Word Sense Disambiguation with Neural Models. ArXiv e-prints (2016)

24. Panchenko, A., Ruppert, E., Faralli, S.: Unsupervised does not mean uninterpretable: the case for word sense induction and disambiguation. In: Proceedings of the 15th Conference of the European Chapter of the Association for Computational Linguistics, pp. 86–98 (2017)

25. Chebil, W., Soualmia, L.F., Darmoni, S.J.: BioDI: a new approach to improve biomedical documents indexing. In: Decker, H., Lhotská, L., Link, S., Basl, J., Tjoa, A.M. (eds.) DEXA 2013. LNCS, vol. 8055, pp. 78–87. Springer, Heidelberg (2013). https://doi.org/10.1007/978-3-642-40285-2_9

26. Chebil, W., Soualmia, L.F., Omri, M.N., Darmoni, S.J.: Extraction possibiliste de concepts MeSH a partir de documents biomedicaux. Revue d Intell. Artificielle **28**(6), 729–752 (2014)

27. Chebil, W., Soualmia, L., Omri, M.N., Darmoni, S.J.: Biomedical documents indexing with Bayesian networks and terminologies. In: 12th International Conference
on Intelligent Systems and Knowledge Engineering (ISKE), pp. 1–6. IEEE, Nanjing
(2017)
28. Bounhas, M., Mellouli, K., Prade, H., Serrurier, M.: Possibilistic classifiers for
numerical data. Soft Comput. **17**(5), 733–751 (2013)
29. De Campos, L.M., Fernandez-Luna, J.M., Huete, J.F.: The BNR model : foundations and performance of Bayesian network-based retrieval model. Int. J. Approximate Reasoning **34**, 265–285 (2003)
30. Hersh, W., Buckley, C., Leone, T.J., Hickam, D.: OHSUMED: an interactive
retrieval evaluation and new large test collection for research. In: Croft, B.W.
(ed.) SIGIR 1994, pp. 192–201. ACM/Springer, Dublin (1994)

On Improving the Prediction Accuracy of a Decision Tree Using Genetic Algorithm

Md. Nasim Adnan[1(⊠)], Md. Zahidul Islam[1], and Md. Mostofa Akbar[2]

[1] School of Computing and Mathematics, Charles Sturt University,
Bathurst, NSW 2795, Australia
{madnan,zislam}@csu.edu.au
[2] Department of Computer Science and Engineering,
Bangladesh University of Engineering and Technology (BUET), Dhaka, Bangladesh
mostofa@cse.buet.ac.bd

Abstract. Decision trees are one of the most popular classifiers used in a wide range of real-world problems. Thus, it is very important to achieve higher prediction accuracy for decision trees. Most of the well-known decision tree induction algorithms used in practice are based on greedy approaches and hence do not consider conditional dependencies among the attributes. As a result, they may generate suboptimal solutions. In literature, often genetic programming-based (a complex variant of genetic algorithm) decision tree induction algorithms have been proposed to eliminate some of the problems of greedy approaches. However, none of the algorithms proposed so far can effectively address conditional dependencies among the attributes. In this paper, we propose a new, easy-to-implement genetic algorithm-based decision tree induction technique which is more likely to ascertain conditional dependencies among the attributes. An elaborate experimentation is conducted on thirty well known data sets from the UCI Machine Learning Repository in order to validate the effectiveness of the proposed technique.

Keywords: Decision tree · Genetic algorithm
Prediction accuracy · Knowledge discovery

1 Introduction

Data mining has entered into our day to day life; we now predict who would be the mayor of our town. This prediction is carried out by classifier(s) based on previously known information. In the same way, classifiers are used in business, science, education, medical, security and many other arena. As classifiers enter such influential and sensitive ambit, the importance of improving their prediction accuracy and knowledge discovery potential is paramount.

There are many different types classifiers in literature such as Artificial Neural Networks [40], Bayesian Classifiers [9], Nearest-Neighbor classifiers [19], Support Vector Machines [11] and Decision Trees [10,30,31]. Among them, the application domain of decision trees are considerably large as they can be readily

© Springer Nature Switzerland AG 2018
G. Gan et al. (Eds.): ADMA 2018, LNAI 11323, pp. 80–94, 2018.
https://doi.org/10.1007/978-3-030-05090-0_7

applied on data sets with categorical, numerical, high dimensional and redundant attributes [28]. More importantly, decision trees can express the patterns that exist in a data set into a set of logic rules (rules) that closely resembles human reasoning [27]. Also, decision trees require no domain knowledge for any parameter setting and therefore more appropriate for exploratory knowledge discovery [15]. Till date, C4.5 [30,31] remains to be one of the most accurate and popular decision tree induction algorithms [24]. In line with the above mentioned facts, we understand that any improvement beyond C4.5 can render significant influence over its large application domain.

Almost all popular decision tree induction algorithms such as C4.5 [30,31] and CART [10] follow the structure of Hunt's Concept Learning System (CLS) [17]. Hunt's CLS is in general a greedy (i.e. nonbacktracking) top-down partitioning strategy that attempts to secure "purer class distribution" in the succeeding partitions. Generally, a greedy strategy picks the locally best attribute that delivers the "purest class distribution" in succeeding partitions as the splitting attribute. However, this locally best attribute may not be the ultimate best attribute selected for that particular partition. The reason is: all impurity measures used for inducing decision trees assume that all non-class attributes are conditionally independent and hence ignore relationships that may have existed among some of the non-class attributes. As a result, greedy strategies may lead to the generation of suboptimal decision trees in applicable cases.

Building the optimal decision tree (in terms of prediction accuracy) by exhaustive search starts by placing each non-class attribute as the splitting attribute for each partition and generate a candidate decision tree from each combination of the splitting attributes. Then from all the generated decision trees, one candidate decision tree with the best prediction accuracy is recognized as the optimal decision tree. However, computing the optimal decision tree by exhaustive search is unrealisable with the increase of non-class attributes as the number of candidate decision trees grows exponentially.

In order to fix the problems of greedy approaches as well as avoid the computational burden of the exhaustive search, one-level look ahead strategies were developed. However, these strategies yielded larger and less accurate decision trees in many occasions [29]. As an obvious solution to these problems, Genetic Algorithm-based techniques can be applied that with high probability can generate optimal/near-optimal decision trees. Despite being computationally intensive, these algorithms are no longer exponential to the number of non-class attributes.

Genetic Algorithm (GA in this paper) is a class of computational framework inspired by evolution [38]. GA was first introduced by John H. Holland as an adaptation of natural evolution (*survival for the fittest*) in computing [16]. GA encodes a potential solution in a simple data structure called chromosome. Typically, the execution of a GA begins with a population of randomly defined chromosomes. Then GA iteratively moves forward by applying genetics-inspired operators/components such as crossover and mutation to create new population(s). Chromosomes of each population are evaluated and chromosomes representing better solution remain in the process to be given more chance to reproduce. The chromosome with the best solution so far is reported as the output of GA. Unlike greedy approaches that have a high chance of being stuck in a local optima for unidirectional search, GA performs a robust search in different directions of the solution space in order to find the optimal/near optimal solution [4].

The size of the solution space can be comparable to the size of an ensemble of decision trees. An ensemble of decision trees or decision forest is a collection of decision trees where an individual decision tree acts as the base classifier. The forest prediction is compiled by taking a vote based on the predictions made by each decision tree [35]. In general, decision forests are generated by inducing different decision trees by perturbing the training data set differently for each decision tree (usually, the training data set is perturbed by excluding some records and/or attributes). As a result of being generated from perturbed data sets, forest trees are compromised in terms of completeness as they partially reflect the training knowledge. Furthermore, forest constituted from decision trees generated for its own purposes closely resembles to a "black box" as the comprehensibility of a single decision tree is lost [8].

Over the last decade or so, a variant of GA called Genetic Programming (GP) has been extensively used for inducing decision trees to overcome different optimization problems such as prediction accuracy and conciseness. The main difference between GA and GP is in the representation of chromosomes. Instead of using a simple data structure like GA, GP uses more complex representation of a chromosome through tree-like structure [13]. However, none of the GP-based techniques proposed so far can effectively address conditional dependencies that may exist among some of the attributes and thus may fall short in generating optimal/near optimal decision tree in terms of prediction accuracy. On the other hand, techniques that adhere to the basic GA philosophy are largely unexplored in constructing decision trees for optimizing the prediction accuracy [8,13]. This inspires us to propose a new, easy-to-implement GA-based decision tree induction technique (NGA-DT) which is more likely to ascertain conditional dependencies among the attributes.

The rest of this paper is organized as follows: In Sect. 2 we provide Background Study that covers a brief introduction to decision tree and some of the major components of GA. In Sect. 3 we discuss some of the well-known and relevant GP-based decision tree induction techniques and their limitations. The proposed GA-based decision tree induction technique is described in Sect. 4. Section 5 discusses the experimental results in detail. Finally, we offer some concluding remarks in Sect. 6.

2   Background Study

2.1   Decision Tree

Hunt's CLS [17] can be credited as the pioneering work for inducing top-down decision trees. According to CLS, the induction of a decision tree starts by selecting a non-class attribute A_i to split a training data set D into a disjoint set of horizontal partitions [30,35]. The purpose of this splitting is to create a purer distribution of class values in the succeeding partitions than the distribution in D. The purity of class distribution in succeeding partitions is checked (using an impurity measure) for all contending non-class attributes and the attribute that gives purer class distribution than others is selected as the splitting attribute. The process of selecting the splitting attribute continues recursively in each subsequent partition D_i until either every partition gets the "purest class distribution" or a stopping criterion is satisfied. By "purest class distribution" we mean the presence of a single class value for all records. A stopping criterion can be the minimum number of records that a partition must contain; meaning that if an splitting event creates one or more succeeding partitions with less than the minimum number of records, the splitting is not considered.

Different decision tree induction algorithms that follow the same structure of CLS usually differs in using impurity measures (for measuring the purity of class distribution) in order to find the splitting attributes. For example, C4.5 [30,31] uses Gain Ratio while CART [10] uses Gini Index as impurity measures.

A decision tree consists of nodes (denoted by rectangles) and leaves (denoted by ovals) as shown in Fig. 1. The node of a decision tree symbolizes a splitting event where the splitting attribute (label of the node) partitions a data set according to its domain values. As a result, a disjoint set of horizontal segments of the data set are generated and each segment contains one set of domain values of the splitting attribute. For example, in Fig. 1 "Trouble Remembering" is selected as the splitting attribute in the root node. "Trouble Remembering" has two domain values: "Y" and "N" and thus it splits the data set into two disjoint horizontal segments in such as way that the records of one segment contain "Y" value for "Trouble Remembering" attribute and the records of another segment contain "N" value. The domain values of the splitting attribute designated for the respective horizontal segments are represented by the labels of edges leaving the node.

2.2   Some of the Major Components of GA

Usually, there are five major components in a GA as described in the following.

Initial Population Selection: We already know, a chromosome is a potential solution encoded in a data structure and a set of chromosomes is termed as "Population" in GA. Hence, the "Initial Population" is realized by encoding the first set of chromosomes.

Crossover: Crossover is crucial for the evolution of new population [32,38]. Generally, crossover operation is applied on a chromosome pair (parents) where they swap segments to form the offspring. In this way, a new set of chromosomes representing new solutions are generated [4,20].

Mutation: In Mutation, chromosomes are arbitrarily changed in order to induce more randomness in search directions of the solution space [4,20].

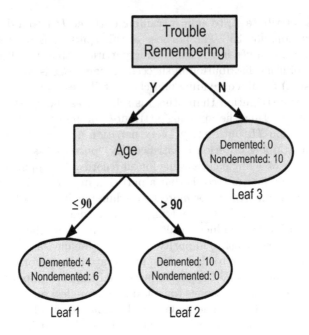

Fig. 1. Decision Tree

Elitist Operation: Elitist operation searches for the best chromosome (the best solution) in a population (from the initial population to a modified population through crossover and mutation operations) [4].

Chromosome Selection for the Next Iteration: After crossover and mutation operations (meaning, after an iteration), the modified population is compared with the immediate previous population (i.e. the population at the beginning of an iteration). If the modified population becomes inferior to the immediate previous population then there is a strong possibility of continuous degradation in subsequent iterations [4]. This may promote the search in wrong directions of the solution space. In order to prevent such scenario, better chromosomes are selected from the modified and the immediate previous populations for the next iteration.

3   Related Works

GP-based decision tree induction techniques that use tree-like chromosomes for encoding decision tree, seem to be the most common in literature [6,8,13,14, 36,41]. Representing a decision tree into a tree-like chromosome gives more flexibility in mimicking rules of a decision tree than a simple data structure such as a string or array. In [6], the authors applied a tree-like chromosome encoding scheme for competitive co-evolution of decision trees. The scheme encodes a binary decision tree by translating the nodes and leaves as 4-tuple:

$node = \{i, N, O, V\}$ where i represents the ID of the attribute tested, N indicates whether it is a node or a leaf and O is the operator used (meaningful for nodes only). V can represent dual values, for nodes it contains the test value and for leaves it contains the binary classification value. Each of the components of the 4-tuple contains numeric value and are subject to modification in the evolution process.

In [33], the authors represented a binary decision tree using a list of 5-tuple: $node = \{t, n, L, R, C\}$. Here, t represents the node number ($t = 0$ at the root), n is the attribute number (meaningful for only nodes), L and R points to left and right children respectively (meaningful for only leaves) and C represents set for counters facilitating the cut point to traverse to left/right child. Similarly, in [36] a decision tree was be represented as 7-tuple in a tree-like chromosome: $node = \{t, label, P, L, R, C, size\}$ where t represents node number ($t = 0$ means the root), $label$ is the class label of a leaf (meaningful for only leaves), P is the pointer to the parent node, L and R represent pointers to left and right children respectively (meaningful for only nodes, for leaves both pointers are $NULL$), C represents a set of registers where for example $C[0]$ stores the ID of the attribute tested and $C[1]$ stores the splitting value for the attribute. Finally, $size$ stores the number of nodes and leaves beneath; such like the size of the root is the size of the whole decision tree while the size of a leaf is 1 [8,36].

In [14], each decision tree of the initial population was generated from a different subset of the training data set in order to stress the decision trees to be as different as possible. All decision trees of the initial population are represented using a list of 4-tuple. This representation allows any node of a decision tree to become the root of a subtree commencing from the node. After initial population selection, two parent decision trees are selected according to roulette wheel technique [25,32] for crossover operation. In crossover, one node is randomly selected from each parent and then subtrees rooted from those nodes are exchanged (see Fig. 2).

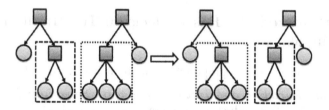

Fig. 2. Crossover

After crossover, mutation operation is applied which involves the exchange of subtrees within a decision tree. In doing so, two nodes within a decision tree are randomly selected and subtrees rooted from those nodes are exchanged (see Fig. 3).

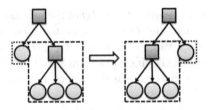

Fig. 3. Mutation

After crossover and mutation operations, it is possible that some decision trees may have logically inconsistent rules. Though, those rules do not affect prediction accuracy (because no records will satisfy logically inconsistent rules), yet the authors [14] opted to prune them. Finally, elitist operation finds the best chromosome (the best decision tree) according to prediction accuracy from the modified population in each iteration.

Despite being flexible in representing decision trees into chromosomes, GP-based techniques have some common limitations. For instance, almost each of the proposed algorithms deals with binary decision tree as it becomes more complicated to represent non-binary decision trees into tree-like chromosomes. The 7-tuple $node = \{t, label, P, L, R, C, size\}$ points to left and right children using two pointers L and R for binary decision tree whereas for non-binary decision trees the number of children may vary in each node. Thus, it becomes more complicated when non-binary decision trees are to be translated into tree-like chromosomes. Evidently, GP is more computationally intensive than GA as GP needs to parse decision trees into complex tree-like chromosomes. Further computational overhead and complications come from applying genetics-inspired operators such as crossover and mutation on those chromosomes. Moreover, with the exchange of subtrees it is difficult to explore conditional dependencies entangled among different sets of attributes in a limited number of iterations.

4   The Proposed GA-Based Decision Tree Induction Technique

The main components of the proposed technique is described as follows.

Chromosome Encoding and Initial Population Selection: The proposed GA-based decision tree induction technique encodes each chromosome (Cr_i) in a one-dimensional array where each cell contains the weight of a non-class attribute. Thus, the length of each chromosome is equal to the number of non-class attributes ($m = \{A_1, A_2, \ldots, A_m\}$) in the training data set. The weights are obtained randomly from a uniform distribution in the interval of $[0, 1]$. As a result, different chromosomes in the initial population obtain different sets of randomly generated weights for the same set of attributes. In the proposed technique, we encode 20 chromosomes to constitute the initial population ($|\mathbf{P}| = 20$) as was done in literature [4] (see Fig. 4).

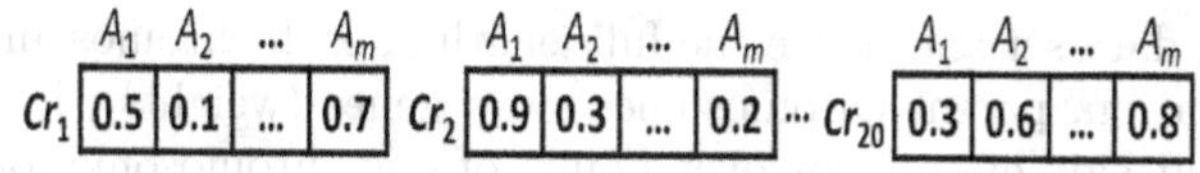

	A_1	A_2	...	A_m		A_1	A_2	...	A_m			A_1	A_2	...	A_m
Cr_1	0.5	0.1	...	0.7	Cr_2	0.9	0.3	...	0.2	...	Cr_{20}	0.3	0.6	...	0.8

Fig. 4. Initial population in the proposed technique

Elitist Operation: We apply elitist operation to find the chromosome (Cr_b) among the initial population ($\mathbf{P}_{Curr}$) from which the best C4.5 decision tree (in terms of prediction accuracy) is generated. As one C4.5 decision tree is generated from a particular chromosome, the weight distribution remains the same for all nodes of a single C4.5 decision tree, but different weight distributions are likely to be exerted for different C4.5 decision trees. When generating a C4.5 decision tree from a chromosome, at each splitting event, merit values of all non-class attributes are calculated by multiplying the value of impurity measure (such as Gain Ratio [30,31]) of each attribute with the respective random weight which is stored in the chromosome. After the merit values of all non-class attributes are calculated, the attribute with the highest merit value is selected as the splitting attribute.

In this way, the use of random weights introduces a random preference to some attributes in the induction process of a C4.5 decision tree. These "some attributes" ($\boldsymbol{S}$) are likely to be different in different chromosomes and hence gives us the opportunity to test conditional dependencies among different $\boldsymbol{S}$. We know that decision trees are considered to be an unstable classifier as a slight perturbation in a training data set can cause significant differences between decision trees generated from the perturbed and original data sets [3,35]. As a result, a population of chromosomes imposing random weights on attributes causing preferences to different $\boldsymbol{S}$ is more likely to help inducing a number of different decision trees incorporated with some of those conditional dependencies. The best decision tree among them is expected to be the best utilizer of such conditional dependencies. The chromosome generating the best C4.5 decision tree (i.e. the best chromosome) is stored as Cr_b.

Crossover and Mutation: For crossover, we select the best chromosome of $\mathbf{P}_{Curr}$ (Cr_b) as the first chromosome of a pair. To select the second chromosome of the pair, we use roulette wheel technique [25,32] where a chromosome Cr_r ($\neq Cr_b$) is selected with a probability $p(Cr_r) = \frac{PA(Cr_r)}{\sum_{i=1}^{|\mathbf{P}_{Curr}|} PA(Cr_i)}$ ($PA(Cr_r)$ is the prediction accuracy of the decision tree generated from chromosome Cr_r) affirming better chromosomes have greater chance of selection over weaker ones. Once a chromosome pair is selected, they are excluded from the process of choosing the upcoming pairs; all pairs are chosen in the same way as described. The motivation behind roulette wheel selection is to induce some randomness in choosing compatible peer from the best available chromosomes in a population. After pairing the chromosomes, standard 1-point crossover operation is applied on them as described in Fig. 5. A single crossover point is selected randomly between 1 and $|m|$ for each parent pair, where $|m|$ is the number of non-class attributes in the training data set and hence the full length of each chromosome. With the crossover operation, parent chromosomes swap genes (weights); left genes (i.e. genes in the left side of the crossover point) of one chromosome join the right genes of another chromosome. After crossover operation all the parent pairs are converted into offspring pairs with same number of genes.

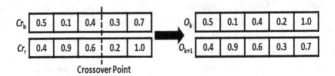

Fig. 5. Crossover in the proposed technique

For mutation, we select one gene randomly from each offspring and then regenerate the weight randomly in the interval of $[0, 1]$. In this way, mutation helps to induce some randomness in search directions. After both crossover and mutation, elitist operation is applied to find the Cr_b.

Chromosome Selection for the Next Iteration: At the end of each iteration, the modified population can be inferior to its immediate previous population. To prevent such degradation, at the end of each iteration we create a pool of chromosomes ($\mathbf{P}_{Pool}$) by adding chromosomes of the modified ($\mathbf{P}_{Mod}$) and its immediate previous population (i.e. the population at the beginning of the iteration, $\mathbf{P}_{Curr}$). Hence, $\mathbf{P}_{Pool}$ consists of 40 chromosomes. We then apply roulette wheel technique to select 20 chromosomes from the 40-chromosome $\mathbf{P}_{Pool}$. The selected chromosomes form the new population ($\mathbf{P}_{Curr}$) for the next iteration. This encourages that good chromosomes representing better solution remain in the process to be given more chance for reproduction. Finally, we obtain the best chromosome (Cr_b) from which we expect to induce an optimal/near optimal C4.5 decision tree.

5   Experimental Results

5.1   Data Set Information and Experimental Setup

We carry out an elaborate experimentation on thirty well known data sets that are publicly available from the UCI Machine Learning Repository [23] covering a variety of areas. The data sets used in the experimentation are described in Table 1. For example, the Car Evaluation data set has six non class attributes, 1728 records distributed in four distinct class values. The data sets are presented in Table 1.

We already know, till date C4.5 [30, 31] remains to be one of the most accurate and popular decision tree induction algorithms [18, 24] and any improvement beyond C4.5 can render significant influence over its large application domain.

Table 1. Description of the data sets

Data Set name (DS)	Non-class attributes	Records	Distinct class values
Abalone (AB)	08	4177	28
Balance Scale (BS)	04	625	3
Breast Cancer (BC)	33	194	2
Car Evaluation (CE)	06	1728	4
Chess (CHS)	36	3196	2
Credit Approval (CA)	15	653	2
Dermatology (DER)	34	358	6
Glass Identification (GI)	09	214	6
Hayes-Roth (HR)	04	132	3
Hepatitis (HEP)	19	80	2
Image Segmentation (IS)	19	2310	7
Ionosphere (ION)	34	351	2
Iris (IRS)	04	150	3
Letter Recognition (LR)	16	20000	26
Libras Movement (LM)	90	360	15
Liver Disorder (LD)	06	345	2
Nursery (NUR)	08	12960	5
Pen-Based Recognition of Handwritten Digits (PD)	16	10992	10
Pima Indians Diabetes (PID)	08	768	2
Seeds (SDS)	07	210	3
Sonar (SON)	60	208	2
Statlog Heart (SH)	13	270	2
Statlog Vehicle (SV)	18	846	4
Teaching Assistant Evaluation (TAE)	05	151	3
Thyroid Disease (TD)	05	215	3
Tic-Tac-Toe (TTT)	09	958	2
Wine (WNE)	13	178	3
Wine Quality (WQ)	11	6497	7
Yeast (YST)	08	1484	10
Zoo (ZOO)	16	101	7

Hence, we apply the proposed GA-based technique (NGA-DT) in order to induce optimal/near optimal C4.5 decision tree. The main purpose of our experimentation is to demonstrate how much improvement (in terms of prediction accuracy) an optimal/near optimal C4.5 decision tree can offer over a regular C4.5 decision tree. Therefore, we use the same settings for generating both NGA-DT and C4.5 decision trees. NGA-DT uses the same impurity measure (Gain Ratio [30,31]) and the entire training data set (not different subsets/bootstrap samples of the training data set) as used in a regular C4.5 decision tree. The minimum Gain Ratio/merit value is set to 0.01 for any attribute to qualify for splitting a node, Each leaf node of a tree contains at least two records and no further post-pruning is applied. In [14], decision trees are generated from different subsets of the training data set and thus compromised in terms of completeness as they partially reflect the training knowledge. Furthermore, those decision trees are not fully grown as logically inconsistent rules are pruned from them. Therefore, a decision tree generated from [14] cannot be regarded as a variant of a regular C4.5 decision tree and hence we exclude [14] from the comparison spectrum.

The experimentation is conducted by a machine with Intel(R) 3.4 GHz processor and 8 GB Main Memory (RAM) running under 64-bit Windows 7 Enterprise Operating System. All the results reported in this paper are obtained using 10-fold-cross-validation (10-CV) [7,21,22] for every data set. In 10-CV, a data set is divided randomly into 10 segments and from the 10 segments each time one segment is regarded as the test data set (out of bag samples) and the rest 9 segments are used for training decision trees. In this way, 10 training and 10 corresponding testing segments are generated. In our experimentation, we generate 20 decision trees from each population and hence a total of 20 × 20 (20 iterations) = 400 decision trees from each training segment and then evaluate their performance on the training and the corresponding testing segments. The best results reported in this paper are stressed through **bold-face**.

5.2   Comparison Between C4.5 and NGA-DT

Prediction Accuracy (PA) is one of the most important performance indicators for any decision tree algorithm [2,5]. In Table 2 we present the PA (in percentage) of C4.5 and NGA-DT on training and testing segments of all data sets considered. From Table 2 we see that NGA-DT performs better than C4.5 on training segments of all thirty data sets. This implies that training segments are more correctly reflected through NGA-DT. Hence, NGA-DT can be more reliable for knowledge discovery compared to C4.5.

It is shown in literature that maximizing the PA on training data set may lead to improving the generalization performance [34]. The results presented in Table 2 validate the proposition as NGA-DT performs better than C4.5 on testing segments of twenty three data sets and for one data set they have a draw. Now, to access the significance of improvement on the testing segments, we conduct a statistical significance analysis using Wilcoxon Signed-Ranks Test [1,39]. Wilcoxon Signed-Ranks Test is said to be more preferable to counting only significant wins and losses for comparison between two classifiers over multiple data sets [12]. We observe that PAs do not follow a normal distribution and thus do not satisfy the conditions for parametric tests. Hence, we perform a non-parametric one-tailed Wilcoxon Signed-Ranks Test [1,39] for $n = 30$ (number of data sets used) with the significance level: $\alpha = 0.005$. Thus, the critical value is

Table 2. Prediction accuracies

DS	C4.5 on training segment	NGA-DT on training segment	C4.5 on testing segment	NGA-DT on testing segment
AB	25.26	**31.95**	18.10	**21.83**
BS	81.57	**82.05**	**65.60**	64.43
BC	81.45	**84.32**	70.55	**72.65**
CE	96.50	**96.60**	**94.10**	94.09
CHS	96.02	**99.78**	95.97	**99.25**
CA	86.67	**87.80**	86.37	**86.90**
DER	71.81	**94.54**	71.87	**93.87**
GI	83.55	**90.70**	65.85	**69.90**
HR	66.84	**70.04**	**46.97**	**46.97**
HEP	94.44	**98.61**	82.50	**87.50**
IS	95.40	**98.56**	94.46	**95.37**
ION	96.30	**97.91**	**92.02**	90.02
IRS	97.48	**98.45**	**95.33**	94.00
LR	75.92	**79.29**	71.05	**73.98**
LM	90.49	**91.98**	64.72	**65.28**
LD	76.03	**77.42**	**67.29**	64.83
NUR	98.81	**99.18**	97.00	**97.06**
PD	97.83	**98.77**	95.03	**95.59**
PID	79.89	**82.12**	72.57	**73.21**
SDS	95.50	**98.62**	91.43	**92.86**
SON	93.33	**98.40**	**72.21**	72.14
SH	90.33	**93.33**	75.55	**78.15**
SV	75.59	**85.99**	65.96	**71.08**
TAE	65.93	**73.29**	47.58	**51.63**
TD	97.73	**99.27**	92.84	**93.90**
TTT	91.11	**92.25**	82.59	**82.60**
WNE	98.88	**99.81**	93.53	**95.67**
WQ	50.55	**59.33**	45.17	**50.33**
YST	60.75	**68.04**	49.39	**55.24**
ZOO	97.14	**97.36**	89.00	**90.00**
Avg.	83.64	*87.52*	75.09	*77.34*

calculated to be: 109 [26,37]. The test statistic for the Wilcoxon Signed-Ranks Test based on the PAs of NGA-DT and C4.5 (on the testing segments) is calculated to be: 70. As the test statistic remains lower than the critical value, we understand that NGA-DT significantly improves the generalization performance of C4.5.

6   Conclusion

In this paper, we propose a new, less complicated GA-based decision tree induction technique called NGA-DT which is more likely to ascertain conditional dependencies among the attributes. From the experimental results, we see that NGA-DT significantly improves the generalization performance of C4.5. It is also shown that NGA-DT can be more reliable for knowledge discovery compared to C4.5. The structure of NGA-DT is developed in such a way that other renowned decision tree induction algorithms (such as CART [10]) can be used instead of C4.5 without any modification.

A major drawback of NGA-DT (which involves many other GA-based techniques) is that it has a large memory and computational overhead. The memory and computational overhead of NGA-DT is roughly 1200 times greater than that of C4.5. Hence, NGA-DT may not be suitable for time-critical applications where quick results are appreciated. However, when additional time, memory and processing power is available, NGA-DT can utilize their full potential. Furthermore, the structure of NGA-DT can be implemented in a parallel environment in order to address higher computational time.

References

1. Abellan, J.: Ensembles of decision trees based on imprecise probabilities and uncertainty measures. Inf. Fusion **14**, 423–430 (2013)
2. Adnan, M.N., Islam, M.Z.: ComboSplit: combining various splitting criteria for building a single decision tree. In: Proceedings of the International Conference on Artificial Intelligence and Pattern Recognition, pp. 1–8 (2014)
3. Adnan, M.N., Islam, M.Z.: Forest CERN: a new decision forest building technique. In: Proceedings of the 20th Pacific Asia Conference on Knowledge Discovery and Data Mining (PAKDD), pp. 304–315 (2016)
4. Adnan, M.N., Islam, M.Z.: Optimizing the number of trees in a decision forest to discover a subforest with high ensemble accuracy using a genetic algorithm. Knowl.-Based Syst. **110**, 86–97 (2016)
5. Adnan, M.N., Islam, M.Z., Kwan, P.W.H.: Extended space decision tree. In: Wang, X., Pedrycz, W., Chan, P., He, Q. (eds.) ICMLC 2014. CCIS, vol. 481, pp. 219–230. Springer, Heidelberg (2014). https://doi.org/10.1007/978-3-662-45652-1_23
6. Aitkenhead, M.J.: A co-evolving decision tree classification method. Expert Syst. Appl. **34**(1), 18–25 (2008)
7. Arlot, S.: A survey of cross-validation procedures for model selection. Stat. Surv. **4**, 40–79 (2010)
8. Barros, R.C., Basgalupp, M.P., de Carvalho, A.C.P.L.F., Freitas, A.A.: A survey of evolutionary algorithm for decision tree induction. IEEE Trans. Syst. Man Cybern. - Part C: Appl. Rev. **42**(3), 291–312 (2012)
9. Bishop, C.M.: Pattern Recognition and Machine Learning. Springer, New York (2008)
10. Breiman, L., Friedman, J., Olshen, R., Stone, C.: Classification and Regression Trees. Wadsworth International Group, Belmont (1985)
11. Burges, C.J.C.: A tutorial on support vector machines for pattern recognition. Data Min. Knowl. Discov. **2**, 121–167 (1998)

12. Demsar, J.: Statistical comparisons of classifiers over multiple data sets. J. Mach. Learn. Res. **7**, 1–30 (2006)
13. Espejo, P.G., Sebastian, S., Herrera, F.: A survey on the application of genetic programming to classification. IEEE Trans. Syst. Man Cybern. - Part C: Appl. Rev. **40**(2), 121–144 (2010)
14. Fu, Z., Golden, B., Lele, S., Raghavan, S., Wasli, E.: Genetically engineered decision trees: population diversity produces smarter trees. Oper. Res. **51**(6), 894–907 (2003)
15. Han, J., Kamber, M.: Data Mining Concepts and Techniques. Morgan Kaufmann Publishers, San Francisco (2006)
16. Holland, J.H.: Adaptation in Natural and Artificial Systems: An Introductory Analysis with Applications to Biology, Control and Artificial Intelligence. MIT Press, Cambridge (1992)
17. Hunt, E., Marin, J., Stone, P.: Experiments in Induction. Academic Press, New York (1966)
18. Kamber, M., Winstone, L., Gong, W., Cheng, S., Han, J.: Generalization and decision tree induction: efficient classification in data mining. In: Proceedings of the International Workshop Research Issues on Data Engineering, pp. 111–120 (1997)
19. Kataria, A., Singh, M.D.: A review of data classification using k-nearest neighbour algorithm. Int. J. Emerg. Technol. Adv. Eng. **3**(6), 354–360 (2013)
20. Kim, Y.W., Oh, I.S.: Classifier ensemble selection using hybrid genetic algorithms. Pattern Recogn. Lett. **29**, 796–802 (2008)
21. Kurgan, L.A., Cios, K.J.: Caim discretization algorithm. IEEE Trans. Knowl. Data Eng. **16**, 145–153 (2004)
22. Li, J., Liu, H.: Ensembles of cascading trees. In: Proceedings of the Third IEEE International Conference on Data Mining, pp. 585–588 (2003)
23. Lichman, M.: UCI machine learning repository. http://archive.ics.uci.edu/ml/datasets.html. Accessed 15 Mar 2016
24. Lim, T.S., Loh, W.Y., Shih, Y.S.: A comparison of prediction accuracy, complexity, and training time of thirty-three old and new classification algorithms. Mach. Learn. **40**, 203–229 (2000)
25. Liu, Y., Shen, Y., Wu, X.: Automatic clustering using genetic algorithms. Appl. Math. Comput. **218**, 1267–1279 (2011)
26. Mason, R., Lind, D., Marchal, W.: Statistics: An Introduction. Brooks/Cole Publishing Company, New York (1998)
27. Murthy, S.K.: On growing better decision trees from data. Ph.D. thesis, The Johns Hopkins University, Baltimore, Maryland (1997)
28. Murthy, S.K.: Automatic construction of decision trees from data: a multidisciplinary survey. Data Min. Knowl. Discov. **2**, 345–389 (1998)
29. Murthy, S.K., Kasif, S., Salzberg, S.S.: A system for induction of oblique decision trees. J. Artif. Intell. Res. **2**, 1–32 (1994)
30. Quinlan, J.R.: C4.5: Programs for Machine Learning. Morgan Kaufmann Publishers, San Mateo (1993)
31. Quinlan, J.R.: Improved use of continuous attributes in C4.5. J. Artif. Intell. Res. **4**, 77–90 (1996)
32. Rahman, M.A., Islam, M.Z.: A hybrid clustering technique combining a novel genetic algorithm with k-means. Knowl.-Based Syst. **71**, 345–365 (2014)

33. Shirasaka, M., Zhao, Q., Hammami, O., Kuroda, K., Saito, K.: Automatic design of binary decision trees based on genetic programming. In: Second Asia-Pacific Conference on Simulated Evolution and Learning. Australian Defense Force Academy, Canberra (1998)
34. Tamon, C., Xiang, J.: On the boosting pruning problem. In: López de Mántaras, R., Plaza, E. (eds.) ECML 2000. LNCS (LNAI), vol. 1810, pp. 404–412. Springer, Heidelberg (2000). https://doi.org/10.1007/3-540-45164-1_41
35. Tan, P.N., Steinbach, M., Kumar, V.: Introduction to Data Mining. Pearson Education, London (2006)
36. Tanigawa, T., Zhao, Q.: A study on efficient generation of decision trees using genetic programming. In: Genetic and Evolutionary Computation Conference (GECCO'2000), pp. 1047–1052. Morgan Kaufmann (2000)
37. Triola, M.F.: Elementary Statistics. Addison Wesley Longman Inc., Reading (2001)
38. Whitley, D.: A genetic algorithm tutorial. Stat. Comput. **4**, 65–85 (1994)
39. Wilcoxon, F.: Individual comparison by ranking methods. Biometrics **1**, 80–83 (1945)
40. Zhang, G.P.: Neural networks for classification: a survey. IEEE Trans. Syst. Man Cybern. **30**, 451–462 (2000)
41. Zhao, H.: A multi-objective genetic programming programming approach to developing pareto optimal decision trees. Decis. Support Syst. **43**(3), 809–826 (2007)

A Genetic Algorithm Based Technique for Outlier Detection with Fast Convergence

Xiaodong Zhu[1], Ji Zhang[2(✉)], Zewen Hu[1], Hongzhou Li[3], Liang Chang[3], Youwen Zhu[4], Jerry Chun-Wei Lin[5], and Yongrui Qin[6]

[1] Nanjing University of Information Science and Technology, Nanjing, China
[2] University of Southern Queensland, Toowoomba, Australia
Ji.Zhang@usq.edu.au
[3] Guilin University of Electronic Technology, Guilin, China
[4] Nanjing University of Aeronautics and Astronautics, Nanjing, China
[5] Western Norway University of Applied Sciences (HVL), Bergen, Norway
[6] University of Huddersfield, Huddersfield, UK

Abstract. In this paper, we study the problem of subspace outlier detection in high dimensional data space and propose a new genetic algorithm-based technique to identify outliers embedded in subspaces. The existing technique, mainly using genetic algorithm (GA) to carry out the subspace search, is generally slow due to its expensive fitness evaluation and long solution encoding scheme. In this paper, we propose a novel technique to improve the performance of the existing GA-based outlier detection method using a bit freezing approach to achieve a faster convergence. Through freezing converged bits in the solution encoding strings, this innovative approach can contribute to fast crossover and mutation operations and achieve an early stop of the GA that leads to more accurate approximation of fitness function. This research work can contribute to the development of a more efficient search method for detecting subspace outliers. The experimental results demonstrate the improved efficiency of our technique compared with the existing method.

1 Introduction

Outlier detection is an important research problem in data mining that aims to find objects that are considerably dissimilar, exceptional and inconsistent with respect to the majority data in an input database [9]. In recent years, we have witnessed a tremendous research interest sparked by the explosion of data collected and transferred in the format of streams. Outlier detection in data streams can be useful in many fields such as analysis and monitoring of network traffic data, web log, sensor networks and financial transactions, etc.

There have been abundant research in outlier detection in the past decade. Most of the conventional outlier detection techniques are only applicable to relatively low dimensional static data [7,10,11,13,14,17,24]. Because they use

© Springer Nature Switzerland AG 2018
G. Gan et al. (Eds.): ADMA 2018, LNAI 11323, pp. 95–104, 2018.
https://doi.org/10.1007/978-3-030-05090-0_8

the full set of attributes for outlier detection, thus they are not able to detect projected outliers. They cannot handle data streams either. Recently, there are some emerging work in dealing with outlier detection either in high-dimensional static data or data streams. However, there have not been any reported concrete research work so far for exploring the intersection of these two active research directions. For those methods in projected outlier detection in high-dimensional space [3, 18–20, 22], they can detect projected outliers that are embedded in subspaces. However, their measurements used for evaluating points' outlier-ness are not incrementally updatable and many of the methods involve multiple scans of data, making them incapable of handling data streams. For instance, [3, 22] use the Sparsity Coefficient to measure data sparsity. Sparsity Coefficient is based on an equi-depth data partition that has to be updated frequently in data stream. This will be expensive and such updates will require multiple scan of data. [18–20, 25] use data sparsity metrics that are based on distance involving the concept of kNN. This is not suitable for data streams either as one scan of data is not sufficient for retaining kNN information of data points. One the other hand, the techniques for tackling outlier detection in data streams [2, 12, 16] rely on full data space to detect outliers and thus projected outliers cannot be discovered by these techniques.

A key observation that motivates our work is that outliers existing in high-dimensional data streams are embedded in some lower-dimensional subspaces. Here, a subspace refers to as the data space consisting of a subset of attributes. These outliers are termed *projected outliers* in the high-dimensional space. The existence of projected outliers is due to the fact that, as the dimensionality of data goes up, data tend to become equally distant from each other [1]. As a result, the difference of data points' outlier-ness will become increasingly weak and thus undistinguishable. Only in moderate or low dimensional subspaces can significant outlier-ness of data be observed.

In order to identify outliers embedded in subspaces in high-dimensional data space, efficient search of the subspaces is required. Evolutionary algorithm, such as genetic algorithm, has been established as an effective method to find the subspaces where outliers are most likely embedded [23]. However, the execution of genetic algorithm is rather time-consuming particularly when dealing with high-dimensional data sets. Acceleration the search space is definitely much-needed to contribute a more efficient detection of subspace outliers.

To this end, this paper proposes an innovative genetic algorithm with the feature of fast convergence using a bit freezing technique to further improve the efficiency of the detection method proposed in [23], which uses the standard genetic algorithm. The major technical contributions of this paper are summarized as follows:

1. First, we propose a new technique to freeze the converged bits in the binary solution encoding strings. Those string bits, each representing an experimental condition, will be frozen to '0' or '1' if their values converge. By doing this, the length of the solution encoding strings can be significantly shortened and the overhead involved in crossover and mutation operations can be reduced;

2. Second, by utilizing the bit freezing technique, we have devised a novel approach to achieve an early stop of GA. This approach is able to automatically determine whether the GA should be continued or can be early terminated anytime during the evolution process. This is a very desired feature for the GA as it can potentially save a substantial portion of the computational overhead incur in the GA;
3. Finally, the experimental results demonstrate the better efficiency of our technique than that of the existing method in detecting subspace outliers.

2 General Genetic Algorithm Framework

Before our bit freezing technique is discussed, we will first introduce the general algorithmic framework for detecting subspace outliers, which has two requirements: (1) find the outlying subspaces of a given data and (2) find the neighboring data points in the subspaces in order to quantify the outlier-ness score for the data. Because there are a large number of condition subsets that we need to potentially evaluate, then the first requirement is much more computationally difficult to tackle. As this research problem is a search problem by nature, thus Genetic Algorithm (GA) is chosen to find the outlying subspaces of the given data. Once the outlying subspaces have been found, the nearest neighbors can be found in a linear complexity order.

Most aspects of the GA design for subspace outlier detection is identical to the standard GA. An initial population of solutions, with each individual in the population representing a particular subspace, are first randomly generated. Fitness of these solutions are then evaluated and good solutions are selected to undergo genetic operations such as crossover and mutation to produce the population of the next generation. This process continues until a specific number of generations are finished. The top solutions amongst all the solutions that have been evaluated in the GA will be retuned as the final solutions.

In the following, a more detailed discussion is presented on the design of solution encoding and fitness function which that are relatively unique for dealing with this problem.

Solution Encoding. Standard binary solution encoding scheme is normally used. In this encoding, all solutions are represented by strings with fixed and equal length M, where M is the number of dimensions (i.e., the number of attributes) of the dataset. Using binary alphabet $\Sigma = \{0, 1\}$ for gene alleles, each bit in the solution will take on the value of "0" and "1", indicating whether or not its corresponding dimension is selected for the subspace, respectively ("0" indicates the corresponding condition is absent and vice versa for "1"). For a simple example, the solution $(1, 0, 0, 1, 0, 1)$ when $M = 6$ means that this solution is a 3-dimensional subspace which contains the 1st, 4th and 6th dimensions.

The subspace search in the GA involves finding those subspaces s that have high fitness defined as:

$$f(d, s) = \frac{1}{k} \sum_{i=1}^{k} |distance(d, d_i)|, d_i \in kNNSet(d, s, D)$$

where $kNNSet(d, s, D)$ denotes the set of kNNs from the database D for the target data d in subspace s.

3   Bit Freezing Technique in GA

In this section, we introduce the bit freezing technique in the Genetic Algorithm. As mentioned earlier, the critical performance bottleneck in the existing approaches for the outlying subspaces of a target data results from the typically high-dimensionality of the dataset, which leads to the long representation for individual solution and the high cost in crossover and mutation operations in the GA.

To better tackle this problem, we developed a novel approach to significantly reduce the overhead of the crossover and mutation by heuristically freezing the bits in the binary solution representation (encoding) strings. This idea is motivated by that fact in subspace outlier detection, some subspaces are critical while others are not important at all with regard to the strength of outlier-ness of the target data. Therefore, some subspaces will be always present (or absent) in the good solutions generated in the populations. This is called convergence in this work. In most cases, we can start to observe such convergence for most, if not all, subspaces in the early stage of the GA. This convergence phenomenon is stable, meaning that once a subspace is converged, then it will not change for the rest of the evolution process of the GA. Because of the stability of convergence, it will be safe to freeze those bits corresponding to these converged subspaces to either '0' or '1' in the corresponding solution encoding strings when the convergence happens. The advantage for doing this is obvious: we can reduce the length of solution encoding strings, and the crossover and mutation operations will only be operated on those unfrozen bits, leading to a fast generation of new solutions in each generation of the GA. This strategy can considerably improve the overall efficiency of the GA without compromising the quality of the detected outlying subspaces.

3.1 w-Moving Average of Bit Value

To quantitatively identify those converged subspaces, we calculate the moving average of the value of each bit in the solution encoding strings. A sliding window with a size of w is used that covers w consecutive generations to compute the moving average. As there are a number of solutions obtained in each generation, thus the bit value in a given generation is actually the average value of that bit for all the solutions in that generation. It is defined as

$$v(b_i, t) = \frac{\sum_{j=1}^{pop\_size} s_j(b_i, t)}{pop\_size}$$

where $s_j(b_i, t)$ denotes the value of ith bit of the jth solution in the tth generation, and we have $1 \leq i \leq M$, $1 \leq j \leq pop\_size$ and $1 \leq t \leq Gen$. pop_size corresponds to the number of solutions generated in each generation, which is a human-specified parameter and normally takes a fixed value for all the generations in the GA. *Gen* is the maximum number of generations the GA will performs.

Definition 1 w **moving average** (w-**MA**). w-MA of a bit b_i at the tth generation in the GA, denoted as $w\_MA(b_i, t)$, is defined as the average bit value in w most recent consecutive generations, that is

$$w\_MA(b_i, t) = \frac{\sum_{j=t-w+1}^{t} v(b_i, j)}{w}$$

It is necessary to keep track of the w moving average of all the bits along the generations of the GA. To this end, we create w moving average vector to store the values of w moving average for the bits. w moving average vector is defined as follows.

Definition 2 w **Moving Average Vector** (w-**MAV**). w-MAV is a $1 \times M$ vector where the ith element of the vector stores the w moving average value for the ith dimension. M is the dimensionality of the given dataset.

When a bit has been frozen to either '0' or '1', its corresponding entry in the w-MAV will no longer be maintained thereafter.

3.2   Bit Freezing Technique for GA

It is very simple and straightforward to design the approach to freeze bits of the representation strings. It involves first specifying the values of two thresholds, T_{low} and T_{high}, used to freeze bits to '0' or '1', respectively, and device the rules for freezing bits. By using these two parameters, the range of $[0.0, 1.0]$, where bit values distribute, is divided into three regions, namely, frozen-to-'0' region ($[0.0, T_{low}]$), unfrozen region ((T_{low}, T_{high})) and frozen-to-'1' region ($[T_{high}, 1.0]$). The rules used to freeze a bit b_i at the tth generation are detailed as follows:

1. **Rule for freezing b_i to '0'**: if $w\_MA(b_i, t) \leq T_{low}$, then b_i is frozen to '0' thereafter;
2. **Rule for freezing b_i to '1'**: if $w\_MA(b_i, t) \geq T_{high}$, then b_i is frozen to '1' thereafter;
3. **Rule for keep b_i active (unfrozen)**: If b_i does not satisfy the above two rules, then b_i continues to remain active (unfrozen).

Even though being human-specified parameters, T_{low} and T_{high} are fairly intuitive to set. A rule-of-thumb is that T_{low} should be relatively close to 0 while T_{high} should be relatively close to 1. For example, $T_{low} \in [0.15, 0.25]$ and $T_{high} \in [0.75, 0.85]$ are the reasonable value ranges for the parameters.

The size of the above regions has an effect on the overall performance of the GA. The smaller the unfrozen region is, the higher speed can be generally achieved, even though there is a higher chance that a small number of true outlying subspaces may be missed.

Under our bit freezing framework, it is necessary to record the state of all the bits in the solution encoding throughout the GA. We use Bit State Vector (BSV) to accomplish this. BSV is defined as follows.

Definition 3 Bit State Vector (BSV). BSV is a $1 \times M$ table where each entry takes one of three symbols, namely '0', '1' or '?'. M corresponds to the number of dimensions of the dataset. The ith entry in BST takes the symbol of '0' (or '1' or '?') if the ith bit in the encoding string is frozen to '0' (or frozen to '1' or remains active).

The construction of BSV is the based on w-MAV and the bit freezing rules that are discussed earlier. Please note that BSV is dynamic by nature as the entries in the vector may be changed from '?' to '0' or '1' sometime in the process of the GA. Specifically, the following rules are followed regarding the state change of bits in the GA.

1. All the entries in a BSV take the symbol of '?' initially at the beginning of the GA;
2. '?' bits may be changed to '0' or '1' sometime in the GA depending upon its w moving average and, once been frozen, remain that state for the rest of the GA;
3. '?' bits may remain active for the whole life span of the GA;
4. '0' bits and '1' bits cannot change to any other states.

Once has been frozen, a bit will no longer participate in the crossover and mutation operations in the subsequent generations of the GA. Those active bits will continue to be involved in the crossover and mutation operations until they are frozen to '1' or '0' or the end of the GA is reached, whichever comes earlier. Figure 1 presents an example showing a possible change of BSV in the GA where $M = 8$ (only the first three generations are shown in the figure).

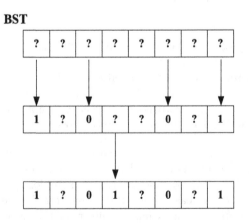

Fig. 1. An example of the change of Bit State Vector (BSV) in the GA

4 Performance Improvement Using Bit Freezing Technique

The immediate improvement that can be achieved using the bit freezing technique is the faster speed for crossover and mutation operations. When they are frozen, bits will no longer participate in the crossover and mutation operations in the subsequent generations. Thus, the *shortened solution encoding* which only contains the active bits from BST can be safely used. Both crossover and mutation on the shortened solution encoding are performed in the same way as on other regular binary strings. The only difference is that the new children generated using shortened solution encoding need to be mapped back to the full-length encoding for fitness evaluation. A mapping between bit index in the shortened solution encoding and that in the full solution encoding is therefore required. To facilitate such mapping/conversion process, we devise Bit Index Mapping (BIM) and partial Effective Bit Vector (EBV). BIM is used for mapping bit index between the shortened and the full solution encodings, while EBV quickly shows which experimental condition(s) participate in the condition subset, an information required in the subsequent fitness function evaluation. BIM and EBV are defined as follows.

Definition 4. A Bit Index Mapping(BIM) is a $l \times 2$ table with $BIM(i, 1)$ (the first entry of each row) being the index of an active bit b in the shortened solution encoding while $BIM(i, 2)$ (the second entry of the same row) being the index of b in BST.

As an example, let us assume that the BST is $(1, ?, 0, ?, ?, 0, ?, 1)$ and a solution in a shortened encoding is $(0, 1, 1, 0)$, then BIM can be constructed as $(1{\rightarrow}2, 2{\rightarrow}4, 3{\rightarrow}5, 4{\rightarrow}7)$, suggesting that the 1st bit in the shortened encoding is mapped to the 2nd bit in the full encoding, and etc.

5 Experimental Results

This section reports some preliminary experimental results on our bit freezing technique. Both efficiency and effectiveness of the algorithm are evaluated. The KDD Cup 99 network intrusion detection dataset is used as a real-life dataset to detect intrusions which manifest as outlying observations. In the experiments, comparative study will be carried out to compare the performance of our bit freezing technique with existing methods. The competitive method we choose in the comparative study are the standard GA without major improvements incorporated.

5.1 Efficiency Study

The speed improvement is an important objective our algorithm needs to achieve. In efficiency study, we investigate the execution time of our bit freezing technique and compare it with the existing methods. The comparison is first conducted under varying number of the dimensions of the dataset. The result is presented in Table 1. The results show that our bit freezing technique is 2–5 times (on average) faster than the standard GA under varying number of the dimensions. The performance improvement is contributed by the boosting strategies of faster genetic operations (crossover and mutation) and early stopping of GA.

Table 1. Time performance of the two methods under varying dimensionality

	5	10	15	20	25	30
Standard GA	8	16	37	38	48	59
GA with Bit freezing	4	6	8	9	10	12

5.2   Effectiveness Study

Effectiveness study investigates how well the GA method can find the good outlying subspaces when bit freezing technique is incorporated. For this study, an appropriate effectiveness metric needs to be devised. *Accuracy* is utilized in this experiment to measure, amongst all the solutions returned by the bit freezing enabled GA, how many of them are really the top solutions. It is defined as follows:

$$Accuracy = \frac{a}{top\_n} \times 100\%$$

where a is the number of true top_n solutions returned by our algorithm. The critical issue involved in calculating accuracy is that we need to first obtain the so-called golden truth result, *i.e.*, the set of the true top_n solutions. Unfortunately, due to the NP nature of the research problem, the set of the true top_n solutions are unknown. However, it is still possible for us to obtain a good approximation of true top_n solutions through the following strategy. The standard GA for detecting outlying subspaces is first executed for multiple times and the top solutions with regard to the target data are then collected, from where the approximated set of top solutions are generated. To ensure that the approximated set of top solutions are as close as possible to the real ones, we continue to run the algorithm until the top solutions returned become stabilized. This stabilized set of solutions are used as the golden truth result for evaluating our bit freezing enabled GA method in the effectiveness study.

Top-n	100	200	300	400	500
Accuracy	86.3%	89.4%	90.5%	93.9%	94.6%

Fig. 2. Accuracy result

We evaluate the accuracy of our method under different values of top_n. top_n denotes the number of top solutions to be obtained at the end of the GA. In the experiment, top_n ranges from 100 to 500 and the corresponding accuracy of our method is shown in Fig. 2. The results reveals that our method performs pretty accurate and bit freezing does not compromise the effectiveness of the method. Furthermore, the accuracy of our method gets better as top_n increases, indicating that our method works particularly well when a relatively large number of top condition subsets are sought.

6   Conclusions

In this paper, we present an innovative bit freezing technique to improve the efficiency of subspace outlier detection from multiple or high dimensional datasets. This bit-freezing technique is motivated by the inherent convergence behavior of bit value for most of the subspaces. Our technique is promising as it offers a number of different strategies to improve efficiency, including speed-up of crossover and mutation operations and early stop the GA whenever it is possible. We have conducted some preliminary experiments for evaluating our method, which show that satisfactory speed and accuracy performance are achieved when bit freezing technique is incorporated in the GA.

Acknowledgment. This research was partially supported by National Key Research and Development Program of China (No. 2017YFB0802300), the National Natural Science Foundation of China (No. 61602240), Guangxi Key Laboratory of Trusted Software (No. kx201615) and Capacity Building Project for Young University Staff in Guangxi Province, Department of Education, Guangxi Province (No. ky2016YB149).

References

1. Aggarwal, C.C., Yu, P.S.: An effective and efficient algorithm for high-dimensional outlier detection. VLDB J. **14**, 211–221 (2005)
2. Aggarwal, C.C.: On abnormality detection in spuriously populated data streams. In: SDM 2005, Newport Beach, CA (2005)
3. Aggarwal, C.C., Yu, P.S.: Outlier detection in high dimensional data. In: SIGMOD 2001, Santa Barbara, California, USA, pp. 37–46 (2001)
4. Aggarwal, C.C., Han, J., Wang, J., Yu, P.S.: A framework for clustering evolving data streams. In: VLDB 2003, Berlin, Germany, pp. 81–92 (2003)
5. Aggarwal, C.C., Han, J., Wang, J., Yu, P.S.: A framework for projected clustering of high dimensional data streams. In: VLDB 2004, Toronto, Canada, pp. 852–863 (2004)
6. Angiulli, F., Pizzuti, C.: Fast outlier detection in high dimensional spaces. In: Elomaa, T., Mannila, H., Toivonen, H. (eds.) PKDD 2002. LNCS, vol. 2431, pp. 15–27. Springer, Heidelberg (2002). https://doi.org/10.1007/3-540-45681-3_2
7. Breuning, M., Kriegel, H.-P., Ng, R., Sander, J.: LOF: identifying density-based local outliers. In: SIGMOD 2000, Dallas, Texas, pp. 93–104 (2000)
8. Guttman, A.: R-trees: a dynamic index structure for spatial searching. In: SIGMOD 1984, Boston, Massachusetts, pp. 47–57 (1984)

9. Han, J., Kamber, M.: Data Mining: Concepts and Techniques. Morgan Kaufman Publishers, Burlington (2000)
10. Knorr, E.M., Ng, R.T.: Algorithms for mining distance-based outliers in large dataset. In: VLDB 1998, New York, NY, pp. 392–403 (1998)
11. Knorr, E.M., Ng, R.T.: Finding intentional knowledge of distance-based outliers. In: VLDB 1999, Edinburgh, Scotland, pp. 211–222 (1999)
12. Palpanas, T., Papadopoulos, D., Kalogeraki, V., Gunopulos, D.: Distributed deviation detection in sensor networks. SIGMOD Rec. **32**(4), 77–82 (2003)
13. Ramaswamy, S., Rastogi, R., Kyuseok, S.: Efficient algorithms for mining outliers from large data sets. In: SIGMOD 2000, Dallas Texas, pp. 427–438 (2000)
14. Papadimitriou, S., Kitagawa, H., Gibbons, P.B., Faloutsos, C.: LOCI: fast outlier detection using the local correlation integral. In: ICDE 2003, Bangalore, India, p. 315 (2003)
15. Pokrajac, D., Lazarevic, A., Latecki, L.: Incremental local outlier detection for data streams. In: CIDM 2007, Honolulu, Hawaii, USA, pp. 504–515 (2007)
16. Subramaniam, S., Palpanas, T., Papadopoulos, D., Kalogeraki, V., Gunopulos, D.: Online outlier detection in sensor data using non-parametric models. In: VLDB 2006, Seoul, Korea, pp. 187–198 (2006)
17. Tang, J., Chen, Z., Fu, A.W., Cheung, D.W.: Enhancing effectiveness of outlier detections for low density patterns. In: Chen, M.S., Yu, P.S., Liu, B. (eds.) PAKDD 2002. LNCS (LNAI), vol. 2336, pp. 535–548. Springer, Heidelberg (2002). https://doi.org/10.1007/3-540-47887-6_53
18. Zhang, J., Lou, M., Ling, T.W., Wang, H.: HOS-miner: a system for detecting outlying subspaces of high-dimensional data. In: VLDB 2004, Toronto, Canada, pp. 1265–1268 (2004)
19. Zhang, J., Gao, Q., Wang, H.: A novel method for detecting outlying subspaces in high-dimensional databases using genetic algorithm. In: ICDM 2006, Hong Kong, China, pp. 731–740 (2006)
20. Zhang, J., Wang, H.: Detecting outlying subspaces for high-dimensional data the new task, algorithms and performance. Knowl. Inf. Syst. (KAIS) **10**, 333–355 (2006)
21. Zhang, T., Ramakrishnan, R., Livny, M.: BIRCH: an efficient data clustering method for very large databases. In: SIGMOD 1996, Montreal, Canada, pp. 103–114 (1996)
22. Zhu, C., Kitagawa, H., Faloutsos, C.: Example-based robust outlier detection in high dimensional datasets. In: ICDM 2005, Houston, Texas, pp. 829–832 (2005)
23. Zhang, J., Gao, Q., Wang, H., Liu, Q., Xu, K.: Detecting projected outliers in high-dimensional data streams. In: Bhowmick, S.S., Küng, J., Wagner, R. (eds.) DEXA 2009. LNCS, vol. 5690, pp. 629–644. Springer, Heidelberg (2009). https://doi.org/10.1007/978-3-642-03573-9_53
24. Zhang, J., Tao, X., Wang, H.: Outlier detection from large distributed databases. World Wide Web J. (WWWJ) **17**(4), 539–568 (2014). https://doi.org/10.1007/s11280-013-0218-4
25. Zhu, X., Zhang, J., Li, H., Fournier-Viger, P., Lin, J.C.-W., Chang, L.: FRIOD: a deeply integrated feature-rich interactive system for effective and efficient outlier detection. IEEE Access **5**, 25682–25695 (2017)

Multivariate Synchronization Index Based on Independent Component Analysis for SSVEP-Based BCI

Yanlong Zhu, Chenglong Dai, and Dechang Pi[✉]

College of Computer Science and Technology,
Nanjing University of Aeronautics and Astronautics,
29 Jiangjun Road, Nanjing 211106, Jiangsu, China
{zhuyanlong, chenglongdai, dc.pi}@nuaa.edu.cn

Abstract. A template-matching approach combined with multivariate synchronization index (MSI) and independent component analysis (ICA) based spatial filtering for steady-state visual evoked potentials (SSVEPs) frequency recognition is proposed in this paper to enhance the performance of SSVEP-based brain-computer interface (BCI). As a type of electroencephalogram (EEG) signals, SSVEPs generated from underlying brain sources is different from other activities and artifacts, this spatial filter has great potential to enhance the signal-to-noise ratio (SNR) of SSVEPs. This study adapted the MSI-ICA based spatial filters to process test data and the averaged training data, and then used the correlation coefficients between them as features for SSVEP classification. Some conventional methods such as canonical correlation analysis (CCA), filter bank-CCA (FBCCA), and ICA based frequency recognition were adapted to do the contrasting experiment, using a 40-class SSVEP benchmark datasets recorded from 35 subjects. The experimental results demonstrate that the MSI-ICA based method outperforms other methods in terms of the classification accuracy and information transfer rate (ITR).

Keywords: Brain computer interface · Steady-state visual evoked potential · Multivariate synchronization index · Independent component analysis

1 Introduction

Brain computer interface (BCI) has achieved unprecedented progress in the past decades, which provides a new communication channel to humans by using the brain activities. Among various BCI modalities, the steady-state visual evoked potential (SSVEP)-based BCI has attracted more attention of researchers due to its high information transfer rate (ITR) and low training requirements. The flow chart of the recognition of SSVEP-based BCI is shown in Fig. 1. First, pre-process the EEG signal and do feature extraction and feature recognition on the signal. Then convert the signal and transmit them through BCI to the external device [1].

© Springer Nature Switzerland AG 2018
G. Gan et al. (Eds.): ADMA 2018, LNAI 11323, pp. 105–115, 2018.
https://doi.org/10.1007/978-3-030-05090-0_9

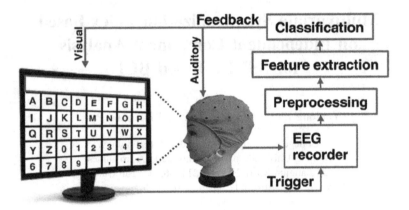

Fig. 1. The process of SSVEP-based BCI

1.1 Motivation

To improve the performance of SSVEP-based BCIs, researchers are working on two main directions: (1) increasing the number of coded target (i.e., visual stimuli), (2) improving frequency recognition algorithms [2].

For the number of targets, plenty of paradigms have been proposed, e.g., multi-frequency coding method [3], frequency-phase mixed coding method [4], joint frequency phase modulation (JFPM) method [5] and etc. For the frequency recognition algorithm, the researchers mainly focus on the multi-channel methods that provides more robust performance than single channel methods.

1.2 Related Works

Various frequency recognition methods have been proposed to accurately detect SSVEPs from electroencephalogram (EEG) signals. The canonical correlation analysis (CCA) methods, which have been widely used to detect the stimulation frequency of SSVEPs due to its ability to enhance the signal-to-noise ratio (SNR) [6]. In recent studies, filter bank has been incorporated into the CCA-based method to improve the frequency detection of SSVEPs [7]. The filter bank analysis can decompose SSVEPs into multiple sub-band components, then fusion the results from all sub-bands to recognize the frequency. Although the CCA method prove robust in detecting SSVEPs, its performance is often affected by the interference from the spontaneous EEG activities [8].

Besides the CCA-based spatial filters have been explored, other spatial filtering methods such as independent component analysis (ICA) are also applied to design spatial filters [9]. ICA has been widely used to decompose multichannel EEG data into statistically independent components (ICs) [10]. However, the correspondence between ICs and source activities is unclear.

1.3   Contributions and Outline

We propose a multivariate synchronization index (MSI) based on ICA method for SSVEP-based BCI. In this method, we aim to estimate the synchronization between the actual EEG signals and the reference signals as a potential index for recognizing the stimulus frequency.

The rest of this paper is organized as follows. We introduce preliminaries of our method in Sect. 2. A MSI-ICA method for SSVEP-based BCI is proposed in Sect. 3. Experimental results are shown in Sect. 4. Finally, the Sect. 5 offers concluding remarks.

2   Preliminaries

2.1   Standard CCA

Canonical correlation analysis (CCA) is a multivariable statistical method for measuring the underlying correlation between two sets of data, and it finds a pair of linear combinations, such that the correlation is maximized. Lin *et al.* [6] firstly applied the CCA method for multi-channel SSVEP detection. When using CCA for frequency recognition [11], we require the reference signals.

Let us denote the EEG signals by a four-way matrix $\chi = (\chi)_{njkh} \in \mathbb{R}^{N_f \times N_c \times N_s \times N_t}$, where n indicates the stimulus index, j indicates the channel index, k indicates the index of sample points, h indicates the number of training trials, N_f is the number of visual stimuli, N_c is the number of channels, N_s is the number of sampling points, and N_t is the number of training trials [9]. The sine-cosine reference signals Y_n for the n-th stimulus f_n are calculated as follows:

$$
Y_n = \begin{bmatrix} \sin(2\pi f_n t) \\ \cos(2\pi f_n t) \\ \vdots \\ \sin(2\pi N_h f_n t) \\ \cos(2\pi N_h f_n t) \end{bmatrix}, \quad t = \frac{1}{F_s}, \frac{2}{F_s}, \cdots, \frac{N_s}{F_s} \tag{1}
$$

where N_h is the number of harmonics ($N_h = 5$ in this study), and F_s is the sampling rate. Considering EEG signal $\chi_{nh} \in \mathbb{R}^{N_c \times N_s}$, reference signal $Y_n \in \mathbb{R}^{2N_h \times N_s}$ and their linear combinations $x = \chi_{nh}^T W_\chi$ and $y = Y_n^T W_Y$, CCA finds the weight vectors, $W_\chi \in \mathbb{R}^{N_c \times N_c}$ and $W_Y \in \mathbb{R}^{2N_h \times N_c}$, which maximizes the correlation between x and y by solving the following problem:

$$
\max_{W_\chi, W_Y} \rho(x, y) = \frac{E\left[W_\chi^T \chi_{nh} Y_n^T W_Y\right]}{\sqrt{E\left[W_\chi^T \chi_{nh} \chi_{nh}^T W_\chi\right] E\left[W_Y^T Y_n Y_n^T W_Y\right]}} \tag{2}
$$

The maximum of ρ_n with respect to W_χ and W_Y is the maximum canonical correlation. We can identify the target stimulus τ by the following equation:

$$\tau = \arg\max_n \rho_n, \quad n = 1, 2, \cdots, N_f \tag{3}$$

2.2   Filter Bank CCA

Filter bank methods have been widely used to analyze signals with multiple sub-band frequency components in signal processing [12]. To better integrate information from the fundamental and harmonic components for SSVEP detection, Chen *et al.* proposed the filter bank canonical correlation analysis (FBCCA) method [7]. FBCCA significantly outperformed the standard CCA method.

The m-th zero-phase Chebyshev Type I Infinite impulse response (IIR) filters are applied to decompose EEG data $\chi_{nh} \in \mathbb{R}^{N_c \times N_s}$ into sub-band components $\chi_{nh}^{(m)} \in \mathbb{R}^{N_c \times N_s}, m = 1, 2, \cdots, N_m$, N_m is the number of sub-bands. So that independent information embedded in the harmonic components can be extracted more efficiently. The sub-bands were designed according to individual harmonic frequency.

In the frequency recognition framework, correlation-based feature values for the m-th sub-band and n-th stimuli can be calculated as:

$$\gamma_n^{(m)} = \rho\left(\chi_{nh}^{(m)T} W_\chi^{(m)}, Y_n^T W_Y^{(m)}\right) \tag{4}$$

where $W_\chi^{(m)}$ and $W_Y^{(m)}$ was calculated in *Eq. 2*. A weighted sum of squares of the combined correlation values corresponding to all sub-band components is calculated as the feature for target identification:

$$\tilde{\rho}_n = \sum_{m=1}^{N_m} a(m) \cdot \left(\gamma_n^{(m)}\right)^2 \tag{5}$$

According to [7], $a(m) = m^{-1.25} + 0.25$. We can identify the target stimulus τ by the follow equation:

$$\tau = \arg\max_n \tilde{\rho}_n, \quad n = 1, 2, \cdots, N_f \tag{6}$$

3   The MSI-ICA Based Frequency Recognition

3.1   Independent Component Analysis

Independent component analysis (ICA) is a computational method for separating a multivariate signal into additive subcomponents. Since Wang *et al.* reported the efficacy of ICA for enhancing the SNR of SSVEPs [13], ICA-based spatial filtering has attracted more attention in the template-based detection method [9]. However, the correspondence between ICs and source activities is unclear.

3.2   Multivariate Synchronization Index

The multivariate synchronization index (MSI) based frequency recognition method is proposed by Zhang *et al.* [14]. When using MSI for frequency recognition, we also require the reference signals described in *Eq.* 1. This method use the S-estimator calculate the synchronization index between the multichannel signals and reference signals as a potential index for recognizing the stimulus frequency. In fact, the S-estimator was designed to measure the amount of synchronization over a single or two regions of the cortex.

Let us denote the EEG signals by a matrix $\tilde{\chi}_n \in \mathbb{R}^{N_c \times N_s \cdot N_t}$ and the reference signals by a matrix $Y_n \in \mathbb{R}^{2N_h \times N_s \cdot N_t}$. Matrix $\tilde{\chi}_n$ and Y_n are z-normalized [14]. Then a correlation matrix C is calculated as:

$$C = \begin{bmatrix} C_{11} C_{12} \\ C_{21} C_{22} \end{bmatrix} = \begin{bmatrix} \frac{1}{N_s \cdot N_t} \tilde{\chi}_n \tilde{\chi}_n^T & \frac{1}{N_s \cdot N_t} \tilde{\chi}_n Y_n^T \\ \frac{1}{N_s \cdot N_t} Y_n \tilde{\chi}_n^T & \frac{1}{N_s \cdot N_t} Y_n Y_n^T \end{bmatrix} \tag{7}$$

We can reduce the influences of autocorrelation using the following linear transformation:

$$U = \begin{bmatrix} C_{11}^{-\frac{1}{2}} & 0 \\ 0 & C_{22}^{-\frac{1}{2}} \end{bmatrix} \tag{8}$$

Then, the transformed correlation matrix is

$$R = UCU^T = \begin{bmatrix} I_{N_c \times N_c} & C_{11}^{-\frac{1}{2}} C_{12} C_{22}^{-\frac{1}{2}} \\ C_{22}^{-\frac{1}{2}} C_{21} C_{11}^{-\frac{1}{2}} & I_{2N_h \times 2N_h} \end{bmatrix} \tag{9}$$

Denote $\lambda_1, \lambda_2, \cdots, \lambda_P$ be the eigenvalues of R, where $P = N_c + 2N_h$. Then, the normalized eigenvalues are calculated as follows:

$$\lambda_i' = \frac{\lambda_i}{\sum_{i=1}^{P} \lambda_i} = \frac{\lambda_i}{tr(R)} \tag{10}$$

Then, we obtain the synchronization index S between $\tilde{\chi}_n$ and Y_n:

$$S = 1 + \frac{\sum_{i=1}^{P} \lambda_i' \log(\lambda_i')}{\log(P)} \tag{11}$$

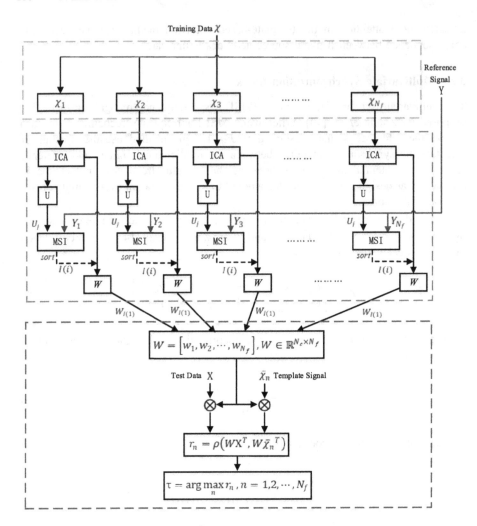

Fig. 2. Flowchart of the proposed frequency recognition method using ICA and MSI

3.3   MSI-ICA Based Frequency Recognition

The proposed method is based on the template-matching framework, in which the correlation coefficient between test data and individual templates after spatial filtering is used as a feature value [2]. Figure 2 shows the flowchart of the proposed method.

The training trials for the n-th stimulus $\chi_n \in \mathbb{R}^{N_c \times N_s \times N_t}$ are concatenated as $\tilde{\chi}_n \in \mathbb{R}^{N_c \times N_s \cdot N_t}$. Then apply the info-max ICA algorithm to concatenated un-maxing matrix $W = (w)_{ij} \in \mathbb{R}^{N_{IC} \times N_c}$ and ICs $U = (u)_{ik} = W\tilde{\chi}_n \in \mathbb{R}^{N_{IC} \times N_s \cdot N_t}$. If an IC contains SSVEPs should be higher than those with ICs unrelated to SSVEP [4]. So, the MSI between $u_i \in \mathbb{R}^{N_s \cdot N_t}$ and $Y_n \in \mathbb{R}^{2N_h \times N_s \cdot N_t}$ were calculated according to 3.2. Then the sorted un-mixing vectors can be denoted as $\tilde{W} = \left[w_{I(1)}; w_{I(2)}; \cdots; w_{I(N_{IC})} \right]$.

The 1-st un-mixing vector $w_{I(1)}$ is stored as a spatial filter $w_n \in \mathbb{R}^{N_c}$ for n-th stimuli. In the method, ensemble spatial filter W is constructed as follows:

$$W = \left[w_1; w_2; \cdots; w_{N_f} \right] \in \mathbb{R}^{N_f \times N_c} \tag{12}$$

The template signal $\bar{\chi}_n \in \mathbb{R}^{N_c \times N_s}$ for the n-th stimulus obtained by averaging all training trials χ_n, the test data $X \in \mathbb{R}^{N_c \times N_s}$. Feature values for the n-th target candidates stimulus can be calculated as follows:

$$r_n = \rho\left(W \bar{\chi}_n^T, W X^T \right) \tag{13}$$

The target stimulus τ can be identified by the follow equation:

$$\tau = \arg \max_n r_n, \quad n = 1, 2, \cdots, N_f \tag{14}$$

4 Experiments and Analysis

This section introduces the dataset, the experimental scheme and the evaluation index. The leave-one-out cross validation is used to estimate the accuracy and ITR of each experiment.

4.1 Dataset

The EEG datasets used in the experiment are provided by Tsinghua University, China, which are collected from thirty-five healthy subjects in an offline 40-target BCI speller BCI experiment [15]. For each subject, the experiment included six blocks, and every one contains 40 trials corresponding to all targets indicated in a random order. Every trial starts with a 0.5 s visual cue indicating a target stimulus. Subjects are asked to shift their gaze to the target as soon as possible within the cue duration. After the cue, all stimuli start to flicker on the screen concurrently and last 5 s. Then, the screen is blank for 0.5 s before the next trial begin. Every trial last 6 s in total.

Moreover, the EEG data are acquired using Synamps2 system at a sampling rate of 1000 Hz from sixty-four channels that are placed on the positions according to an extended 10–20 system to record whole-head EEG.

The EEG data have been segmented into 6 s epochs. The epochs are subsequently down-sampled to 250 Hz. A notch filter at 50 Hz is applied to remove the power-line noise in recording. More detail information about this dataset can be found in [15]. Considering the 140 ms latency delay in the visual system [7, 15], the used data are extracted in [0.64 s 0.64 + d s] from each epoch, where d is the time window length for frequency recognition. According to [7, 11], nine channels over the parietal and occipital areas (O1, O2, Oz, PO7, PO8, POz, P3, P4 and Pz) are used in this study.

4.2   Performance Evaluation

Classification accuracy and information transfer rate (ITR) [16] are wildly used for evaluating BCI performance. We can estimate ITR (in bits/min) as follows:

$$ITR = \left(log_2 M + P log_2 P + (1 - P)log_2 \left[\frac{1-P}{M-1}\right]\right) \times \frac{60}{T} \tag{15}$$

where M is the number of classes, P is the accuracy of target identification, and T (seconds/selection) is the average time for a selection, the gaze shifting time should be added to T in the calculation.

The leave-one-out cross validation is used to estimate the classification accuracy. In each of 6 rounds, cross-validation is performed using 5 blocks for training and 1 block for testing.

4.3   Compared Methods with Various Subject

In this experiment, the number of harmonics N_h is 5, the number of channels N_c is 9, the latency delay is 140 ms. Table 1 lists the classification accuracy and ITRs for each subject with time window length of 0.5 s by CCA, FBCCA, ICA-based and MSI-ICA based methods.

Table 1. Accuracy (%) and ITR (bits/min) for each subject with 0.5 s data length

ID	Accuracy (%)				ITR (bits/min)			
	CCA	FB-CCA	ICA	MSI-ICA	CCA	FB-CCA	ICA	MSI-ICA
S1	11.25	22.08	**89.17**	**89.17**	8.38	26.8	**256.77**	**256.77**
S2	12.08	27.92	77.5	**80.83**	8.91	41.27	202.22	**216.84**
S3	28.75	53.33	97.08	**97.5**	42.71	111.8	299.48	**302.48**
S4	30	44.58	90.83	**92.5**	45.29	84.36	264.7	**274.24**
S5	11.25	27.08	73.33	**79.58**	7.68	38.33	185.16	**211.46**
S6	7.08	14.17	55.42	**59.17**	3.17	13.36	119.44	**132.03**
S7	10	15.83	46.25	**53.33**	7.62	15.2	91.52	**113.66**
S8	6.67	10	52.92	**60**	2.93	6.28	111.09	**134.55**
S9	10.83	19.17	43.33	**47.08**	8.31	21.35	82.13	**93.6**
S10	9.58	17.5	51.25	**61.25**	6.51	18.7	106.23	**141.98**
S11	5.42	7.08	10.42	**12.92**	2.21	3.27	7.19	**9.19**
S12	12.08	22.08	52.08	**58.33**	9.48	26.95	108.54	**130.64**
S13	15	12.5	75.83	**77.92**	13.31	9.64	196.24	**204.51**
S14	16.67	31.67	77.5	**82.08**	16.8	50.01	202.58	**223.02**
S15	6.67	7.5	45	**46.25**	3.13	3.48	85.92	**90.28**
S16	3.33	6.25	35	**39.17**	0.63	2.55	57.65	**69**
S17	7.08	8.75	55	**61.25**	4.7	6.1	118.65	**139.27**
S18	16.25	29.17	74.17	**75.83**	16.01	43.06	188.52	**195.95**
S19	4.58	5.42	**12.08**	**12.08**	1.92	2.98	**9.66**	**9.66**

(continued)

Table 1. (*continued*)

ID	Accuracy (%)				ITR (bits/min)			
	CCA	FB-CCA	ICA	MSI-ICA	CCA	FB-CCA	ICA	MSI-ICA
S20	6.25	12.92	50.83	**52.5**	2.02	11	107.63	**112.89**
S21	8.33	10.83	29.17	**34.17**	6.09	8.42	43.31	**55.48**
S22	15.83	33.75	85.42	**89.17**	15.88	54.55	237.58	**255.83**
S23	13.75	30.42	64.17	**73.33**	12.45	47.11	150.43	**185.78**
S24	12.5	23.33	77.08	**80**	10.79	29.55	201.28	**212.8**
S25	17.92	23.33	75.42	**78.33**	19.06	29.87	193.64	**205.78**
S26	23.33	27.5	74.58	**79.58**	29.78	39.9	191.93	**212.48**
S27	9.58	21.25	**67.92**	67.08	6.51	25.56	**163.82**	161
S28	10.83	21.67	81.25	**82.5**	8.12	26.84	219.22	**224.45**
S29	3.75	9.17	47.5	**49.58**	1.1	6.14	94.28	**101.31**
S30	10	11.25	**57.08**	53.33	6.15	8.28	**124.69**	111.94
S31	24.17	40	92.92	**95**	31.14	73.03	275.1	**286.87**
S32	18.75	29.17	82.08	**82.92**	23.06	44.82	222.18	**226.7**
S33	5	8.75	15	**18.33**	1.26	5.81	13.74	**20.48**
S34	17.08	20.83	75	**77.92**	17.66	24.42	195.98	**207.28**
S35	5.42	5.83	31.67	**37.08**	1.73	2.11	49.96	**64.09**
Avg	12.20	20.34	60.60	**63.91**	11.5	27.51	147.95	**159.55**

CCA and FBCCA show lower performance at short time window length. CCA-based method is often affected by the interference from the spontaneous EEG activities. The MSI-ICA based approach achieves the highest accuracy and ITR for most subjects, which means ICA combine with MSI is effective. And MSI-ICA has the highest ITR at 302.48 bits/min for S3 with time window length is 0.5 s. However, all the methods show lower performance for S11, S19 and S33. This might be because the amplitude of the SSVEP is not the same for different subjects.

To summary MSI-ICA has higher robustness could potentially enhance the ITR of BCI systems.

4.4   Compared Methods with Various Time Windows

Figure 3 shows the average accuracies and ITRs across all subjects obtained by CCA, FBCCA, ICA-based and MSI-ICA based methods at various time windows (i.e. 300–700 ms with an interval of 100 ms).

Increasing the time windows length can increase the ITR, which will improve the recognition accuracy, as shown in Fig. 3(a) and (b). CCA reaches the peak at 2 s (Acc 63.46%, ITR 60.81 bits/min), FBCCA reaches the peak at 1 s (Acc 65.13%, ITR 108.89 bits/min), ICA reaches the peak at 0.7 s (Acc 76.16%, ITR 173.83 bits/min) and MSI-ICA reaches the peak at 0.7 s (Acc 79.13%, ITR 184.41 bits/min).

The result indicates that the MSI-ICA approach can get higher recognition accuracy than other approach at various time window lengths.

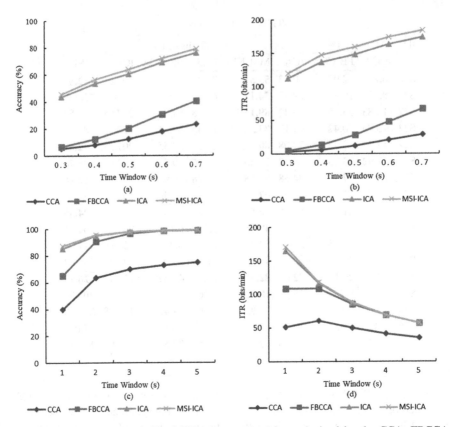

Fig. 3. The average accuracies and ITRs across all subjects obtained by the CCA, FBCCA, ICA-based and MSI-ICA based methods

5   Conclusion

We proposed a new frequency recognition method based on MSI-ICA and verified its efficiency with offline real EEG data. The results indicated that the MSI performed better than CCA, FBCCA and ICA. The standard CCA method, which does not require any calibration data, showed the lowest detection performance. The other three methods, which incorporated individual calibration data in SSVEP detection, all showed significantly improved performance. The MSI not only has the advantages of no calibration data, no channel selection and no parameter optimization, but also have been demonstrated higher detection accuracy and robustness, which could potentially enhance the ITR of BCI systems. To optimize the proposed method, we will focus on the configuration of electrodes in the future work.

Acknowledgments. The research work is supported by National Natural Science Foundation of China (U1433116) and the Fundamental Research Funds for the Central Universities (NP2017208).

References

1. Chen, X., Wang, Y., Nakanishi, M., et al.: High-speed spelling with a noninvasive brain–computer interface. Proc. Natl. Acad. Sci. **112**(44), E6058–E6067 (2015)
2. Nakanishi, M., Wang, Y., Wang, Y.T., et al.: A high-speed brain speller using steady-state visual evoked potentials. Int. J. Neural Syst. **24**(06), 1450019 (2014)
3. Zhang, Y., Xu, P., Liu, T., et al.: Multiple frequencies sequential coding for SSVEP-based brain-computer interface. PLoS ONE **7**(3), e29519 (2012)
4. Jia, C., Gao, X., Hong, B., et al.: Frequency and phase mixed coding in SSVEP-based brain–computer interface. IEEE Trans. Biomed. Eng. **58**(1), 200–206 (2011)
5. Chen, X., Chen, Z., Gao, S., et al.: A high-ITR SSVEP-based BCI speller. Brain-Computer Interfaces **1**(3–4), 181–191 (2014)
6. Lin, Z., Zhang, C., Wu, W., et al.: Frequency recognition based on canonical correlation analysis for SSVEP-based BCIs. IEEE Trans. Biomed. Eng. **53**(12), 2610–2614 (2006)
7. Chen, X., Wang, Y., Gao, S., et al.: Filter bank canonical correlation analysis for implementing a high-speed SSVEP-based brain–computer interface. J. Neural Eng. **12**(4), 046008 (2015)
8. Wang, Y., Nakanishi, M., Wang, Y.T., et al.: Enhancing detection of steady-state visual evoked potentials using individual training data. In: Proceedings of 36th Annual International Conference of the IEEE Engineering in Medicine and Biology Society, pp. 3037–3040 (2014)
9. Nakanishi, M., Wang, Y., Hsu, S.H., et al.: Independent component analysis-based spatial filtering improves template-based SSVEP detection. In: 2017 39th Annual International Conference of the IEEE Engineering in Medicine and Biology Society (EMBC), pp. 3620–3623. IEEE (2017)
10. Delorme, A., Sejnowski, T., Makeig, S.: Enhanced detection of artifacts in EEG data using higher-order statistics and independent component analysis. Neuroimage **34**(4), 1443–1449 (2007)
11. Bin, G., Gao, X., Yan, Z., et al.: An online multi-channel SSVEP-based brain–computer interface using a canonical correlation analysis method. J. Neural Eng. **6**(4), 046002 (2009)
12. Ang, K.K., Chin, Z.Y., Zhang, H., et al.: Filter bank common spatial pattern (FBCSP) in brain-computer interface. In: IEEE World Congress on Computational Intelligence Neural Networks, IEEE International Joint Conference on IJCNN 2008, pp. 2390–2397. IEEE (2008)
13. Wang, Y., Wang, R., Gao, X., et al.: A practical VEP-based brain-computer interface. IEEE Trans. Neural Syst. Rehabil. Eng. **14**(2), 234–240 (2006)
14. Zhang, Y., Xu, P., Cheng, K., et al.: Multivariate synchronization index for frequency recognition of SSVEP-based brain–computer interface. J. Neurosci. Methods **221**, 32–40 (2014)
15. Wang, Y., Chen, X., Gao, X., et al.: A benchmark dataset for SSVEP-based brain–computer interfaces. IEEE Trans. Neural Syst. Rehabil. Eng. **25**(10), 1746–1752 (2017)
16. Chen, W., Wang, S., Zhang, X., et al.: EEG-based motion intention recognition via multi-task RNNs. In: Proceedings of the 2018 SIAM International Conference on Data Mining. Society for Industrial and Applied Mathematics, pp. 279–287 (2018)

Big Data

Forecasting Traffic Flow in Big Cities Using Modified Tucker Decomposition

Manish Bhanu[1]($\boxtimes$), Shalini Priya[1], Sourav Kumar Dandapat[1],
Joydeep Chandra[1], and João Mendes-Moreira[2]

[1] Indian Institute of Technology, Patna, India
{manish.pcs16,shalini.pcs16,sourav,joydeep}@iitp.ac.in
[2] LIAAD - INESC TEC, Faculty of Engineering, University of Porto,
Rua Dr. Roberto Frias s/n, 4200-465 Porto, Portugal
jmoreira@fe.up.pt

Abstract. An efficient traffic-network is an essential demand for any smart city. Usually, city traffic forms a huge network with millions of locations and trips. Traffic flow prediction using such large data is a classical problem in intelligent transportation system (ITS). Many existing models such as ARIMA, SVR, ANN etc, are deployed to retrieve important characteristics of traffic-network and for forecasting mobility. However, these methods suffer from the inability to handle higher data dimensionality. The tensor-based approach has recently gained success over the existing methods due to its ability to decompose high dimension data into factor components. We present a modified Tucker decomposition method which predicts traffic mobility by approximating very large networks so as to handle the dimensionality problem. Our experiments on two big-city traffic-networks show that our method reduces the forecasting error, for up to 7 days, by around 80% as compared to the existing state of the art methods. Further, our method also efficiently handles the data dimensionality problem as compared to the existing methods.

Keywords: ODM · CP decomposition · Tucker · Time-series · CUR Traffic flow

1 Introduction

Over the last few decades, a large body of flow prediction techniques have been developed to facilitate the traffic organizations in controlling and improving the transportation efficiency—ranging from driver assistance to vehicle routing, forecasting and signal coordination [2,17]. Traffic-network is generally studied with Origin-Destination matrix (ODM). An ODM represents traffic volume in a given time duration. Many stacked ODMs for consecutive time duration represent time series of traffic volume of the sources and destinations. Given a large volume of origin-destination data (identified by the GPS locations), collected over a period of time, the traffic prediction problem is to forecast the traffic volume between locations at any given time in the future.

© Springer Nature Switzerland AG 2018
G. Gan et al. (Eds.): ADMA 2018, LNAI 11323, pp. 119–128, 2018.
https://doi.org/10.1007/978-3-030-05090-0_10

A large body of works exists that deals with traffic forecasting. Forecasting techniques can be broadly categorized as parametric and non-parametric techniques. Examples of parametric techniques include linear and nonlinear regression, smoothing techniques [16] and auto-regressive process by time-series analysis [5], to list a few. On the other hand the non-parametric techniques include non-parametric regression [3] and neural network based techniques [11,12]. Existing approaches for forecasting the traffic flow using time series approaches like, ARIMA [9], SARIMA [8] and Vector ARMA [7] are used to study the characteristics of the ODM and forecasting the same. Due to peak and off-peak time, traffic flow unveils a regular pattern, i.e. it presents daily seasonality. According to many previous studies, in order to predict better seasonal pattern SARIMA model is proved to perform better than the models constructed using support vector regression, historical average, and simple ARIMA [9]. Some machine learning techniques like ANN [6], SVR [1], KNN [18] and Bayesian network [14] have also been used. These methods generally perform well when modeling and prediction is done for each OD pair separately. Further, most of these approaches predict each time series independently. Forecasting each trip is considered isolated and prediction of thousands of trips goes separately. Thus these works do not capture the existing inter-relationship among the trips [4,19]. With widening popularity of deep learning techniques, Lv et al. [11] applied the same for traffic flow prediction using several traffic features. Experimental results indicate that these methods obtain much improved performance over traditional methods. However, these single point prediction methods are inefficient when dealing with larger traffic networks. For example, a traffic network containing $N \times N$ source-destination pair requires $O(n^2)$ time-series to predict, which is difficult to process when N exceeds thousands.

Tensor based methods on the other hand can deal with these problems better by factorizing the ODM into essential components. Tensor decomposition can yield dimensionality reduction; the reduced dimensional components are easy to be operated. This helps handling complete network as one unit. Tensors have found wide applications in forecasting traffic [13,15]. Most of these techniques aim to derive traffic approximations using low rank tensor decompositions. However, the major drawback of this approach is the requirement of huge historic data to predict the traffic flow from one source to a destination. There are two major scopes of improvement to these techniques, (a) improving the prediction accuracy and (b) maintaining higher accuracy in face of missing traffic data. Our paper is directed towards fulfilling these goals.

In this paper we propose a tensor decomposition based traffic forecasting technique that addresses both these requirements. The proposed approach uses three and four dimensional tensors to predict complete city traffic flow at once using modified Tucker tensor decomposition. Our modified Tucker decomposition technique uses CUR factorization, that gives the advantage of restoring the original values of matrix during approximation for prediction. The Frobenius norm of the difference between original and approximated matrix is very less when CUR is used over other matrix factorization methods like SVD, QR etc.

For our analysis, we use two datasets, one from New York city and the other from Thessaloniki city. Both contain traffic flow from many sources to destinations in the city and we predict the traffic flow (volume) of complete city. Experimental results indicate that the proposed approach yields significant forecasting accuracy even for sparse network data. The forecasting accuracy of our approach when compared to the recent techniques indicates an improvement of around 80%. Further, the proposed approach also effectively handles higher dimensionality of the traffic data.

The organization of the paper is as follows. We discuss the problem definition in Sect. 2. The proposed approach is discussed in Sect. 3 followed by the experimental procedure in Sect. 4. The experimental results are discussed in Sect. 5. Finally, we draw our conclusions in Sect. 6.

2   Problem Statement

For a big city, the traffic network consists of millions[1] of trips with GPS information of start and end locations along with the trip-timestamps. The ODM of these cities contains millions of time-series data. Using such a large ODM in any predictive model would incur huge computational cost. To tackle the problem of predicting traffic flows in large networks, considering the inter-dependence relationships of the time-series, we need to address the following research questions:-

> **RQ1:** How to create an efficient ODM for a big city traffic network with the desired number of influencing factors (components) of mobility?
> **RQ2:** How to handle the interdependence of the trip time-series to forecast the traffic flow in the entire network?
> **RQ3:** How to handle higher dimensionality of the traffic data (influencing factors or components) that can affect the forecasting results?

We formulate the problem of large city traffic forecasting as a two-step process: (i) decomposing (unfolding) the traffic data T into components A_i (for i^{th} component) and (ii) approximating components with factorized matrices (F_i) for which the Frobenius norm of the difference is minimum under low-rank constraints [10,15]. The problem lies in finding approximated components A_i, F_i and reconstructing T after prediction.

$$\underset{A_i}{\operatorname{argmin}} \sum_i \|A_i - F_i\|_F^2$$
$$\text{subject to}: unfold(T, i) = A_i$$
$$rank(A_i) = k < rank(T_i) \tag{1}$$

3   Methodology

In this section, we outline our methodology adopted to address each of the issues described in the previous section.

---

[1] in our examples: New York 2.6M, Thessaloniki 1.7M; M: million trips in 3 months.

RQ1 (ODM creation): The solution to the first question is the multidimensional ODM which can be easily represented with Tensor. The OD-tensor can be represented as a three-dimensional tensor $\mathcal{T} \in \mathbb{R}^{T \times S \times D}$, where T, S and D, represents the time periods, sources and destinations respectively. Likewise, we also use OD-tensor with four components $(T \times S \times D \times W)$ where W is the day of a week. Each entry of an OD-tensor cell is the volume of traffic from source S to destination D for given time T on a particular weekday W. We cannot create an ODM directly from GPS locations as the number of unique positions would be in millions. An efficient approach is to use clustering algorithms to group a set of nearby positions to represent a particular area of the city, that can act as an origin/destination in the ODM. We use a grid based strategy in our work as (a) it's faster, (b) density-based clustering techniques like DBSCAN caused uneven distributions of GPS locations resulting in few clusters to have approximately 90% of all locations while (c) hierarchical-based techniques need heavy calculations of Haversine distance between each location.

RQ2 (Forecasting): OD-tensor of the traffic network contains time-series data for each pair of origin-destination. The number of time-series for four components (c-4) will increase multiplicatively with the increase in the number of components in tensor. During decomposition we gain dimensionality reduction by decomposing the tensor with dimension $(T \times S \times D \times W)$ to smaller dimensional components, $T \times K_1$, $S \times K_2$, $D \times K_3$ and $W \times K_4$ respectively along with a tensor (core) with dimension $K_1 \times K_2 \times K_3$, where $K_1 < T$, $K_2 < S$, $K_3 < D$ and $K_4 < W$. We use a generalization of HOSVD decomposition called *Tucker decomposition*[2]. Tensor decomposition is also used in missing data problem rather than forecasting [10]. However, modelling the same for forecasting would require large historic traffic data. (we present details in Subsect. 5.2). However CP decomposition has no flexibility to assign different ranks to each components separately. Hence we use Tucker decomposition with CUR factorization owing to the advantages of CUR over SVD, (1) retaining real-valued data of time-series, (2) maintaining sparsity in components and (3) lower Frobenius norm difference than that of SVD.

Proposed Approach (Modified Tucker-CUR): Given an OD-tensor $\mathcal{T}$, we initially perform Tucker decomposition with CUR factorization to approximate the component matrices (Algorithm 1). The selection of best columns and rows in CUR is done on the basis of top higher L_1 values of columns and rows. We use ALS to approximate A_i (Algorithm 2).

---

[2] https://iksinc.online/2018/05/02/understanding-tensors-and-tensor-decompositions -part-3/.

Algorithm 1. Tucker Decomposition

1: **procedure** (Tucker) $(\mathcal{T}, R1, R2, ..., RN)$
2:     **for** n=1,2,...,N **do**
3:         $A^{(n)} \leftarrow C_n \ CUR(\mathcal{T}_n)$
4:     **end for**
5:     $G \leftarrow \mathcal{T} X_1 A^{(1)T} X_2 A^{(2)T} ... X_N A^{(N)T}$
6:     **return** $G, A^{(1)}, A^{(2)}, ..., A^{(N)}$
7: **end procedure**

Algorithm 2. Sampled CUR factorization

Input: $A(i)_{m \times n}$.
Output: CUR
1: **for** i:n **do**
2:     $r_{l2} = \|A_{(i)}(i,:)\|$
3:         $R \Leftarrow select\_top\_k_1\_rows: C_{l2}$
4: **end for**
5: **for** i:n **do**
6:     $c_{l2} = \|A_{(i)}(i,:)\|$
7:         $C \Leftarrow select\_top\_k_2\_columns: r_{l2}$
8: **end for**
9: $U \Leftarrow pseudo-inverse(R \cap C)$
10: $U \Leftarrow (U^+U)^{-1}U^+$
11: **return** C,U,R

For forecasting we use the approximated temporal component matrix, which would be of dimension $T \times K_1$, where T is the maximum days and K_1 is the number of columns in the low rank matrix (K_1 is 100 or less in our experiments). Thus this matrix produces K_1 time series with data for 1 to T days. We use these time series values to predict additional values up to δ additional days (we use a maximum value of $\delta = 7$), using ARIMA and LSTM. Applying prediction model on the temporal component (size $T \times K_1$) is computationally less expensive than applying on the entire tensor ($S \times D \times W$). We subsequently obtain a new matrix of size $(T + \delta) \times K_1$ by combining the predicted rows to the original temporal component matrix and then use the same to reconstruct the new tensor T' with dimension $(T + \delta) \times S \times D \times W$. The new entries in the T' for the δ additional days represent the forecasted values of the traffic. The steps of the proposed approach is outlined in Algorithm 3.

RQ3 **Higher Dimensionality:** The proposed approach in Algorithm 3 can effectively handle higher dimensionality. To investigate the same we conducted forecasting experiments with 3 and 4 components ($T \times S \times D$ and $T \times S \times D \times W$ respectively) to see the effect of more components on the predictive results.

We next discuss the dataset used for our experiments followed by the experimental results.

Algorithm 3. Tucker-CUR: Proposed algorithm

Input: $X_{old}, k_1, k_2, k_3, k_4, \delta$.
Output: $X_{predicted}$
 1: $G, factors \leftarrow Tucker(X_{old}, k_1, k_2, k_3, k_4)$
 2: $Factors : C_{T*K_1}, C_{S*K_2}, C_{D*K_3}, C_{W*K_4}$
 3: $C_{(T+\delta)*k_1} \leftarrow predict(C_{T*K_1}, \delta)$
 4: $X_{predicted} \leftarrow G_{K1 \times K2 \times K3} X_1 C_{(T+\delta) X_2 K_1} X_2 C_{S \times K2} X_3 C_{D \times K3} X_4 C_{W \times K4}$
 5: **return** $X_{predicted}$

4   Experimental Procedure

For our experiments, we used New York City Green Taxi data(NYC) from January to March 2014 available online[3]. The other data is from Thessaloniki(THS) private taxi company for January to March 2015 which is provided in famous "taxi fare challenge kaggle competition, 2017". For Green Taxi data, each trip has GPS data of pickup location, drop-off location and time duration of the trip. The same information were obtained for the THS data, with an assumption that the startup time of the next trip being same as the drop-off time of the previous trip.

- For both the cities, the city GPS locations were clustered in grid size of 2 and 5 km². Then we created OD-tensor with components $T \times S \times D$ (c-3) and $T \times S \times D \times W$ (c-4) for all the grided city network. We obtained eight city traffic-networks: NYC02(c-3, c-4) & NYC05(c-3, c-4) and THS02(c-3, c-4) & THS05(c-3, c-4)[4]. Each grid is considered as a node of the OD-tensor. We only used the node-pairs whose traffic volume were equal or greater than 60 for this time duration.
- For 300 locations, the OD-tensor time series count is 90,000. We apply Algorithm 1 to obtain the component matrices and use the temporal component matrix as stated in Algorithm 3 to derive the predicted OD-tensor.
- During prediction we used 80% of temporal component data as train-set and rest as test-set. The best models are obtained at ARIMA(7,2,1) and at LSTM (nodes_per_Layer = 200, epoch = 100, batch size = 3, hidden layers = 4, optimizer = "adam"), validation_split = 0.1. Data were normalized before prediction by LSTM. Sigmoid activation and multivariate of window size 7 was used along with time steps up to 7. We tuned other parameters of the models to get the best results on networks using HOSVD (one of baseline approach) and then the same models were used for Tucker-CUR (Algorithm 3) during prediction.

---

[3] http://www.nyc.gov/html/tlc/html/about/trip_record_data.shtml.
[4] (Newyork/Thessaloniki city with grid size 02/05 and components 3/4, c stands for components).

5   Results and Discussions

We next outline the evaluation strategies to analyze the performance of the proposed method and subsequently outline the results obtained.

5.1   Evaluation Strategies

We compare our approach with existing methods proposed for traffic-network predictions [13, 15]. We compare the efficiency of our proposed approach in terms of the data requirement and prediction accuracy. Dynamic Tensor Completion (DTC) [15] DTC is a recent state of the art approach proposed for traffic prediction, the prediction model of DTC is not directly comparable to our proposed method due to its large data requirement. We subsequently compare the prediction accuracy of our approach with two state-of-the-art tensor decomposition methods proposed in [13]:

- **CP Decomposition:** In [13], the authors have used CP decomposition which is a tensor decomposition method that uses a random assignment of initial decomposed matrices.
- **HOSVD:** We also compare our work with the HOSVD decomposition which uses SVD as factorization approach. We use the HOSVD[5] (Tucker SVD) approach to obtain the temporal component matrix for prediction.

5.2   Data Requirement

In [15], the authors proposed a dynamic tensor completion technique for predicting short term traffic data. However the successful implementation of the technique requires satisfying two major conditions of the ODM:

- A matrix with missing data must not have a complete row or column missing;
- A matrix must be a low rank matrix.

The second constraint, however, is applicable for almost every approach as very few factors (low-rank) are responsible for determining the traffic between locations. But the first constraint enforces heavy data requirement, i.e. a large volume of data is required to predict few data values. For an $M \times N$ matrix the predicted values cannot exceed max(M, N). We compared the maximum predicted values of our proposed approach with the DTC model. Giving a brief representation of data requirement in each approach, when ODM $M \times N$ is used, maximum predicted values in DTC is max(M-1, N-1) while it is cN (multiple columns) in our proposed approach where c is a positive integer. Both approaches used 90 days (3 months) for experimentation, hence "predicted values to data required" ratio for DTC is $1/M$ and for our approach it's c/M. We predicted up to 7 days, hence for c = 7 and M = 90, DTC and proposed approach have "predicted to data required" ratio as 0.01 and 0.07 respectively. This shows when modeling missing data problem as prediction problem like in DTC [15], the model demands more data requirement than proposed approach for same count of predicted values.

---

[5] k assignment at good prediction result: NYC-THS02 (50, 100, 100, 5) NYC-THS05 (50, 30, 30, 5) & same for Tucker-CUR.

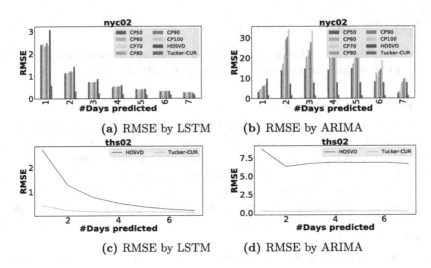

Fig. 1. LSTM performs much better than ARIMA. When the complete city network is predicted.

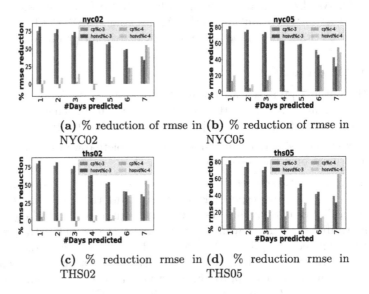

Fig. 2. % Reduction in RMSE using LSTM over both the city networks using (Eq. 2). Tucker-CUR is compared to CP70 (best among CP) and HOSVD in both (c-3, c-4).

5.3   Prediction Accuracy

Choosing the Time Series Model: We consider two major techniques for prediction of the time series: LSTM and ARIMA. Figure 1 shows the RMSE values of the predicted time series using LSTM and ARIMA. The range of values of RMSE in case of LSTM (Figs. 1a and c) is much lower as compared to

ARIMA (Figs. 1b and d). This is because traffic-network time series can be non-linear. LSTM captures the non-linearity of the data much better as compared to ARIMA. So, we use LSTM in our future experiments.

Comparison of Prediction Accuracy. We predicted eight traffic-networks using LSTM for 1 to 7 days using our approach as well as the CP and HOSVD based decomposition and calculate the RMSE values. The reduction in RMSE values is calculated using the following equation:

$$(OtherApproach_{RMSE} - TuckerCUR_{RMSE}) * 100/TuckerCUR_{RMSE} \quad (2)$$

It is observed that our approach has much lower RMSE as compared to CP (with different $K = 50$, 60, 70, 80, 90, 100) and HOSVD (with different k approximations) Fig. 2. In most cases, as shown in Fig. 2, our proposed method achieves around 80% reduction in RMSE, except in very few cases where CP($k = 70$) performs better.

6   Conclusion

In this paper, we proposed an OD-tensor decomposition based method for forecasting large city traffic. The proposed method overcomes the problem of traditional time-series based prediction methods and allows addition of new dimensions if required. We have gained reduction in prediction error in terms of RMSE up to 80% compared to the state of the art approach. Our Tucker-CUR algorithm is efficient in sense of using real time-series data and reduced time complexity over CP, HOSVD. We also showed the effect of spatial and component changes in OD-tensor and this leaves scope for further analysis in this direction.

Acknowledgment. This work is funded by Project "TEC4Growth - Pervasive Intelligence, Enhancers and Proofs of Concept with Industrial Impact/NORTE-01-0145-FEDER-000020", a project financed by North Portugal Regional Operational Programme (NORTE 2020), under the PORTUGAL 2020 Partnership Agreement.

References

1. Ahn, J.Y., Ko, E., Kim, E.: Predicting spatiotemporal traffic flow based on support vector regression and Bayesian classifier. In: 2015 IEEE Fifth International Conference on Big Data and Cloud Computing (BDCloud), pp. 125–130. IEEE (2015)
2. Bhanu, M., Chandra, J., Mendes-Moreira, J.: Enhancing traffic model of big cities: network skeleton & reciprocity. In: 2018 10th International Conference on Communication Systems & Networks (COMSNETS), pp. 121–128. IEEE (2018)
3. Davis, G.A., Nihan, N.L.: Nonparametric regression and short-term freeway traffic forecasting. J. Transp. Eng. **117**(2), 178–188 (1991)

4. Gabrielli, L., Rinzivillo, S., Ronzano, F., Villatoro, D.: From tweets to semantic trajectories: mining anomalous urban mobility patterns. In: Nin, J., Villatoro, D. (eds.) CitiSens 2013. LNCS (LNAI), vol. 8313, pp. 26–35. Springer, Cham (2014). https://doi.org/10.1007/978-3-319-04178-0_3

5. Ghosh, B., Basu, B., O'Mahony, M.: Bayesian time-series model for short-term traffic flow forecasting. J. Transp. Eng. **133**(3), 180–189 (2007)

6. Jun, M., Ying, M.: Research of traffic flow forecasting based on neural network. In: Second International Symposium on Intelligent Information Technology Application, IITA 2008, vol. 2, pp. 104–108. IEEE (2008)

7. Kamarianakis, Y., Prastacos, P.: Forecasting traffic flow conditions in an urban network: comparison of multivariate and univariate approaches. Transp. Res. Rec.: J. Transp. Res. Board (1857), 74–84 (2003)

8. Kumar, S.V., Vanajakshi, L.: Short-term traffic flow prediction using seasonal arima model with limited input data. Eur. Transp. Res. Rev. **7**(3), 21 (2015)

9. Lee, S., Fambro, D.: Application of subset autoregressive integrated moving average model for short-term freeway traffic volume forecasting. Transp. Res. Rec.: J. Transp. Res. Board (1678), 179–188 (1999)

10. Liu, J., Musialski, P., Wonka, P., Ye, J.: Tensor completion for estimating missing values in visual data. IEEE Trans. Pattern Anal. Mach. Intell. **35**(1), 208–220 (2013)

11. Lv, Y., Duan, Y., Kang, W., Li, Z., Wang, F.-Y.: Traffic flow prediction with big data: a deep learning approach. IEEE Trans. Intell. Transp. Syst. **16**(2), 865–873 (2015)

12. Priya, S., Bhanu, M., Dandapat, S.K., Ghosh, K., Chandra, J.: Characterizing infrastructure damage after earthquake: a split-query based IR approach. In: 2018 IEEE/ACM International Conference on Advances in Social Networks Analysis and Mining (ASONAM), pp. 202–209. IEEE (2018)

13. Ren, J., Xie, Q.: Efficient OD trip matrix prediction based on tensor decomposition. In: 2017 18th IEEE International Conference on Mobile Data Management (MDM), pp. 180–185. IEEE (2017)

14. Sun, S., Zhang, C., Zhang, Y.: Traffic flow forecasting using a spatio-temporal bayesian network predictor. In: Duch, W., Kacprzyk, J., Oja, E., Zadrożny, S. (eds.) ICANN 2005 Part II. LNCS, vol. 3697, pp. 273–278. Springer, Heidelberg (2005). https://doi.org/10.1007/11550907_43

15. Tan, H., Wu, Y., Shen, B., Jin, P.J., Ran, B.: Short-term traffic prediction based on dynamic tensor completion. IEEE Trans. Intell. Transp. Syst. **17**(8), 2123–2133 (2016)

16. Tan, M.-C., Wong, S.C., Xu, J.-M., Guan, Z.-R., Zhang, P.: An aggregation approach to short-term traffic flow prediction. IEEE Trans. Intell. Transp. Syst. **10**(1), 60–69 (2009)

17. Wang, J., Zhang, L., Zhang, D., Li, K.: An adaptive longitudinal driving assistance system based on driver characteristics. IEEE Trans. Intell. Transp. Syst. **14**(1), 1–12 (2013)

18. Xiaoyu, H., Yisheng, W., Siyu, H.: Short-term traffic flow forecasting based on two-tier k-nearest neighbor algorithm. Procedia-Soc. Behav. Sci. **96**, 2529–2536 (2013)

19. Yuan, Y., Raubal, M.: Extracting dynamic urban mobility patterns from mobile phone data. In: Xiao, N., Kwan, M.-P., Goodchild, M.F., Shekhar, S. (eds.) GIScience 2012. LNCS, vol. 7478, pp. 354–367. Springer, Heidelberg (2012). https://doi.org/10.1007/978-3-642-33024-7_26

A Sparse and Low-Rank Matrix Recovery Model for Saliency Detection

Chao Wang, Jing Li$^{(\boxtimes)}$, KeXin Li, and Yi Zhuang

College of Computer Science and Technology,
Nanjing University of Aeronautics and Astronautics,
Nanjing 210016, Jiangsu, China
{wangchao159,lijing,kexinli,zhuangyi}@nuaa.edu.cn

Abstract. The previous low-rank matrix recovery model for saliency detection have a large of problem that the transform matrix obtained on the open datasets may not be suitable for the detecting image and the transform matrix fails to combine the low-level features of the image. In this paper, we propose a novel salient object detection model that combines sparse and low-rank matrix recovery (SLRR) with the adaptive background template. Our SLRR model using a selection strategy is presented to establish the adaptive background template by removing the potential saliency super pixels from the image border regions, and the background template is obtained. And the sparse and low rank matrix recovery model solved by Inexact Augmented Lagrange Multiplier (ALM). Both quantitative and qualitative experimental results on two challenging datasets show competitive results as compared with other state-of-the-art methods. In addition, a new datasets which saliency object on the edge (SOE), containing 500 images is constructed for evaluating saliency detection.

Keywords: Low rank · Saliency detection · Matrix recovery
Inexact augmented lagrange multiplier

1 Introduce

In saliency detection, the task is to identify saliency object and to suppress the redundant background informs from visual scenes [1–3]. Saliency detection is the use of computer vision method in an image by suppressing the image of the redundant background information extracted from people interested in the object. In recent years, the study of saliency detection has attracted the interest of many scholars, and many saliency detection methods have been proposed [6], and it has been widely applied to image segmentation [7], target recognition retrieval [10] and image classification [11] and other fields.

In sparse and low rank recovery model, the salient object are indicated by the sparse noises. In order to determine the position of the saliency object, we introduce a novel multi-scale segmentation strategy to handle size variation of salient objects. The saliency value of a neighbor's pixel will have an effect on itself. If the pixel's feature is similar to the low-level feature of the neighboring pixel, the greater the influence of the neighbor pixel.

G. Gan et al. (Eds.): ADMA 2018, LNAI 11323, pp. 129–139, 2018.
https://doi.org/10.1007/978-3-030-05090-0_11

Motivated by the above discussion, we proposed sparse and low-rank matrix recovery model base on adaptive background template (SLRR). We use a new selection strategy based on contrast principle and the connecting principle to generate adaptive background template. Then, we apply multi-scale analysis with multiple segmentations to handle size variation of salient objects. The salient map is obtained by solving sparse and low rank recovery model.

The main contributions of this paper are listed as follows. We optimize the method of obtaining the transform matrix in the sparse and low rank matrix models by using the adaptive background template, and presented a new spatial prior to integrate the saliency detection results by aggregating two complementary measures such as image center preference and the background template exclusion in order to get adaptive background template. This sparse and low-rank matrix recovery is solved by Inexact Augmented Lagrange Multiplier (ALM), and it incorporates a background template to make sparse and low-rank matrix recovery more perfect.

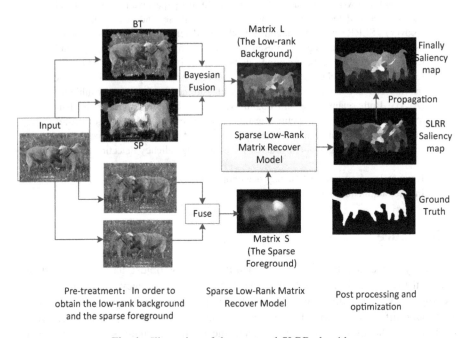

Fig. 1. Illustration of the proposed SLRR algorithm

2 Sparse Low-Rank Matrix Recover for Saliency Object Detection

The framework of saliency detection method based on sparse and low rank matrix recovery model (SLRR) is shown in Fig. 1, which contains three parts: pre-treatment, sparse and low rank matrix recover and post processing and optimization. We first analyze the image sparse and low rank matrix recovery model in Sect. 2.1. Section 2.2

introduce a lots of priori knowledge is used in the background template to ensure that the background of the image is low-rank subspace. Finally, Sect. 2.3 describes how to used ALM method solved the SLRR problem.

2.1 Sparse Low-Rank Matrix Recover Model

An image can be represented as a low-rank matrix plus a sparse noises in a feature space. The input image $I(F$ representing the I's feature space) can be decomposed into the sum of the low rank matrix L (representing the background region) and the sparse matrix S (representing the saliency object), which is $F = L + S$. The saliency is obtained by matrix decomposition, and the low rank matrix recovery model can be expressed by Eq. 1.

$$\arg \min_{L,S}(rank(L) + \lambda\|S\|_0) \tag{1}$$

$$s.t. \quad F = L + S$$

where $rank(.)$ is used to solve the rank of the matrix, $\|.\|_0$ indicates l_0-norm. But this problem is NP-hard problem, it is necessary to be converted to Eq. 2.

$$\arg \min_{L,S}(\|L\|_* + \lambda\|S\|_1) \tag{2}$$

$$s.t. \quad F = FL + S$$

On the basis of Liu [19], we propose a model of sparse and low rank matrix recover for saliency object detection. A background prior matrix and foreground sparse based on multi-scale saliency detection are applied to let the low rank (LR) model more accurate. We propose the sparse and low rank matrix recovery model for saliency object detection that can be described as Eq. 3

$$\arg \min_{\substack{L_1,\ldots,L_K \\ S_1,\ldots,S_K}} \left(\sum_{i=1}^{k} \|L_i\|_* + \lambda\|S'\|_1 \right) \tag{3}$$

$$s.t. \quad W_i * F_i = F_i * L_i + S_i, i = 1,\ldots,K$$

W is the feature space transformation matrix. L is the scale of SLIC. F is the characteristic subspace at each scale (L). And then obtain the foreground sparse matrix by multi-scale segmentation. We obtain the saliency map by solving the new low rank matrix recovery model (as shown Eq. 3). Finally, the final saliency map is obtained integrating the saliency map based on the propagated saliency map.

2.2 Background Low-Rank Matrix Based on Adaptive Background Temple and Spatial Prior

First, the original image is segmented into multiple super pixels. Based on the assumption that the regions located in image boundary are most likely to be the background, and this background regions are inevitably contains the saliency super pixels when the saliency object appears near the image border. In order to remove the salient patches in and increase the robustness of saliency detection methods based on background prior, we design a new selection strategy based on the contrast principle and the connecting principle.

The salient score S_i^{bt} of super pixels sp_i is defined as the weighted sum of super pixels sp_i color contrasts to other super pixels in background template N_{bt}.

$$S_i^{bt} = \sum_{k=1,k\neq i}^{N_{br}} \exp\left(\frac{-d_s(sp_i,sp_k)}{\sigma_s^2}\right)(d_c(sp_i,sp_k)) \tag{4}$$

Where σ_s controls the strength of spatial weights, $d_s(sp_i,sp_k)$ is the spatial distance and $d_c(sp_i,sp_k)$ is the color distance. Our spatial prior S_i^{sp} of pixel p_i belonging to super pixels sp_j is calculated as follow:

$$S_i^{sp} = \exp\left\{-\left[\frac{1}{N}\sum_{i=1}^{N}\left((x_i-\mu_x)^2+(y_i-\mu_y)^2\right)+\frac{\eta}{N_{br}}\sum_{k=1,k\neq i}^{N_{br}}\left((x_i-x_k)^2+(y_i-y_k)^2\right)^{-1}\right]\right\} \tag{5}$$

Where (x_i, y_i) and (x_k, y_k) denote the coordinates of pixel p_i and p_k, (μ_x, μ_y) denotes the image center and η controls the relative weight between the image center preference and the image border excluded. The low-rank background matrix score W_i of pixel p_i is calculated by integrating the spatial prior and background template saliency map in Eq. 6.

$$W_i = S_j^{bt} * S_i^{sp} \tag{6}$$

The sparse low-rank matrix recover model can obtain the transformed matrix W in Eq. 3 by the adaptive background temple.

2.3 Solving SLRR Model by Inexact ALM

This SLRR problem can be solved by the ALM method, which minimizes the following augmented Lagrange function:

$$\gamma = \arg \min_{\substack{L_1,...,L_K \\ S_1,...,S_K}} \left(\sum_{i=1}^{K}\|L_i\|_* + \lambda\|S'\|_1\right) + \sum_{i=1}^{K}\left(tr(Y_1^T(W_i * F_i - F_i * L_i - S_i)\right)$$

$$+ \frac{\mu}{2}\left(\sum_{i=1}^{K}\left(\|W_i * F_i - F_i * L_i - S_i\|_F^2\right)\right) \tag{7}$$

So it can be minimized with respect to L_i and S respectively, by fixing the other variables, and then updating the Lagrange multipliers Y_1, where $\lambda > 0$ is a penalty parameter. The inexact ALM method, also called the alternating direction method (Table 1).

Table 1. Solving problem by inexact ALM

Algorithm 1 Solving Problem by Inexact ALM

Input: data matrix W, parameter β, λ

Initialize: $L_i = S = 0$, $Y_1 = 0$, $\mu = 10^{-6}$, $\mu max = 10^6$, $\rho = 1.1$, and $\varepsilon = 10^{-8}$

while not converged **do**

1. fix the others and update L_i by

$$L_i = \arg\min \frac{1}{\mu} \left| L_{i(m',n')} \right|_* + \frac{1}{2} (\| W_i * F_i - F_i * L_i - S_i \|_F^2)$$

2. fix the others and update S by

$$S' = \arg\min \frac{\lambda}{\mu} \| S' \|_1 + \left\| S_i - \frac{(W_i * F_i - F_i * L_i)}{\mu} \right\|_F^2$$

3. update the multipliers

$$Y_1 = Y_1 + \mu (W_i * F_i - F_i * L_i - S_i)$$

4. update the parameter μ by $\mu = min(\rho\mu, \mu max)$.

5. check the convergence conditions:

$$\| W_i * F_i - F_i * L_i - S_i \|_\infty < \varepsilon$$

end while

---

3 Experimental Result

The proposed method is evaluated with eleven state-of-the-art algorithms on two datasets. The datasets and evaluation criteria are introduced firstly, followed by comparing the experimental results in a quantitative way. In the experiments of this paper, we shall focus on analyzing the essential aspects of SLRR, under the context of adaptive background template and foreground sparse matrix.

We also consider for comparison some previous saliency detection methods, including Frequency-Tuned saliency (FT) [12], Spectral Residual saliency (SR) [19], Dense and low-rank Gaussian (DLRG) [14], Region-based Contrast (RC) [13], Hierarchical Saliency (HS) [20], Dense and Sparse Reconstruction (DSR) [6], Spatiotemporal Cues (LC) [4], Histogram-based Contrast (HC) [13], Low Rank Matrix Recovery (LR) [9], Markov Chain saliency (MC) [17]. Whenever they are available, we use the author-provided results. Results of FT, SR, HC and RC are generated by using the codes provided by [13], and adopt the public implementations from the original authors for DLRG, LR, HS, DSR, MC. Note that the saliency maps of all methods are mapped to the range [0, 255] by the same max-min normalization method for the further evaluation.

3.1   Results on HKU_IS

HKU_IS contains 4447 images selected by the G Li. Each image contains at least one salient objects, and the color information of the objects and the background is more complex. At the same time, it provides a real image manually labeled.

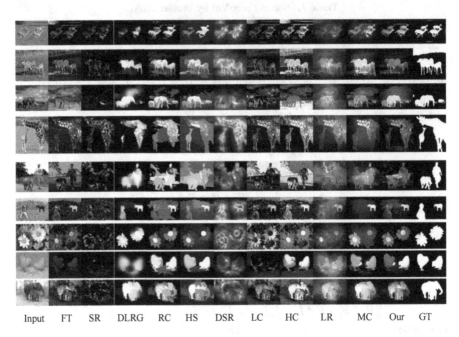

Input   FT   SR   DLRG   RC   HS   DSR   LC   HC   LR   MC   Our   GT

Fig. 2. Visual comparison of saliency maps of some state-of-the-art methods on HKU_IS

Visual Comparison. Figure 2 shows that our model generates more accurate saliency maps with uniformly highlighted foreground and well suppressed background in a variety of challenging cases of HKU-IS. For example, as shown on the first and second lines, there are two salient targets in the image, messy background in the third row, four to six has salient object locates in the image border, saliency objects with complex structure with the seventh and ninth rows and low contrast between object and background with the seventh and eighth rows. Experimental results show that the proposed SLRR method not only effectively suppresses the background region but also can completely detect saliency objects.

For precision-recall curves, saliency maps are binarized at each possible threshold within range [0, v255]. F-Measure is defined as $F = \frac{(\beta^2 + 1)PR}{\beta^2 P + R}$ where we set $\beta^2 = 0.3$ to emphasize precision [14].

PR curve and F-measure Metrics. The quantitative comparison results are given in Fig. 3. For the HKU-IS datasets, our proposed method SLRR consistently outperforms the compared methods such as CA. It can be seen from the comparative analysis in Fig. 6 that the proposed method is equivalent to the MC and RC algorithms on the P-R curve and the F-measure histogram, and is superior to the SR, CA and LC algorithms. The algorithm proposed in this paper is more suitable for complex multi-target situations.

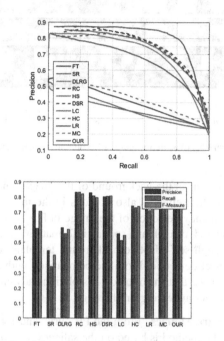

Fig. 3. Quantitative results of the precision/recall curves and F-measure metrics on HKU_IS

3.2 Results on SOE

Saliency Object on the Edge (SOE) contains 300 images collected from four datasets SED2, PASCAL-1500, MSRA1000, THUR15000. The datasets are more challenging with complex objects and include multiple salient objects in each image. The new datasets is constructed by choosing images from the above four datasets accord with at least one of the criteria: (i) the saliency object on the edge. (ii) Saliency objects with rich structure. (iii) The images have a complex background.

Visual Comparison. Figure 4 shows the visual comparison of saliency maps of some state-of-the-art methods on SOE. Compared with Fig. 2, the salient object of the images in Fig. 4 appear at the edges of the image. The contrast effect of the salient map

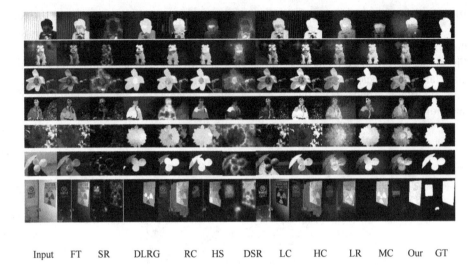

Input   FT   SR   DLRG   RC   HS   DSR   LC   HC   LR   MC   Our   GT

Fig. 4. Visual comparison of saliency maps of some state-of-the-art methods on SOE

shows that the method of our in the background of the image (lines 2 and 5 in Fig. 4). The SLRR saliency detection effect is also obvious in the more complex case. Compared with Fig. 2, the saliency detection of the image in Fig. 4 are located on the edge of the image. This method can obtain spatial information of adaptive background template effectively through self-adaptive background template, which guarantees the integrity of saliency object detection. From the comparison with the saliency maps of other algorithms, we can see that this method has obvious advantages in the detection effect of the saliency object appearing on the edge.

PR Curve and F-Measure Metrics. From the PR curve and F-measure histogram in Fig. 5, we can see that the method is based on the saliency map of the algorithm when the saliency object on the edge (SOE) PR curve and F-measure value histogram Other self-algorithms have relatively obvious advantages. This method has great P-R curves with both RC and MC. On the F-measure value, the proposed method is 7.7% higher than MC. The saliency object of the image or background information in the HKU_IS dataset is relatively simple. On the contrary, for the dataset SOE whose image saliency object located at the edge of the image, general algorithms are spatially weighted based on the image center, and this method can obtain spatial distribution of salient targets through adaptive background templates, and sparse and low rank matrix recovery, so that the detection effect is more accurate. For example, the saliency object is located at the edge of the image, the salience object contains multiple regions with obvious contrast, etc.

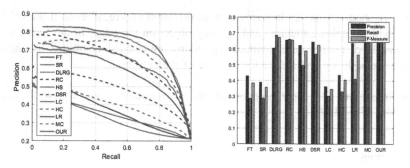

Fig. 5. Quantitative results of the precision/recall curves and F-measure metrics on SOE

3.3 Mean Absolute Error

It is defined as $MAE = \frac{1}{W \times H} \sum\limits_{x=1}^{W} \sum\limits_{y=1}^{H} |S(x,y) - G(x,y)|$, which $S(x,y)$ represent saliency map and $G(x,y)$ represent ground truth. Figure 6 shows the MAE metric of SLRR and other methods on four datasets. For the HKU_IS data set, the MAE value of our method is 0.1373, which is equivalent to the algorithm MC algorithm whose MAE is 0.1427. On the data set SED, the average absolute error (MAE) of all algorithms has been increased. The MAE of the OUR algorithm is 0.1453, and the MAE of the RC and MC are 0.1636 and 0.1989, respectively. Compared with other comparison algorithms, the MAE has more obvious advantages. This is consistent with the above analysis of the P-R curve and the F-measure histogram.

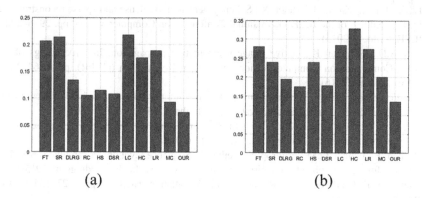

(a)   (b)

Fig. 6. (a) HKU_IS, (b) SOE dataset, MAE histogram of different saliency detection methods.

4 Conclusions

In this paper, we have proposed a sparse and low-rank matrix recovery model for saliency detection based on adaptive background template. By evaluating our method on two datasets and comparing with previous works, the experimental results indicate that our method is superior in terms of both precision and recall even if the original images with complex structures and saliency object on the edge. In the future work, more low-level features and optimal algorithms are needed combined to improve accuracy consistency and reduce the runtime of the detection.

Acknowledgement. This paper is supported by the Fundamental Research Funds for the Central Universities (NS 2015092).

References

1. Borji, A., Cheng, M.M., Jiang, H., et al.: Salient object detection: a benchmark. IEEE Trans. Image Process. **24**(12), 5706–5722 (2015)
2. Yuan, Y., Li, C., Kim, J., et al.: Reversion correction and regularized random walk ranking for saliency detection. IEEE Trans. Image Process. **27**(3), 1311–1322 (2018)
3. Li, X., Lu, H., Zhang, L.: Saliency detection via dense and sparse reconstruction. In: Proceedings of the 2013 IEEE International Conference on Computer Vision (ICCV), pp. 2976–2983. IEEE, Sydney (2013)
4. Wang, K., Lin, L., Lu, J., Li, C., Shi, K.: PISA: pixelwise image saliency by aggregating complementary appearance contrast measures with edge-preserving coherence. IEEE Trans. Image Process. **24**(10), 3019–3033 (2015)
5. Wang, L., Lu, H., Ruan, X., Yang, M.H.:Deep networks for saliency detection via local estimation and global search. In: Proceedings of the 2015 IEEE Conference on Computer Vision and Pattern Recognition (CVPR). pp. 3183–3192. IEEE, Boston (2015)
6. Chen, W., et al.: EEG-based motion intention recognition via multi-task RNNs. In: Proceedings of the 2018 SIAM International Conference on Data Mining, pp. 279–287. Society for Industrial and Applied Mathematics (2018)
7. Li, X., Lu, H., Zhang, L., Ruan, X.: Saliency detection via dense and sparse reconstruction. In: Proceedings of the IEEE International Conference on Computer Vision, pp. 2976–2983. IEEE, Sydney (2013)
8. Jain, S.D., Grauman K.: Active image segmentation propagation. In: Proceedings of the IEEE Conference on Computer Vision and Pattern Recognition (CVPR), pp. 2864–2873. IEEE, Nevada (2016)
9. Yu, G., Yuan, J., Liu, Z.: Propagative Hough voting for human activity detection and recognition. IEEE Trans. Circuits Syst. Video Technol. **25**(1), 87–98 (2015)
10. Yue Lin, Weitong Chen, Xue Li, Wanli Zuo, and Minghao Yin.: A survey of sentiment analysis in social media. Knowl. Inf. Syst. 1–47 (2018)
11. Horbert, E., García, G.M., Frintrop, S., Leibe, B.: Sequence level object candidates based on saliency for generic object recognition on mobile systems. In: Proceedings of 2015 IEEE International Conference on Robotics and Automation (ICRA), pp. 127–134. IEEE, Washington (2015)
12. Yang, X., Qian, X., Xue, Y.: Scalable mobile image retrieval by exploring contextual saliency. IEEE Trans. Image Process. **24**(6), 1709–1721 (2015)
13. Lei, B., Tan, E.L., Chen, S., Ni, D., Wang, T.: Saliency-driven image classification method based on histogram mining and image score. Pattern Recognit. **48**(8), 2567–2580 (2015)
14. Chandra S, Usunier N, Kokkinos I.:Dense and low-rank Gaussian CRFs using deep embeddings[C]//International Conference on Computer Vision (ICCV), pp. 5113–5122. IEEE, Venice (2017)

15. Yao, X., Han, J., Zhang, D., et al.: Revisiting co-saliency detection: a novel approach based on two-stage multi-view spectral rotation co-clustering. IEEE Trans. Image Process. **26**(7), 3196–3209 (2017)
16. Jiang, B., Zhang, L., Lu, H., Yang, C., Yang, M.H.: Saliency detection via absorbing Markov chain. In: Proceedings of the 2013 IEEE International Conference on Computer Vision (ICCV), pp. 1665–1672. IEEE, Sydney (2013)
17. Lu, H., Li, X., Zhang, L., Ruan, X., Yang, M.H.: Dense and sparse reconstruction error based saliency descriptor. IEEE Trans. Image Process. **25**(4), 1592–1603 (2016)
18. Ishikura, K., Kurita, N., Chandler, D.M., et al.: Saliency detection based on multiscale extrema of local perceptual color differences. IEEE Trans. Image Process. **27**(2), 703–717 (2018)

Instruction SDC Vulnerability Prediction Using Long Short-Term Memory Neural Network

Yunfei Liu, Jing Li$^{(\boxtimes)}$, and Yi Zhuang

College of Computer Science and Technology,
Nanjing University of Aeronautics and Astronautics,
Nanjing 210016, Jiangsu, China
18351936098@163.com, {lijing,zhuangyi}@nuaa.edu.cn

Abstract. Silent Data Corruption (SDC) is one of the serious issues in soft errors and it is difficult to detect because it can cause erroneous results without any indication. In order to solve this problem, a new SDC vulnerability prediction method based on deep learning model is proposed. Our method predicts the SDC vulnerability of each instruction in the program based on the inherent and dependent features of each instruction in the Lower Level Virtual Machine (LLVM) intermediate. Firstly, the features are extracted from benchmarks by LLVM passes and feature selection is performed. Then, LLVM Based Fault Injection Tool (LLFI) is used to get SDC vulnerability labels to obtain the SDC prediction data set. Long Short-Term Memory (LSTM) neural network is applied to classification of SDC vulnerability. Finally, compared with the model based on SVM and Decision Tree, the experiment results show that the average accuracy of LSTM in classification of SDC vulnerability is 11.73% higher than SVM, and 10.74% higher than Decision Tree.

Keywords: LSTM · Silent data corruption · Fault injection · Prediction

1 Introduction

The transient failure of a semiconductor circuit caused by Single event upset (SEU) is called soft error [1]. Soft errors occur intermittently, and they may cause the running program to fail by affecting the signal transmission and changing the stored values.

In recent decades, processor designers have been improving the performance of processors by shrinking feature sizes, lowering voltage levels, and reducing noise margins [2]. However, these advances make processors increasingly vulnerable to soft errors [2]. Soft errors caused by SEU affect running programs in three ways [3, 4]: (1) they may not have any effect on the running program (benign/masked), and (2) they may cause the program to crash or hang, or (3) they may lead to erroneous output results, called Silent Data Corruption (SDCs). SDCs are more hidden because they never give any indication of failure. We have found that SDCs have a high probability of occurring in many programs, especially those that focus on mathematical operations. Many prior researches have broadly focused on crash and hang [4], so we want to focus on the software-level detection techniques of SDCs in applications by machine learning.

© Springer Nature Switzerland AG 2018
G. Gan et al. (Eds.): ADMA 2018, LNAI 11323, pp. 140–149, 2018.
https://doi.org/10.1007/978-3-030-05090-0_12

To address the SDCs caused by SEU, designers introduce hardware-level redundancy [5–8], but the hardware-level redundancy is too expensive. Software-level redundancy provides a lower-cost and more flexible alternative [2]. Selective software-level redundancy is clearly better because it can reduce time and space cost while reducing the SDC vulnerability of the program. The prediction of SDC vulnerability of each instruction is the most critical step before redundancy. Most of the previous works try to find out the SDC vulnerability of instructions by a large number of time-consuming fault injections, especially for large programs. [10] have used SVM to learn code instructions in scientific codes that must be protected to avoid Silent Output Corruption (SOC), but it doesn't consider the mask instructions in propagation path. [4] also have used Classification and Regression Tree to forecasting the SDC vulnerability of each instruction, but it mainly focuses on *store* and *cmp* instructions without paying more attention to other instructions. Their results indicate that the inherent information and propagation path also have an effect on SDC vulnerability of each instruction. Long Short-Term Memory (LSTM) neural network outperforms traditional machine learning in many ways, such as speech and text recognition and language modeling [11–14]. Due to the sequential nature of instruction execution, the ability to predict SDC vulnerability of LSTM is worth exploring.

With the above challenges, this paper proposes a LSTM-based instruction SDC vulnerability prediction model without a large number of fault injections. Firstly, we use the LLVM framework to extract the features related with SDC vulnerability of instruction. Then, we get the SDC vulnerability of an instruction by fault injection. After that, we perform feature selection on the data set and introduce our LSTM model for classification of SDC vulnerability. In the subsequent prediction of SDC vulnerability, we no longer need to perform a large number of time-consuming fault injections. Finally, we compare the performance of LSTM on predication of SDC vulnerability with Support Vector Machine (SVM) and Decision Tree.

In summary, the contribution of this paper is as follows:

We use the LLVM compilation framework to extract the inherent and dependent features of intermediate code instructions through LLVM pass. LLFI injection tools is used to get SDC vulnerability of each instruction. Then, we propose an LSTM-based instruction SDC vulnerability classification model, which takes advantage of the memory function of LSTM. This model combines the inherent features of the LLVM instructions with the features of the instruction propagation path to forecast SDC vulnerability of each instruction.

2   Motivation

2.1   Definition of SDC Vulnerability

When a program fails, the fault may occur in the source operands or destination operands of the instruction. Then it propagates to the basic block and function. Finally, it propagates to propagation path and reaches the program's output. Based on this basic principle, we extract the features related to instruction, basic block, function and propagation of instruction to predict the SDC vulnerability.

SDC vulnerability of instruction is defined as Eq. (1).

$$P_{SDC}(I_i) = \frac{1}{W_i} \times \sum_{t=1}^{W_i} \frac{S_i}{T_i} \qquad (1)$$

I_i represents the ith instruction in the program, W_i represents the width of the destination register of the ith instruction, S_i represents the number of SDC errors of the program after the fault injections on one bit, T_i represents the number of fault injections on one bit. In our experiment, $T_i = 100$.

2.2   Overall Framework

Figure 1 shows the overall framework of our method. It mainly includes three parts: (1) LLVM IR level feature extraction using LLVM passes and fault injections using LLFI. (2) Selecting the most effective features by feature selection. (3) Training, testing and evaluation of instruction SDC vulnerability classification model using LSTM, SVM, Decision Tree respectively.

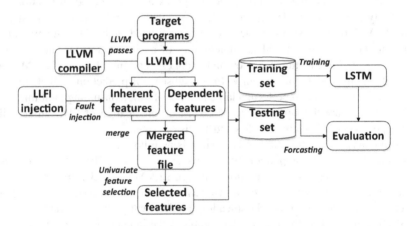

Fig. 1.   Overall framework of our method

3   Feature Extraction and Selection

Previous studies have shown that the SDC problem is caused by a relatively small part of the instructions in programs [9], and the SDCs are caused by the errors propagated to the program's output, so the propagation path (dependent features) of an instruction is critical to the SDC vulnerability. The SDC vulnerability of an instruction is not only related to the propagation path, but also related to some features of the instruction itself, we call them inherent features, such as the instruction type. Therefore, we extract features related to the SDC vulnerability of an instruction from (1) the inherent features and (2) dependent features.

3.1   Inherent Features

The inherent features of the instruction can indicate the characteristics of the instruction itself. In the execution of the instructions, the instruction type, the basic block and function where it is located can reflect the vulnerability of the instruction to some extent [4].

When the instruction belongs to the instructions which involve address calculation, the SDC vulnerability of the instruction will be greatly reduced because they are more inclined to crash or hang.

The tendency of arithmetic calculation instructions to generate SDC errors is significantly higher than that of other instructions. The vulnerability of instructions in *basicmath* (a testing program with a large number of arithmetic operations) is significantly higher than that of other programs.

The loop depth of the instruction also has a certain influence on the SDC vulnerability. The deeper the loop depth is, the more critical the instruction is to the program, so the SDC vulnerability of the instruction may also increase. In addition to these, information such as the size of the basic block and the call number of function are all related to the SDC vulnerability of the instruction. These Inherent features are shown in the Table 1.

Table 1. Inherent features for model building.

Inherent features	Description
type of inst	Type of instruction such as arithmetic instruction, branch instructions, etc.
bb length	The number of static instructions in the basic block that contains the instruction
bb remaining length	The number of remaining static instructions in the basic block starting from the instruction
function length	The number of static instructions in the function that contains the instruction
function remaining length	The number of remaining static instructions in the function starting from the instruction
in loop	Whether the instruction is in loop
loop depth	The loop depth of the structure where the instruction is located
dynamic count of execution	The counts of dynamic execution of the instruction
dynamic count of function calls	The counts of dynamic calling of the function where the instruction is located
in global	Whether the instruction changes a globally defined variable
inst func execution ratio	Dynamic counts of the instruction divided by the dynamic counts of the function
data width	The number of bits of the result of the instruction

3.2 Dependent Features

We need to find relevant features related to SDC from the instruction's propagation path. The propagation path of the instruction is defined as Eq. (2).

$$Path(I_i) = \{I_{i+1}, \ldots, I_k, \ldots, I_K | \forall (i + 1 \leq k \leq K), I_k \delta I_{k+1}, I_K = I_{end}(I_i)\} \qquad (2)$$

$Path(I_i)$ indicates the propagation path of the ith instruction, $I_k \delta I_{k+1}$ indicates that there is a data dependency relationship between the instruction I_k and the instruction I_{k+1}. I_K is the end instruction of the propagation path, so I_k to I_{k+1} constitutes the propagation path of instruction I_i. The end instructions in the propagation path are the *store*, *br*, and *call* instructions. Because the *store* and *br* have no destination registers, and the *call* instructions will generate a new stack frame, all these instructions will terminate the data propagation in the program. The propagation related features: the mask instructions and the address calculation instructions.

Mask Instructions. The mask instructions include logic operation instructions (such as *and*, *or*) and displacement operation instructions (such as *shl*, *lshr*). These instructions may mask errors in the instruction operands and reduce the SDC vulnerability of the instructions. For example, performing an AND operation with 0 or performing an OR operation with 1 may not affect the result, although an error has occurred before. Table 2 shows the mask instructions.

Table 2. Mask instructions.

Types	Mask instructions
Logic operation instructions	*and, or, xor*
Displacement operation instructions	*shl, lshr, ashr*

Address Calculation Instructions. If there is an error in the address calculation instruction (such as *load*, *getelementptr*) in the propagation path, segment errors are often more likely to occur, causing the program to directly crash, and therefore the SDC vulnerability of the instruction is reduced.

The features of the propagation path we extracted are shown in the Table 3.

Table 3. Dependent features for model building.

Dependent features	Description
is mask	Whether there is a mask instruction on the propagation path
number of mask inst	Number of mask instructions on the propagation path
type of operand of mask inst	Type of the operands of the mask instructions on the propagation path
is addr calc	Whether there is an address calculation instruction on the propagation path
number of addr calc inst	Number of address calculation instructions on the propagation path
type of operand of addr calc inst	Type of operands of the address calculation instructions on the propagation path

3.3   Feature Selection

We select the univariate feature selection method to select features. The univariate feature selection method, which is independent of the classification algorithm, tests each feature and calculates a statistical indicator for each feature to measure the relationship between the features and the prediction category. According to our SDC classification task, we perform ANOVA on the data set, and obtain the P-value of each feature based on the F-value calculated before and calculate the importance score of each feature according to the formula as shown in Eqs. (3) and (4).

$$score_i' = -\log_{10}(P - value_i) \tag{3}$$

$$score_i = \frac{score_i'}{\max(\{score_i'|j = 1, 2, \ldots, n\})} \tag{4}$$

P-value is the P-value of the ith feature, n is the number of features, $score_i$ represents the importance score of the ith feature.

3.4   SDC Vulnerability Classification Model Based on LSTM

The LSTM is an improved version of the RNN and is also developed to solve problem of gradient explosion [11, 12]. LSTM solves the problem of long-term dependence by adding a unit state C that allows it to store long-term states, known as cell states. Its capability in SDC vulnerability forecasting is worth exploring due to the sequential nature of instructions execution. Therefore, we choose the LSTM model as our classification model.

In order to complete the 4-classification of SDC vulnerability, we designed a LSTM-based classifier with the following details.

Input Features. The input is the feature values and the SDC vulnerability categories. The data set can be represented as D = {X1, X2, ...Xi..., Xd}, d means the size of data set, Xi = {x1, x2, ..., xi, ..., xn, y} means each piece of data, xi means each feature from feature extraction and selection, y∈ {0,1,2,3} corresponds to 4 categories of SDC vulnerability. Before training, we divide the entire dataset into k equal parts based on timestep, where timestep * k = d. Every time we train, we send timestep data to LSTM.

Output. We added a fully connected layer at the end of the LSTM model to complete the classification. Each time we send the timestep data to the LSTM model for training, we take the output of the last time step as the input of the full connection layer. We get the normalized probability value of each label $\tilde{y} \in [0, 1]$ by softmax.

Loss Function. When training the model parameters, we use the cross-entropy loss function as our loss function. The cross-entropy loss function is shown in Eq. (5). yi represents the true value, and $\tilde{y}_i$ is the predicted value by LSTM.

$$J = -\sum_{j \in D} y_i \log(\widetilde{y}_i) \qquad (5)$$

4 Experiments and Results

4.1 Benchmarks

The benchmarks used in this paper are sourced from Mi-Bench, the open and free programs set. The selected programs for training and testing set in this paper: qsort (quick sort), FFT (fast Fourier Transform), jpeg (JPEG encoding decoding program), bitcount (count number of bit which is 1), lame (MP3 conversion), CRC32 (CRC32 calculation tool), basicmath (mathematical calculation test), patricia (leaf sparse tree structure), stringsearch, susan (image recognition tools), dijkstra (shortest path algorithm), sha (SHA hash algorithm). The first six are selected as training sets, and the last six are selected as test sets. The training and testing set are different because we want to maintain the integrity of the training and testing programs and observe performance of the model on new programs.

4.2 Fault Injection Experiment

LLFI is an LLVM based fault injection tool that injects faults into the LLVM intermediate representation (IR) of the application source code. The faults can be injected into specific program points, and the effect can be easily tracked back to the source code.

Fault Model. Using LLFI to simulate the effect of single event upset on the program, we only consider the effect of single-bit upset by inverting one bit in the destination register of the LLVM IR instruction. During the injection process, we perform 100 fault injections on each bit of data in the destination register, so the SDC vulnerability of an instruction with 32-bit operand requires 3,200 injections. We average the SDC vulnerability value of each injected bit as the final SDC vulnerability value of the instruction. Finally, we classified the SDC vulnerability into four categories based on the actual vulnerability of the injection results: highly vulnerable (80%–100%), vulnerable (50%–80%), relatively vulnerable (20%–50%), not vulnerable (0–20%).

4.3 Feature Extraction and Selection Experiment

The LLVM Project is a collection of modular and reusable compiler and toolchain technologies. Despite its name, LLVM has little to do with traditional virtual machines. LLVM uses static single assignment (SSA) in LLVM IR, so each instruction is equal to its result. We use LLVM to implement feature extraction.

We extracted a total of 18 features in the feature extraction process, and then performed univariate feature selection on these 18 features. The score for each feature was obtained, as shown in the Fig. 2. We found that some features have very low scores, so we removed less important features whose score is less than 0.1 from the feature set. The features we removed include the number of mask inst, loop depth, type of operand of mask inst, and dynamic count of execution.

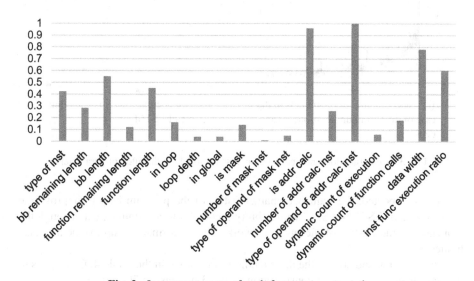

Fig. 2. Importance score of each feature we extracted

4.4 LSTM-Based Classification Model Experiment

In the experiment, we compared our LSTM model with traditional machine learning model, using the same data set.

In the SVM experiment, we find the best SVM model where C = 10, kernel = rbf, gamma = 0.001. The highest accuracy of SVM is 0.77, and the average accuracy is 0.7285. For Decision Tree model under the same data, the highest accuracy rate is 0.748, but the average accuracy can reach 0.735 and is higher than the SVM. Therefore, the classification performance of SVM and Decision Tree is quite close.

Taking into account the influence of the number of network layers on the learning ability of network, we set up different number of network layers and cells for comparison. Figure 3a shows the convergence process of loss function of the models in the case of different number of layers and cells (only part of the experimental results is shown in the Fig. 3a). When the number of layers is 2 and the number of cells is 10, the model loss function decreases fastest and LSTM network has the best learning capabilities.

The best accuracy of LSTM on the same data set is shown in the Fig. 3b. With the increasing number of iterations, the final accuracy can reach 0.8253, and the average accuracy can reach 0.814, which is 11.73% higher than the average accuracy of SVM and 10.74% higher than Decision Tree. It shows that the classification accuracy of LSTM is obviously better than SVM and Decision Tree. The parameters of the optimal LSTM model obtained by training: 2 layers, 10 cells, learning rate 0.001, droupout rate 0.8, sequence 5, top accuracy 0.8253.

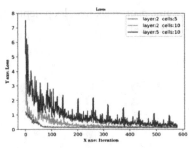
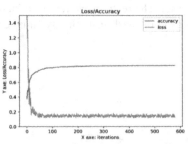

(a)   loss function of the models in the case of different number of layers and cells

(b) best accuracy and loss function curve of LSTM

Fig. 3. Loss function and accuracy curves of models with different parameters

As we expected, the sequence characteristics of the program help to improve the performance of LSTM on SDC vulnerability prediction to a certain extent. Though the sequence characteristics of the program works, the performance improvement is also limited.

The accuracy and cost of the three methods are shown in the Table 4. Cost refers to the forecasting average time consumption of the 3 models on 2000 data.

Table 4. Accuracy and time consumption of 3 methods

Method	Top accuracy	Average accuracy	Cost
SVM	0.77	0.7285	2.65 s
Decision tree	0.748	0.735	2.89 s
LSTM	0.8253	0.814	4.93 s

The accuracy of 3 methods in different testing programs is shown in the Fig. 4. LSTM surpasses SVM and Decision Tree in the accuracy of all testing programs, and the performance is relatively stable.

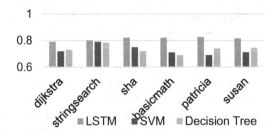

Fig. 4. The accuracy of 3 methods in different testing programs

5 Conclusion

This paper proposes an LSTM-based instruction SDC vulnerability prediction classification model based on the information of instruction itself and its propagation. Inherent features and dependent features are used in our model and help LSTM achieve better performance.

Acknowledgement. This paper is supported by the Fundamental Research Funds for the Central Universities (NS 2015092).

References

1. Walter, J.P., Zick, K.M., French, M.: A practical characterization of a NASA spacecube application through fault emulation and laser testing. In: IEEE International Conference on Dependable Systems and Networks, pp. 1–8 (2013)
2. Reis, G.A., Chang, J., Vachharajani, N., Rangan, R.: SWIFT: software implemented fault tolerance. In: International Symposium on Code Generation and Optimization, pp. 243–254 (2005)
3. Hari, S.K.S., Adve, S.V., Naeimi, H.: Low-cost program-level detectors for reducing silent data corruptions. In: IEEE/IFIP International Conference on Dependable Systems and Networks, pp. 1–12 (2012)
4. Lu, Q., Pattabiraman, K., Gupta, M.S.: SDCTune: a model for predicting the SDC proneness of an application for configurable protection. In: International Conference on Compilers, Architecture and Synthesis for Embedded Systems, pp. 1–10 (2015)
5. Farnsworth, C., Clark, L.T., Gogulamudi, A.R.: A soft-error mitigated microprocessor with software controlled error reporting and recovery. IEEE Trans. Nuclear Sci. **63**(4), 2241–2249 (2016)
6. Elnozahy, E.N., Plank, J.S.: Checkpointing for peta-scale systems: a look into the future of practical rollback-recovery. IEEE Trans. Dependable Secur. Comput. **1**(2), 97–108 (2004)
7. Bernick, D., Bruckert, B., Vigna, P.D.: NonStop® advanced architecture. In: International Conference on Dependable Systems and Networks, pp. 12–21 (2005)
8. Chang, J., Reis, G.A., August, D.I.: Automatic instruction-level software-only recovery. IEEE Micro **27**(1), 36–47 (2007)
9. Thomas, A., Pattabiraman, K.: Error detector placement for soft computation. In: 43rd Annual IEEE/IFIP International Conference on Dependable Systems and Networks, pp. 1–12 (2013)
10. Laguna, I., Schulz, M., Richards, D.F.: IPAS: intelligent protection against silent output corruption in scientific applications. In: IEEE/ACM International Symposium on Code Generation and Optimization (CGO), pp. 227–238 (2016)
11. Zhao, J., Qu, H., Zhao, J.: Towards traffic matrix prediction with LSTM recurrent neural networks. Electron. Lett. **54**(9), 566–568 (2018)
12. Chen, W.T., et al.: EEG-based motion intention recognition via multi-task RNNs. In: Proceedings of the 2018 SIAM International Conference on Data Mining, Society for Industrial and Applied Mathematics, pp. 279–287 (2018)
13. Yue, L., Chen, W.T., Li, X., Zuo, W.L., Yin, M.H.: A survey of sentiment analysis in social media. Knowl. Inf. Syst. 1–47 (2018)
14. Greff, K., Srivastava, R.K., Koutník, J.: LSTM: a search space odyssey. IEEE Trans. Neural Netw. Learn. Syst. **28**(10), 2222–2232 (2015)

Forecasting Hospital Daily Occupancy Using Patient Journey Data - A Heuristic Approach

Shaowen Qin[✉] and Dale Ward

College of Science and Engineering, Flinders University, Tonsley, SA 5042, Australia
{shaowen.qin,dale.ward}@flinders.edu.au

Abstract. Hospitals are dynamic environments that involve many stochastic processes. Each day, some patients are discharged from hospital, emergency patients arrive and require admission, and a variable number of elective admissions are planned for the day. The ability to forecast hospital occupancy will assist hospital managers to balance the supply and demand on inpatient beds on a daily basis, which in turn will reduce the risk of hospital congestion. This study employed a heuristic approach to derive a forecasting model based on hospital patient journey data. Instead of using estimated overall length of stay (LOS) for each patient, the forecasting model relies on daily evaluation of the probabilities of staying or being discharged based on a patient's current LOS. Patients' characteristics are introduced as additional model parameters in an incremental manner to balance model complexity and prediction accuracy. It was found that a model without enough details can provide indications of overall trends in terms of the mean occupancy. However, more parameters, such as day of the week, must be considered in order to capture the extremes present in the data. Of course, as more parameters are introduced, less data become available for meaningful analysis. This proof-of-concept study provides a demonstration of a heuristic approach to determine how complex a model needs to be and what factors are important when forecasting hospital occupancy.

Keywords: Forecast model · Hospital occupancy
Probability of discharge · Patient journey

1   Introduction

There has been a growing demand for hospital services in Australia. Most public hospitals often operate at or close to 100% of their capacity. The ability to forecast hospital occupancy on a daily basis will assist hospital managers to balance

This work was supported by the ARC linkage grant LP130100323, jointly awarded to Flinders University, the Southern Adelaide Health Service (Flinders Medical Centre) and the Central Adelaide Local Health Network (Royal Adelaide Hospital).

© Springer Nature Switzerland AG 2018
G. Gan et al. (Eds.): ADMA 2018, LNAI 11323, pp. 150–159, 2018.
https://doi.org/10.1007/978-3-030-05090-0_13

the supply and demand on their servicing resources, which in turn will reduce the risk of hospital congestion. However, given the uncertainties involved in demand for hospital services, such as the number of emergency patient arrivals, planned admissions, and diverse range of medical conditions that require treatment processes of different durations, forecasting future hospital occupancy with desirable accuracy is a challenging task. In particular the patient turnover (number of discharges per day vs arrivals) can result in significant shifts in occupancy from day to day.

There have been many studies addressing the need to forecast hospital occupancy, including a selected more relevant few referenced in this study [1–4]. Health care data analysis is traditionally done using statistical techniques in order to report and hopefully forecast health care service performance and demand [1,2]. With the advancement of computing technology, new data analysis and modelling approaches, such as data-mining [3], machine learning [4], and combinations of these techniques, have been applied to health care service management related studies. Computer simulation of hospital operations has also proven to be an effective approach to facilitate hospital managers identifying process improvement opportunities and testing the impact of their decisions both short and long term [5,6]. However, due to the nature of hospital services, e.g., the maximum number of patients a hospital can serve each year, the amount of data available for predictive analysis of hospital occupancy has remained more or less the same, which is what we have to rely on for building forecasting models that can help hospital managers to make resource planing decisions on a day-to-day basis.

The art, as well as the science, of mathematical modelling is to construct the simplest model that captures the salient features of the system. In the words of Einstein, "a scientific theory should be as simple as possible, but no simpler". For this to occur, there is a requirement that the choice of the model being used is appropriate. In particular, modelling must strike the right balance between fit-for-purpose and complexity [7,8]. In this study we employed a heuristic approach to consider whether a simple forecast model might be developed to approximate the occupancy of the hospital at some future time step based on the occupancy and its composition of patient types at the present time. Furthermore, we aim to limit the amount of necessary case-specific information required for forecasting and try to use simple measures freely available to the hospital.

2   Modelling Occupancy Using Patient Journey Data

Let us define occupancy as the number of patients present within the hospital at some time t. Note this is different from admitted occupancy or inpatients assigned to a hospital bed. The occupancy is composed of the number of both emergency and elective patients present within the hospital at time t. For proof-of-concept, however, the study will focus on the emergency patients $X(t)$ only.

Given the occupancy $\mathbf{X}(t)$ at time t we wish to estimate the occupancy at time $t + 1$. That is we want a function f s.t.

$$\mathbf{X}(t + 1) = f(\mathbf{X}(t), \mathbf{A}(t + 1)) + \epsilon(t + 1), \qquad (1)$$

where $\mathbf{A}$ is the number of patients that arrive at the emergency department (ED) over time t to $t + 1$ and ϵ is an approximation error. Note that the number of patients discharged over the same period is a function of $X(t)$.

With access to a well-maintained Patient Journey Database (PJD) containing historical records of the various movements of every patient from arrival at ED to discharge, a data-driven approach can be applied to Eq. (1). The data set used in this study was from a large Australian tertiary hospital that includes information about every patient that arrived from January 1, 2012, to April 1, 2015. The details in the data cover: type and time of arrival, triage score, admission decision, admission and discharge wards/units (if the patient was admitted), timestamps for major events (being seen by a doctor/nurse in ED, discharged from ED, admitted to hospital, discharged). As part of the ethics approval process, information related to patient's identity was removed before the data was provided to the researchers.

Most of the existing work starts with estimating total LOS for patient groups with similar characteristics, and work out daily occupancy by checking whether a patient's LOS is expired on that day [1–3]. Instead of using this approach, this study attempts to conduct daily evaluation of the probabilities of staying or being discharged based on a patient's current LOS. This approach at least appears to better resemble the actual decision making process where the knowledge of total LOS of a patient is usually not available. An incremental model development process is also applied, which offers the advantage of growing the model complexity only when additional parameters deemed important and necessary are identified. Sections 3 and 4 present the implementation of the described heuristic approach.

3   A Basic Model: LOS and Triage

Let us introduce factors τ and k to subdivide emergency patients based on their conditions and treatment progress, so that $X_{\tau,k}(t)$ is the number of emergency patients with triage τ and length of stay $LOS \geq k$ (has spent k days in hospital) at time t:

$$\mathbf{X}(t) = \begin{pmatrix} \cdots \\ X_{\tau,k}(t) \\ \cdots \end{pmatrix}. \qquad (2)$$

Note that LOS is the number of nights since arrival to time t, i.e., $LOS = 0$ means a patient has spent 0 nights in hospital, and total LOS is the number of nights since arrival to discharge.

From time t to $t + 1$ we consider two actions, that is a patient is discharged (leaves hospital) or stays. For the purposes of this analysis we do not distinguish discharge types (i.e., leaves hospital for home, transfer, etc) and whether staying changes the status or treatment of the patient. Based on this simple two case model, we denote the probability that a patient with triage τ and current LOS of k days $STAYING$ as $\alpha(\tau, k)$ and the probability they are $DISCHARGED$ as $1 - \alpha(\tau, k)$ (as probabilities must sum to one), as indicated below.

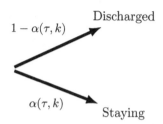

Given $X_{\tau,k}(t)$, if we know $\alpha(\tau, k)$ $\forall \tau = \{1, 2, \cdots, 6\}$ and $k \geq 0$ then

$$X_{\tau,k}(t+1) = \begin{cases} \underline{\alpha(\tau, k)} \cdot X_{\tau,k-1}(t); \ k \geq 1, \\ \underline{\alpha(\tau, 0)} \cdot \underline{A_\tau(t+1)}; \ k = 0. \end{cases} \tag{3}$$

The above relationship can be used to approximate the number of emergency patients within the hospital tomorrow based on the number of patients present today and the arrivals tomorrow. Underlined terms represent unknown factors that need to be modelled. The term $A(t+1)$ representing the number of emergency arrivals tomorrow can be modelled using a variety of different techniques including a time series, empirical or fitted distribution etc. We instead focus on calculating the α terms.

We note that

$$\alpha(\tau, k) = P_\tau(LOS \geq (k+1)|LOS \geq k), \tag{4}$$

that is the probability of staying given a current $LOS = k$ and triage score τ is equivalent to the probability of having $LOS > k$ given already spending k days in hospital and having triage score τ. Therefore,

$$\alpha(\tau, k) = \frac{\sum_{m \geq k+1} N(\tau, m)}{\sum_{m \geq k} N(\tau, m)} \in [0, 1], \tag{5}$$

where $N(\tau, m)$ is the number (count) of patients with triage τ and discharged with $LOS = m$.

Using Eq. (5) the values of $\alpha(\tau, k)$ can be calculated using data extracted from PJD database, i.e, arrival date, discharge date and triage score. Due to limitations with data prior to 2012, patient data from the 1st of January 2012 to 31st of December 2014, which includes 225,596 patient journeys in total, is used to approximate $\alpha(\tau, k)$. The results of the calculation are displayed in Fig. 1.

As we have yet to define an arrival model A_τ we focus on the first term of Eq. (3) (case $k \geq 1$), which includes estimates of the probability of patients

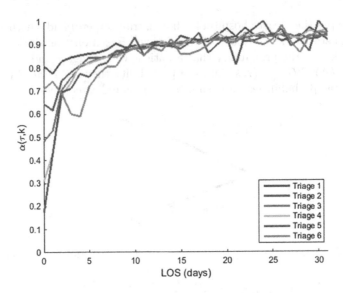

Fig. 1. Probability of staying in hospital, $\alpha(\tau, k)$. Each coloured curve represents a different triage category. LOS limited to first 30 days. (Color figure online)

staying in hospital between t to $t + 1$ and the number of patients who were already present at time t. We test our forecasting model using the probability of staying in hospital α calculated in Fig. 1 and snapshots of both the 2014 data and additional data from the 1st of January 2015 to the 1st of April 2015 that was not used in modelling.

For purposes of simplifying implementation we present the results of the number of emergency patients discharged for $k \geq 1$ between each day rather than the actual occupancy. However, no fundamental changes to the methodology presented above are required. The results are presented in Fig. 2.

It is clear that our forecast model utilising τ and k provides a poor estimate of the number of emergency discharges. Certainly a key issue is that the model is not capturing the extremes present in the data. However, it can be noted that the forecast does a reasonable job in capturing the overall trends present most of the time.

A quick analysis of the data clearly indicates there are strong weekly patterns existing in the data that the current model is unable to address. This seems to suggest that additional factors are also influencing the discharge behaviour of patients than just their condition and treatment time. Such observations appear intuitive since the hospital has structural and behavioural practices that will likely influence the timing of a patient being discharged. As such we should introduce additional parameters to refine the model.

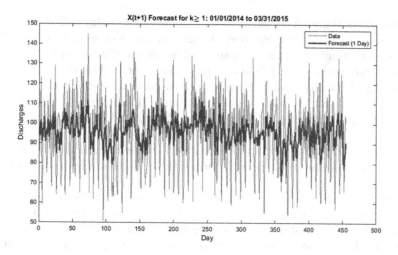

Fig. 2. Number of discharged emergency patients forecast vs actual data. Period is from 1st of Jan 2014 to 1st of April 2015.

4   A Refined Model: LOS, Triage and Day of the Week

The model is refined by introducing day of the week into our discharge probabilities. That is we define a unique $\alpha(t, k)$ for each day of the week $d \in \{1, \cdots 7\}$, where $d = 1$ is Sunday and $d = 7$ is Saturday. The factoring of weekdays into the model can be loosely interpreted as incorporating the hospital operational behaviour implicitly into the model. Intuitively one might expect the hospital to operate differently across weekdays based on ease of access to facilities and specialist equipment, different staffing levels or even different hospital policy in regards to discharge.

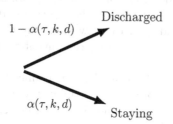

Now,

$$\alpha(\tau, k, d) = P_\tau(LOS \geq (k+1)|LOS \geq k, D = d), \tag{6}$$

$$= \alpha(\tau, k, d) = \frac{\sum_{m \geq k+1} N(\tau, m, \delta(m, k, d))}{\sum_{m \geq k} N(\tau, m, \delta(m, k, d))} \in [0, 1], \tag{7}$$

where $\delta(m, k, d)$ is the function

$$\delta(m, k, d) = (d - 1 + (m - k)) \mod 7 + 1 \in \{1, \cdots, 7\}, \tag{8}$$

that calculates the weekday $m - k$ days from current weekday d. The δ function is crucial in Eq. (7) as it cycles over the days of the week as m increases. This made sure that only relevant cases are used in our calculation, for instance, if a patient is not discharged on Wednesday the day after m days then he/she may be discharged the day after on Thursday or two days after on Friday. However, it is not possible for them to be discharged on a Monday $m + 1$ days. As such only discharges of the relevant LOS and weekday should be used in calculating $\alpha(\tau, k, d)$.

Similar to the previous τ and k case the α values are calculated, this time using the refined expression in Eq. (7) using the PJD data. The results are presented in Fig. 3.

The plots in Fig. 3 suggest that different discharge behaviours might exist across different weekdays. In particular, the weekend days (Sat and Sun) have significantly higher probability of staying in hospital than those for weekdays (Mon–Fri). This is most apparent for triage scores τ ranges from 2 to 4 where the probability of staying, for $LOS > 20$, is around 0.9 on weekdays and around 0.95 on weekends. The large volatility in α for triage 1, 5, 6 (particularly for large k) is due to the fact that there are only a small number of such patients treated in the hospital. The additional subdivision means that for higher k there are less observations to obtain robust estimates of α.

It should be noted that despite the inclusion of an additional factor the expression for $X(t+1)$ given in Eq. (3) remains essentially unchanged. In fact it is calculated almost entirely the same with $\alpha(\tau, k)$ being replaced with $\alpha(\tau, k, \hat{d})$ where $\hat{d}$ is the day of week at time $t + 1$. The new approximation therefore is given by

$$X_{\tau,k}(t+1) = \begin{cases} \alpha(\tau, k, \hat{d})X_{\tau,k-1}(t); & k \geq 1, \\ \alpha(\tau, 0, \hat{d})A_{\tau,\hat{d}}(t+1); & k = 0. \end{cases} \tag{9}$$

Note that we also adjusted A to include a subscript for $\hat{d}$ in case different arrival rates for different days of the week is to be incorporated.

Again we test the validity of the new model by forecasting the number of discharges for each day and comparing the output with the actual data. The results are presented in Fig. 4.

It can be seen that the refined model is a significant improvement upon the LOS and Triage model. The forecast now exhibits the strong weekly patterns that were noted previously in the data and the model better captures the extremes shown in the actual data. Based on appearance alone the model seems to typically under-estimate the number of discharges. This under-estimation could be an artifact of truncation when calculating the series whereby the forecast model under-estimates the number of LOS outliers, i.e., $k > 30$, being discharged. The error between data and the forecast are displayed in Figs. 5a and b. The histogram and normal fit indicate that the errors are not normal, particularly around the tails. This supports what was observed in the data where the model

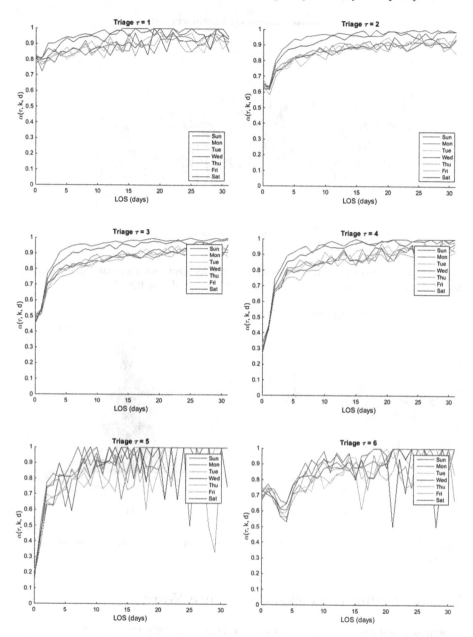

Fig. 3. Probability of staying in hospital on different day of the week, $\alpha(\tau, k, d)$

was unable to capture the full extent of the extremes in discharges. Some of these large scale errors could be reduced by separately considering periods around public holidays which typically demonstrate different behaviour.

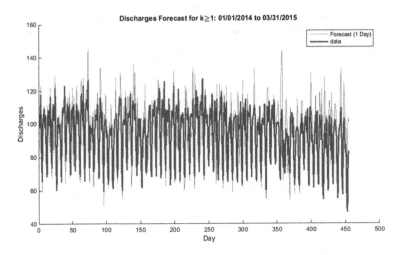

Fig. 4. Number of discharged emergency patients forecast vs actual data incorporating day of the week. Period is from 1st of Jan 2014 to 1st of April 2015.

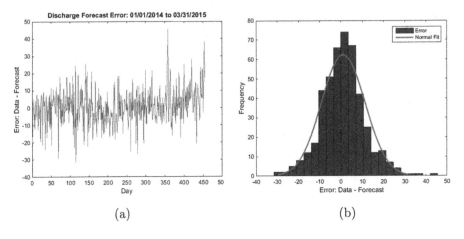

Fig. 5. Modelling error (a) and Histograms of error with normal fit (b) of data - forecast.

5   Concluding Remarks and Future Work

A heuristic approach is employed to investigate the effectiveness of forecasting hospital occupancy based on daily evaluation of the probabilities of staying or being discharged for each patient. This is different from most existing work where total LOS is estimated for each patient upfront, and daily occupancy is then worked out by checking whether a patient's LOS is expired. It appears to better resemble the decision making process which does not involve the knowledge of total LOS of a patient. An incremental model development process is applied, which offers the advantage of growing model complexity in a controlled and necessary manner that allows learning of the new parameter's importance at the same time. It was found that a model without enough details can provide indications of overall trends in terms of the mean occupancy. However, more parameters, such as day of the week, must be considered in order to capture the extremes present in the data. Of course, as more parameters are introduced, less data become available for meaningful analysis.

Future work includes investigating the impact of other parameters, such as diagnosis-related-group (DRG); integrating the model in our simulation study; as well as using machine learning approaches to conduct such analysis and compare the accuracy of predictions obtained from different approaches.

References

1. Van Walraven, C., Forster, A.: The TEND (Tomorrow's Expected Number of Discharges) model accurately predicted the number of patients who were discharged from the hospital the next day. J. Hosp. Med. **13**(3), 158–163 (2017)
2. Fuhs, P., Martin, J., Hancock, W.: The use of length of stay distributions to predict hospital discharges. Med. Care **17**(4), 355–368 (1979)
3. Azari, A., Janeja, V.P., Mohseni, A.: Healthcare data mining: predicting hospital length of stay (PHLOS). Int. J. Knowl. Discov. Bioinform. **3**(3), 44–66 (2012)
4. Xu, H., Wu, W., Nemati, S., Zha, H.: Patient flow prediction via discriminative learning of mutually-correcting processes. IEEE Trans. Knowl. Data Eng. **29**(1), 157–171 (2017)
5. Barado, J., Guergué, J., Esparza, L., Azcárate, C., Mallor, F., Ochoa, S.: A mathematical model for simulating daily bed occupancy in an intensive care unit. Crit. Care Med. **40**(4), 1098–1104 (2012)
6. Ben-Tovim, D., Filar, J., Hakendorf, P., Qin, S., Thompson, C., Ward, D.: Hospital event simulation model: arrivals to discharge - design, development and application. Simul. Model. Pract. Theory **68**, 80–94 (2016)
7. Mackay, M., Lee, M.: Choice of models for the analysis and forecasting of hospital beds. Health Care Manag. Sci. **8**(3), 221–230 (2005)
8. Hastie, T., Tibshirani, R., Friedman, J.: The Elements of Statistical Learning - Data Mining, Inference and Prediction, 2nd edn. Springer, New York (2009). https://doi.org/10.1007/978-0-387-84858-7

An Airport Scene Delay Prediction Method Based on LSTM

Zhongbin Li[✉], Haiyan Chen, Jiaming Ge, and Kunpeng Ning

College of Computer Science and Technology,
Nanjing University of Aeronautics and Astronautics,
Nanjing 210016, Jiangsu, China
lizbl994@163.com, nuaagjm@163.com,
wjh291371205@163.com, chenhaiyan@nuaa.edu.cn

Abstract. Due to the highly dynamic nature of flight operations, the prediction for flight delay has been a global problem. At the same time, existed traditional prediction models have difficulty capturing sequence information of delay, which may be caused by the subsequent transmission of delay. In this paper, a delay prediction method based on Long Short-Term Memory Model (LSTM) is proposed firstly. Furthermore, the relevant features are selected and we divide the delay levels. Then we cross-contrast performances of the model based on different hyper parameters on the actual dataset. Finally, the optimal prediction model of the scene delay is obtained. Experimental results show that compared with the traditional prediction model whose average accuracy is 70.45%, the proposed prediction model has higher prediction accuracy of 88.04%. In addition, the proposed model is verified to be robust.

Keywords: Airport scene delay · Long Short-Term Memory (LSTM)
Deep learning · Data sequence

1 Introduction

Over the past decades, flight delays led by the increasing flight volume has become more prominent. Delay may occur at every stage of the flight operation. According to the general rule of a flight, the delay is roughly divided into the scene delay and route delay. Scene delays means that the actual departure time of the flight of the airport lags the planned departure time while the latter is the delay caused by the flight on its route, that is, the actual flight time exceeds the planned flight time.

Since the cost of waiting on the ground is much lower than waiting in the air, scene delays often occur compared to less frequent air delays. Scene delays may be led by multiple causes. According to statistics, the delay of airport scene of a certain period is huge if delays in the pre-order period have not been eliminated in time, which may lead to accumulation continuously. Among the scene delays of various periods, there is a spread from upstream to downstream and accumulative relationship. Therefore, when an airport is experiencing scene delays for some reason, it is indispensable to make predictions of scene delay based on the current situation and future flight plans in time to reduce the downward spread of delays and avoid more flights from being affected.

© Springer Nature Switzerland AG 2018
G. Gan et al. (Eds.): ADMA 2018, LNAI 11323, pp. 160–169, 2018.
https://doi.org/10.1007/978-3-030-05090-0_14

The scene operation command center makes scheduling decisions based on the prediction results to control the overspread of delays.

This paper focuses on the propagation and accumulating relationship among scene delays in an operating cycle, makes full use of the characteristics of the LSTM model, and proposes an airport scene delay prediction model based on LSTM and verifies its capability.

2 Related Work

Scholars have tried different ways to predict airport delay [1]. In terms of non-machine methods, Wieland etc. [2] used a queue model to simulate and predict delays in his article. Rong [3] et al. combined Bayesian network with Gauss's expectation model to put forward the influence factor model of random flight point to delay, and finally to predict aircraft delay. Wong et al. [4] used the survival model to predict the delay propagation in the national aviation system. Mueller et al. [5] proposed the normal distribution and Poisson distribution model to simulate the arrival delay and takeoff delay of the flight.

As a hot method of machine learning in recent years, deep learning has achieved good results applicant for many fields. In voice recognition [6], speech recognition [7] and natural language processing [8], sentiment analysis [9], deep learning has achieved great development and relevant scholars have applied deep learning methods to the field of civil aviation.

Venkatesh et al. [10] proposed the model based on neural network and deep confidence network in machine learning. He took the distance and the weather conditions as features. Then, he predicted the delay of the specific flight and repeated training to achieve the best effect, and the final accuracy reflected the advantage of deep learning on the prediction of delay of flight. However, the proposed model has limited practical effect in that its predicting results is in the binary form.

Kang et al. [11] suggested long-short term memory recursive neural network (LSTM) to predict traffic flow, and comparatively analyzed the effects of different inputs on performance of short-term traffic flow prediction. Finally, the authors found better results can be obtained with traffic flow. This approach carried out embodies the advantages of LSTM in traffic flow prediction.

As a branch of deep neural network, recurrent neural network can capture the sequence information of data. Considering the advantages of deep neural network, this paper suggests the long-short term memory networks to predict the delay of airport scenes.

3 Long Short-Term Memory (LSTM)

LSTM is an improved version of the recurrent neural network [12–15], whose internal structure is shown in Fig. 1.

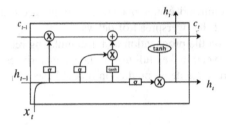

Fig. 1. LSTM internal structure

Oblivion gate Equation:

$$f_t = \sigma(W_f \bullet [h_{t-1}, x_t] + b_f) \qquad (1)$$

Equation (1) shows the Oblivion Gate realized through the activation layer. $[h_{t-1}, x_t]$ represents the merge matrix of the cell state at the previous time step and the input of the current step. The σ is the activation function.

Input layer gate output Equation is showed in Eq. (2)

$$i_t = \sigma(W_i \bullet [h_{t-1}, x_t] + b_i) \qquad (2)$$

$$c_t = \tanh(W_c \bullet [h_{t-1}, x_t] + b_i) \qquad (3)$$

Where W_i represents the weight between the input layer and the hidden layer. Equation (3) creates a candidate vector c_t, which will be added to the current cell state. Equation (4) shows the equation of the update of cell states.

$$c_t = f_t \bullet c_{t-1} + i_t \bullet c_t \qquad (4)$$

f_t and i_t represents the rate while c_{t-1} is the old cell state. Output gate Equation is as follows:

$$o_t = \sigma(W_o \bullet [h_{t-1}, x_t] + b_0) \qquad (5)$$

$$h_t = o_t \bullet \tanh(c_t) \qquad (6)$$

The output gate determines the output of a new cell state, and then passes the cell state through the active layer [16] and combines with the output of the output gate.

4   LSTM-Based Scene Delay Prediction Model

4.1   Structure of Scene Delay Prediction Model

The method proposed in this paper is to take LSTM to predict the delay time of the airport based on the information related to the pre-order departure delay. To train the

Table 1. Input and output of delay prediction model

Attributes of flight data (input variables)	Day of a week (DOW), Period (15 min as a period)
Attributes of pre-order departure delay data (input variables)	Actual inbound flights (AIF), actual departure flights (ADF), planned departures flights (PDF), departure delayed flights (delayed previously) (DDFP), departure delayed flights (not delayed previously) (DDF), departure delayed time (DDT)
Output variables	Class of delays (0,1,2, 3)

model, data from the airport's historical on-time data for the Guangdong Baiyun Airport in 2016 and the departure delays were used.

By applying the experience threshold to the delay time of this period we can obtain no delays, slight delays, moderate delays, severe delays in four delays. Table 1 shows the details of the flight data and the attributes of pre-order delay used by the model for forecasting.

In the divided sequence, the delay of the previous several periods may cause delay in the subsequent period. The concrete conceptual map of the model is shown in Fig. 2. Based on the flight data and prior-order delay data of the current period, the LSTM cells in the model not only calculate and output the delay level of the current period, but also transmit the data information through the network to the next LSTM cell.

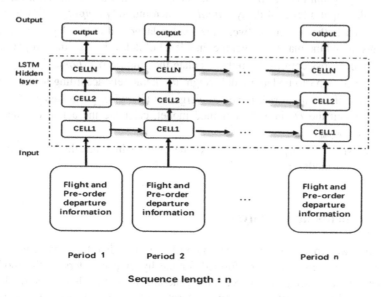

Fig. 2. Structure of LSTM-based scene delay prediction model

We let a sequence of 8 scene features mentioned before to pass the hidden layer of model, and apply truncation in case of excess. Finally the model can calculate and output the delay level of subsequent periods based on part of the flight information and the pre-order departure delay information in the current period.

4.2   The Process of Scene Delay Prediction Model

After the correct configuration of the proposed LSTM model, we train the model with the airport flight information and the pre-order delay status information as the input of the model. The main steps of the LSTM scene delay prediction algorithm are as follows:

Step 1   Divide the training data set divided by 9:1 into equal parts according to the time sequence *timestep*, and construct the prediction training dataset $D_1 = \{T_1, T_2, \ldots, T_{timestep}, T_{timestep+1}, \ldots, T_{2*timestep}, \ldots, T_{m*timestep}\}$ and the test dataset $D_2 = \{T_1, T_2, \ldots, T_{timestep}, T_{timestep+1}, \ldots, T_{2*timestep}, \ldots, T_{n*timestep}\}$, represented with $T_i = (x_{i1}, x_{i2}, \ldots, x_{id}, y_i)$ and d as the number of features. According to the candidate model's initialization sequence length, LSTM layer number, cell number, Dropout probability and learning rate, set model set $M = \{m_1, m_2, \ldots, m_k\}$.

Step 2   Take a model m_j of M, set and initialize the network weights of model m_j with $\theta_{m_j} = \{W, b\} \sim N(0, 1)$.

Step 3   Transmit training *timestep* samples into the utilized LSTM model, then in the $i(1 \le i \le timestep)$ step the current tuple T_i is input to the hidden layer which will output a level of delay $output_{i-1}$, then multiply $output_{timestep-1}$ and W_{m_j}, which is the weights between the hidden layer and output layer and add the result and the bias b_{m_j} to get the final level of delay through the Dropout layer. By the cross-entropy formula calculate the empiric risk $Loss_{m_j}$.

Step 4   Repeat step3 to update model weight θ_{m_j} iteratively according to the learning rate a if empirical risk is not convergent. Otherwise, go to step 5.

Step 5   Transmit the D_2 into the obtained model, calculate the actual test accuracy $Accuracy_{m_j}$.

Step 6   Let $M = M\setminus\{m_i\}$, if $M = \emptyset$, take the model which has the best accuracy as the best model, and otherwise go to step 2.

5   Experiments and Analysis

In this paper, we propose the model based on LSTM to predict the delay of the airport, cross-contrast experiments are performed for the hyper parameters of the model to obtain the optimal model, and to access the generalization capability of the model, the dataset is divided into training set and test set, and test results were recorded and analyzed.

5.1   Preparation of Dataset

The dataset of experiment in this article is based on the actual operation data of Baiyun Airport in 2017. With 15 min as a period, the actual inbound flights, actual departure flights, planned departure flights, departure delayed flights (delayed previously), departure delayed flights (not delayed previously), departure delayed time are extracted and calculated for each day of the year.

Binning Operation. We observe that the distribution of delay time has a certain relationship with the growth of the period, and there are some very prominent points in the distribution chart whose value is obviously much larger than the average. Boxplot is used to analyze the quartile values of the standardized data and filter out abnormal points in the data. The box bitmap is demonstrated in Fig. 3.

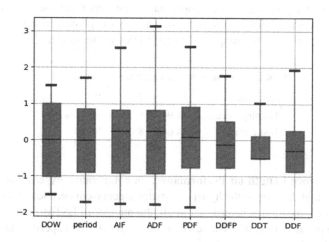

Fig. 3. Boxplot of the standardized delay data

Discretization of Features. In our project, we need to discretize the time for continuous delay. We use the Chi Merge method and the key steps are as follows:

Step 1   Instances are sorted by attributes to be discrete: each instance belongs to a range.

Step 2   Consolidate adjacent intervals, including two steps:
   (1)   Calculate the chi-squared value for each pair of adjacent intervals
   (2)   Combine a pair of intervals with the smallest chi-square value, and the

Step 3   Repeat Step 2 until all pairs of intervals have X^2 values exceeding the X^2 threshold (described below):

$$X^2 = \sum_{i=1}^{2} \sum_{j=1}^{2} \frac{(A_{ij} - E_{ij})^2}{E_{ij}} \tag{7}$$

Where A_{ij} represents the number of instances of class j in interval i, and E_{ij} means the expectation probability of A_{ij}, which is calculated as $\frac{M_i * N_j}{M}$, where M is the total number of samples, M_i means the number of samples in group i and N_j represents proportion of the sample of class j.

Taking industry experience and method mentioned above into consideration, scene delays can be divided into four warning levels, which are converted into corresponding category labels $c_0, \ldots, c_3$ by judging which level y_i belongs to. The value of scene delay is also embedded to obtain a one-hot matrix. In this way, the final training and test dataset is obtained as $D' = \{(x_t, c_t)\}$, $X_i = (x_{i1}, x_{i2}, \ldots, x_{id})$, $C_i = (c_0, c_1, c_2, c_3)$.

5.2   Analysis

The LSTM model proposed in this paper has some hyper parameters needed to be defined during training. For each hyper parameter, this paper has set up corresponding comparative experiments to compare the capabilities of the model.

Table 2. The results of comparative experiment 1

Sequence	5	7	9
Training accuracy	84.94%	85.97%	86.18%
Test accuracy	84.66%	85.68%	85.84%

Effect of Sequence Length on Performance. In comparative experiment 1, the data set uses the actual data set, in the hyper parameter aspect (sequence length is 5, 7, and 9 for comparison), the hidden layer is set to 20, the number of LSTM model layer is 15, and the keep probability is 0.8. The learning rate is 1e−3. The experimental results are shown in Table 2.

The experimental results of Comparative Experiment 1 show that the model has the highest accuracy when the sequence length is 9, but the advantage is not obvious to 7. In addition, the difference between the test results and the training accuracy is stable at 0.3%, which shows that the robustness of the proposed model is good.

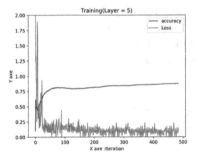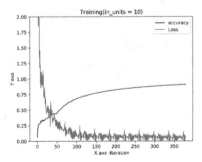

Fig. 4. Optimal loss and accuracy of the optima models in Experiment 2 and Experiment 3

Effect of Hidden Layers on Performance. In comparative experiment 2, the length of the sequence is set to 7 and the number of neurons in the hidden layer is set to 1, 2, 5, and 10, and others are consistent with experiment 1. The experimental results show that the number of 10 appears best but the choice of the optimal number of neurons often needs to consider the actual model calculation cost and actual accuracy. In this experiment, the optimal neuron is 5, and the corresponding loss graph and accuracy rate are shown in Fig. 4. The model's Loss (marked in blue) converges to about 0, and the test accuracy rate (marked in red) reaches 86.4%.

Effect of LSTM Cells on Performance. In the comparative experiment 3, the sequence length is set to 7, and the number of LSTM cells is set to 1, 2, 5, 10 and 15 respectively. The others are the same as those of the comparative experiment 1. In the deep learning model, the number of neurons and network layers in a model often lead to greater calculations and possibility of over fitting at the same time. Therefore, the optimal number of neuron layers is 3, the corresponding test accuracy rate reached 86.3%, the corresponding loss map and accuracy rate diagram is shown in Fig. 5.

Table 3. Accuracy of different dropout probability

Keep-prob	0.7	0.8	0.9
Test accuracy	84.66%	86.8%	85.84%

Effect of Dropout Probability on Model Performance. In comparison experiment 4, the sequence length is set to 7, and the dropout probability is set to 0.7, 0.8, and 0.9. The others are the same as those of experiment 1. Since the larger data information used, it also brings about increasing of the number of iterations, but not obvious. The optimal probability in comparison experiment 4 is 0.8. The corresponding loss and accuracy rate chart is shown in Table 3, and the test accuracy rate reaches 86.8%.

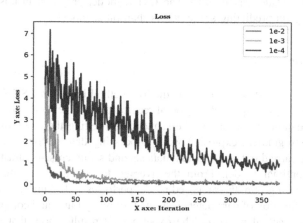

Fig. 5. Losses of trainings with different learning rate. (Color figure online)

Effect of Learning Rate on Model Performance. For the comparison of experimental 5, the learning rate was set at 1e−4, 1e−3, and 1e−2 for comparison. The length of the sequence was 7, and the other hyper-parameters were consistent with the experiment 1. The learning rate is a parameter that controls the speed of model loss reduction in the deep learning model. In this experiment, when the learning rate is set too large (1e−2, marked in blue), the Loss curve will often miss the extreme point. However, if it is set too small as 1e−4 (marked in red), it will lead to slow movement on the error curve, difficult to converge, resulting in wasting a lot of computing resources and time. This project sets the above three values separately. The experimental result is shown in Fig. 5. It can be concluded that the test accuracy rate is 86.7% at 1e−3 (green). The Loss eventually converges to around 0, indicating that the model has a good training result, and the final test accuracy rate is about 87%.

Comparison Experiment with BPNN. After cross-validation of the above process for the five groups of hyper-parameters, an optimal LSTM model is obtained. The corresponding learning rate is 1e−3, the number of neurons is 20, the number of model layers is 5, the sequence length is 9, and keep probability is 0.8. The optimal model was trained and tested three times. The accuracy rate obtained was 88.04%.

Table 4. Results of comparative experiment

	1	2	3	Average
LSTM	87.54%	88.57%	88.03%	88.04%
BPNN	68.63%	72.34%	70.39%	70.45%

The experiment also shows the performance of the Back Propagation Neural Network Model (BPNN) on the same data set. For the BPNN model parameters, we take on the same cross-contrast experiments to obtain the best model, which is then compared with the LSTM in appearance. The accuracy comparison chart is shown in Table 4. The LSTM performance was better than the traditional neural network, and the latter average value was 70.45%. The LSTM model proposed in this paper has a greater advantage in predicting scene delays than other models.

6   Conclusion

In this paper, the airport scene delay prediction method based on LSTM is proposed. The LSTM model is an improved version of recurrent neural network model, which can store the information of the data in the sequence by setting the cell unit to solve the problem of RNN gradient explosion while processing data of long sequence. Then experiments are carried out on the actual data set and set up cross-contrast experiments to obtain the optimal model. From the average test accuracy of the comparison experiments in each group, the proposed model performs well. Finally, the model with the optimal combination of parameters is trained for many times and compared with the back propagation neural network. The experimental results show that the proposed model performs better than the prediction model based on BPNN, which proves the accuracy of the proposed model for airport scene delay prediction.

The method proposed in this paper performs well on prediction for the delay of airport scene, which can capture the context information of the delay data well. However, in actual airport applications, the delay data is affected by many factors. It is an important task for future work that how to capture more information in data when dealing with actual airport cases, take more appropriate models for prediction and analysis, and apply this method to data with greater magnitude of delays.

References

1. Sternberg, A., Soares, J., Carvalho, D., et al: A Review on Flight Delay Prediction (2017)
2. Wieland, F.: Limits to growth: results from the detailed policy assessment tool [air traffic congestion]. In: Digital Avionics Systems Conference (1997)
3. Rong, F., Qianya, L., Bo, H., Jing, Z., Dongdong, Y.: The prediction of flight delays based the analysis of Random flight points. In: 2015 34th Chinese Control Conference (CCC), Hangzhou (2015)
4. Wong, J.-T., Tsai, S.-C.: A survival model for flight delay propagation. J Air Transp. Manag. **23**, 5–11 (2012)
5. Mueller, E., Chatterji, G.: Analysis of aircraft arrival and departure delay characteristics. AIAA J. (2008)
6. Trianto, R., Tai, T.C., Wang, J.C.: Fast-LSTM acoustic model for distant speech recognition. In: 2018 IEEE International Conference on Consumer Electronics (ICCE), Las Vegas, NV, pp. 1–4 (2018)
7. Lu, N., Wu, Y., Feng, L., Song, J.: Deep learning for fall detection: 3D-CNN combined with LSTM on video kinematic data. IEEE J. Biomed. Health Inform
8. Chen, W., et al.: EEG-based motion intention recognition via multi-task RNNs. In: Proceedings of the 2018 SIAM International Conference on Data Mining, pp. 279–287. Society for Industrial and Applied Mathematics (2018)
9. Yue, L., Chen, W., Li, X., Zuo, W., Yin, M.A.: A survey of sentiment analysis in social media. Knowl. Inf. Syst., 1–47 (2018)
10. Venkatesh, V., Arya, A., Agarwal, P., et al.: Iterative machine and deep learning approach for aviation delay prediction. In: 2017 4th IEEE Uttar Pradesh Section International Conference on Electrical, Computer and Electronics (UPCON), Mathura, pp. 562–567 (2017)
11. Kang, D., Lv, Y., Chen, Y.Y.: Short-term traffic flow prediction with LSTM recurrent neural network. In: 2017 IEEE 20th International Conference on Intelligent Transportation Systems (ITSC), Yokohama, pp. 1–6 (2017)
12. Graves, A.: Generating sequences with recurrent neural networks. Computer Science (2013)
13. Graves, A., Schmidhuber, J.: Framewise phoneme classification with bidirectional LSTM and other neural network architectures. Neural Netw. **18**(5–6), 602–610 (2004)
14. Kalchbrenner, N., Danihelka, I., Graves, A.: Grid long short-term memory. Computer Science (2015)
15. Bengio, Y., Simard, P., Frasconi, P.: Learning long-term dependencies with gradient descent is difficult. IEEE Trans Neural Netw. **5**(2), 157–166 (1994)
16. Bottou, L.: Large-scale machine learning with stochastic gradient descent. In: Lechevallier, Y., Saporta, G. (eds.) Proceedings of COMPSTAT 2010. Springer, Heidelberg (2010). https://doi.org/10.1007/978-3-7908-2604-3_16

DSDCS: Detection of Safe Driving via Crowd Sensing

Yun Du[1], Xin Guo[1], Chenyang Shi[1], Yifan Zhu[1], and Bohan Li[1,2(✉)]

[1] College of Computer Science and Technology,
Nanjing University of Aeronautics and Astronautics, Nanjing 211106, China
bhli@nuaa.edu.cn
[2] Collaborative Innovation Center of Novel Software Technology
and Industrialization, Nanjing 210016, China

Abstract. Traffic safety plays an important role in smart transportation, and it has become a social issue worthy of attention. For detection of safe driving, we focus on the collection, processing, distribution, exchange, analysis and utilization of information, and aim at providing diverse services for drivers and passengers. By adopting crowdsourcing and crowd-sensing, we monitor the extreme driving behavior during the process of driving, trying to reduce the probability of traffic accidents. The smartphones are carried by passengers, which can sense the driving state of the vehicles with our proposed incentive mechanism. After the data is integrated, we are able to monitor the driving behavior more accurately, and finally secure the public transit. Finally, we developed a safe driving App for monitoring and evaluation.

Keywords: Crowd sensing · Detection of safe driving
Crowdsourcing incentive

1 Instruction

As we know, public transport plays a very important role in life. With the increase in the use of shuttle buses, the frequency of accidents is getting higher and higher. As almost every accident can cause major casualties and property losses, the security of public transport is of great significance. The road safety situation in China is very serious. The number of deaths due to road traffic safety accidents exceeds 200,000 each year. Statistics on traffic accidents in China indicate that more than 90% of accidents are caused by human factors and dangerous driving behavior. The driver is the manipulator of the car. If we can train the driver's driving literacy by detecting extreme driving behavior, many accidents will be avoided. Traditional technical means used professional equipment to detect extreme driving behavior. This method costs a lot and

This work is supported by National Natural Science Foundation of China (61672284, 41301407), Funding of Security Ability Construction of Civil Aviation Administration of China (AS-SA2015/21), Fundamental Research Funds for the Central Universities (NJ20160028, NT2018028, NS2018057).

© Springer Nature Switzerland AG 2018
G. Gan et al. (Eds.): ADMA 2018, LNAI 11323, pp. 170–177, 2018.
https://doi.org/10.1007/978-3-030-05090-0_15

is difficult to popularize. Some researchers proposed to detect the driver's eye fatigue by installing a camera to detect the driver's eye movement. However, the image captured by the camera is not accurate enough and thus may easily lead to misjudgment. The sensor on the smartphone has high accuracy and no additional cost, and thus can be applied to traffic detection scenes. In addition, we also integrate the ideas of crowdsourcing. Drivers and passengers jointly sense data, and each cell phone can only be equivalent to one sensor node. Using the cooperative advantages of group perception, the data detected by crowd-sensing data can improve the sensing precision and enhance the credibility of the data through cleaning and fusion. In addition, due to differences in the location of the vehicle and the location of the smartphone, exploring the complementary relationship between the data in each group can further reduce the data error.

Crowdsourcing is a distributed problem-solving mechanism open to the public on the Internet. It integrates computers and public on the Internet to accomplish tasks that are difficult for computers to accomplish. According to the different forms of public crowdsourcing, crowdsourcing is divided into collaborative crowdsourcing and competitive crowdsourcing. The task of collaborative crowdsourcing is the need for public collaboration to complete, and the people who complete the task usually don't get rewards; and the task of competitive crowdsourcing is usually done by the individual independently, and the individuals who complete the task will receive corresponding rewards (such as money Remuneration) [1]. At present, there are many successful applications which use crowdsourced methods to solve image recognition and entity analysis which is difficult for computers to accomplish. The multi-sensing capabilities (geo-location, light, movement, audio and visual sensors, among others) of smartphones, provide a new variety of efficient means for opportunistic data transfer new crowdsourcing applications. A macro crowdsourcing task needs to be subdivided into many microscopic tasks, and the microscopic task is highly accepted by the user, so that the user is willing to assume the task. In this project, after the vehicle is over, the data obtained by each sensor node will be uploaded through the network. After that, the user needs to evaluate or label the trip. Since crowd-sourced data perception is based on human beings, crowd-sourced data collection can not only provide large-scale data for analysis, but also acquire subjective perceptions of passengers when detecting extreme driving behavior. Comprehensive objective data and subjective perceptions make the test results more accurate and user-friendly.

2 Related Work

2.1 Extreme Driving Behavior Detection System

Guo et al. [2] proposed "crowdsafe", which is a mobile crowd sensor system that utilizes the cooperative ability of bus passengers to improve detection of extreme driving behavior. In particular, they proposed the basic idea of solving the following three problems and made corresponding experimental results, including recognition of the passenger environment (for example, locating a passenger in a bus, detecting the position of a smartphone, such as a hand or a pocket), based on Multi-sensor fusion of extreme driving behavior detection, and strategies for investigating conscious behavior based on group collaboration and conflict resolution. Khaisongkram et al. [3] studied the impact of the development of automotive autopilot assist systems and sensor technology on personal driving behavior. Therefore, the data stored by the on-vehicle device can be analyzed based on the driving state described by the driver's maneuver and driving environment. The precondition is that the classifier that identifies the driving state is modeled by the push sequence mark method (BSLM). By analyzing the recognition accuracy of each driving state, the BSLM-trained classifier is verified. As a result, the accuracy of recognizing the braking and deceleration conditions is normal, but the cruising accuracy of the parking state is high.

2.2   Incitation Mechanism of Mobile Crowd Sensing

Ogie [4] proposed that mobile crowd sensing is an emerging concept that allows intelligent cities to utilize the sensing capabilities of mobile devices and ubiquitous features to capture and mark phenomena of common interest. Therefore, whether or not it is possible to fully and effectively motivate participants to determine the most appropriate incentive mechanism is a major challenge for whether mobile crowd-funding applications can be widely used and adopted.

In addition to the existing work, researchers also studied complex event processing methods based on driving behavior detection which is based on smart phones [5], using five-factor models to predict driving behavior [6], and integrated driving behavior model [7], Bayesian network model scenarios and non-situation driving behavior assessment [8]. With regard to the practical application of crowdsourcing ideas, they studied a crowdsourcing perception-improving network model based on the incentive game [9], based on the study of the duration of the weighted scheduling framework for mobile sensor data acquisition [10], dynamic data-driven crowd sensing tasks assignment [11], a blacklist based on mobile crowd sensing anonymous authentication scheme [12] and a dynamic and quality-enhanced incentive mechanism for mobile crowd-sourcing [13]. Therefore, under the consideration of the idea of crowdsourcing and the adoption of a long-term incentive mechanism, we study the detection of extreme driving behavior based on crowdsourcing ideas.

3   Safe Driving Model of Sensing Fusion

3.1   The Basic Concept of Crowdsourcing

"Crowdsourcing" was first proposed by the American journalist Jeff Howe. He believes that crowdsourcing is the act of a company or organization that has outsourced its tasks to its indefinite public network. When work is synergistic, crowdsourcing or mass production occurs; but crowdsourcing is often undertaken by individuals. Crowdsourcing is an important prerequisite for open call-to-action and potentially broad labor networks. To be precise, crowdsourcing is based on network technology and is a publicly called way to make full use of the Internet's individual user resources to outsource activities to the public's practice model.

Spatio-temporal crowdsourcing is based on the Internet of Things as a platform and individual terminal equipment as a unit. It gathers crowd-sourced participants on the Spatio-temporal platform, distributes various tasks through the crowdsourcing platform's engine, and controls the quality of crowd-sourced participants so that they can complete tasks in the spacetime of the physical world and receive certain rewards. Among them, the physical Spatio-temporal position of crowd-sourced participants generally has to meet the time-space constraints of crowd-sourcing tasks. Spatio-temporal crowdsourcing aims to use the Internet of Things platform to find people who are free, have a certain amount of knowledge and related skills, have a certain interest, and have corresponding computing capabilities and battery capacity. Spatio-temporal crowdsourcing uses their human intelligence and the terminal's meta-computing capabilities to organize their online offline physical world to complete some of the challenges that are too challenging for the machine and to make effective use of online and offline resources.

Extracting the features described above, we conclude that crowdsourcing is a distributed problem solving model. It assigns a work task to the mass network in a free and voluntary form through decomposition, which is a group of wisdom.

3.2 Extreme Driving Behavior Detection Based on Position Correlation

Differences in the location of passengers in the car will have an impact on the collection of data on the driving of the smartphone. The top view of the vehicle is shown in Fig. 1, when the shuttle is turning left, the centrifugal force felt by passengers in Area C is greater and the sensor accuracy is the highest. For the right turn, the sensor accuracy of the D-zone passenger's smartphone is the highest. When the shuttle changes lanes, since the front wheels turn more than the rear wheels when turning, the accuracy in the front area of the shuttle is higher than that in the rear area, so the accuracy of the mobile phones in the front areas is the highest. For emergency braking and rapid acceleration, the accuracy of the mobile phone at the rear of the vehicle will be higher. In order to detect the driving condition of the vehicle, it is necessary to use data provided by a plurality of passengers sitting at different positions. In order to determine the position of the passengers, the passengers need to select the area before the start of the driving test to label the journey accordingly. In addition, these data are collected from multiple sensors and require the fusion of different characteristics of the data. Here, we use the tri-axis gyroscope sensor and tri-axis acceleration sensor configured in most smartphones. The tri-axis gyroscope sensor measures the angular velocity in six directions ($x\theta+$, $x\theta-$, $y\theta+$, $y\theta-$, $z\theta+$, $z\theta-$), tri-axis acceleration sensor is used to sense the linear acceleration in six directions ($xa+$, $xa-$, $ya+$, $ya-$, $za+$, $za-$). The data of the two sensors respectively reflect the different vehicle states. Crossover comparisons and compensations have been implemented to achieve the detection of extreme driving behaviors with multi-sensor fusion.

It is easier to judge the extreme driving behavior by using single sensor data. After collecting the data, it is judged that the overall angular velocity and linear acceleration are within the limited range, and the angular velocity is the same.

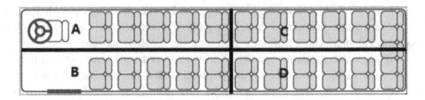

Fig. 1. Schematic diagram of space distribution in bus

To solve this problem that the detection results obtained from different locations. In various decisions, we summarizing the previous papers can get Bayesian decision theorem to have higher accuracy, so the Bayesian formula is adopted [3]. Assume that n passengers {u1, u2, u3,..., un} have detected a driving behavior with an accuracy of {a1, a2, a3,..., an}. The probability that the detected final result r ∈ R is correct can be calculated as:

$$P(r|\Omega) = \frac{P(\Omega|r)P(r)}{P(\Omega)} = \frac{P(\Omega|r)P(r)}{\Sigma_{r_i \in R}P(\Omega|r_i)P(r_i)}$$

The distribution of the detection result is represented by Ω, and r_i indicates each result within the area of the result w represents the user's usage coefficient. In this way, we can Calculate the weight of the result provided by each passenger:

$$c_j = w * \ln\frac{a_j}{1 - a_j}$$

As a result, we can calculate the weight of each result provided by passengers and use the highest credibility result as the final result. $f(u_i)$ indicates the result of passenger u_i detection:

$$P(r) = P(r|\Omega) = \frac{e^{\Sigma_{f(u_j)=rc_j}}}{\Sigma_{r_i \in R}\left(e^{\Sigma_{f(u_j)=r_i}C_j}\right)}$$

3.3 Decision Tips Based Machine Learning

In fact, crowd-based extreme driving behavior detection often encounters the following situation: For the same extreme driving behavior, we may use mobile phones at different locations in the vehicle to obtain different results. There are systematic errors in the data of passengers, including instrumental errors, operational errors, etc. of the sensors. Since the systematic error is always too large or too small, it cannot be reduced by averaging over many experiments. Crowdsourcing can solve this problem. After taking into account the data differences caused by location factors, we can compare the data uploaded by all passengers on a journey and use the fusion of data to compensate to reduce systematic errors in the stage of data preprocessing. It is especially important when using crowdsourced data to analyze driving behaviors. Based on the character-istics of the data collected by the two sensors, the passenger's evaluation was quan-tified, and the neural network algorithm was used to train the machine learning model using the preprocessed data. Then use Bayesian voting theory to make group decisions, and give appropriate weights based on the results given by the previous model. The final result is given considering the number of votes for each extreme driving behavior and the passenger's position information.

We assume that each passenger has uploaded all crowd-sourced data for the trip and given accurate evaluation tags and scores. The evaluation tag matrix is [Asses-sLabel] and the evaluation score matrix is [Score]. We set up the sensor to collect data once every 0.2 s. In the experiment, one trip is about 30 min. The uploaded 9000 sets of data contain the time matrix [Time] and the two sensor's triaxial parameters, which are respectively recorded as the acceleration parameter matrix [Acc] and angular velocity parameter matrix [Ang], the three matrices are combined into [data] matrix. This matrix of 9000 * 7 size was passed to the neural network for training after invalid data and null data were removed. In addition, after the ride is over, the tags and scores given by the passengers are combined into a [Label] matrix as an index to reduce the loss function.

$$Acc = [AccX, AccY, AccZ] \quad Ang = [AngX, AngY, AngZ] \quad Time = [T]$$
$$Data = [Acc \ Ang \ T] \quad Label = [Score \ AssessLabel]$$

This neural network includes an input layer, an output layer, and three hidden layers. The first two hidden layers learn the change of the [Acc] matrix, the [Ang] matrix in one travel time series, and the third hidden layer learns the [Time] matrix changes in the multiple travels. All three hidden layers use a Gradient Descent algo-rithm to reduce the loss function and optimize the model parameters. Gradient Descent algorithm is a kind of iterative method, which is used to iteratively solve the minimum value of loss function to get the minimum loss function and model parameter value. Its iteration formula is:

$$A_{k+1} = A_k + P_k S^{(k)}$$

$S^{(k)}$ represents the negative direction of the gradient, P_k represents the search step length in the gradient direction, and gradient direction can be obtained by deriving the function. We determine the step size by a linear search algorithm. The closer to the target value, the smaller the step and the slower the progress.

3.4   Incentive Measures

Incentive measures are important tools of crowdsourcing data acquisition. There are many ways to motivate the situation, including entertainment games, rewards, virtual points, social relationship incentives, and so on. How to improve the availability and efficiency of crowdsourced data. This requires designing appropriate incentives for crowdsourced awareness task requirements. Combined with relevant counts, it also satisfies the core issues of maximizing the respective benefits of the platform and the participants. For example, maximum incentives and top-k incentives have differences advantages and disadvantages. They differ in different situations.

We consider taking the passengers' uploading data as a condition to obtain the right to use the free Wi-Fi in the car and giving reward cards after the evaluation interface is completed, which can be used to redeem travel codes. First of all, we use anonymity to ensure the privacy of passengers. The incentive condition is that for each upload of good ride data, we give a reward card with a ride code, which can be used to offset the shuttle ride cost. The higher the level, the higher and lower the limit of the ridership code attached to the reward card, which can inspire the user to use the cycle. In order to introduce new users, we can make greater rewards to new users. When uploading the first good data, you can directly redeem a larger number of ride codes. In addition, if there is a condition for on-vehicle Wi-Fi in the shuttle, it can also provide the right to use Wi-Fi after the user uploads crowdsourced data.

4   Implementation of the Platform

The Fig. 2 show the gyroscope data and acceleration sensor data during the ride. The horizontal axis is time and the longitudinal axis is the value perceived by the sensor. And we can check shuttle bus information like the Fig. 3, Departure location, arrival location, departure date for your choice, and then you can check the corresponding bus information. The Fig. 4 is a reward card with a different amount of travel code, which can be used to offset the ride cost.

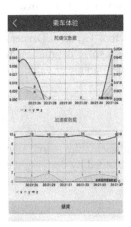

Fig. 2. Sensor data   **Fig. 3.** Bus information   **Fig. 4.** Reward code

5   Conclusion and Future Work

Crowdsourcing is one of the most efficient ways to obtain data and process data so far. Both of these are crucial to the time with information explosion. In this paper, we try to build a safe driving platform based on crowdsourcing, using the smart phone sensors of passengers to achieve real-time detection of driving and implementing incentive mechanisms for passengers. However, it remains to be seen whether there will be problems in the implementation process or whether the incentive mechanism is effective. We plan to apply our methods to intelligent car like unmanned driving in the future.

References

1. Chatzimilioudis, G., Konstantinidis, A., Laoudias, C., ZeinalipourYazti, D.: Crowdsourcing with smartphones. IEEE Internet Comput. **16**, 36–44 (2012)
2. Guo, Y., Guo, B., Liu, Y., Wang, Z., Ouyang, Y., Yu, Z.: CrowdSafe: detecting extreme driving behaviors based on mobile crowdsensing. In: Proceedings of the 14th IEEE International Conference on Ubiquitous Intelligence and Computing (UIC 2017), 4–8 August 2017, San Francisco, California, USA (2017)
3. Khaisongkram, W., Raksincharoensak, P., Shimosaka, M., Mori, T., Sato, T., Nagai, M.: Automobile driving behavior recognition using boosting sequential labeling method for adaptive driver assistance systems. In: Dengel, A.R., Berns, K., Breuel, T.M., Bomarius, F., Roth-Berghofer, T.R. (eds.) KI 2008. LNCS (LNAI), vol. 5243, pp. 103–110. Springer, Heidelberg (2008). https://doi.org/10.1007/978-3-540-85845-4_13
4. Ogie, R.I.: Adopting incentive mechanisms for large-scale participation in mobile crowdsensing: from literature review to a conceptual framework. Hum.-Cent. Comput. Information Sci. **6**(1), 24 (2016)
5. Vasconcelos, I., Oliveira Vasconcelos, R., Olivieri, B., Roriz, M., Endler, M., Colaço Junior, M.: Smartphone-based outlier detection: a complex event processing approach for driving behavior detection. J. Internet Serv. Appl. **8**(1) (2017)
6. Herzberg, P.Y.: Beyond "accident-proneness": Using Five-Factor Model prototypes to predict driving behavior. J. Res. Pers. **43**(6), 1096–1100 (2009)
7. Toledo, T., Koutsopoulos, H.N., Ben-Akiva, M.: Integrated driving behavior modeling. Transp. Res. Part C: Emerg. Technol. **15**(2), 96–112 (2007)
8. Zhu, X., Yuan, Y., Hu, X., Chiu, Y.-C., Ma, Y.-L.: A Bayesian network model for contextual versus non-contextual driving behavior assessment. Transp. Res. Part C: Emerg. Technol. **81**, 172–187 (2017)
9. Liu, X., Ota, K., Liu, A., Chen, Z.: An incentive game based evolutionary model for crowd sensing networks. Peer-to-Peer Netw. Appl. **9**(4), 692–711 (2016)
10. Thejaswini, M., Rajalakshmi, P., Desai, U.B.: Duration of stay based weighted scheduling framework for mobile phone sensor data collection in opportunistic crowd sensing. Peer-to-Peer Netw. Appl. **9**(4), 721–730 (2016)
11. Pournajaf, L., Xiong, L., Sunderam, V.: Dynamic data driven crowd sensing task assignment. Procedia Comput. Sci. **29** (2014)
12. Li, H., Jia, K., Yang, H., Liu, D., Zhou, L.: Practical blacklist-based anonymous authentication scheme for mobile crowd sensing. Peer-to-Peer Netw. Appl. **9**(4), 762–773 (2016)
13. Guo, B., et al.: TaskMe: toward a dynamic and quality-enhanced incentive mechanism for mobile crowd sensing. Int. J. Hum.-Comput. Stud. **102**, 14–26 (2016)

Power Equipment Fault Diagnosis Model Based on Deep Transfer Learning with Balanced Distribution Adaptation

Kaijie Wang[✉] and Bin Wu[✉]

Beijing Key Laboratory of Intelligent Telecommunications
Software and Multimedia, School of Computer Science,
Beijing University of Posts and Telecommunications, Beijing, China
WKJ1995@bupt.edu.cn

Abstract. In recent years, an increasing popularity of deep learning models has been widely used in the field of electricity. However, in previous studies, it is always assumed that the training data is sufficient, the training and the testing data are taken from the same feature distribution, which limits their performance on the imbalanced tasks. So, in order to tackle the imbalanced data distribution problem, this paper presents a new model of deep transfer network with balanced distribution adaptation, aiming to adaptively balance the importance of the marginal and conditional distribution discrepancies. By conducting comparative experiments, this model is proved to be effective and have achieved a better performance in both classification accuracy and domain adaptation effectiveness.

Keywords: Transfer learning · Domain adaptation
Balanced distribution adaptation · Deep transfer network · Power data analysis

1 Introduction

In recent years, certain achievements have been made in the work of fault diagnosis, load forecasting, energy saving and emission reduction, etc. The existing classification, clustering, and association analysis methods are able to satisfy the mining and processing of existing grid data information. However, with the continuous expansion of practical applications, traditional data mining methods will face some new problems. The traditional data mining methods require the source data information to be very full, so it may seem powerless for data sparseness and other issues. Moreover, in the process of power data collection, there will be an objective lack or subjective lack of large amounts of data, which will bring great interference to the analysis and processing of power data. Due to the strong correlation between power data, the transfer learning may

This research is supported by the National Key R&D Program of China (No. 2018YFC0831500), the National Social Science Foundation of China (No. 16ZDA055), and the Special Found for Beijing Common Construction Project.

© Springer Nature Switzerland AG 2018
G. Gan et al. (Eds.): ADMA 2018, LNAI 11323, pp. 178–188, 2018.
https://doi.org/10.1007/978-3-030-05090-0_16

play an important role in the analysis and processing of power data. Figure 1 is a brief framework of transfer learning.

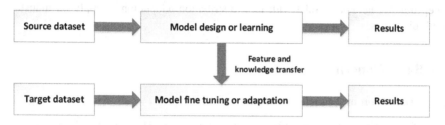

Fig. 1. Brief framework of transfer learning

Section 1 in this paper introduces our work briefly. Some related work of the paper is included in Sect. 2. Basic concepts of our work are introduced in Sect. 3. Section 4 shows the detail implement the training strategy of the model we proposed. Experiments and results are described in Sect. 5, and Sect. 6 summarizes the conclusion and future works of this model.

Our major contribution includes:

1. We propose a deep transfer network model for classification in different domain, namely Deep Transfer Network with Balanced Distribution Adaptation. This model uses balanced distribution adaptation to tackle the domain adaptation problem and convolutional neural network to tackle the classification problem.
2. As we currently know, we apply the Deep Transfer Network with Balanced Distribution Adaptation on the power data analysis for the first time. Due to the strong correlation and imbalanced distribution of the power data, it has been proved that the model we proposed has obvious advantages over other methods.

2 Related Work

In recent years, transfer learning has been viewed in many different areas. A variety of comprehensive surveys has been made. In the surveys, transfer learning has been categorized into four branches, Instance based Transfer Learning, Feature based Transfer Learning, Model based Transfer Learning and Relation based Transfer Learning [1, 2]. Li proposed to jointly select feature and preserve structural properties [5]. Long et al. proposed joint distribution adaptation method (JDA) to match both marginal and conditional distribution between domains [3]. Others extended JDA by adding structural consistency, domain invariant clustering, and target selection. To investigate the importance of each distribution, Wang proposed balanced distribution adaptation method (BDA) to adaptively adjust the marginal and conditional distribution according to specific situations [10].

With the rapid development of deep learning, more and more researchers begin to use deep neural networks for transfer learning. Compared to the traditional transfer learning methods, the deep transfer network can automatically extract more expressive features and meet end-to-end requirements in practical applications. In 2014, Yosinski *et al.* first proposed the feasibility study of deep transfer network. [7] Han proposed a deep transfer network model with joint distribution adaptation to apply on industrial fault diagnosis.

3   Basic Concept

3.1   Definition of Transfer Learning

Transfer learning is novel machine learning framework. There are two basic concepts in transfer learning, domains and tasks. The definition is as followed.

Definition 1 (Domain). A domain D is composed of two components: a feature space X and a marginal probability distribution $P(X)$, where $X = \{x_1, \ldots, x_n\}$ is a particular training dataset, i.e., $D = \{X, P(X)\}$.

Definition 2 (Task). A task T consists of two parts, a label space y and a predictive function $f(X)$, which can be learned from the instance set X, i.e., $T = \{y, f(X)\}$. Also, $f(X) = Q(Y|X)$ is the conditional probability distribution.

Definition 3 (Transfer learning). Given a source domain D_s with a learning task T_s and a target domain D_t with a learning task T_t, transfer learning aims to facilitate the learning process of target predictive function $f_t(X)$ in D_t by using the related information or knowledge in D_s and T_s, where $D_s \neq D_t$, or $T_s \neq T_t$.

3.2   Maximum Mean Discrepancy

MMD is an index to empirically estimate the discrepancy of two distributions. [10] Given two dataset x_s, x_t, $P_s(x_s) \neq P_t(x_t)$ and a nonlinear mapping function ϕ in a reproducing Kernel Hilbert space H(RKHS), the formulation of MMD can be defined as:

$$\text{MMD}_H(x_s, x_t) = \left\| \frac{1}{n_s} \sum\nolimits_{i=1}^{n_s} \emptyset(x_i^s) - \frac{1}{n_t} \sum\nolimits_{i=1}^{n_t} \emptyset(x_i^t) \right\|_H \tag{1}$$

Where n_s and n_t are the numbers of samples in the two datasets respectively.

4   Deep Transfer Network with Balanced Distribution Adaptation

4.1   Balanced Distribution Adaptation

Balanced distribution adaptation aims to find a transformation A to make the distance between the transformed $P(A^T x_s)$ and $P(A^T x_t)$ and the distance between $P(y_s|A^T x_s)$ and $P(y_t|A^T x_t)$ as small as possible. So, balanced distribution adaptation can be divided into three steps [3].

Marginal Distribution Adaptation. First, we need to minimize the distance between $P(A^T x_s)$ and $P(A^T x_t)$. MMD is as followed.

$$DISTANCE(x_s, x_t) = \|P(x_s) - P(x_t)\| \tag{2}$$

Conditional Distribution Adaptation. Second, we need to find another transformation A to make the distance between $P(y_s|A^T x_s)$ and $P(y_t|A^T x_t)$ small enough, we first need to train a simple classifier to predict the label y_t. The MMD distance can be described as followed.

$$DISTANCE(D_s, D_t) \approx \|P(y_s|x_s) - P(y_t|x_t)\| \tag{3}$$

So, joint distribution adaptation can be described as followed.

$$DISTANCE(D_s, D_t) \approx \|P(y_s|x_s) - P(y_t|x_t)\| + \|P(x_s) - P(x_t)\| \tag{4}$$

In some cases, marginal distribution and conditional distribution may not be equally important. To solve this problem, we use a balance factor to dynamically adjust the distance between two distributions. The equation is as followed [10].

$$DISTANCE(D_s, D_t) \approx \mu \|P(y_s|x_s) - P(y_t|x_t)\| + (1 - \mu)\|P(x_s) - P(x_t)\| \tag{5}$$

Where $\mu \in [0, 1]$ represents the balance factor.

In order to compute the marginal and conditional distribution divergences in (5), we adapt maximum mean discrepancy (MMD) to empirically estimate both distribution discrepancies. Formally speaking, (5) can be represented as

$$DISTANCE(D_s, D_t)$$

$$\approx (1 - \mu)\left\|\frac{1}{n}\sum_{i=1}^{n} X_{s_i} - \frac{1}{m}\sum_{j=1}^{m} X_{t_i}\right\|_H^2$$

$$+ \mu \sum_{c=1}^{C} \left\|\frac{1}{n_c}\sum_{X_{s_i} \in D_s^{(C)}} X_{s_i} - \frac{1}{m_c}\sum_{X_{t_i} \in D_t^{(C)}} X_{t_i}\right\|_H^2 \tag{6}$$

Where H means RKHS (the reproducing kernel Hilbert space), c is the class label, n, m denotes the number of samples in the source and target domain, $D_s^{(C)}$ and $D_t^{(C)}$ denote the samples belonging to class c in source and target domain. (8) can be formalized as followed:

$$\min tr\left(A^T X\left((1 - \mu)M_0 + \mu \sum_{c=1}^{C} M_c\right)X^T A\right) + \lambda\|A\|_F^2 \tag{7}$$
$$\text{s.t. } A^T XHX^T A = I, \ 0 \leq \mu \leq 1$$

To obtain the optimal transformation matrix **A**, we use Lagrange function to solve (7). First, we denote $\emptyset = (\emptyset_1, \emptyset_2, \ldots, \emptyset_d)$ as Lagrange multiplier. Then, we can get

$$L = \min tr\left(A^T X\left((1-\mu)M_0 + \mu \sum_{c=1}^{C} M_c\right) X^T A\right) + \lambda \|A\|_F^2 + tr\left((I - A^T XHX^T A)\emptyset\right) \quad (8)$$

Assume that $\partial L/\partial A = 0$, we get the equation as followed.

$$\left(X\left((1-\mu)M_0 + \mu \sum_{c=1}^{C} M_c\right) X^T + \lambda I\right)A = XHX^T A \quad (9)$$

The detail of BDA can be summarized in Table 1.

Table 1. BDA: Balanced Distribution Adaptation

Input: Source and target feature matrix X_s and X_t source label vector y_s, #dimension d, balance factor μ, regularization parameter λ

Output: Transformation matrix A and classifier f

1: begin

2: Train a base classifier on X_s and apply prediction on X_t to get its soft labels y_t. Construct $X = [X_s, X_t]$, initialize M_0 and M_c by Eq. (5) and (6)

3: repeat

4: Solve the feature decomposition problem in (8) and use d smallest eigenvectors to build A

5: Train a classifier f on $\{A^T x_s, y_s\}$

6: Update the soft labels of D_t: $\hat{y}_t = f(A^T x_t)$

7: Update matrix M_c using Eq. (6)

8: until convergence

9: return Classifier f

4.2   Deep Transfer Network

Generally, we can train a CNN model on the sufficient source data from scratch with the optimization task. The cross-entropy l_{ce} between estimated probability distribution and true label is served as the loss function. When applying the pre-trained CNN model to domain adaptation, a new objective function is redefined by integrating the l_{ce} and regularization term of BDA, rewritten as:

$$L(\theta) = l_{ce} + \lambda D_H(J_s, J_t) \quad (10)$$

Where $\theta = \{W_i, b_i\}_{i=1}^{l}$ is the parameter collection of a CNN with l layers and λ is non-negative regularization parameter.

By minimizing (14), the pre-trained CNN model can be further transferred and adapted for target tasks. The mini-batch stochastic gradient descent (SGD) and backpropagation algorithm are used for network optimization. The gradient of objective function with respect to network parameters is

$$\nabla_{\theta^l} = \frac{\partial l_{ce}}{\partial \theta^l} + \lambda (\nabla D_H(J_s, J_t))^T \left(\frac{\partial \emptyset(x)}{\partial \theta^l} \right) \tag{11}$$

Where $\frac{\partial \emptyset(x)}{\partial \theta^l}$ are the partial derivatives of the output of $(l-1)$th layer with network parameters. The detailed formulations of $\nabla D_H^2(J_s, J_t)$ are described as:

$$\nabla D_H^2(J_s, J_t) = \nabla MMD_H^2(P_s, P_t) + \sum_{c=1}^{C} \nabla MMD_H^2\left(Q_s^{(c)}, Q_t^{(c)}\right) \tag{12}$$

Where:

$$\nabla MMD_H^2(P_s, P_t) = \begin{cases} \frac{2}{n_s}\left(\frac{1}{n_s}\sum_{i=1}^{n_s}\emptyset(x_i^s) - \frac{1}{n_t}\sum_{j=1}^{n_t}\emptyset(x_j^t)\right) x \in D_s \\ \frac{2}{n_t}\left(\frac{1}{n_t}\sum_{j=1}^{n_t}\emptyset(x_j^t) - \frac{1}{n_s}\sum_{i=1}^{n_s}\emptyset(x_i^s)\right) x \in D_t \end{cases} \tag{13}$$

$$\nabla MMD_H^2\left(Q_s^{(c)}, Q_t^{(c)}\right) = \begin{cases} \frac{2}{n_s^{(c)}}\left(\frac{1}{n_s^{(c)}}\sum_{x_i^s \in D_s^{(c)}}\emptyset(x_i^s) - \frac{1}{n_t^{(c)}}\sum_{x_j^t \in D_t^{(c)}}\emptyset(x_j^t)\right) x \in D_s \\ \frac{2}{n_t^{(c)}}\left(\frac{1}{n_t^{(c)}}\sum_{x_j^t \in D_t^{(c)}}\emptyset(x_j^t) - \frac{1}{n_s^{(c)}}\sum_{x_i^s \in D_s^{(c)}}\emptyset(x_i^s)\right) x \in D_t \end{cases} \tag{14}$$

4.3 Training Strategy

The complete training procedure of DTN with BDA is summarized in Table 2.

Table 2. Training procedure of DTN with BDA

Input: Given the dataset $D_s = \{x_i^s, x_i^s\}_{i=1}^{n_s}$ in source domain, unlabeled dataset $D_t = \{x_i^t\}_{i=1}^{n_t}$ in target domain, the architecture of deep neural network, the trade-off parameters λ

Output: Transferred network and predicted labels for target samples

1: begin

2: Train a base deep network on the source dataset Ds

3: Predict the pseudo labels $\hat{Y}_0 = \{y_i^t\}_{i=1}^{n_t}$ for target samples with base network

4: repeat

5: $j=j+1$

6: Compute the regularization term of BDA according to (8)

7: Network optimization with respect to (9)

8: Update the pseudo labels $\hat{Y}_j$ with optimized network

9: until convergence or $\hat{Y}_j = \hat{Y}_{j-1}$

10: Check the diagnosis performance of transferred network on other target samples

---

5   Experiments

5.1   Data Description

To demonstrate the efficiency, superiority as well as practical value of proposed transfer framework, we conduct experiments on three datasets. The first dataset contains the six kinds of dissolved gas concentration of the oil, the temperature of the oil in the transformer and the label if the transformer fails from January, 2017 to April, 2018 with a monitoring period of 4 h. The second dataset contains the pressure, the temperature, the moisture content and the gas density in the switch and the label if the switch fails from July, 2016 to March, 2018 with a monitoring period of 1 h. The detail of datasets is shown in the Table 3.

Table 3. Datasets description

No.	Name	Samples
1	Gas concentration datasets	2916
2	Pressure datasets	12206

To conduct the experiment properly, we perform the experiments under 5 different temperature conditions ranging from -10 °C to 40 °C (Temp. 1–5). As for pressure datasets, we divide it into 6 parts based on temperature conditions ranging from -10 °C to 50 °C (Temp. 6–11). Temperature differences of each condition range 10 °C. The details of the tasks de-signed are listed in Table 4.

Table 4. Diverse transfer tasks across different conditions

Transfer tasks	Source domain	Target domain
A → B	Temp. 1–2	Temp 3
B → A	Temp 3	Temp.1–2
C → D	Temp. 2–3	Temp. 4–5
D → C	Temp. 4–5	Temp. 2–3
E → F	Temp. 1	Temp. 5
F → E	Temp. 5	Temp. 1
G → H	Temp. 8	Temp. 9
H → G	Temp. 9	Temp. 8
I → J	Temp. 6–8	Temp. 9–11
J → I	Temp. 9–11	Temp. 6–8

5.2   Balance Factor Settings

In this part, we evaluate the effectiveness of the balance factor (Fig. 2). We run the model with $\mu \in \{0, 0.1, 0.2, \ldots, 1.0\}$, the results are shown in Fig. 3.

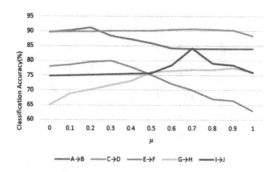

Fig. 2. The classification accuracy (%) on different balanced factor

From the figure we can conclude that different tasks have different optimal. For the transfer task A → B, the optimal is 0.2, for the transfer task C → D, the optimal is 0.7, For the transfer task E → F, the optimal is 0.3, For the transfer task G → H, the optimal is 0.9, For the transfer task I → J, the optimal is 0.7. So it's necessary to balance the marginal and conditional distribution between domains. Therefore, BDA is more capable of achieving good performance.

5.3   Results and Analysis

We choose five other methods as comparison. (1) SVM (2) RF (Random Forest) (3) CNN (4) DTN with TCA (5) DTN with JDA. Among these methods, the first three are traditional machine learning classification algorithms while the other two are previous transfer learning approaches.

Classification Accuracy. The results are shown as followed (Table 5).

Table 5. Classification results (%) on 10 tasks

Task	SVM	RF	CNN	DTN with TCA	DTN with JDA	DTN with BDA
A → B	74.2	77.9	90.8	89.7	85.9	91.3
B → A	76.7	68.3	78.3	83.1	80.9	85.4
C → D	79.8	88.8	82.4	89.6	90.1	90.6
D → C	85.7	91.5	93.6	94.7	95.8	98.1
E → F	62.5	77.3	73.4	78.1	75.3	79.9
F → E	59.8	64.0	67.6	65.5	66.7	67.4
G → H	47.7	70.3	58.9	65.2	76.1	77.5
H → G	58.3	50.7	67.3	87.7	90.0	94.2
I → J	66.9	61.8	70.9	74.9	75.8	84.1
J → I	82.0	67.1	86.9	80.3	81.5	82.0
Ave.	69.4	71.8	77.0	80.9	81.8	85.0

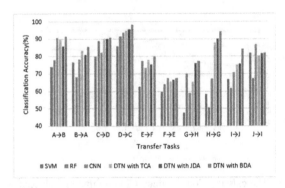

Fig. 3. Classification results (%) on 10 tasks

From the results above, we can draw the conclusion that (a) the model we proposed performs differently in different cases due to the different variability between source domains and target domains. (b) DTN with BDA performs the best in average classification accuracy. (c) In most cases, DTN with BDA performs best compared to other methods and the maximum classification accuracy can reach 98.1%. However, in some cases, the result may not be as good as other methods possibly due to negative transfer and the solutions still need to be investigated.

Effectiveness of Domain Adaptation. Furthermore, we test the distribution adaptation of BDA and compare with other two distribution adaptation methods: TCA and JDA. We evaluate the performance of distribution adaptation with MMD distances calculated using (1) (Fig. 4).

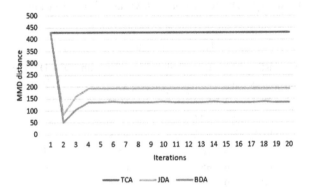

Fig. 4. MMD distance with increasing iterations

The figures above show the MMD distance of BDA, TCA and JDA when increasing iteration. Based on the results, we can draw a conclusion that (a) MMD distances of all methods can be reduced with the increasing iterations. This indicates the effectiveness of TCA, JDA and BDA; (b) MMD distance of TCA is not reduced largely as it only adapts the marginal distribution distance and requires no iteration; (c) MMD distance of JDA is obviously larger than BDA, since BDA could balance the importance of marginal and conditional distribution via balance factor μ; (d) BDA achieves the best performance in domain adaptation.

6 Conclusion

To tackle the imbalanced data distribution problem and to make full use of the existing model in the field of electricity, this paper proposed a new model of Deep Transfer Network with Balanced Distribution Adaptation and perform it on the actual power data. By conducting experiments, it is proved that the model has achieved better results when comparing to other existing methods. In the future, we will continue the exploration in these two aspects: by developing more strategies to leverage the distributions and handle the parallelization of the model for better efficiency.

References

1. Pan, S.J., Yang, Q.: A survey on transfer learning. IEEE Trans. Knowl. Data Eng. **22**(10), 1345–1359 (2010)
2. Weiss, K., Khoshgoftaar, T.M., Wang, D.: A survey of transfer learning. J. Big Data **3**, 9 (2016)
3. Long, M., Wang, J., Ding, G., Sun, J., Yu, P.S.: Transfer feature learning with joint distribution adaptation. In: IEEE International Conference on Computer Vision, pp. 2200–2207 (2013)
4. Pan, S.J., Tsang, I.W., Kwok, J.T., Yang, Q.: Domain adaptation via transfer component analysis. IEEE Trans. Neural Netw. **22**(2), 199–210 (2011)
5. Li, J., Zhao, J., Lu, K.: Joint feature selection and structure preservation for domain adaptation. In: International Joint Conferences on Artificial Intelligence (2016)
6. Long, M., Wang, J., Ding, G., Sun, J., Yu, P.S.: Transfer joint matching for unsupervised domain adaptation. In: IEEE Conference on Computer Vision and Pattern Recognition, pp. 1410–1417 (2014)
7. Yosinski, J., Clune, J., Bengio, Y., Lipson, H.: How transferable are features in deep neural networks. Eprint Arxiv **27**, 3320–3328 (2014)
8. Long, M., Zhu, H., Wang, J., Jordan, M.I.: Deep transfer learning with joint adaptation networks. In: International Conference on Machine Learning (2016)
9. Hsu, T.M.H., Chen, W.Y., Hou, C.-A., Tsai, Y.-H.H., Yeh, Y.-R., Wang, Y.-C.F.: Unsupervised domain adaptation with imbalanced cross-domain data. In: IEEE International Conference on Computer Vision, pp. 4121–4129 (2015)
10. Wang, J., Chen, Y., Hao, S., Feng, W., Shen, Z.: Balanced distribution adaptation for transfer learning. In: IEEE International Conference on Data Mining, pp. 1129–1134 (2017)
11. Costilla-Reyes, O., Scully, P., Ozanyan, K.B.: Deep neural networks for learning spatio-temporal features from tomography sensors. IEEE Trans. Industr. Electron. **65**, 645–653 (2018)
12. Khatami, A., Babaie, M., Tizhoosh, H.R., Khosravi, A., Nguyen, T., Nahavandi, S.: A sequential search-space shrinking using CNN transfer learning and a Radon projection pool for medical image retrieval. Expert Syst. Appl. **100**, 224–233 (2018)

13. Wei, Y., Zhang, Y., Yang, Q.: Learning to transfer. Eprint arxiv, arXiv:1708.05629 [cs.AI] (2017)
14. Qureshi, A.S., Khan, A., Zameer, A., Usman, A.: Wind power prediction using deep neural network based meta regression and transfer learning. Appl. Soft Comput. **58**, 742–755 (2017)
15. Lu, W., Liang, B., Cheng, Y., Meng, D., Yang, J., Zhang, T.: Deep model based domain adaptation for fault diagnosis. IEEE Trans. Industr. Electron. **64**, 2296–2305 (2017)
16. Zhang, J., Li, W., Ogunbona, P.: Joint geometrical and statistical alignment for visual domain adaptation. In: IEEE Conference on Computer Vision and Pattern Recognition (2017)
17. Li, S., Song, S., Huang, G.: Prediction reweighting for domain adaptation. IEEE Trans. Neural Netw. Learn. Syst. **99**, 1–14 (2016)

Event Extraction with Deep Contextualized Word Representation and Multi-attention Layer

Ruixue Ding[✉] and Zhoujun Li[✉]

School of Computer Science and Engineering,
Beihang University, Beijing 100191, China
{ruixue_ding,lizj}@buaa.edu.cn

Abstract. One common application of text mining is event extraction. The purpose of an event extraction task is to identify event triggers of a certain event type in the text and to find related arguments. In recent years, the technology to automatically extract events from text has drawn researchers' attention. However, the existing works including feature based systems and neural network base models don't capture the contextual information well. Besides, it is still difficult to extract deep semantic relations when finding related arguments for events. To address these issues, we propose a novel model for event extraction using multi-attention layers and deep contextualized word representation. Furthermore, we put forward an attention function suitable for event extraction tasks. Experimental results show that our model outperforms the state-of-the-art models on ACE2005.

Keywords: Event extraction · Muti-attention layer
Deep contextualized word representation

1 Introduction

The event extraction is a challenging task in text mining. It aims at identifying event triggers of a certain event type in the text (Event Detection) and finding related arguments with different roles (Argument Identification). For example, in the sentence "He died in the hospital", the event detection system is supposed to detect an event *Die* with the trigger word "*died*" and the argument identification system is expected to identify "*He*" and "*hospital*" as event arguments with roles respectively *Person* and *Place*.

Existing methods include feature based systems and neural network models. Traditional methods [1–5] use crafted features to represent text. There are two categories of features which are lexical features and contextual features. Lexical features include NER features, POS features and some other features like lemma and capitalization. Contextual features are often task specified and required many efforts to find them. Recently, with the development of neural network, low feature models outperform these traditional models and need much less feature engineering. Chen et al. [6] used dynamic multi-pooling convolutional neural network (CNN) to extract features.

© Springer Nature Switzerland AG 2018
G. Gan et al. (Eds.): ADMA 2018, LNAI 11323, pp. 189–201, 2018.
https://doi.org/10.1007/978-3-030-05090-0_17

Nguyen et al. [7] proposed a joint event extraction model via recurrent neural networks to predict event triggers and arguments for sentences simultaneously. Tozzo et al. [8] used a recurrent neural network enhanced with structural features and adopted curriculum learning to train models.

In spite of these advances, there are still some issues which are not well solved. Firstly, the same trigger word can lead to different event types in different contexts. For example, in the following two events:

Obama beats McCain.
Tyson beats his opponent.

The event type in the first sentence is *Elect*, while the event type in the second sentence is *Attack*, even though their event triggers are the same word "*beats*".

Secondly, it is hard to exploit deep semantic information to infer relationship between arguments and the event. For example:

In Baghdad, a cameraman died when an American tank fired on the Palestine Hotel.

In this sentence, there are two trigger words which are "*died*" and "*fired*" implying event type *Die* and *Attack* respectively. We can directly infer some simple relations like that the cameramen is the *Victim* of event *Die*, the Palestine Hotel is the *Target* of event *Attack*. Based on the deep semantic information that the event *Die* is the result of event *Attack*, we can conclude that the cameramen is also the victim of event *Attack*.

In the paper, we use deep contextualized word representation to tackle the first problem and use multi-attention layers to address the second problem. In the word representation, the contextual information needs to be encoded into the word to differentiate it in different contexts. In feature-based systems, contextual features are often task specified and require many efforts to find them. In neural network models, Chen et al. [6] utilized context tokens to represent the context information, which doesn't perform well in distinguishing words' meanings in different contexts, because it only takes into account the left and right tokens of the candidate word. We propose a deep contextualized word representation based on language model and can embed more context information into word representation. In deep semantic information inference, the sentence representation needs to be well modeled so that deep semantic information can be extracted. Chen et al. [6] used CNN to represent sentence features, Nguyen et al. [7] used RNN to model relations between elements. Both of them can't deal with longer dependency relations. We put forward a new model with multi-attention layers to address deep semantic information extracting problem. By using attention network, no matter how long the sentence is, the relation between words and event related words can always be well modeled. Our experiments are conducted on ACE2005. It is a widely used dataset for event extraction tasks and the results show that our model outperforms the state-of-the-art models.

To sum up, our main contributions are:

- We use a contextualized word representation based on language models to represent words with context information. It has a better performance in distinguish the different meanings of the same word in different contexts, compared to previous model which just used crafted features or context tokens.

- In the context modeling layer, we present multi-attention layers to extract relations in sentence which can extract deep semantic information. Besides, we propose an attention function suitable for event extraction tasks.
- We conduct experiments on ACE2005 to assess the effectiveness of the deep contextualized word representation and multi-attention layer. The experimental results show that they are very informative and effective in the task of event extraction tasks.

2 Preliminary

2.1 Word Representation

To represent the meaning of a word in a computer, one common solution is to use a resource containing lists of synonym sets and hypernyms so that the relations between words can be built. But there are some problems. For example, "proficient" can be a synonym for "good" in some contexts and can be "bad" in the others. And it's difficult to keep the dictionary up-to-date because it requires much human labor to do this. Besides, there isn't efficient way to calculate the similarity. To address these problems, in traditional NLP, we use one-hot vectors to represent word and the vector dimension equals the number of words in vocabulary. But it still can't model the similarity between words. To encode similarity in the vectors themselves, semantic vector space models are proposed and have been used in a lot of applications, like the information retrieval [9], document classification [10], question answering [11] and parsing [12]. Later, Word2vec [13] was introduced. The core idea is that a word's meaning is given by the words that frequently appear close by. When a word appears in a text, its context is the set of words that appear nearby so we can use the many contexts of a word to build up a representation of it. Recently, deep contextualized word representation was introduced [14]. Each token in the context is represented by a function of entire input sentence. It can be easy implemented into models and is proved to outperform state-of-the-art models in lots of natural language processing tasks. Vectors are trained from de BiLSTM language model on a large text corpus so it can represent deep semantic information because they are a function of all the internal layers of the model.

2.2 Recurrent Neural Network

Recurrent Neural Network [15] was proposed to process variable input length and make cells have memory. Long-short term memory [16] reduced the problem of long term dependency. Recently, RNNs have outperformed state-of-the-art models in many tasks such as language modeling [17], machine translation [18], and speech recognition [19]. The idea behind RNNs is to make use of sequential information. In a traditional neural network, we assume that all inputs (and outputs) are independent of each other. But for many tasks that's a very bad idea. If you want to predict the next word in a sentence you better know which words came before it. RNNs are called recurrent because they perform the same task for every element of a sequence, with the output being depended on the previous computations. Another way to think about RNNs is that they have a "memory" which captures information about what has been calculated so far. Although RNNs improve a lot performance, it takes a lot of time to train and predict because it is time dependent. Therefore, a gated recurrent unit (GRU) was proposed by Chung et al. [20] to make each recurrent unit to adaptively capture dependencies of different time scales. Similar to the LSTM unit, the GRU has gating units that modulate the flow of information inside the unit, however, without having a separate memory cell.

3 System Architecture

Our model can be divided into three parts as shown in Fig. 1. System architecture. The first is a feature rich word encoder which uses deep contextualized word representation and other features to encode words into a vector with semantic and contextual information from raw sentences. The second is a multi-attention layer which uses attention networks to generate sentence representation. And the last part of the model is a classifier, it gives a confidence score on each candidate type.

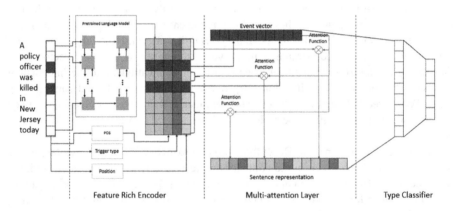

Fig. 1. System architecture

Besides, in the event extraction task, the event argument detection is depended on the result of trigger detection, so there are usually two types of event extraction system, one is to firstly detect the trigger word and then find related arguments, the other is to tag the trigger and its argument simultaneously. We propose to perform them separately because the event type is an important feature in argument role detecting. They share similar architectures so we will first describe argument detection model in detail and then show modifications when it is applied to event trigger detection.

3.1   Feature Rich Word Encoder

To represent the syntax and semantic information of a word, we use deep contextualized word representation that models both complex characteristics of word use (e.g., syntax and semantics), and how these uses vary across linguistic contexts (i.e., to model polysemy) by pretraining a deep language model and linearly combining its hidden states in each layer. Other features like POS, trigger position, and event type are also gathered to form the word representation.

Traditional word representation often uses a fixed pretrained word embedding such as Glove or Word2Vec along with a CNN or RNN character embedding. However, this can encode little context information into word. So we replace pretrained word embedding with a pretrained language model to generate word representation so that context information can be encoded into it. Language models [21, 22] is used to model probability distribution over sequences of words. Given a sequence of N words $t_1, \ldots, t_i, \ldots, t_N$, it assigns a probability $P(t_1, \ldots, t_N)$. According to *Bayes' theorem*, we have:

$$P(t_1, \ldots, t_N) = \prod_{i=1}^{N} P(t_i | t_1, \ldots, t_{i-1})$$

In a neural language model, $P(t_{i+1} | t_1, \ldots, t_i)$ is represented by a RNN cell at time step i which takes t_i and last hidden state h_{i-1} as input and outputs the current hidden state h_i then this hidden state will be translated to a RNN output at time step i. The probability distribution is calculated by taking softmax over RNN output. The loss function is the cross entropy. The RNN in language model can be replaced by LSTM, GRU or a bidirectional one and multiple layers can also be added to it. The higher-level hidden state usually captures context information and the low-level hidden state usually captures syntax information. In our model, we take advantage of both of them and use their linear combination to represent contextualized word $w_{e,i}$ at position i:

$$w_{e,i} = \sum_{l=1}^{L} \gamma_l h_{i,l}$$

Where L is the number of layers, $\gamma_l \in \mathbb{R}$ is the weight of each layer learned by neural network and $h_{i,l} \in \mathbb{R}^{d_e}$ is the hidden state of each layer, d_e is the hidden size.

In addition, we consider some other low features. The part of speech is also very important when identifying role of argument. For example, the victim of the Attack event is usually a noun. So, we use POS information when considering word embedding. We give every POS tag a unique id and note it as a one-hot vector $e_{p,i}$ for word at position i. And the event type of trigger word heavily influences the detection of role argument. So, it is also included in the features. We note it as a one-hot vector $e_{d,i}$. The distance from current word to trigger word influences words' role of argument since the longer the distance is, the weaker the relation is. So, we add it as a feature to indicate spatial relation between current word and trigger word and note it as $e_{s,i}$. For

$e_{p,i}$, $e_{d,i}$ and $e_{s,i}$, they will be mapped into vectors $w_{p,i}$, $w_{d,i}$, $w_{s,i}$ with corresponding dimensions d_p, d_s and d_s by using mapping matrixes M_p, M_d and M_s.

$$w_{p,i} = M_p e_{p,i}$$

$$w_{d,i} = M_d e_{d,i}$$

$$w_{s,i} = M_s e_{s,i}$$

After applying mapping, we get embeddings $w_{p,i}$, $w_{d,i}$, $w_{s,i}$ for those features.

Finally, our feature rich word encoder concatenates all these embeddings together and mapping it to a d_e dimension word representation w_i using matrix M_w.

$$w_i = M_w \left[w_{e,i}; w_{p,i}; w_{d,i}; w_{s,i} \right]$$

Where w_i is the word embedding of i-th word. We use a bidirectional GRU extract higher contextual information over feature rich word embedding. Let w_i be the i-th word representation, the i-th output of Bi-GRU is obtained by:

$$r_i = \sigma(W_r w_i + U_r h_{i-1} + b_r)$$

$$z_i = \sigma(W_z w_i + U_z h_{i-1} + b_z)$$

$$\widehat{h}_i = \tanh(W w_t + r_i U h_{i-1} + b)$$

$$h_i = (1 - z_i)\widehat{h}_i + z_i h_{i-1}$$

Where σ is a sigmoid function, W, U and b are parameters learnt by model, h_i, h_{i-1} are output vectors at i and $i - 1$, r_i is the reset gate vector and z_i is the update gate vector.

3.2   Multi-attention Layer

In previous works [6, 23–25], CNNs were directly used to extract features and then do prediction based on them. However, different triggers may cause different roles of arguments. So, the sentence features are highly related to the trigger word and candidate arguments. It's important to use the information while computing sentence representation. In our word, we use attention neural network instead of simple CNNs to extract sentence features.

Attention mechanisms [26, 27] were firstly applied in machine translation and then outperformed state-of-art models in many NLP tasks. It is usually applied to encode a sequence of vectors into a fixed length sentence representation. However, in our work, to address the problem that one sentence contains more than one event and the same argument may play different roles according to different trigger words. It's necessary to emphasize the change of the trigger word. In our work, we propose a multi-attention mechanism to explicitly encode variation of trigger word into sentence representation.

Let o_i be the output of word encoder for word w_i, now we can use these word embeddings to generate the event vector q, which are composed of extracted event trigger and candidate argument, and a sentence representation u.

For the event vector, Chen et al. [6] used the word embeddings of candidate words (candidate trigger and candidate argument) and the context tokens (left and right tokens of the candidate words). In our work, since the contextual information has already been encoded into o_i by Bi-GRU, so we only use the word representation of candidate words. Let i_t be the position of trigger word and i_c be the position of candidate argument.

$$q = [o_{i_t}; o_{i_c}]$$

The sentence representation u is computed by concatenating three sub representations which are split by the trigger word and the candidate argument. We can suppose that $i_t < i_c$:

$$u = [att(o, 1, i_t); att(o, i_t + 1, i_c); att(o, i_c + 1, n)]$$

$$att(o, a, b) = \sum_a^b \alpha_i o_i$$

The weight α_i of each word is computed by:

$$\alpha_i = \frac{\exp(e_i)}{\sum_j \exp(e_j)}$$

Where

$$e_j = a(o_j, q)$$

Is an alignment model which scores how well the event vector and the word at position j match $a(x, y)$ is an attention scoring function and we propose the following formulations:

1. Additive attention (MLP) [27]: $s^T \tanh(W_1 x + W_2 y)$.
2. Multiplicative attention: $x^T U^T V y$.
3. Scaled multiplicative attention [28]: $\frac{1}{\sqrt{k}} x^T U^T$, where k is the attention hidden size.
4. Scaled multiplicative with nonlinearity: $\frac{1}{\sqrt{k}} f(Ux)^T f(Vy)$.
5. Symmetric multiplicative attention: $x^T U^T D U y$, where D is diagonal.
6. Symmetric multiplicative with nonlinearity: $f(Ux)^T D f(Uy)$.

We propose to use the scaled multiplicative with nonlinearity scoring function because it takes into account the hidden size and the nonlinearity makes function more expressive. Our assumption is proved in the Experiments part.

3.3   Type Classifier

The input v of classifier is obtained by concatenating event vector and sentence representation.

$$v = [q; u]$$

$v \in \mathbb{R}^{5d_o}$ where d_o is the output dimension of word encoder and q is event vector, u is sentence representation. The classifier can be formulated by:

$$\hat{y} = W_s v + b_s$$

$\hat{y} \in \mathbb{R}^m, W_s \in \mathbb{R}^{m \times 5d_o}, b_s \in \mathbb{R}^m$, where m is the number of argument role including "None role", W_s and b_s are parameters learnt by model and $\hat{y}$ is the output of classifier, it gives a confidence score to each argument role.

3.4   Training

Let Θ be the set of parameters used in out model. We use the negative log probabilities of the classifier output as our loss function:

$$L(\Theta) = -\frac{1}{n} \sum_{i=1}^{n} y_i \log(\hat{y}_i)$$

Where y_i equals 1 when it denotes the true argument role and 0 if not.

3.5   Event Trigger Detection

Event trigger detection has similar architecture to argument role detection except that the sentence is only split by trigger word so it's a 2-attention layer instead of a 3-attention layer. And the event vector will only be composed of trigger embedding vector. The rest of the model is the same as argument role detection's.

4   Experiments

4.1   Dataset

ACE2005 is a widely used corpus in event extraction tasks. The ACE 2005 corpus includes 6 different domains:broadcast conversation (bc), broadcast news (bn), telephone conversation (cts), newswire (nw), usenet (un) and webblogs (wl). We also use this dataset and follow the same data split schema as previous works [1, 4–6, 29] to make it compatible. We randomly selected 30 articles from different genres as the development set, and subsequently conducted a blind test on a separate set of 40 ACE 2005 newswire documents. We used the remaining 529 articles as our training set.

4.2   Model Settings

We use 50 dimensions for POS, event type and position embeddings, 1024 dimensions for deep contextualized word representation, 400 dimensions for the GRU hidden layer. The dropout is set to 0.4, and the batch size is set to 50.

The implementation of model is based on pytorch. We use standfordNLP to get POS and dependency of words and for deep contextualized word representation, we use AllenNLP's pre-trained ELMo representations. During training, Adam optimizer is adapted to update model parameters.

4.3   Results

We compared our model with current state-of-art models. Li's baseline [1] use lexical features, basic features and syntactic features to implement a feature-based system. Liao's cross-event [4] uses document level information to improve performance. Hong's cross-entity [3] uses cross-entity inference to improve event extraction model and achieves the best-reported feature-based system score in the literature on using gold standards argument candidates. Li's structure [1] improved performance by using structure prediction and is the best-reported structure-based system. DMCNN [6] uses multi-pooling convolutional neural networks. JRNN [7] did event extraction in a joint framework with bidirectional recurrent neural networks. In our results, P, R means precision and recall respectively, while F1 score (F) is calculated by:

$$F = 2\frac{PR}{P+R}$$

Table 1 shows the performance of state-of-art models and our model. From the results we can see that our model achieves the best performance on both trigger classification and argument role classification. For the F1 score, our model outperformances the state-of-art model by 1.9% on trigger classification and by 1.2% on argument role classification which shows the effectiveness of our model. Compared to JRNN which also use recurrent neural network, our model improving a lot precision and maintains the recall in the same time. It can be interpreted that our rich feature word encoder represents better words and deep contextual information helps improve the precision. Besides, the attention mechanism does well in modeling relations between words. For example, in the sentence:

I would think, probably, various facilities as well as major government buildings that they need to control to set up da new government

Start-Org was recognized as event type by both JRNN and our model. But JRNN failed to extract "government" as an argument, which, by the help of attention network, was successfully found out by our model.

Table 1. Overall performance

Model	Trigger identification (%)			Trigger identification + classification (%)			Argument identification (%)			Argument role (%)		
	P	R	F	P	R	F	P	R	F	P	R	F
Li's baseline	76.2	60.5	67.4	74.5	59.1	65.9	74.1	37.4	49.7	65.4	33.1	43.9
Liao's cross-event	N/A			68.7	68.9	68.8	50.9	49.7	50.3	45.1	44.1	44.6
Hong's cross-entity	N/A			72.9	64.3	68.3	53.4	52.9	53.1	51.6	45.5	48.3
Li's structure	76.9	65.0	70.4	73.7	62.3	67.5	69.8	47.9	56.8	64.7	44.4	52.7
DMCNN	80.4	67.7	73.5	75.6	63.6	69.1	68.8	51.9	59.1	62.2	46.9	53.5
JRNN	68.5	75.7	71.9	66.0	73.0	69.3	61.4	64.2	62.8	54.2	56.7	55.4
Our model	73.6	76.2	**74.9**	70.3	72.2	**71.2**	64.7	65.0	**64.8**	57.4	55.8	**56.6**

4.4 Attention Function Comparison

We train models with different attention functions to justify our choice of the scaled multiplicative with nonlinearity scoring function and the results are shown in Table 2.

From the results we can see that the model which takes scaled multiplicative with ReLU as attention function achieves the best performance. And non-linear functions are better than linear functions, multiplicative functions are better than additive functions.

Table 2. Attention function comparison

Model	Trigger identification (%)			Trigger identification + classification (%)			Argument identification (%)			Argument role (%)		
	P	R	F	P	R	F	P	R	F	P	R	F
Additive	72.1	75.4	73.7	68.9	71.3	70.1	64.1	64.5	64.3	56.6	55.2	55.9
Multiplicative	73.2	75.7	74.4	69.5	71.9	70.7	64.2	64.7	64.4	57.1	55.3	56.2
Scaled multi	73.3	75.7	74.5	69.8	71.8	70.8	**64.8**	64.9	64.8	57.2	55.5	56.3
Scaled multi (ReLU)	73.6	**76.2**	**74.9**	70.3	**72.2**	**71.2**	64.7	65.0	**64.8**	**57.4**	**55.8**	**56.6**
Sym multi	73.5	74.8	74.1	70.6	71.3	70.9	63.9	65.2	64.5	56.9	55.6	56.2
Sym multi (ReLU)	**73.9**	75.2	74.5	**70.8**	71.6	71.2	64.2	**65.4**	64.8	57.2	55.7	56.4

4.5 Ablations

To show contributions of each part in networks, we do ablation test and the results are shown in Table 3.

Deep contextualized word representation improves model's F1 score by 1.9% in trigger identification and by 1.6% in argument role identification, it improves a lot the precision and also contributes in recall. The attention networks improve model's F1 score by 0.8% in trigger identification and by 0.9% in argument role identification. Results below proves the effectiveness of word representation and attention networks.

Table 3. Ablation

Model	Trigger identification (%)			Trigger identification + classification (%)			Argument identification (%)			Argument role (%)		
	P	R	F	P	R	F	P	R	F	P	R	F
Our model	73.6	76.2	74.9	70.3	72.2	71.2	64.7	65.0	64.8	57.4	55.8	56.6
- Contextualized representation	72.3	75.4	73.8	69.2	71.1	70.1	62.2	63.8	63.0	55.6	54.4	55.0
- Attention	72.9	75.8	74.3	69.3	71.6	70.4	63.5	63.9	63.7	56.6	54.9	55.7

5 Related Work

Earlier research of event extraction concentrates on pattern-based methods which are formed by predicates, event triggers and other important contexts [30]. Later, feature-based approach is adopted to this task and the performance relies heavily on feature sets [1, 3, 5, 31–35]. Recently, neural networks have been introduced into event extraction. Nguyen and Grishman [29] use CNNS to study domain adaptation and event extraction. Chen et al. [6] uses multi-pooling CNNs and use a pipelined framework. Nguyen et al. [7] uses RNN to do joint event extraction. None of them uses multi-attention layer and deep contextualized word representation in the model.

6 Conclusion

We present a model with deep contextualized word representation and multi-attention layers to resolve the lack of context information problem and extract deep semantic information. Our evaluation of the proposed model demonstrates the effectiveness of our method. In the future, we plan to explore how active learning can help us reduce the demand of training dataset.

Acknowledgement. This work was supported in part by the National Natural Science Foundation of China (Grand Nos. U1636211, 61672081, 61370126), Beijing Advanced Innovation Center for Imaging Technology (No. BAICIT-2016001), and the National Key R&D Program of China under Grant 2016QY04W0802.

References

1. Li, Q., Ji, H., Huang, L.: Joint event extraction via structured prediction with global features. In: Proceedings of the 51st Annual Meeting of the Association for Computational Linguistics, pp. 73–82 (2013)
2. Li, Q., Ji, H., Hong, Y., Li, S.: Constructing information networks using one single model. In: Proceedings of the 2014 Conference on Empirical Methods in Natural Language Processing (EMNLP), pp. 1846–1851 (2014)
3. Hong, Y., Zhang, J., Ma, B., Yao, J., Zhou, G., Zhu, Q.: Using cross-entity inference to improve event extraction. In: Proceedings of the 49th Annual Meeting of the Association for Computational Linguistics: Human Language Technologies, vol. 1, pp. 1127–1136. Association for Computational Linguistics, June 2011
4. Liao, S., Grishman, R.: Using document level cross-event inference to improve event extraction. In: Proceedings of the 48th Annual Meeting of the Association for Computational Linguistics, pp. 789–797. Association for Computational Linguistics, July 2010
5. Ji, H., Grishman, R.: Refining event extraction through cross-document inference. In: Proceedings of ACL 2008: HLT, pp. 254–262 (2008)
6. Chen, Y., Xu, L., Liu, K., Zeng, D., Zhao, J.: Event extraction via dynamic multi-pooling convolutional neural networks. In: Proceedings of the 53rd Annual Meeting of the Association for Computational Linguistics and the 7th International Joint Conference on Natural Language Processing, pp. 167–176 (2015)
7. Nguyen, T.H., Cho, K., Grishman, R.: Joint event extraction via recurrent neural networks. In: Proceedings of the 2016 Conference of the North American Chapter of the Association for Computational Linguistics: Human Language Technologies, pp. 300–309 (2016)
8. Tozzo, A., Jovanovic, D., Amer, M.: Neural event extraction from movies description. In: Proceedings of the First Workshop on Storytelling, pp. 60–66 (2018)
9. Larson, R.R.: Introduction to information retrieval. J. Am. Soc. Inf. Sci. Technol. **61**(4), 852–853 (2010)
10. Sebastiani, F.: Machine learning in automated text categorization. ACM Comput. Surv. (CSUR) **34**(1), 1–47 (2002)
11. Tellex, S., Katz, B., Lin, J., Fernandes, A., Marton, G.: Quantitative evaluation of passage retrieval algorithms for question answering. In: Proceedings of the 26th Annual International ACM SIGIR Conference on Research and Development in Information Retrieval, pp. 41–47. ACM, July 2003
12. Socher, R., Bauer, J., Manning, C.D.: Parsing with compositional vector grammars. In: Proceedings of the 51st Annual Meeting of the Association for Computational Linguistics, pp. 455–465 (2013)
13. Mikolov, T., Sutskever, I., Chen, K., Corrado, G.S., Dean, J.: Distributed representations of words and phrases and their compositionality. In: Advances in Neural Information Processing Systems, pp. 3111–3119 (2013)
14. Peters, M.E., et al.: Deep contextualized word representations. arXiv.org. arXiv:1802.05365 (2018)
15. Goller, C., Kuchler, A.: Learning task-dependent distributed representations by backprop-agation through structure. In: 1996 IEEE International Conference on Neural Networks, pp. 347–352. IEEE, June 1996
16. Hochreiter, S., Schmidhuber, J.: Long short-term memory. Neural Comput. **9**(8), 1735–1780 (1997)
17. Mikolov, T., Kombrink, S., Burget, L., Černocký, J., Khudanpur, S.: Extensions of recurrent neural network language model. In: 2011 IEEE International Conference on Acoustics, Speech and Signal Processing (ICASSP), pp. 5528–5531. IEEE, May 2011

18. Cho, K., et al.: Learning phrase representations using RNN encoder-decoder for statistical machine translation. arXiv.org. arXiv:1406.1078 (2014)
19. Graves, A., Mohamed, A.R., Hinton, G.: Speech recognition with deep recurrent neural networks. In: 2013 IEEE International Conference on Acoustics, speech and signal processing (ICASSP), pp. 6645–6649. IEEE, May 2013
20. Chung, J., Gulcehre, C., Cho, K., Bengio, Y.: Empirical evaluation of gated recurrent neural networks on sequence modeling. arXiv.org. arXiv:1412.3555 (2014)
21. Jozefowicz, R., Vinyals, O., Schuster, M., Shazeer, N., Wu, Y.: Exploring the limits of language modeling. arXiv.org. arXiv:1602.02410 (2016)
22. Melis, G., Dyer, C., Blunsom, P.: On the state of the art of evaluation in neural language models. arXiv.org. arXiv:1707.05589 (2017)
23. Collobert, R., Weston, J., Bottou, L., Karlen, M., Kavukcuoglu, K., Kuksa, P.: Natural language processing (almost) from scratch. J. Mach. Learn. Res. 12(Aug), 2493–2537 (2011)
24. Kim, Y.: Convolutional neural networks for sentence classification. arXiv.org. arXiv:1408.5882 (2014)
25. Zeng, M., et al.: Convolutional neural networks for human activity recognition using mobile sensors. In: 2014 6th International Conference on Mobile Computing, Applications and Services (MobiCASE), pp. 197–205. IEEE, November 2014
26. Bahdanau, D., Cho, K., Bengio, Y.: Neural machine translation by jointly learning to align and translate. arXiv.org. arXiv:1409.0473 (2014)
27. Chorowski, J.K., Bahdanau, D., Serdyuk, D., Cho, K., Bengio, Y.: Attention-based models for speech recognition. In: Advances in Neural Information Processing Systems, pp. 577–585 (2015)
28. Vaswani, A., et al.: Attention is all you need. In: Advances in Neural Information Processing Systems, pp. 6000–6010 (2017)
29. Nguyen, T.H., Grishman, R.: Event detection and domain adaptation with convolutional neural networks. In: Proceedings of the 53rd Annual Meeting of the Association for Computational Linguistics and the 7th International Joint Conference on Natural Language Processing, pp. 365–371 (2015)
30. Grishman, R., Westbrook, D., Meyers, A.: NYU's English ACE 2005 system description, vol. 5. ACE, (2005)
31. Ahn, D.: The stages of event extraction. In: Proceedings of the Workshop on Annotating and Reasoning about Time and Events, pp. 1–8. Association for Computational Linguistics, July 2006
32. Gupta, P., Ji, H.: Predicting unknown time arguments based on cross-event propagation. In: Proceedings of the ACL-IJCNLP 2009 Conference Short Papers, pp. 369–372. Association for Computational Linguistics, August 2009
33. Patwardhan, S., Riloff, E.: A unified model of phrasal and sentential evidence for information extraction. In: Proceedings of the 2009 Conference on Empirical Methods in Natural Language Processing, pp. 151–160. Association for Computational Linguistics, August 2009
34. Liao, S., Grishman, R.: Acquiring topic features to improve event extraction: in pre-selected and balanced collections. In: Proceedings of the International Conference Recent Advances in Natural Language Processing 2011, pp. 9–16 (2011)
35. McClosky, D., Surdeanu, M., Manning, C.D.: Event extraction as dependency parsing. In: Proceedings of the 49th Annual Meeting of the Association for Computational Linguistics: Human Language Technologies, pp. 1626–1635. Association for Computational Linguistics, June 2011

A Novel Unsupervised Time Series Discord Detection Algorithm in Aircraft Engine Gearbox

Zhongyu Wang, Dechang Pi$^{(\boxtimes)}$, and Ya Gao

College of Computer Science and Technology,
Nanjing University of Aeronautics and Astronautics, 29 Jiangjun Road,
Nanjing 211106, Jiangsu, People's Republic of China
1109763309@qq.com, 703165600@qq.com, nuaacs@126.com

Abstract. Aircraft engine discord detection is an important way to ensure flight safety. Unsupervised algorithms will be relatively effective due to the lack of models and tagged discord data. For the aircraft time series data collected from sensors, this paper proposes a Trend Featured Dynamic Time Wrapping for J Distance Discord Discovery algorithm based on the J-Distance Discord anomaly definition which combined with the trend information of the data. The experiments on aircraft engine gearbox data show that the TFDTW for JDD Discovery algorithm is better than the normal J-Distance Discord Discovery algorithm and also better than some other classic time series data discord detection algorithms.

Keywords: Time series data · Unsupervised learning · Abnormal detection
Aircraft engine

1 Introduction

In the modern aviation industry, the safety of aircraft equipment is crucial because it has relation with the crews' life and property. In many aircraft equipment, the engine is one of the most important equipment of the aircraft. The failure of an aircraft engine can cause catastrophic consequences. In order to avoid this situation, it is necessary to accurately detect the anomaly of the aircraft engine, and send warning information before fatal engine failure.

In general, many sensors are installed in the engine components to detect the operating conditions of each component. An anomaly detection method for the engine is analyzing the data collected by the sensors. The data collected by the sensor is in the form of time series data. How to detect the anomaly of the time series data becomes key factor of detecting the sensor's failure [1].

An inherent feature of time series is that there are mutual dependencies between adjacent observations, and this mutual dependence often contains a large amount of information of observed things or phenomena in a specific environment or at a specific time. Research and analysis of this mutual dependence have great practical value. Time series analysis is an analytical technique that studies the interdependence of such observations. Among all of them, the analysis of so called 'discords' in time series data

© Springer Nature Switzerland AG 2018
G. Gan et al. (Eds.): ADMA 2018, LNAI 11323, pp. 202–210, 2018.
https://doi.org/10.1007/978-3-030-05090-0_18

is an important part of time series analysis. Discords are defined as specific patterns in data that are inconsistent with pre-defined normal patterns. A time series discord is a sub-sequence of time that is inconsistent with the normal pattern in the context of the time series. Existing discord detection methods can be divided into two methods from the perspective of data use: 1. Model-driven approach; 2. Data-driven approach.

The input of the model-driven discord detection method is a training data set and a test data set, the output is a model established from the training data set, and anomalies in the test data. The use of the model is divided into two phases: training and testing. The data sets used in the two phases of training and testing are often different. The model-driven method is either the Supervised method or the Semi-supervised method. Common model-driven discord detection methods include artificial neural network ANN [2], SVM [3], Bayesian network [4], rule-based detection methods, and so on the model-driven methods have a faster detection process, but at the same time, they also bring many disadvantages to the detection method. The quality of the model is the key of the effectiveness of the detection method. However, for aircraft engine data, due to the lack of prior knowledge, it is difficult to find an efficient model for discord detection. Besides, in aircraft engine data, abnormal data are very scarce compared to normal data, and it is difficult for supervised methods to deal with such data imbalances.

The input of data-driven discord detection method is a test data set, the output is the most unusual data instance in the test data set and their anomaly degree, there is no training process. The principle is based on key assumptions:

1. Aircraft engine operation is normal for most of the time, and abnormal time accounts for a very small part;
2. The data in the abnormal time will be different from the normal data.

With the continuous development of sensor technology, the types of time series in monitoring data are continuously increasing. It is difficult to find a suitable model in a short time. Therefore, model-driven abnormal detection methods cannot adapt to the requirements of abnormal detection. Therefore, this paper uses data-driven unsupervised discord detection method to abnormally detect aircraft engine time series data.

In the study of unsupervised sequential data discord detection, Keogh proposed the definition of discord and first gave the naïve algorithm. [5] Subsequently, Keogh proposed the HOTSAX algorithm and achieved higher efficiency and accuracy. [6] However, Keogh's discord definition relies on the distance of the discord sequence from its normal neighboring sequence. However, in the industrial field, machine anomalies are likely to occur continuously in a short period of time. The definition will have the twin monster Freak Twins problem. In order to solve this problem, Huang [7] proposed a new exception definition JDD, and proposed the JDD Discovery algorithm for anomaly detection. However, JDD Discovery calculates distances between subsequences based on the Euclidean distance. It is not suitable for aircraft engines with large noise, which will result in a decrease in accuracy. However, if SAX is used for data compression, data details will be lost. Therefore, this paper proposed the TFDTW for JDD Discovery algorithm based on the Trend Feature of time series data and dynamic time warping (DTW) algorithm. We made detection on the data of aircraft engine gearbox, experiments' result shows that our algorithm achieved higher recognition accuracy compared with the existing ones.

The following organizational structure of this paper is as follows. Section 2 proposes the TFDTW for JDD Discovery algorithm and Sect. 3 describes the experimental data. The experimental results and their analysis are in Sect. 4. Finally, Sect. 5 summarizes the experiments' results.

2 Algorithm

2.1 Definition of Discord Data in Time Series JDD

Time series data is a specific type of data. Therefore, before defining the abnormal time series data, we give definitions of time series and their subsequences first, as Definitions 1 and 2.

Definition 1 Time Series. The time series is an ordered combination of m real numbers, expressed as $T = (t_1, t_2, \ldots, t_m)$, $t_i \in R$.

Definition 2 Subsequence. Given a time series T of length m, the subsequence $C_{p,n}$ is a subset of T obtained from the position p consecutively sampled n times, expressed as $C_{p,n} = T[p : n]$.

In the unsupervised case, there has not enough abnormal data to train the model to define the anomaly. Therefore, the anomaly can only be defined by the difference between normal and abnormal. In this case, a clear definition of the exception needs to be made. Keogh proposed an exception definition, as defined in Definition 3.

Definition 3 Time Series Discord. Given a time series T, C_d is a length n subsequence of T whose starting point is d. If C_d has the largest nearest neighbor distance, then C_d is a discord in T. i.e. for any subsequence C_o of T, $|d - o| \geq n$, we have nearest neighbor distance $nnDist(C_d) > nnDist(C_o)$.

Definition 3 quantifies the degree of abnormality of a subsequence as the distance between the sequence and its neighboring sequences, providing the basis for detecting abnormality of the time series data. On this basis, Keogh proposed the naive algorithm and the HOTSAX algorithm. However, sometimes, the same kind of system anomaly may appear several times in a short time. This phenomenon is known as the "Twin Freak" problem. The phenomenon is especially common in industrial data. When an exception occurs in a component, it causes a continuous sequence of abnormalities, as shown in Fig. 1.

The Twin Freak problem is indefinable by Definition 3, because its abnormal subsequence appears nearby and thus cannot be the subsequence that is most dissimilar to other subsequences, and thus cannot be recognized normally.

In Definition 3, the abnormality of a time series depends on the distance between it and non-overlapping nearest neighbors. When an anomaly occurs multiple times in a time series, these anomalies are the nearest neighbors to each other. The distance

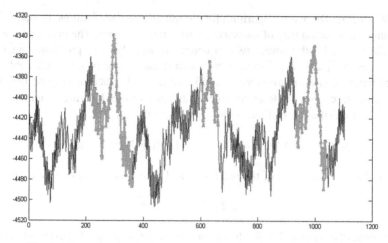

Fig. 1. Example of twin freak problem in aircraft engine gearbox data

between shape-like anomalies may be small, which makes them impossible to capture by the original definition of time series discord. The new discord definition should be able to perceive the presence of multiple similar discords. To solve this problem, Huang proposed J-distance discord. The J-distance is the distance between a time subsequence and its Jth non-overlapping nearest neighbor. A normal subsequence has at least J neighbors that are very similar to it. The discord with J distance as the criterion is the subsequence which has the largest J distance in time series.

Definition 4 (J-distance discord). Given a time series T, C_d is a subsequence of T of length n whose starting point is d. If C_d is T's J-distance discord, C_d has the longest J-distance. i.e., for any subsequence C_o of T, $|d - o| \geq n$, we have nearest neighbor distance $nnDist(C_d) > nnDist(C_o)$.

From Definition 4, it can be seen that the definition of the original time series discord is a special case of the new discord definition, that is, the case of J = 1. According to this definition, Huang proposed the J-Distance Discord Discovery algorithm, which achieved good results in ECG discord recognition task. However, ECG data is usually irregular, and there are few problems such as offset. In aircraft engine data, the data offset is more serious, so the JDD Discovery algorithm does not perform well.

2.2 Trend Featured Dynamic Time Wrapping for JDD Discovery

The calculation of JDD depends on the calculation of the time series distance. Accurate time series calculation method will contribute to the accuracy of the algorithm. In the time series data provided by the industrial sensors, the lengths of the two time series that need to calculate the similarity may not be equal, and the offset situation may occur. Therefore, based on the JDD Discovery algorithm, this paper proposes the

TFDTW for JDD Discovery algorithm for aircraft engine data features. Dynamic Time Wrapping is a common way of measuring time series distances. The principle is to find the smallest curved path between two sequences through dynamic programming. Given two time series T_1 and T_2. Construct an $n \times m$ distance matrix, every element of the distance matrix can be represented as $d(x1_i, x2_j)$, the distance usually be defined as Euclidean distance. Set the total cumulative distance of the dynamic planning path to $g(i, j)$, which calculation method is shown as formula (1).

$$g(i, j) = d(x1_i, x2_j) + \min\{g(i, j-1), g(i-1, j-1), g(i-1, j)\} \qquad (1)$$

Then the DTW distance of time series T_1 and T_2 can be calculated by formula (2).

$$dist_{dtw}(T_1, T_2) = g(n, m) \qquad (2)$$

However, the simple DTW algorithm only considers the distance between the values of each time point in the time series, and does not take the change trend of the time series into consideration. This may cause deviations in the time series distance metrics. At the same time, the dynamic programming algorithm will cause the algorithm to have high time complexity, and the speed cannot be tolerated in the case of a long time series. Therefore, this paper proposes to extract the features of the original data into the trend feature sequence by using trend features, and to calculate the distance by the dynamic time warping algorithm of the trend feature sequence. The distance thus obtained will be more accurate.

Given time series $T = (x_1, x_2, \ldots x_n)$, calculate its derivative sequence $\hat{T} = (\hat{x}_1, \hat{x}_2, \ldots \hat{x}_n)$, the method is shown as formula (3).

$$\hat{x}_h = \begin{cases} x_2 - x_1 & h = 1 \\ \frac{1}{2}(x_{h+1} - x_{h-1}) & h \in [2, 3, \ldots n-1] \\ x_n - x_{n-1} & h \in n \end{cases} \qquad (3)$$

Then the trend distance between the two sequence numerical points is $\bar{d}(x1_i, x2_j)$, the calculation formula is shown as formula (4).

$$\bar{d}(x1_i, x2_j) = \sqrt{\mu \cdot (\hat{x}1_i - \hat{x}2_j)^2 + \lambda(x1_i - x2_j)^2} \qquad (4)$$

Based on the trend distance combined with formula (1) and (2), the trended DTW distance $dist_{TFDTW}(T_i, T_j)$ for any two time series data can be calculated.

We combine the TFDTW algorithm and the JDD Discovery algorithm into a TFDTW for JDD Discovery algorithm, as shown in Algorithm 1.

Algorithm 1: Trend Featured DTW for J-Distance Discord Discovery

---

Input: Time series T, length of T m, length of sliding window n, J

Output: Anomaly start index and the anomaly distance

01. anomalyIndex = 0, anomalyDistance = maxInt
02. *For i in 0 to m-n //start from first sliding window*
03.   *Jdist(C_i) = maxInt //define the J distance of this subsequence as infinity*
04.   *For j in 0 to m-n ordered by random //randomly generate the subsequence*
05.   *If not abs(i-j) < n then //ensure not overlapping*
06.   *distance = distTFDTW(C_i,C_j) //calculate the TFDTW distance*
07.   *Update Jdist(C_i) with distance*
08.   *End if*
09.   *End for*
10.   *If anomalyDistance< Jdist(C_i) then*
11.   *anomalyDistance = Jdist(C_i)*
12.   *anomalyIndex = i*
13.   *End if*
14. *End for*

According to Algorithm 1, we can get the starting position of a certain time series' discord and the distance between the discord sequence and its surrounding sequence. This distance reflects the degree of discord. Comparing the obtained discord sequence with the actual discord sequence, the actual effect of the algorithm can be obtained.

3   Data Set and Performance Evaluation

3.1   Time Series Dataset of Aircraft Engine Gearbox

We use aeroengine gearbox data to verify the accuracy of the Trend Featured Dynamic Time Wrapping for JDD Discovery algorithm. The gearbox data is collected by the gearbox bearing sensor and filtered by the filter. The data format is shown in Table 1.

Table 1. Time series discord detect score

Failure type	Length of time series	Number of time series	Number of failure time period
Flaking failures	120 min	1500	721
Tooth broken failure	120 min	1500	971
Wear-out failure	120 min	1500	843
Tooth broken failure + wear-out failure	30 min	800	454
Flaking failures + wear-out failure	30 min	800	397

The data contains three types of single faults: pitting, broken teeth, and wear. It also includes two types of mixed faults, broken teeth + wear and pitting + wear. At the same time, each time series records the value of the filtered sensor signal per second and the time interval of the fault occurrence interval. An example of the data is shown in Fig. 2, where the red part represents the discord interval.

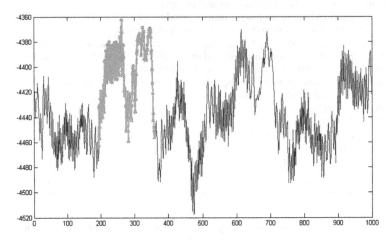

Fig. 2. Example of flaking failures in aircraft gearbox data (Color figure online)

3.2   Performance Evaluation

In this paper, both the detection results and the actual results are time intervals, and it is impossible to use a simple absolute accuracy rate to measure the detection results. Therefore, the time period of the detection results needs to be compared with the actual time period. The more common parts there are, the better the algorithm is detected.

Let the detected discord result set is $S_d = \{T_{d1}, T_{d2}, \ldots T_{dn}\}$, the real discord interval set is $S_r = \{T_{r1}, T_{r2}, \ldots T_{rm}\}$. Define the repetition rate of the two discord intervals as $Overlap(T_i, T_j)$, which is the percentage of overlap between two sections over their total length, the calculation formula is shown as Formula 5.

$$Overlap(T_i, T_j) = \frac{Overlaplen(T_i, T_j)}{len(T_i) + len(T_j)} \tag{5}$$

According to the definition of the overlap, the discord detection score can be definition as Formula 6.

$$score = \sum_{i=1}^{n} \max(Overlap(T_i, T_j), T_i \in S_d, T_j \in S_r) \tag{6}$$

It can be seen from Formula 6 that the higher the score is, the more coincident the detected discord interval and the actual interval are, and the better the detection result is.

4   Experimental Results and Analysis

This section uses the proposed TFDTW for JDD Discovery algorithm to detect the discord occurred in the aircraft engine gearbox.

Software environment: Intel(R) Core(TM) i5-3470 3.2 GHZ CPU, 8 G main memory, 1T hard disk and Microsoft Windows7 system. Experiment platform: Matlab. Language: Matlab.

The result of experiment is shown in Table 2.

Table 2. Time series discord detect score

Dataset	Naïve [5]	HOTSAX [6]	APCA-JDD discovery	SAX-JDD discovery [7]	TFDTW JDD discovery
Flaking failures	372.3	658.4	1132.3	1382.1	**1401.9**
Tooth broken failure	225.9	442.3	1377.2	1287.5	**1463.3**
Wear-out failure	311.1	398.8	2031.5	**2534.9**	2510.0
Tooth broken failure + wear-out failure	190.3	215.2	900.7	1107.2	**1329.1**
Flaking failures + wear-out failure	121.2	154.6	392.8	566.3	**782.5**

From Table 2, it can be seen that the JDD Discovery algorithm is significantly better than the simple definition of naïve and HOTSAX algorithms, while the TFDTW JDD Discovery algorithm is generally better than the APCA and SAX-based JDD Discovery algorithm.

5   Conclusion

This paper presents the Trend Feature based Dynamic Time Wrapping algorithm, which optimizes the original JDD Discovery algorithm from the perspective of similarity measure. The TFDTW algorithm considers the global numerical similarity and local trend similarity between trajectories at the same time, which makes the similarity measure of time series become more accurate and improves the effect of JDD Discovery discord detection algorithm. Experiments proved that the algorithm is more accurate than the naïve and HOTSAX algorithms for discord intervals, and it is also superior to JDD Discovery based on APCA and SAX, we come to the conclusion that our algorithm has higher accuracy and practicability.

Acknowledgements. The research work is supported by National Natural Science Foundation of China (U1433116), the Fundamental Research Funds for the Central Universities (NP2017208) and Fundation of Graduate Innovation Center in NUAA (kfjj20171603).

References

1. Wang, T., Yu, J., Siegel, D., et al.: A similarity-based prognostics approach for remaining useful life estimation of engineered systems. In: International Conference on Prognostics and Health Management, PHM 2008, pp. 1–6. IEEE (2008)
2. Ryan, J., Lin, M.J., Miikkulainen, R.: Intrusion detection with neural networks. In: Advances in neural information processing systems, pp. 943–949. Morgan Kaufmann Publishers (1998)
3. Mukkamala, S., Janoski, G., Sung, A.: Intrusion detection using neural networks and support vector machines. In: Proceedings of the 2002 International Joint Conference on Neural Networks, IJCNN 2002, vol. 2, pp. 1702–1707. IEEE (2002)
4. Friedman, N., Geiger, D., Goldszmidt, M.: Bayesian network classifiers. Mach. Learn. **29**(2–3), 131–163 (1997)
5. Keogh, E., Lin, J., Lee, S.H., Van Herle, H.: Finding the most unusual time series subsequence: algorithms and applications. Knowl. Inf. Syst. **11**(1), 127 (2007)
6. Keogh, E., Lin, J., Fu, A.: Hot SAX: efficiently finding the most unusual time series subsequence. In: Fifth IEEE International Conference on Data Mining, pp. 8. IEEE, November 2005
7. Huang, T., Zhu, Y., Wu, Y., et al.: J-distance discord: an improved time series discord definition and discovery method. In: IEEE International Conference on Data Mining Workshop, pp. 303–310. IEEE (2015)

A More Secure Spatial Decompositions Algorithm via Indefeasible Laplace Noise in Differential Privacy

Xiaocui Li[1], Yangtao Wang[1], Xinyu Zhang[2], Ke Zhou[1(✉)], and Chunhua Li[1]

[1] Wuhan National Laboratory for Optoelectronics,
Huazhong University of Science and Technology, Wuhan, China
{LXC,ytwbruce,k.zhou,li.chunhua}@hust.edu.cn
[2] School of Computer, Wuhan University, Wuhan, China
zhangxinyu@whu.edu.cn

Abstract. Spatial decompositions are often used in the statistics of location information. For security, current works split the whole domain into sub-domains recursively to generate a hierarchical private tree and add Laplace noise to each node's points count, as called differentially private spatial decompositions. However Laplace distribution is symmetric about the origin, the mean of a large number of queries may cancel the Laplace noise. In private tree, the point count of intermediate nodes may be real since the summation of all its descendants may cancel the Laplace noise and reveal privacy. Moreover, existing algorithms add noises to all nodes of the private tree which leads to higher noise cost, and the maximum depth h of the tree is not intuitive for users. To address these problems, we propose a more secure algorithm which avoids canceling Laplace noise. That splits the domains depending on its real point count, and only adds indefeasible Laplace noise to leaves. The ith randomly selected leaf of one intermediate node is added noise by $\frac{(\beta-i+1)+1+\beta}{(\beta-i+1)+\beta}Lap(\lambda)$. We also replace h with a more intuitive split unit u. The experiment results show that our algorithm performs better both on synthetic and real datasets with higher security and data utility, and the noise cost is highly decreased.

Keywords: Indefeasible Laplace noise · Low noise cost
Differential privacy · Spatial decompositions

1 Introduction

In the era of Big Data, A variety of data mining algorithms and prediction strategies [1–3] were developed to analyze users' behavior habits, which brings heavily privacy threats to users. In many investigations, the statistic of location information is needed for different academic research, such as the distribution of endangered species, the distribution of hotel occupancy in a certain city, the trip distribution of occupied taxis, and so on. That can help biological scientists,

© Springer Nature Switzerland AG 2018
G. Gan et al. (Eds.): ADMA 2018, LNAI 11323, pp. 211–223, 2018.
https://doi.org/10.1007/978-3-030-05090-0_19

government, business decision-makers etc. make the correct and effective decision through some recommendation algorithms [4,5].

The existing approaches obtain statistic of location information via spatial decomposition. The process of spatial decompositions is that given a data set D of tuples in domain Ω, recursively decompose Ω into a set of sub-domains if the point count of current domain is larger than the given threshold θ. When the recursive termination condition is reached, a Hierarchical spatial decompositions tree is generated. Through retrieving this tree, the location information in a certain region can be obtained. Actually, we only hope users get the statistic information of database but not individual as individual's location may leak one's privacy.

However the spatial decompositions tree may reveal individuals' privacy, simply, some leaves of this tree may only contain one tuple. Some purpose oriented adversary can filch most individuals' privacy by various technical means such as data mining algorithms via retrieving the spatial decompositions tree. When publishing database, the data owner needs to perturb the information to achieve preserving privacy [6], which is known as preserving privacy data publishing [7,8]. It is hoped that users get the statistical information of the database as a whole but not individuals' information when users query the database.

Dwork et al. first gave the notion of differential privacy [9]: deleting an element from a statistical database should not substantially increase the risk of the record owner's privacy. Consequently, Dwork proposed a theoretical framework [10–12] called ε-differential privacy, and proved that the *Laplace mechanism* [13] can achieve differential privacy for numerical queries. In the last decade, differential privacy was applied in many algorithms, such as histogram-based data publishing [14], batch query [15], decision tree [16] and so on. In 2016, Apple Inc. was planning to adopt the differential privacy to preserve the users' privacies, and it is the first time that the differential privacy algorithm has been applied in practical applications.

In 2012, Cormode first applied the differential privacy to spatial decompositions quadtree [17], which took the approach of adding Laplace noise to each node of the spatial decompositions tree and publishing the noisy private tree to achieve the purpose of privacy preserving. Thereafter more excellent algorithms [19–21] emerged for spatial decompositions based on differential privacy, of which the PrivTree proposed by Zhang 2016 was especially noteworthy. In the algorithms proposed by Granham and Zhang, a private tree was generated through spatial decompositions. Each split of a node will generate 2^d children, where d is the dimension of the dataset, which finally outputs a 2^d-tree. For preserving privacy, each node of the 2^d-tree was added by Laplace noise. Users can obtain the answers to queries of the point count of a given area by retrieving the private tree.

1.1 Motivation and Contributions

In 2008, Dwork indicated that preserving privacy by adding Laplace noise into the true answer is delicate [10], as Laplace noise is symmetric about of the origin and the same question is asked many times, the responses may be averaged, canceling out the noise. In spatial decompositions private tree, each intermediate node's point count equals the summation of all its descendants' point counts, that may reveal the intermediate node's privacy for the summation of all its descendants Laplace noise may be canceled.

Another problem is that the private tree with every node added Laplace noise makes domain splitting imprecise for the noise snowballs from root to leaf. Moreover, the depth of the private tree h, also the maximum depth of recursion should be predefined when splitting Ω. However, the choice of h can be challenging, since a smaller h will cause coarse splitting of Ω, while a larger h will lead to the depth of the private tree too high, as a result the noise added to the private tree will increase.

To address the limitations of above, we propose a more secure spatial decompositions algorithm via indefeasible Laplace noise in differential privacy. We first propose a more secure spatial decompositions algorithm via indefeasible Laplace noise in differential privacy, which only adds noise to leaves but not to intermediate nodes. We use a more intuitive threshold u, the minimal split unit to limit the maximum depth of the private tree. We add indefeasible noise to the ith randomly selected leaf child of each intermediate node of the private tree through multiplying Laplace noise by coefficient of $\frac{(\beta-i+1)+\beta+1}{(\beta-i+1)+\beta}$. It satisfies ε-differential privacy and the answer to a query will be not real because the added noise cannot be canceled. It is also proved that the noise cost is lower than existing algorithms. Finally, we conduct extensive experiments to demonstrate the performance of our algorithm.

2 Preliminaries

In this section, we give the background of differential privacy and introduce the problem of spatial decomposition.

2.1 Differential Privacy

Definition 1 *(Neighboring Datasets). The dataset D and D' are neighbors if D and D' differ in at most one element.*

Definition 2 *(ε-Differential Privacy). The randomized function F satisfies ε-differential privacy if, for any two neighboring datasets D and D' and for all output $S \in Range(F)$,*

$$\frac{Pr\left[F\left(D\right) \to S\right]}{Pr\left[F\left(D'\right) \to S\right]} \le e^{\varepsilon} \tag{1}$$

where Pr[·] denotes the probability of an event.

Definition 3 *(Sensitivity). Let f be a function that maps a dataset D to a vector of real numbers. The global sensitivity of f is defined as:*

$$S\left(f\right) = \max_{D,D'} \frac{\|f\left(D\right) - f\left(D'\right)\|_1}{dis_{Ham(D,D')}} \tag{2}$$

where $\|\cdot\|_1$ denotes the $L1$ norm, and the $dis_{Ham}(\cdot)$ denotes the hamming distance.

The *Laplace Mechanism* is the fundamental algorithm used in numerical function through adding *i.i.d.* noise into each output. The noise obeys *Laplace distribution* with the following probability density function:

$$Pr(x) = \frac{1}{2\lambda} e^{-\frac{|x|}{\lambda}} \tag{3}$$

2.2　Spatial Decompositions

Spatial decomposition was usually classified into data-dependent decomposition and data-independent decomposition. The data-dependent decomposition indicates that the partition of space is dependent on the input data. The most common data-dependent decomposition are KD-tree [22] and R-tree [23]. The data-independent decomposition indicates that the partition of the spatial nodes is independent of the input data. It is computed by splitting the space into two average parts on each coordinate. The best known is quadtree in two dimensions and 2^d-tree [24–26] in higher dimensions. In our algorithm, we use data-independent decomposition as the spatial decomposition.

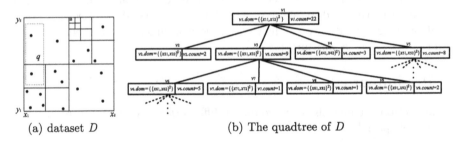

(a) dataset D　　　　　　　　　(b) The quadtree of D

Fig. 1. Spatial decomposition quadtree of a dataset D in 2-dimensional space Ω

Let D be a dataset consisting of data points in a d-dimensional space Ω. A spatial decomposition of D generates a 2^d-tree, which decomposes Ω into its sub-domains, along with partitioning of its data points into the leaves of the decomposition tree. In Fig. 1, (a) gives a 2- dimensional dataset D containing 22 points, it recursively divides into four regions until the number of its data points is less than a given threshold. (b) is the generation of a 2^d-tree (quadtree with d=2) of D. The root of the quadtree v_1 corresponds to the region that covers the entire domain Ω, it has four child nodes v_2, v_3, v_4 and v_5 respectively corresponding to the four sub-domains of Ω. Each node of the quadtree consists of its corresponding domain and the number of data points contained in its region.

The 2^d-tree is widely used for querying the spatial region through up-down traversing the quadtree from the root.

3   Related Work

In this section we first review private spatial decompositions algorithms, then discuss their disadvantages and the dilemmas.

Cormode et al. first proposed the differentially private spatial decompositions through quadtree in 2012, as called DPSD. It takes the solution of publishing the perturbed version of the spatial decomposition quadtree, which adds Laplace noise to each node of quadtree to achieve ε-differential privacy. Algorithm 1 presents the main approach of DPSD. The input contains two thresholds h and θ, which respectively represent the maximum recursion depth and the decision value of split as discussed in Sect. 2.2, and $\overline{v_i.count}$ is the noisy version of the $v_i.count$.

Algorithm 1 DPSD(D,λ,θ,h)

1: initialize a quadtree T with a root node v_1;
2: set $v_1.dom = \Omega$, and mark v_1 as visited ;
3: **while** there exists an unvisited node v_i **do**
4:    mark v_i as visited ;
5:    compute the number $v_i.count$ of points in D that are contained in $v_i.dom$;
6:    compute the noisy version of $v_i.count$, $v_i.\overline{count} = v_i.count + Lap(\lambda)$;
7:    **if** $\overline{v_i.count} > \theta$ and $depth(v_i) < h-1$ **then**
8:       split v_i, and add its children to T;
9:       mark the children of v_i as unvisited;
10: **return** T;

Assume that D and D' are neighboring datasets. The quadtrees of D and D' have at most h nodes with different point count, and these h nodes form a path from the root to one leaf. That indicates that the sensitivity of quadtree is h, so the Algorithm 1 satisfies ε-differential privacy if $\lambda \geq \frac{h}{\varepsilon}$. In Algorithm 1 there is n Laplace noise added to the private tree, the noise cost of the private tree is $\sum_{i=1}^{n} |Lap(\lambda)|$. The main limitation of Algorithm 1 is that the privacy cost depends on the recursive depth h.

Zhang et al. in 2016 proposed PrivTree, in which biased count of v_i was given, $b_i.count = \max\{\theta - \delta, v_i.count - depth(v_i) \cdot \delta\}$. In PrivTree input h was replaced by a new parameter θ and the splitting only depends on the noisy version of $b_i.count$.

Both of these two algorithms consider the split problem by the noisy version of each node of the private tree, which will lead to higher privacy cost of the private tree. In PrivTree, the choice of δ in PrivTree is a difficulty as $b_i.count$ at least equals $\theta - \delta$. If δ is very small, $b_i.count$ plus $Lap(\lambda)$ may easily be larger than δ, which results in unnecessary splitting.

4   A More Secure Indefeasible Laplace Noise Spatial Decompositions Algorithm

In this section, we propose a more secure spatial decompositions algorithm via indefeasible Laplace noise in differential privacy, the InLN_DPSD which can avoid the Laplace noise to be canceled and has lower noise cost.

4.1   Low Noise Cost Private Tree for Spatial Decompositions

According to the above-mentioned analysis, we propose a low noise cost private tree based differential privacy algorithm.

We modified Algorithm 1 with a new parameter minimal split unit u instead of h, which is more intuitive for split termination, and whether a node should be split depends on its real point count $v_i.count$ instead of $\overline{v_i.count}$. We replace the $\overline{v_i.count} > \theta$ and $depth(v_i) < h - 1$ by $v_i.count > \theta$ and $\frac{\Omega}{2^{depth(v_i)*d}} > u$ of Algorithm 1 in line 7, which indicates that if $v_i.dom$, $\frac{\Omega}{2^{depth(v_i)*d}}$ is smaller than u, even if $v_i.count > \theta$, stop splitting v_i, where d is the dimensionality of Ω and $depth(v_i)$ denotes the depth of v_i in the private tree. The private tree T generated by the modified algorithm has all leaves with Laplace noise, while all intermediate nodes do not. For privacy preserving data publishing, we also get the point count of intermediate nodes by summing up the count of the leaves under it.

In order to compare with Algorithm 1, we make the same assumption that D and D' are neighboring datasets differing only by one element. There are at most $\max(depth(v_i))$ nodes with different point count in quadtrees of D and D', and these nodes must form a path from the root to the one leaf. In particular, we let $\max(depth(v_i))$ equal h. Then,

$$ln\frac{Pr[D \rightarrow T]}{Pr[D' \rightarrow T]} = \sum_{i=1}^{h-1} ln\frac{Pr[v_i.count + Lap(\lambda) > \theta]}{Pr[v_i'.count + Lap(\lambda) > \theta]}$$

$$+ln\frac{Pr[v_h.count + Lap(\lambda) = \overline{v_h.count}]}{Pr[v_h'.count + Lap(\lambda) = \overline{v_h.count}]}$$

Although the first h-1 intermediate nodes do not add Laplace noise, the D and D' can also approximately output a same T satisfying ε-differential privacy if $\lambda > \frac{h}{\varepsilon}$. Every node v_i for any $i \in [1, h - 1]$ is the ancestor of v_h and $v_h.dom \subset v_i.dom$, that means $v_i.count$ has the same Laplace noise with the leaf's.

For any $i \in (1, h - 1)$ with $v_i.count \leq \theta$, $ln\frac{Pr[v_i.count+Lap(\lambda)>\theta]}{Pr[v_i'.count+Lap(\lambda)>\theta]}$ equals $\frac{1}{\lambda}$. Otherwise it is less than $\frac{1}{\lambda}$, which has been proved by Zhang et al. in [19]. So,

$$\sum_{i=1}^{h-1} ln\frac{Pr[v_i.count + Lap(\lambda) > \theta]}{Pr[v_i'.count + Lap(\lambda) > \theta]} \leq \frac{h-1}{\lambda} \qquad (4)$$

$$ln\frac{Pr[v_h.count + Lap(\lambda) = \overline{v_h.count}]}{Pr[v_h'.count + Lap(\lambda) = \overline{v_h.count}]} = \frac{1}{\lambda} \qquad (5)$$

$$ln\frac{Pr[D \rightarrow T]}{Pr[D' \rightarrow T]} \leq \frac{h}{\lambda} \leq \varepsilon \tag{6}$$

Assume that there are n points in T, m is the number of the intermediate nodes, t is the number of the leaf nodes, and β is the fanout of the T. The total noise added to this private tree T is $\sum_{i=1}^{t} |Lap(\lambda)|$. It then can be seen that the noise cost of modified algorithm is smaller than the private tree generated by Algorithm 1, which is $\sum_{i=1}^{n} |Lap(\lambda)|$.

4.2 The Full Private β-tree with Indefeasible Laplace Noise

There is a deficiency of the low noise cost private tree, in which the privacy of the intermediate node is delicate. The intermediate nodes are noisy with the integrated Laplace noises of the leaves under it. The noise of an intermediate node which has β leaves equals $\sum_{i=1}^{k} Lap(\lambda)$ which may be 0 with a very high probability as Laplace is symmetric about the origin. As a result the intermediate node's point count approximately to be real.

We make the same assumption as the modified algorithms. In particular, we assume a private tree T is a full β-tree and each leaf has $\beta - 1$ brother leaves. Let $L = \{l_1, l_2, \cdots, l_\beta\}$ be the set of leaves of one intermediate node, $|L| = \beta$. Each time we randomly select one leaf l_i from the set L and add Laplace noise $\frac{i+1+\beta}{i+\beta} Lap(\lambda)$ into its point count, then update $L = L \setminus \{l_i\}$. Repeat the above operation until $L = \Phi$. The noise of intermediate node which has β leaves equals $\sum_{i=1}^{k} \frac{i+1+\beta}{i+\beta} Lap(\lambda) \neq 0$, therefore the noise of intermediate node will not be cancelled out.

$$\overline{v_i.count} = v_i.count + \frac{i+1+\beta}{i+\beta} Lap(\lambda) \tag{7}$$

Lemma 1.

$$ln\frac{Pr[D \rightarrow T]}{Pr[D' \rightarrow T]} = \sum_{i=1}^{h-1} ln\frac{Pr[v_i.count + \frac{i+1+\beta}{i+\beta} Lap(\lambda) > \theta]}{Pr[v'_i.count + \frac{i+1+\beta}{i+\beta} Lap(\lambda) > \theta]}$$
$$+ln\frac{Pr[v_h.count + \frac{i+1+\beta}{i+\beta} Lap(\lambda) = \overline{v_h.count}]}{Pr[v'_h.count + \frac{i+1+\beta}{i+\beta} Lap(\lambda) = \overline{v_h.count}]} < \frac{h}{\lambda}$$

The Laplace noise is proportional to $S(f)$ and inversely proportional to ε. While $S(f)$ and ε are fixed values, increasing the noise with proportion of $\frac{i+1+\beta}{i+\beta}$ can also satisfy the ε-differential privacy.

Lemma 2. *The total noise cost of the private tree generated by improved algorithm is $\frac{t}{\beta} \sum_{i=1}^{\beta} |\frac{i+1+\beta}{i+\beta} Lap(\lambda)|$, smaller than the noise cost of private tree generated by Algorithm 1, which is $\sum_{i=1}^{n} |Lap(\lambda)|$.*

4.3 The General Private β-tree with Modified Indefeasible Laplace Noise

In practice, the spatial decomposition tree is not a full β-tree, but a general β-tree which each intermediate node has the same fanout β. We modify our strategy by adding noise into the ith selected leaf child with $\frac{(\beta-i+1)+1+\beta}{(\beta-i+1)+\beta}Lap(\lambda)$, causing its first selected leaf child with noise of $\frac{\beta+1+\beta}{\beta+\beta}Lap(\lambda)$.

It can be computed that $m = (t-1)/(\beta-1)$. We define the function $f(k) = \frac{k+\beta+1}{k+\beta}$, for any $k \in [1,\beta]$. $f(k)$ is a monotonically decreasing function, that is $f(i) > f(i+1)$ for any $i \in [1, \beta-1]$.

Lemma 3. *Assume that there are x_k leaves which are kth selected and add noises with $f(k)Lap(\lambda)$ in the β-tree, for any $k \in [1,\beta]$. The total noise cost of the private tree $\sum\limits_{k=1}^{\beta} |x_k f(k) Lap(\lambda)| = (x_1 f(1) + \cdots x_\beta f(\beta))|Lap(\lambda)|$ (and $\sum\limits_{k=1}^{\beta} x_k = t$) is smaller than the noise cost of the full β-tree, which has the same leaves count and intermediate nodes count with the general β-tree. Besides, the noise cost of full β-tree can be symbolically simplified as $\frac{t}{\beta}\sum\limits_{k=1}^{\beta}|f(k)Lap(\lambda)|$.*

4.4 The Spatial Decomposition Algorithm InLN_DPSD

Our technique for private spatial decompositions is presented in Algorithm 2. We first generate a β-tree applying low noise cost private tree for spatial decompositions(line 1$\sim$8). We split the node according to the real point count $v_i.count$ and the algorithm terminates depending on the minimum unit u, which is more intuitive than the maximal depth of recursion h(line 6).

Algorithm 2 InLN_DPSD(D,λ,θ,u)

1: initialize a quadtree T with a root node v_1;
2: set $v_1.dom = \Omega$, and mark v_1 as visited ;
3: **while** there exists an unvisited node v_i **do**
4:     mark v_i as visited ;
5:     computer the number $v_i.count$ of points in D that are contained in $v_i.dom$;
6:     **if** $v_i.count > \theta$ and $\frac{\Omega}{2^{depth(v_i)*d}} > u$ **then**
7:        split v_i, and add its children to T;
8:        mark the children of v_i as unvisited;
9: **for** each $i \in [1, n]$ **do**
10:     **if** $isleaf(v_i) == 0$ **then**
11:        initialize a children set $L = \phi$;
12:        add all v_i's children to L;
13:        **for** each $j \in [1, \beta], k = \beta$ **do**
14:           randomly select v_j from L;
15:           **if** $isleaf(v_j) == 1$ **then**
16:              $\overline{v_j.count} = v_j.count + \frac{k+\beta+1}{k+\beta} * Lap(\lambda)$;
17:              $k--$;
18:           $L = L \setminus v_j$;
19: **return** T;

Then, add noise to the leaves of the β-tree (line 9~18), which is the biggest innovation of InLN_DPSD. We first give the notion of indefeasible Laplace noise, which multiplies Laplace noise with coefficient $\frac{(\beta-i+1)+\beta+1}{(\beta-i+1)+\beta}$, where i indicates the ith leaf randomly selected from the intermediate node's children, and β is the fanout of the private tree.

In the process of adding indefeasible noise, we check all nodes of the β-tree. If the node is not a leaf, add noise to its children who are leaves. An intermediate node's child leaves are added to Laplace noise multiplied with coefficient $\frac{k+\beta+1}{k+\beta}$, where k decreases from β to 1. Usually k may be greater than 1 for not all the β children are leaves, and k only is decreased by 1 when a child leaf is processed(line 15~17). It has been proved in Sect. 4.2 that when the private tree was added to Laplace noise multiplied $\frac{k+\beta+1}{k+\beta}$ where k decreases from β, its total noise cost is less than the DPSD algorithm. Otherwise if the k increases from 1 to β, the noise cost is larger than existing algorithms'.

5   Experiment

This section we evaluate InLN_DPSD against the state-of-the-art algorithms of spatial decompositions based on differential privacy.

5.1   Competing Methods and Testing Datasets

Competing Methods. To evaluate the efficacy of the proposed approaches, we compare InLN_DPSD with the DPSD and PrivTree.

Testing Datasets. In the experiments we employ one synthetic two-dimensional dataset and two real spatial datasets. SD contains 1 million location information; Beijing[1] is two-dimensional real dataset which contains 15 million records of pickup locations of Beijing taxis; NYC[2], a four-dimensional real dataset containing one hundred million records of pickup and drop-off locations of NYC taxis in 2013. The distribution of these datasets is shown in Fig. 2.

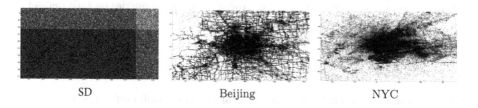

SD   Beijing   NYC

Fig. 2. The distribution of each dataset

---

[1] http://research.microsoft.com/apps/pubs/?id=152883.
[2] http://publish.illinois.edu/dbwork/open-data/.

5.2   Evaluation Measures

We run each algorithm on every dataset to evaluate their performances. We respectively generate ten thousand queries on the region covering [0.1%, 1%), [1%, 10%), [10%, 100%) of the dataset domain and get their answers with each algorithm. For evaluating the query accuracy, we define the relative error RE [17,18] as the measure accuracy of a perturbed answer $\overline{q(D)}$ to a query q by its real answer $q(D)$.

$$RE(\overline{q(D)}) = \frac{|\overline{q(D)} - q(D)|}{\max\{q(D), \Delta\}}$$

where Δ is a smoothing factor [27] set to 1% of the dataset cardinality n. We repeat each experiment 1000 times and report the average relative error of each method for each query set.

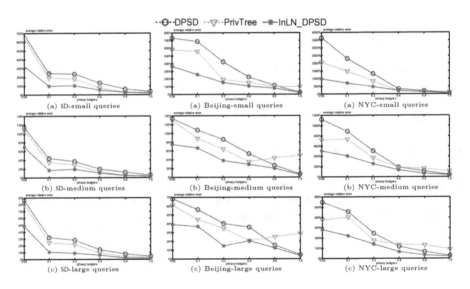

Fig. 3. The results of average relative error of range count queries on each dataset

5.3   Results and Analyses

Figure 3 shows the average relative error of the three different range queries on each dataset applying every algorithm with different privacy budget ε. From the results we can see:

(1) The relative error of InLN_DPSD algorithm is always smaller than the other two algorithms. That shows our algorithm outperforms.
(2) The relative error of NYC dataset is much larger than SD and Beijing datasets. The more skewed the dataset is, the larger relative error it has.

(3) The relative error increases as privacy budget ε increases. Since the noise added to dataset is inversely proportional to ε.

(4) The relative error is inversely proportional to the query area. That indicates the larger the query area is, the more accurate the answer is. If the query area is much smaller, even to one tuple's location, the relative error will be very large, which preserves the individual's privacy.

Table 1. The number of Laplace noises added to the private tree

	DPSD	PrivTree	InLN_DPSD
SD	736449	755261	227677
Beijing	405269	371705	145216
NYC	130209	109041	63527

Table 1 shows the number of noises added to the private tree in different datasets when applying these three algorithms. From the table we can see that the InLN_DPSD algorithm adds fewer noises to the private tree.

6   Conclusion

In this paper, we study the problem of spatial decompositions based differential privacy. The existing algorithm's total noise cost is too high and the recursion depth h of the private tree is hard to choose. Most important of all, the traditional differential privacy is delicate as the Laplace distribution is symmetric about the origin, the sum of several Laplace noises may be 0 so that the added noises may be canceled and the privacy may be compromised. We take the strategy of only adding indefeasible Laplace noises to leaves, the noises will not be canceled and the domain splitting will be more accurate which leads to higher data utility and the total noise cost of the private tree is highly decreased. The predefined h is replaced by minimal split unit u, which is more intuitive for users. Based on this strategy we propose InLN_DPSD, a more secure spatial decompositions algorithm via indefeasible Laplace noise in differential privacy. The experiment results show that InLN_DPSD outperforms the DPSD and PrivTree both in a synthetic dataset and real dataset.

References

1. Yin, H., Chen, H., Sun, X., et al.: SPTF: a scalable probabilistic tensor factorization model for semantic-aware behavior prediction. In: IEEE International Conference on Data Mining, pp. 585–594. IEEE Press, New Orleans (2017)
2. Chen, H., Yin, H., Wang, W., et al.: PME: projected metric embedding on heterogeneous networks for link prediction. In: 24th ACM SIGKDD International Conference on Knowledge Discovery & Data Mining, pp. 1177–1186. ACM Press, London (2018)
3. Chen, T., Yin, H., Chen, H., et al.: TADA: trend alignment with dual-attention multi-task recurrent neural networks for sales prediction. In: IEEE International Conference on Data Mining. IEEE Press, Singapore (2018)
4. Yin, H., Wang, W., Wang, H., et al.: Spatial-aware hierarchical collaborative deep learning for POI recommendation. IEEE Trans. Knowl. Data Eng. **29**(11), 2537–2551 (2017)
5. Yin, H., Sun, Y., Cui, B., et al.: LCARS: a location-content-aware recommender system. In: 19th ACM SIGKDD International Conference on Knowledge Discovery & Data Mining, pp. 221–229. IEEE Press, Chicago (2013)
6. Friedman, A., Schuster, A.: Data mining with differential privacy. In: 16th International Conference on Knowledge Discovery and Data Mining, pp. 493–502. ACM Press, Washington (2010)
7. Fung, B.C.M.: Privacy-preserving data publishing. ACM Comput. Surv. **42**(4), 1–53 (2010)
8. Hardt, M., Ligett, K., Mcsherry, F.: A simple and practical algorithm for differentially private data release. In: Advances in Neural Information Processing Systems, pp. 2339–2347 (2010)
9. Dwork, C.: Differential privacy. In: Bugliesi, M., Preneel, B., Sassone, V., Wegener, I. (eds.) ICALP 2006. LNCS, vol. 4052, pp. 1–12. Springer, Heidelberg (2006). https://doi.org/10.1007/11787006_1
10. Dwork, C.: Differential privacy: a survey of results. In: Agrawal, M., Du, D., Duan, Z., Li, A. (eds.) TAMC 2008. LNCS, vol. 4978, pp. 1–19. Springer, Heidelberg (2008). https://doi.org/10.1007/978-3-540-79228-4_1
11. Dwork, C.: A firm foundation for private data analysis. Commun. ACM **54**(1), 86–95 (2011)
12. Dwork, C., Roth, A.: The algorithmic foundations of differential privacy. Found. Trends Theor. Comput. Sci. **9**(3–4), 211–407 (2014)
13. Dwork, C., McSherry, F., Nissim, K., Smith, A.: Calibrating noise to sensitivity in private data analysis. In: Halevi, S., Rabin, T. (eds.) TCC 2006. LNCS, vol. 3876, pp. 265–284. Springer, Heidelberg (2006). https://doi.org/10.1007/11681878_14
14. Xu, J., Zhang, Z., Xiao, X., et al.: Differentially private histogram publication. In: 29th IEEE International Conference on Data Engineering, pp. 32–43. IEEE Press, Brisbane (2013)
15. Xiao, X., Wang, G., Gehrke, J.: Differential privacy via wavelet transforms. In: 26th IEEE International Conference on Data Engineering, pp. 225–236. IEEE Press (2010)
16. Mohammed, N., Chen, R., Fung, B.C.M., et al.: Differentially private data release for data mining. In: ACM SIGKDD International Conference on Knowledge Discovery and Data Mining, pp. 493–501. ACM press (2011)
17. Cormode, G., Procopiuc, C., Srivastava, D., et al.: Differentially private spatial decompositions. In: 28th IEEE International Conference on Data Engineering, pp. 20–31. IEEE Press, Washington (2012)

18. Li, N., Yang, W., Qardaji, W.: Differentially private grids for geospatial data. In: 28th IEEE International Conference on Data Engineering, pp. 757–768. IEEE Press, Washington (2012)
19. Zhang, J., Xiao, X., Xie, X.: PrivTree: a differentially private algorithm for hierarchical decompositions. In: 35th ACM Conference on Management of Data, pp. 155–170. ACM Press, San Franciso (2016)
20. Zhang, J., Cormode, G., et al.: PrivBayes: private data release via Bayesian networks. In: 33th ACM Conference on Management of Data, pp. 1423–1434. ACM Press, Utah (2014)
21. Zhang, J., Cormode, G., et al.: Private release of graph statistics using ladder functions. In: 34th ACM Conference on Management of Data, pp. 731–745. ACM Press, Melbourne (2015)
22. Miller, F.P., Vandome, A.F., Mcbrewster, J.: KD-tree (2009)
23. Guttman, A.: R-trees: a dynamic index structure for spatial searching. In: International Conference on Management of Data 1984, pp. 47–57. ACM Press, Massachusetts (1984)
24. Bodlaender, H.L.: A linear-time algorithm for finding tree-decompositions of small treewidth. In: The 25th ACM Symposium on Theory of Computing, pp. 226–234 (1993)
25. Demaine, E.D., Mozes, S., Rossman, B., et al.: An optimal decomposition algorithm for tree edit distance. ACM Trans. Algorithms 6(1), 1–19 (2007)
26. Li, B., et al.: Dynamic reverse furthest neighbor querying algorithm of moving objects. In: Li, J., Li, X., Wang, S., Li, J., Sheng, Q.Z. (eds.) ADMA 2016. LNCS (LNAI), vol. 10086, pp. 266–279. Springer, Cham (2016). https://doi.org/10.1007/978-3-319-49586-6_18
27. Xiao, X., Wang, G., Gehrke, J.: Differential privacy via wavelet transforms. IEEE Trans. Knowl. Data Eng. 23(8), 1200–1214 (2011)

Towards Geological Knowledge Discovery Using Vector-Based Semantic Similarity

Majigsuren Enkhsaikhan[1]([⊠])(iD), Wei Liu[1](iD), Eun-Jung Holden[1](iD),
and Paul Duuring[2](iD)

[1] University of Western Australia, Perth, Australia
majigsuren.enkhsaikhan@research.uwa.edu.au,
{wei.liu,eun-jung.holden}@uwa.edu.au
[2] Department of Mines, Industry Regulation and Safety, Perth, Australia
paul.duuring@dmirs.wa.gov.au

Abstract. It is not uncommon for large organisations and corporations to routinely produce various kinds of reports indefinitely. Apart from archiving them and the occasional retrieval of some, very little can be done to take advantage of these massive resources for valuable knowledge discovery. The under-utilised unstructured data written in natural language text is often referred to as part of the "dark data". The good news is, recent success of learning distributed representation of words in vector spaces, especially, the similarity and analogy queries enabled by the so-learned word vectors drive a paradigm shift from "document retrieval" to "knowledge retrieval". In this paper, we investigated how representational learning of words can affect the entity query results from a large domain corpus of geological survey reports. Extensive similarity tests and analogy queries have been performed. It demonstrated the necessity of training domain-specific word embeddings, as pre-trained embeddings are good at capturing morphological relations, but are inadequate for domain specific semantic relations. Carrying out entity extractions prior to word embedding training will further improve the quality of analogy query results. The framework developed in this paper can also be readily applied to other domain specific corpus.

Keywords: Word embedding · Word2Vec · FastText · Word analogy
Cosine similarity · Geological domain

1 Introduction

Word embeddings or distributed representations of words in a vector space [1] have been shown in many Natural Language Processing (NLP) tasks to be able to capture valuable semantic information. They are used as a replacement of lexical dictionary such as WordNet [2] for semantic meaning expansion, leveraging the linguistic background knowledge. These include but are not limited to document classification [3,4], named entity recognition [5], and word sense disambiguation [6].

© Springer Nature Switzerland AG 2018
G. Gan et al. (Eds.): ADMA 2018, LNAI 11323, pp. 224–237, 2018.
https://doi.org/10.1007/978-3-030-05090-0_20

A word embedding framework contains a set of language modeling and feature learning techniques of NLP and it transforms words or phrases into vectors of real numbers. Popular embedding frameworks include Word2Vec [1,7], GloVe [8] and the very recent FastText [9,10]. Generating word embeddings may require a significant amount of time and effort by collecting large-scale data, pre-processing the data, machine learning, evaluating the results and tuning hyper-parameters for performance improvement [11,12]. Therefore, using pre-trained word vectors learned from billions of words is a cost-effective solution with potential better performance due to the massive amount of data used in obtaining these pre-trained embeddings. Although the pre-trained Word2Vec[1], GloVe[2], and FastText[3,4] embeddings are readily available for various NLP tasks, they are trained on the general domain only. Training domain-specific terminologies using domain corpus can create embeddings that capture more meaningful semantic relations in the application domain.

The most popular test for demonstrating the effectiveness of distributed representations of words, using real-valued vectors, is the analogy test [13]. The analogy test is often in the form of the relations between the first pair of words is equal to that of the second pair. A famous example to showcase the valid semantic relations captured by word embeddings is:

$$king : man :: queen : woman$$

In other words, `king` is related to `man` as to `queen` is related to `woman`.

The analogy relations can be broadly classified as morphological relations and semantic relations. For general purpose corpus, the semantic relations might include purpose, cause-effect, part-whole, part-part, action-object, synonym/antonym, place, degree, characteristics, sequence, and etc. When it comes to a special domain, such as the geological survey domain we are interested in, domain specific relations such as commodity and its mineralisation environment, locations, and host rock types.

All existing analogy test data are for general purpose only. Assuming word embeddings have been trained sufficiently to capture semantic relations between words, such distributed representations would become an invaluable tool in querying real-world textual data to support knowledge discoveries from the wide variety of reports available. Take the Western Australian Mineral Exploration Reports (WAMEX) data for example, a query of `Kalgoorlie-Gold+Iron ore=?` may help discover a selected number of words related to the `Iron ore` in the same way as `Gold` is related to `Kalgoorlie`, potentially capturing the location relation to `Iron ore`.

In this paper, a framework that includes pre-processing, domain dictionary construction, entity extraction, clustering for exploratory study of text reports

---

[1] https://code.google.com/archive/p/word2vec/.

[2] https://nlp.stanford.edu/projects/glove/.

[3] https://fasttext.cc/docs/en/english-vectors.html.

[4] https://github.com/facebookresearch/fastText/blob/master/pretrained-vectors.md.

and finally analogy queries is developed to support domain-specific information retrieval. 33,824 geological survey reports are used to train and obtain geological word embeddings. Our experiment compared these geological domain-specific embeddings against pre-trained word embeddings in answering analogy queries that are of interests to the domain experts.

Our results have been confirmed by geological experts, Wikipedia and even Google Map on the effectiveness of how meaningful these word embeddings are in capturing domain specific information. This warrants more future work in the direction of developing word embedding enabled production of information retrieval systems for domain specific textual data. Our framework is designed in a modular fashion so that it can be readily applicable to other domains, despite that our experimental results are for geological survey reports.

This initial success of the prototype system built using this framework on the geological survey reports left a lot to be desired in applying this to real world document collections. It provides hope in addressing the significant yet challenging problem of learning from the massive amount of "dark corporate data" stored in the less accessible textual format.

2   Literature Review

2.1   Embedding Architectures

Three state-of-the-art neural network architectures for learning are discussed here, namely, Word2Vec, GloVe and FastText. They have been shown to perform well on various NLP tasks and on large scale corpora of billions of words [14].

Word2Vec [1,7] supplies two predictive model architectures to produce a distributed representation of words: Continuous Bag-of-Words (CBOW) or Continuous Skip-Gram (CSG). The CBOW model predicts the target word from a window of surrounding context. The order of surrounding words does not influence prediction, so the name is bag-of-words. The CSG model uses the target word to predict the context in the window. The CSG model weighs nearby context words higher than distant context words. It is reported that CBOW is faster, but CSG performs better for infrequent words and is more accurate on a large corpus. An optimisation method such as Hierarchical Softmax or Negative Sampling optimizes the computation of the output layer, to speed up the training of a model [15]. Some studies [1,7] reported that the Hierarchical Softmax works better for infrequent words, while Negative Sampling performs better for frequent words and with low dimensional vectors.

Global Vectors (GloVe) learns by constructing a word co-occurrence matrix that captures the frequency of word appearance in a context. GloVe needs a factorizing matrix to reduce the dimension of that large co-occurrence matrix [8].

While Word2Vec and GloVe treat each word as the smallest unit to train on, **FastText** uses n-gram characters as the smallest unit [9,10]. For example, one word vector could be broken down into several vectors to represent multiple n-gram characters. The benefit of using FastText is generating word embeddings for rare or unseen words during training, because the n-gram character vectors can be shared by multiple words.

2.2 Word Relations

Words relate to each other differently. Gladkova et al. [16] introduced the Bigger Analogy Test Set, which explained the types of word relations: inflectional and derivational morphology, and lexicographic and encyclopedic semantics.

Morphological relations can be inflectional or derivational. Inflectional morphology is the modification of a word to express different grammatical categories including numbers, tenses and comparatives. For example, *rock:rocks, occur:occurred*, and *hard:harder*. Derivational morphology is a word formation by changing syntactic category or by adding new meaning to a word. For example, *able:unable, produce:reproduce*, and *employ:employment, employer, employee, employable*.

Semantic relations can be lexicographic and encyclopedic. Lexicographic relations includes hypernymy (superordinate relation), hyponymy (subordinate relation), meronymy (a part-of relation), synonymy (same meaning) and antonymy (opposite meanings). For example, *apple:fruit, color:blue, member:team, talk:speak*, and *up:down*. Encyclopedic semantics define closely related words without considering grammatical changes. For example, *Australia:English* (country:language), *sky:blue* (thing:color) and *dog:puppy* (adult:young).

2.3 Algorithms for Solving Analogy Test

Analogy tests are performed on morphological or semantic similarities of words including countries and their capital cities, countries and their currencies, modification of words to express different grammatical categories such as tenses, opposites, comparatives, superlatives, plurals, and gender inflections. For example, *France:Paris::Italy:Rome; go:gone::do:done; cars:car::tables:table; wife:woman::husband:man* and *better:good::larger:large*.

Performing analogy tests depends on an embedding method, its parameters, specific word relations [16], and a method of solving analogies [17]. Analogies solved by one method may not be solved by another method on the same embedding. Therefore, generating embeddings are useful for exploring rather than evaluating the underlying dataset. **Pair-based methods** such as 3CosAdd [13] and 3CosMul [18] perform analogical reasoning based on the offset of word vectors. 3CosAdd performs a linear sum to normalize and ignore the lengths or embedding vectors, unlike the Euclidean distance. 3CosMul method amplifies the differences between small quantities and reduces the differences between larger ones by using the vector multiplication instead of the addition. **Set-based methods** [17] include 3CosAvg and LRCos. 3CosAvg works on vector offset averaged over multiple pairs. LRCos incorporates a supervised learning of the target class and the cosine similarity. **Pair–Pattern matrix method** called Latent Relational Analysis (LRA) [19] takes word pairs, and constructs a matrix to find the relational similarity between word pairs, by deriving patterns automatically from a large corpus with synonyms. **Dual-Space method** depends on direction and ignores spatial distance between word vectors [20].

3   Methodology

3.1   Architectural Overview

An overview of our study is presented in Fig. 1. The system consists of *a pre-processing module, a dictionary construction module* which supports *an annotation and filtering module* for entity extraction, *a word embedding training module* to learn the embeddings from other pre-processed text or extracted entities, *a similarity module* that implements multiple similarity tests, and *an analogy test module* with various solvers.

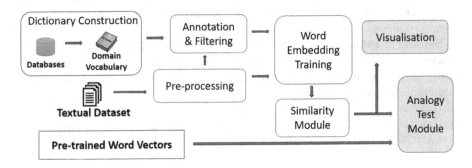

Fig. 1. An overview

Pre-processing module includes cleaning, tokenization and lemmatization processes using Natural Language Toolkit (NLTK) [21]. The geological corpus data are cleaned by removing the stop words, numbers and delimiter characters. The stop word list includes 353 common words such as *the, a, an,* and *is.* After cleaning, we tokenise the words by breaking up sentences into lists of words. The remaining words are then lemmatised to remove inflectional endings and reduced to their base form, using NLTK WordNetLemmatizer.

Annotation. To target domain specific entities and avoid noise, a dictionary-based named entity extraction method is implemented. Only the words of interest from the geological corpus are collected for learning embeddings. A geological vocabulary of 5623 terms is created, which contains minerals, commodity names, geological eras, rocks, stratigraphic units, and mineralisation styles, as well as geographical information, such as location names, mines, tectonic setting names and regions. Figure 2 shows some terms in the vocabulary. The sources of these geological terminologies are Wikipedia[5], Geographical Locations of Western Australia (WA)[6] and WA Stratigraphic Units Database[7]. The data are annotated

---

[5] https://en.wikipedia.org.

[6] http://www.geonames.org.

[7] http://dbforms.ga.gov.au/www/geodx.Stratigraphic_Units_Reports.states_ext.

using this domain vocabulary. Once annotated, the textual data are transformed into the format shown in Fig. 3.

list_commodities	list	135	['aluminum', 'amethyst', 'andalusite', 'antimony', 'asbestos', 'barit
list_eras	list	45	['archaean', 'archean', 'neoarchean', 'mesoarchean', 'paleoarchean',
list_gold_mineralogical_terms	list	33	['microscopic gold', 'submicroscopic gold', 'surface gold', 'gold all
list_locations	list	577	['western australia', 'murchison', 'pilbara', 'kayes well', 'kimberle
list_minerals	list	1067	['acerila', 'achroite', 'acmite', 'adamine', 'adularia', 'aegyrine',
list_rocks	list	187	['adakite', 'adamellite', 'alkali feldspar granite', 'altered granite
list_stratigraphy	list	3231	['81 mile vent lamproite', 'abattoir east gabbro', 'abattoir west gab
list_wa_cratons_basins_creeks	list	96	['albany fraser province', 'albany', 'amadeus basin', 'arafura basin'
list_wa_mine_names	list	190	['abydos', 'agnew project', 'alec mairs', 'argyle', 'athena', 'atlas
list_wa_regions	list	33	['central west', 'central wheat belt', 'esperance', 'eucla', 'gascoyn

Fig. 2. Domain vocabulary

The area is underlain predominantly by low grade metamorphic sedimentary rocks belonging to the Ashburton Formation, the uppermost stratigraphic unit of the Wyloo Group.

The dominant lithologies in the Ashburton Formation are chloritic mudstone and siltstone, with subordinate immature lithic sandstones and minor conglomerate.

area underlain predominantly low grade metamorphic [sedimentary rock,ROCK] belonging [ashburton formation,STRATIGRAPHIC_UNIT] uppermost stratigraphic unit [wyloo group,STRATIGRAPHIC_UNIT]

dominant lithology [ashburton formation,STRATIGRAPHIC_UNIT] chloritic [mudstone,ROCK] [siltstone,ROCK] subordinate immature lithic [[[sandstone,COMMODITY],ROCK],LOCATION] minor [[conglomerate,COMMODITY],ROCK]

Fig. 3. Raw text to annotated text

formation, wyloo group, ashburton formation, mudstone, sandstone, conglomerate, banded iron, ashburton formation, quaternary, laterite, metal, gold, sedimentary rock, ashburton formation, wyloo group, ashburton formation, dolomite, edmund group, ashburton formation, mudstone, sandstone, conglomerate, banded iron, quartz, ashburton formation, quartz, ashburton formation, laterite, ashburton formation, siltstone, quartz, sandstone, conglomerate, greywacke, felsic volcanic rock,

Fig. 4. WAMEX entities for training

Filtering. During the filtering process, all entities (words and phrases) are extracted based on the annotation using our geological dictionary. All documents are filtered and only dictionary terms are kept for the embedding learning process. The whole document contains only dictionary-specific terminologies and is then used for embedding training. Figure 4 shows two example sentences appear that in the filtered text; entities from those two sentences are underlined.

Embedding. Embedding models use two types of context: linear context refers to the positional neighbours of the target word, while dependency based context uses syntactic neighbours of the target word based on a dependency parse tree using part-of-speech labeling. Word representations can be bound or unbound. Bound context representation considers sequential positions of context with the target word. Unbound context representation treats all words within the chosen context window as the same, irrespective of their positions from the target word. Linear context is sufficient for comparing topical similarity compared to dependency based context, according to Li et al. [14], who compared Word2Vec GSG, Word2Vec GBOW and GloVe models. They stated that word analogies are most effective with unbound representation. Therefore, for this study we choose Word2Vec models of unbound representations with linear context.

3.2 Data Clustering and Visualisation Using t-SNE

t-distributed Stochastic Neighbour Embedding (t-SNE) [22] is a popular dimensionality reduction technique that projects high dimensional vectors onto a low dimensional plane, while preserving the distances and similarities of the data as much as possible. We use t-SNE to visualise semantic closeness of words in the various embeddings obtained in this paper. Two main similarity measures for text clustering are *Cosine similarity* and *Euclidean distance*.

Cosine similarity of two given n-dimensional vectors $A = (a_1, a_2, \ldots, a_n)$ and $B = (b_1, b_2, \ldots, b_n)$ is calculated as the cosine of the angle between them, where the vectors represent a pair of words, phrases, sentences, documents or corpora. When we have two documents with similar contents, but one is several times bigger in size, cosine similarity defines how similar they are to each other in terms of the context, not of the size. The similarity is measured in the range of 0 to 1 in cosine distance, where 0 means the most different, near 1 means the highly similar and 1 is for identical. Cosine similarity is defined as:

$$\cos(A, B) = \frac{AB}{||A||||B||} = \frac{\sum_{i=1}^{n} a_i \cdot b_i}{\sqrt{\sum_{i=1}^{n} a_i^2} \sqrt{\sum_{i=1}^{n} b_i^2}} \tag{1}$$

Euclidean distance is defined by any non-negative value. Euclidean distance of two n-dimensional vectors $A = (a_1, a_2, \ldots, a_n)$ and $B = (b_1, b_2, \ldots, b_n)$ is defined as:

$$distance(A, B) = \sqrt{\sum_{i=1}^{n} (a_i - b_i)^2} \tag{2}$$

The cosine similarity deals with relative difference between words, instead of absolute frequency difference. For example, the vector A = (2, 3) has the highest similarity with B = (4, 6), because they have the same angle, although the latter vector is longer, while Euclidean distance between them is 3.6 units. We prefer the cosine similarity for relative difference between words, thus vector length is ignored. t-SNE displays words closer in the visualisation if their high-dimensional vectors are similar, distant if dissimilar.

3.3 Analogy Investigation

A proportional analogy holds between two word pairs: $A{:}B{::}C{:}D$, which means A is to B as C is to D. Mikolov et al. [13] first reported that word embeddings capture relational similarities and word analogies. Analogy tasks answer the questions such as what is the word X that is similar to *woman* in the same way that *King* is similar to *man*? The answer is expected to be *Queen*, if the model is trained well: $King + (woman - man) = Queen$

An analogy query is answered by performing algebraic operations over the word vectors to find the angular distance for the query. For example, in the embedding space, cosine similarity can be used to find a word X from *woman* in the same distance and angle as *King* from *man* using vector analysis. The embedding vectors are all normalized to unit norm. X is the continuous space representation of the word for the answer. If no word is found in that exact position, the word vector with the greatest cosine similarity to X is the answer.

Analogical reasoning method *3CosAdd* [13], which is based on the offset of word vectors, is used in this study, to find the answer to an analogy query and to show how semantically meaningful terms are related. To maintain consistency, we use this cosine similarity measure for all tests on our embeddings.

4   Results

The WAMEX dataset contains unannotated geological text reports obtained from Geological Survey of Western Australia (GSWA)[8]. The dataset contains 33,824 geological reports with 42.6 million tokens, while after filtering using our domain dictionary, the *WAMEX terms dataset* is fifteen times smaller than the WAMEX dataset and only contains the words that are valid mineralisation system terms. The number of tokens is reduced to 2.8 million.

Six sets of embeddings as shown in Table 1 are prepared for this research. Two geological embeddings are trained on the pre-processed WAMEX dataset using Word2Vec and FastText models, respectively. The Word2Vec embedding is named *Word2Vec raw embedding* and the FastText embedding is named *FastText raw embedding*. Another two geological embeddings are trained by Word2Vec and FastText models on the WAMEX terms dataset which only contain terms representing geological entities that of interests to the mineralisation process. The so learned embeddings are named *Word2Vec terms embedding* and *FastText terms embedding* respectively. These four embeddings are created with the CSG model using the GenSim package[9]. The following hyper-parameters are used for the

---

[8] http://www.dmp.wa.gov.au/WAMEX-Minerals-Exploration-1476.aspx.
[9] https://radimrehurek.com/gensim/.

training: dimensionality of vectors is set to 100, window size is set to 5 (window of neighbouring five words), negative sampling size is set to 5, and minimum count for frequency of words is set to 300. In addition, two pre-trained embeddings are downloaded: *Word2Vec pre-trained embedding* (see footnote 1) and *FastText pre-trained embedding* (see footnote 3). They are pre-trained on Google news of 100 billions words and web crawl of 600 billion words, respectively. The vocabularies include general knowledge, including geological terminologies.

Table 1. Embeddings

Data	Word2Vec embedding	No. of vec.	FastText embedding	No. of vec.
WAMEX dataset	*Word2Vec raw*	8,562	*FastText raw*	8,730
WAMEX terms	*Word2Vec terms*	837	*FastText terms*	838
General domain	*Word2Vec pre-trained*	3 mil.	*FastText pre-trained*	2 mil.

Note: Word2Vec Raw and FastText Raw embeddings are trained on the pre-processed WAMEX dataset.

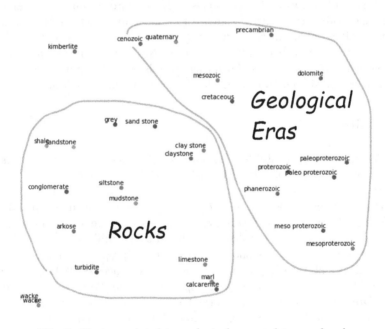

Fig. 5. Clusters related to geological eras and types of rocks

4.1   t-SNE Clustering and Visualisation

We visualise *Word2Vec Terms Embedding*, which is created by Word2Vec model on dictionary-based terms, in order to explore geological terms and their relations in WAMEX dataset. We use *t-SNE* for visualisation and validation of the trained vectors in a 2D vector space. About 840 unique geological entities are visualised. Figure 5 shows clusters of entities related to geological eras and types of rocks. The geological eras in our visualisation are mentioned in the Wikipedia page about the Geologic time scale[10]. The rocks appear in Fig. 5 are mentioned in the Wikipedia page about Sedimentary Rocks[11].

This result shows the groupings of different types of geological information, i.e. entity groups such as geological eras, rock types are effectively clustered using t-SNE. Similar meaningful results are also obtained for entities related to *iron ore*. The relevance are confirmed by Wikipedia. Interestingly, the location name distance on t-SNE have great resemblance to their locations on GoogleMap.

4.2 Similarity Query

To compare and understand what types of semantic information are captured through the representational learning, extensive similarity queries are conducted on all six sets of word embeddings. Table 2 shows the top ten similar tokens given two commodities as query inputs, one for gold and one for iron ore.

The query results demonstrate that the geological domain specific embeddings provide more relevant and useful information than the pre-trained Google or FastText vectors, despite the much smaller training corpus. In particular, dictionary-based entity embeddings retrieved more relevant terms than other embeddings. The largest endowment of gold in Western Australia is located in the Kalgoorlie gold camp. Whereas, iron ore, the Hamersley Province in the Pilbara region of Western Australia contains world class iron ore deposits. Banded iron-formations are a dominant host to the iron ores. The similarity query results contain relations between minerals and their associated locations, host rock types and geological eras. These provide more critical information for mineral explorers operating in WA using our domain/region (WA) specific word vectors than the pre-trained vectors by Google and Facebook.

To validate our *Geological entities' corpus* further, let's take a look at another example on the geological entity ashburton formation using *Word2vec Terms embedding*. The query for most similar entities for ashburton formation returned top five most similar entities as follows: wyloo group, mount minnie group, mount mcgrath formation, june hill volcanics, and capricorn group. This result is checked using the Explanatory Notes System (ENS) in Fig. 6, available from the GSWA, which stores relevant geological descriptions (e.g. formal names, rock compositions and age) and interpretations between major rock groups in Western Australia. These explanatory notes include the stratigraphic unit description of the Ashburton Formation, which contains collected field observations. The use of this independent data source from the training corpus, provides an unbiased assessment of the embedding analysis results. Stratigraphic information is important for mineral explorers as most mineral

---

[10] https://en.wikipedia.org/wiki/Geologic_time_scale.

[11] https://en.wikipedia.org/wiki/Sedimentary_rock.

Table 2. Similarity test

Embedding	Term	Top 10 most similar words
Word2Vec raw	*gold*	*au, copper, precious, nickel, metal, antimony, arsenic, copper-gold, tantalum, tungsten*
	iron ore	*dso, iron, hematite, cid, hypogene, windarling, bid, did, undercover, ore*
FastText raw	*gold*	*au, copper, nickel, antimony, metal, precious, arsenic, tungsten, zinc, historically*
	iron ore	*iron, ironcap, brockman, ironstones, marra, manganese, bif, ironstone, wittenoom, nammuldi*
Word2Vec terms	*gold*	*mineralisation, surface_gold, gold_mineral, kalgoorlie, mineralization, metal, archaean, greenstone_belt, western_australia, nickel*
	iron ore	*iron, hematite, west_angelas, marandoo, martite, hamersley, mount_jackson, banded_iron_formation, windarling, western_australia*
FastText terms	*gold*	*mineralisation, surface-gold, gold-mineral, metal, kalgoorlie, mineralization, archaean, western-australia, greenstone-belt, nickel*
	iron ore	*iron, hematite, west-angelas, banded-iron-formation, windarling, mount-jackson, marandoo, wickham, hamersley, western-australia*
Word2Vec pre-trained	*gold*	*silver, precious_metal, doomed_Moussa, Nastia_deserved, Gold, Oludamola_stripped, precious_metals, Cuba_Yumileidi Cumba_Jay, bullion, Safiullah_seated*
	iron ore	*Iron_ore, coking_coal, manganese_ore, iron_ores, ore, thermal_coal, Iron_Ore, iron_ore_exporter, steelmaking_ingredient, steelmakers*
FastText pre-trained	*gold*	*silver, Gold, Silver, GOLD, gold-, gold., gold.Gold, platinum, bronze, diamond*
	iron ore	*Iron-ore, bauxite, ferrochrome, Gindalbie, ferro-alloy, chromite, steel-making, zinc-lead, base-metals, bauxites*

Note: WAMEX abbreviations: *dso* - direct shipping ore, *cid* - channel iron deposit, *bid* - bedded iron deposit, *did* - detrital iron deposit, and *bif* - banded iron-formation

deposits are controlled by structures (e.g. geological faults) and/or stratigraphic relationships (e.g. banded iron-formations that host iron ores).

These findings are further confirmed by the domain experts that domain-specific embeddings learned from geology related data provide more targeted knowledge for geological applications.

4.3   Analogy Test

As shown in the similarity query results in Table 2, more morphological relations are present in the Pre-trained embeddings, for example, `iron ore` and `Iron_Ore`, whereas more semantic relations are captured by the purposely trained embeddings. For example, we can see commodity and geochemical name (e.g. `gold:au`),

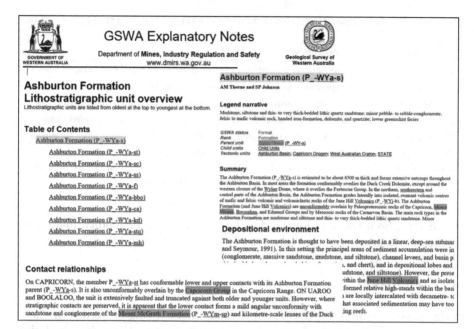

Fig. 6. A GSWA explanatory note extract for Ashburton Formation

commodity and geological era (e.g. gold:archaean). The domain experts confirmed that the most relations that are critical to the understanding of mineralisation systems are present in the embeddings learnt from the WAMEX Terms dataset.

Take the most intuitive Commodity:Location relation as example, we perform an analogy test, with results shown in Table 3. Our geological terms vectors from WAMEX dataset reflect detailed information such as town names for the *iron ore* in WA, while Google pre-trained vectors represent the general knowledge. For example, the query Kalgoorlie + (iron ore - gold) should return terms related to iron ore in the same way as Kalgoorlie relates to gold.

The result from Google news vectors return the location names of Pilbara, Port_Hedland and Karratha, which are closely associated with commodity-related locations. Our vectors trained on WAMEX reports return hematite - the most important ore of iron, martite - a type of iron ore, marandoo - the Marandoo iron ore mine in the Pilbara region, iron, west_angelas - the West Angelas iron ore mine in the Pilbara region, windarling - Windarling Iron Ore Mine, mount_newman_member - Australian stratigraphic unit in Hamersley Basin.

The pre-trained Google vectors return the same types of entities, while our data shows all highly related entities, but with different types. For example, when we query associated terms for a location, Google vectors return results that are all location names, while our results return related entities of locations and minerals. More data improves these results and helps to return only entities

Table 3. Relation Commodity:Location. Query: `kalgoorlie` + `iron-ore` − gold

Embedding	Top answers
Word2Vec raw	*pannawonica, windarling, mmif, ravensthorpe, karratha, newman*
FastText raw	*esperance, geraldton, bunbury, hyden, karratha*
Word2Vec terms	*martite, marandoo, iron, west_angelas, hematite, windarling*
FastText terms	*hematite, west-angelas, windarling, cunderdin, marandoo, iron*
Word2Vec pre-trained	*Pilbara, Pilbara_region, Port_Hedland, Pilbara_iron_ore, Karratha*
FastText pre-trained	*Pilbara, Karratha, Oakajee, Middlemount, Nullagine*

Note: The blue text marks a location name and the underlined text marks a commodity name. *mmif* is the abbreviation for Marra Mamba Iron Formation.

of that type. So if the question word was a location or a commodity, more data during the training helps to return only locations or commodity types.

5   Conclusion

In this paper, we investigated how representational learning of words can affect the entity retrieval results from a large domain corpus. Extensive similarity tests and analogy queries have been performed, which demonstrated the necessity of training domain-specific word embeddings. Pre-trained embeddings are good at capturing morphological relations, but are inadequate for domain specific semantic relations. This seemed to only confirmed the obvious, but we also demonstrated that the importance of entity extraction. A dictionary based entity extraction filter is used to create the entity-only datasets, with the sentence structure completely removed. The embeddings trained over the large sequence of entities using Word2Vec and FastText provide meaningful domain-specific semantic relations better than the embeddings based on the raw data. All results are confirmed by multiple sources, such as Wikipedia, relevant external datasets (e.g. GSWA Exploratory Notes), and more importantly, domain experts.

The success of this initial investigation confirmed the feasibility of using vector representations of words for concept or entity retrieval. Other types of embeddings such as those generated from non-linear context should also be investigated. Different analogy solvers such as knowledge graph models are also currently under investigation.

References

1. Mikolov, T., Sutskever, I., Chen, K., Corrado, G.S., Dean, J.: Distributed representations of words and phrases and their compositionality. In: Advances in Neural Information Processing Systems, pp. 3111–3119 (2013)
2. Miller, G.A.: WordNet: a lexical database for English. Commun. ACM **38**(11), 39–41 (1995)

3. Kusner, M., Sun, Y., Kolkin, N., Weinberger, K.: From word embeddings to document distances. In: International Conference on Machine Learning, pp. 957–966 (2015)
4. Lai, S., Xu, L., Liu, K., Zhao, J.: Recurrent convolutional neural networks for text classification. In: AAAI, vol. 333, pp. 2267–2273 (2015)
5. Lample, G., Ballesteros, M., Subramanian, S., Kawakami, K., Dyer, C.: Neural architectures for named entity recognition. arXiv preprint arXiv:1603.01360 (2016)
6. Iacobacci, I., Pilehvar, M.T., Navigli, R.: Embeddings for word sense disambiguation: an evaluation study. In: Proceedings of the 54th Annual Meeting of the Association for Computational Linguistics (Volume 1: Long Papers), vol. 1, pp. 897–907 (2016)
7. Mikolov, T., Chen, K., Corrado, G., Dean, J.: Efficient estimation of word representations in vector space. arXiv preprint arXiv:1301.3781 (2013)
8. Pennington, J., Socher, R., Manning, C.: GloVe: global vectors for word representation. In: Proceedings of the 2014 Conference on Empirical Methods in Natural Language Processing (EMNLP), pp. 1532–1543 (2014)
9. Bojanowski, P., Grave, E., Joulin, A., Mikolov, T.: Enriching word vectors with subword information. arXiv preprint arXiv:1607.04606 (2016)
10. Mikolov, T., Grave, E., Bojanowski, P., Puhrsch, C., Joulin, A.: Advances in pre-training distributed word representations. arXiv preprint arXiv:1712.09405 (2017)
11. Mikolov, T., Dean, J., Le, Q., Strohmann, T., Baecchi, C.: Learning representations of text using neural networks. In: NIPS Deep Learning Workshop, pp. 1–31 (2013)
12. Google archive: Word2vec (2013). https://code.google.com/archive/p/word2vec/. Accessed 01 March 2018
13. Mikolov, T., Yih, W.T., Zweig, G.: Linguistic regularities in continuous space word representations. In: Proceedings of the 2013 Conference of the North American Chapter of the Association for Computational Linguistics: Human Language Technologies, pp. 746–751 (2013)
14. Li, B., et al.: Investigating different syntactic context types and context representations for learning word embeddings. In: Proceedings of the 2017 Conference on Empirical Methods in Natural Language Processing, pp. 2421–2431 (2017)
15. Rong, X.: Word2vec parameter learning explained. arXiv preprint arXiv:1411.2738 (2014)
16. Gladkova, A., Drozd, A., Matsuoka, S.: Analogy-based detection of morphological and semantic relations with word embeddings: what works and what doesn't. In: Proceedings of the NAACL Student Research Workshop, pp. 8–15 (2016)
17. Drozd, A., Gladkova, A., Matsuoka, S.: Word embeddings, analogies, and machine learning: beyond king-man + woman = queen. In: Proceedings of COLING 2016, the 26th International Conference on Computational Linguistics: Technical Papers, pp. 3519–3530 (2016)
18. Levy, O., Goldberg, Y.: Linguistic regularities in sparse and explicit word representations. In: Proceedings of the Eighteenth Conference on Computational Natural Language Learning, pp. 171–180 (2014)
19. Turney, P.D.: Similarity of semantic relations. Comput. Linguist. **32**(3), 379–416 (2006)
20. Turney, P.D.: Domain and function: a dual-space model of semantic relations and compositions. J. Artif. Intell. Res. **44**, 533–585 (2012)
21. Bird, S., Klein, E., Loper, E.: Natural Language Processing with Python: Analyzing Text with the Natural Language Toolkit. O'Reilly Media Inc., Newton (2009)
22. van der Maaten, L., Hinton, G.: Visualizing data using t-SNE. J. Mach. Learn. Res. **9**(Nov), 2579–2605 (2008)

Keep Calm and Know Where to Focus: Measuring and Predicting the Impact of Android Malware

Junyang Qiu[1(✉)], Wei Luo[1], Surya Nepal[2], Jun Zhang[3], Yang Xiang[3], and Lei Pan[1]

[1] School of Information Technology, Deakin University, Geelong, Australia
{qiuju,wei.luo,l.pan}@deakin.edu.au
[2] Data61, CSIRO, Melbourne, Australia
Surya.Nepal@data61.csiro.au
[3] Digital Research and Innovation Capability Platform,
Swinburne University of Technology, Melbourne, Australia
{junzhang,yxiang}@swin.edu.au

Abstract. Android malware can pose serious security threat to the mobile users. With the rapid growth in malware programs, categorical isolation of malware is no longer satisfactory for security risk management. It is more pragmatic to focus the limited resources on identifying the small fraction of malware programs of high security impact. In this paper, we define a new research issue of measuring and predicting the impact of the detected Android malware. To address this issue, we first propose two metrics to isolate the high impact Android malware programs from the low impact ones. With the proposed metrics, we created a new research dataset including high impact and low impact Android malware samples. The dataset allows us to empirically discover the driving factors for the high malware impact. To characterize the differences between high impact and low impact Android malware, we leverage features from two sources available in every Android application. (1) the readily available *AndroidManifest.xml* file and (2) the disassembled code from the compiled binary. From these characteristics, we trained a highly accurate classifier to identify high impact Android malware. The experimental results show that our proposed method is feasible and has great potential in predicting the impact of Android malware in general.

Keywords: Android malware · Research malware dataset
High impact malware · Low impact malware · Machine learning
Static analysis

1 Introduction

Android has become the most popular mobile operating system, with 86.1% of the world's mobile phones running Android in Q1, 2017 [1]. The popularity

© Springer Nature Switzerland AG 2018
G. Gan et al. (Eds.): ADMA 2018, LNAI 11323, pp. 238–254, 2018.
https://doi.org/10.1007/978-3-030-05090-0_21

of Android, in combination with its openness, causes the number of Android malware skyrocketed in both the official and third-party Android application stores. In Q1, 2017 alone, over 750,000 new Android malware were detected by G DATA security experts [21]. That means that almost 8,400 new malware samples were discovered every day. At the same time, the number of Android malware families have reached 295 by 2016, according to the statistic report produced by McAfee [25]. The growing Android malware has caused severe harms to both end users as well as service providers.

To maintain a healthy ecosystem for Android, as expected, the research communities and security corporations have proposed various techniques to detect Android malware, e.g., software engineering based technique [4,6] and machine learning based technique [2,3]. Traditionally, according to whether executing the application, the existing techniques can be categorized into static analysis, dynamic analysis, or hybrid analysis. Among these proposed methods, almost all the current works focus on determining whether the candidate Android application is malicious or benign. However, with thousands of Android malware emerge every day, but only a small part of them will cause great damage to mobile users and human society. Thus it is more important and meaningful to predict whether the detected malware is of high impact.

In this paper, we address the issue of how to predict the impact of the detected Android malware. To the best of our knowledge, there is no similar work towards classifying the detected malware into high impact or low impact. The current works are all related to the detection of Android malware. To investigate the solutions to our proposed research question, we are facing three key challenges. The first challenge is how to measure the impact of Android malware. The second challenge lies that there is no dataset available which contains both high impact and low impact malware samples. The final challenge is how to predict the impact of the newly detected malware.

Towards addressing these identified challenges, our paper makes the following contributions:

Firstly, we propose two metrics (one is the number of compromised Android devices, and the other is the number of infected countries) to measure the impact of Android malware families. We obtain the metrics values for each family from mobile security reports, blogs or news released by security-related corporations or labs.

Secondly, according to the proposed metrics, we set the thresholds to filter the high-impact Android malware. Then we create a new dataset including both high impact and low impact Android malware samples based on four open source Android malware datasets. Besides, to facilitate the future research work in the area of malware impact prediction, we will release the dataset to the research community.

Next, to characterize the intrinsic nature of high-impact and low-impact Android malware, we extract various features from *AndroidManifest.xml* file and disassembled code file, e.g., requested permissions or suspicious API calls. The discriminative power of the extracted features is visualized using t-SNE feature reduction method. Then we perform the classification experiments to validate the feasibility of our proposed approach.

Finally, from the experiments, we find that the requested permissions, intent filters and API calls play an important role in discriminant high impact and low impact malware. Besides, the features extracted from *AndroidManifest.xml* have greater representation power than features from disassembled code.

The rest of this paper is structured as follows: Sect. 2 presents the related work of Android malware detection. Our proposed methodology is discussed in Sect. 3. In Sect. 4, the evaluation metrics are introduced. We show the experiments results and detailed analysis in Sect. 5. The limitations of our work and some conclusions and future research directions are given in Sects. 6 and Sect. 7, respectively.

2   Related Work

2.1   Static Analysis Techniques

Static analysis technique is performed by disassembling and inspecting the source code of the given Android application without executing it. The static analysis technique can be categorized into three methods. The first is named signature-based method which generates a robust signature using the specific strings or semantic patterns in the source code [14]. If an application's signature matches with existing malware's signatures, then it will be classified as a malware. AndroSimilar [13], DroidAnalytics [32] and ASTROID [15] are all signature-based methods. The second method is permission-based and focuses on analyzing the requested permissions in the *AndroidManifest.xml* file to identify malware. Most notably permission based analysis works are Puma [24], Droidmat [28] and [20]. The third method is based on Dalvik bytecode. The key building block of this method is to recover the source code. A lot of works attempt to decompile the bytecode into the source code [5,10,12,22]. After the bytecode has been decompiled into source code, the semantic information, e.g., API call graph, control flow graph, or data flow graph can be extracted to capture the patterns of malicious and benign applications [2,26,30].

2.2   Dynamic Analysis Techniques

Dynamic analysis technique monitors the behaviors of Android application during execution. The first dynamic analysis system for Android was TaintDroid, an efficient, system-wide dynamic taint tracking and analysis system capable of simultaneously tracking the flow of sensitive data [11]. Another notable dynamic analysis work was DroidScope which can reconstruct both the Linux OS level and Java Dalvik level semantic information simultaneously and seamlessly [29]. In 2013, a dynamic framework for automatically analyzing Android applications, AppsPlayground, was proposed. AppsPlayground integrated various automatic detection and exploration techniques (e.g., taint tracing tool [11], kernel level system call monitoring) for large-scale Android applications analysis [23]. Zhang *et al.* designed a dynamic analysis platform named VetDroid, which can reconstruct sensitive behaviors in Android applications from a new permission use perspective [31]. In 2014, a client-side approach that leveraged light-weight operating system level virtualization was presented to enhance the security of Android platform and facilitate the defense capability against Android malware [27].

2.3 Machine Learning Based Techniques

The low efficiency and difficulty of manually constructing and updating detection patterns for Android malware detection has motivated the application of machine learning. Several approaches have been proposed to automatically analyze applications using machine learning methods. For example, Drebin is a classic machine learning based method that enables detecting Android malware on the device [3]. The features for Drebin are extracted from both *AndroidManifest.xml* as well as *classes.dex* files. Besides, the methods DroidMat [28], Crowdroid [7], Adagio [16], MAST [8], and DroidAPIMiner [2] analyze features statically extracted from Android applications using machine learning techniques. Generally, the key of machine learning based technique is to extract informative and robust features for malware detection. The features used can be categorized into statistic features (e.g., requested permissions, API calls [3]), tree features (e.g., abstract syntax tree) and graph features (e.g., control flow graph, API call graph [19], data flow graph).

3 Methodology

In this section, we present our proposed framework in Fig. 1. Firstly, we collect the Android malware data, and then label the malware data into high impact and low impact. Secondly, we extract the semantic features from the *AndroidManifest.xml* as well as disassembled code to characterize each Android malware sample, the feature selection task can also be performed to select the informative and robust feature set. Thirdly, with the extracted features, the model can be trained to classify the Android malware as high impact or low impact. In the rest of this section, we present each of these steps in detail.

3.1 Data Acquisition

The key challenge of our proposed framework is data collection and data preprocessing. Currently, there are many open source Android malware datasets available for the research community. However, to the best of our knowledge, there is no dataset presenting the high impact or low impact groundtruth. Thus, we need to create a dataset containing the high impact or low impact labels for our proposed framework.

In our work, the created dataset is based on four open source Android malware datasets:

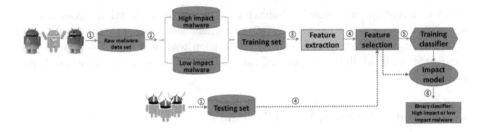

Step 1: ① collect Android malware APKs   ② data preprocessing (filtering the high impact malware based on 2 metrics)

Step 2: ③ feature extraction (**semantic features** based on reverse engineering)

Step 3: ④ feature selection

Step 3: ⑤ model training (SVM, Random Forest)

Step 4: ⑥ detection/classification and results explanation

Fig. 1. The mainly framework of our proposed method

1. contagion mobile (http://contagiominidump.blogspot.com.au/)
   This is a very influential open source malware dataset release site. Contagio mobile mini-dump offers an upload dropbox for everyone to share their mobile malware samples. And people can also download any samples. Currently, this site contains labeled (family label) Android malware data from 2011 till now.
2. https://github.com/fs0c131y/Android-Malwares (Accessed on 7 November, 2017)
   This repository is a collection of Android malware samples. For each sample, we can find the link with the analysis. Currently, there are 292 Android malware samples belonging to 33 malware families.
3. https://github.com/ashishb/android-malware (Accessed on 7 November, 2017)
   This open source site provides a collection of android malware samples. Everyone can share their Android malware samples in this site. Currently, this site contains 144 Android malware samples belonging to 26 families
4. Android Malware Dataset (http://amd.arguslab.org/)

This is a carefully-labeled and well-studied dataset that includes comprehensive profile information of malware. Currently, the dataset contains 24,553 samples, categorized in 135 varieties among 71 malware families ranging from 2010 to 2016.

To simplify the problem, the high impact or low impact of Android malware data is collected based on malware family-grained. In other words, the Android malware samples belong to the high impact Android malware families are considered as high impact malware samples, and Android malware samples belong to other Android malware families are collected as low impact malware samples. The base Android malware families list can be accessed through the link https://forensics.spreitzenbarth.de/android-malware/.

3.2   Impact Metrics of Malware

Now the key problem is how to divide the Android malware families into high impact or low impact. Intuitively, we can employ many metrics to evaluate the impact of malware, e.g., the financial loss caused by the given malware. However, in reality, we can not obtain the accurate metrics value due to the resources limitation. Thus, in this work, to simplify the research problem, two metrics are proposed to measure the impact of the malware families.

1. *The number of infected mobile devices that the malware family has caused*
2. *The number of countries that the malware family has compromised.*

Table 1. The statistical information of the collected high impact Android malware data

Year	Family label	Number of samples	Number of infected countries	Number of infected devices
2012	Plankton	25	Null	Over 5 million
2012	Carberp	3	Over 10	Null
2012	MMarketPay	14	Over 1	Over 100,000
2012	Counterclank	5	Null	5 million
2013	Opfake	10	97	over 1 million
2013	Zitmo	24	Over 2	Over 3.5 million
2014	Koler	69	30	Over 200,000
2015	Fakeinst	2172	Over 130	Over 5 million
2015	Simple Locker	173	Over 1	Over 150,000
2016	Copycat	9	Over 5	Over 14 million
2016	Fusob	1277	Over 100	Null
2016	Hummingbad	590	Over 100	85 million
2016	Pokemongo	1	3	Over 7.5 million
2016	Svpeng	34	Over 23	Over 318,000
2016	Ztorg	28	Null	Over 1 million
2016	Viking Horde	7	13	Over 100,000
2017	Chrysaor	3	Over 11	Null
2017	Judy	14	Null	Over 36 million
2017	Xavier	6	Over 6	Over 100 million

For each malware family, we attempt to obtain the above two metrics values from various information sources through the Google Browser. In this paper, the information sources can be cyber security white papers (reports) or security news released by security related corporations, laboratories, e.g., *CheckPoint*[1], *Symantec*[2], *Kaspersky*[3] and *McAfee*[4]. For example, from the 2016 research report of *CheckPoint*, we can find a high impact Android malware family named *Copycat*, it has infected over 14 million Android phones, rooted approximately 8 million of them, and earned the hackers behind the campaign approximately $1.5 million in fake ad revenues in two months[5].

---

[1] https://www.checkpoint.com/products-solutions/mobile-security/.

[2] https://www.symantec.com/.

[3] https://www.kaspersky.com/.

[4] https://www.mcafee.com/us/index.html.

[5] https://www.checkpoint.com/downloads/resources/copycat-research-report.pdf.

Table 2. The statistical information of the collected low impact Android malware data

Year	Family label	Number of samples	Year	Family label	Number of samples
2010	FakePlayer	21	2011	DroidKungfu	546
2011	GingerMaster	128	2011	GoldDream	53
2012	Boxer	44	2012	FakeAngry	10
2012	FakeDoc	21	2012	FakeTimer	12
2012	Fjcon	16	2012	Lotoor	329
2012	MobileTX	17	2012	Nandrobox	76
2012	Penetho	18	2012	SmsZombie	9
2012	SpyBubble	10	2012	Steek	12
2012	UpdtKiller	24	2013	AndroidRat	45
2013	Bankun	70	2013	Boqx	215
2013	FakeUpdates	5	2013	Gumen	145
2013	Ksapp	36	2013	Kyview	175
2013	Lnk	5	2013	Monimob	203
2013	Mseg	235	2013	Mtk	67
2013	Obad	9	2013	SmsKey	165
2013	Spambot	15	2013	Stealer	25
2013	Tesbo	5	2013	Vidro	23
2013	Vmvol	13	2013	Winge	19
2014	Airpush	7840	2014	Andup	45
2014	Aples	20	2014	Cova	17
2014	Erop	46	2014	FakeAV	5
2014	Finspy	9	2014	Jiust	560
2014	Ramnit	8	2014	Univert	10
2014	Utchi	12	2015	BankBot	724
2015	Dowgin	3377	2015	Fobus	4
2015	Gorpo	37	2015	Kemoge	14
2015	Kuguo	1199	2015	Leech	127
2015	Mecor	1820	2015	Ogel	6
2015	Roop	48	2015	SlemBunk	174
2015	Youmi	1301	2016	RuMMS	311
2016	Triada	210			

In this paper, the malware family will be categorized into high impact malware if one of the following two conditions is met. Otherwise this family will be considered as low impact malware.

Condition 1: infected more than 100,000 mobile devices
Condition 2: infected more than 10 countries.

The detailed information of the collected high impact Android malware data is listed in Table 1, while the low impact Android malware data information is in Table 2.

3.3   Feature Engineering and Embedding

After the training dataset has been created, the next issue is extracting the informative and robust features to characterize the Android malware so as to better distinguish the high impact and low impact malware samples.

In this work, the semantic features based on the *AndroidManifest.xml* and disassembled code are extracted. The information stored in *AndroidManifest.xml* file, e.g., requested hardware components, requested permissions, app components and filtered intents can be efficiently extracted to characterize the impact of Android malware. We also disassemble the bytecode and extract all API calls and strings included in the application.

The detailed features information can be found in Table 3. The former 4 feature sets are extracted from *AndroidManifest.xml*, while the latter 4 are from the disassembled code. The first column is feature source, while the second column is feature type. In column three we present the description of each feature type. It is noted that the size of each feature type is determined by the training dataset, the number of features in column 4 is calculated based on the experiment in Subsects.5.2 and 5.3. The impact of Android malware can be represented in specific patterns and combinations of the extracted features. After the string features for each application have been extracted, we need to map all the extracted features into a joint vector space. Then an application a is mapped to this vector space. For each feature f contained in a the respective dimension is set to 1 and all the other dimensions are set to 0. By mapping each Android application to the joint vector space, those applications sharing similar features will lie close to each other in this representation, while applications with different features will be separated by large distances. However, this representation method will produce high dimensional vector. To improve efficiency and reduce storage memory, in this work, the generated feature matrix of samples are represented using Compressed Sparse Column matrix.

3.4   Model Training and Testing

In this step, the classification task is performed, i.e., labeling Android malware as either high impact or low impact. To this end, we test the proposed framework using Support Vector Machine [17] classification algorithms. Each model is trained using the feature vectors extracted from the Android malware training samples. The detailed experimental results and analysis are presented in Sect. 4.

Table 3. The detailed description of the extracted features

Features	Detailed information about the features	Number of features
Requested permissions	An application will request permissions from users to access sensitive resources at installation	1495
Application components	Each Android application can state several components, e.g., Activity, Service, ContentProvider, Broadcastreceiver	42767
Filtered intents	The intent filter of a component specifies the action it can perform and the data type it is able to operate	3065
Hardware components	Requesting certain hardware or a combination of specific hardwares have potentially security implications	31
Critical API calls	The critical calls can be extracted to deeply reveal the sensitive functionality of an application	209
Suspicious API calls	The suspicious API calls are extracted to reflect suspicious behaviors of malware	45
Used permissions	The critical API calls can be employed to determine and match both requested and actually used permissions	40
Network addresses	Some of network addresses may be involved in botnets or other malicious sites	459

4   Experimental Analysis

In the experiment setup part, to fully addressing the proposed research issue, we attempt to answer the following 3 research questions:

- RQ1: How can we better characterize the high impact and low impact Android malware employing *AndroidManifest.xml* information (e.g., requested permissions, application components)?
- RQ2: How to characterize the high impact and low impact Android malware using the knowledge extracted from disassembled code (e.g., critical API calls or suspicious API calls)?
- RQ3: Is it possible to predict the impact of newly detected Android malware?

4.1   Experimental Setting

In the feature extraction phase, we utilize the open source tool *Androguard,desnos2011androguard*. *Androguard* is a full python tool to play with Android files, e.g., Android's binary xml, disassemble DEX bytecodes. We employ the traditional SVM (Support Vector Machine) algorithm to perform the classification tasks. It is well known that SVM is an interpretable classification algorithm. It can clearly deliver the weight score of different features, which is important for us to understand and explain the classification results. The range of parameters of SVM are set as follows: $C = [1e3, 5e3, 1e4, 5e4, 1e5], gamma = [1e-4, 5e-4, 1e-3, 5e-3, 1e-2, 1e-1]$. We use grid search method to find the best parameters for each classification algorithm. 50% of the dataset serves as the training set and the rest 50% is the testing set.

4.2   Characterize the High Impact and Low Impact Android Malware Employing *AndroidManifest.xml* Information

To answer the first research question, in this subsection, we perform experiment based on the features from *AndroidManifest.xml* files.

To see the discriminative power of the extracted features, we present the visualization of the data using t-Distributed Stochastic Neighbor Embedding (t-SNE) method [18], t-SNE is a technique for dimensionality reduction that is particularly well suited for the visualization of high-dimensional datasets.

Figure 2(a) presents the visualization of the training data. It can be seen that the high impact and low impact malware is obviously distinct in *AndroidMani-fest.xml* feature space. In other word, the features from *AndroidManifest.xml* can better discriminant high impact and low impact Android malware. Figure 2(b) presents the visualization of the testing data. Compared Fig. 2(b) with Fig. 2(a), we can see that the distribution of testing data is similar to the training data, which will contribute to a good classification performance.

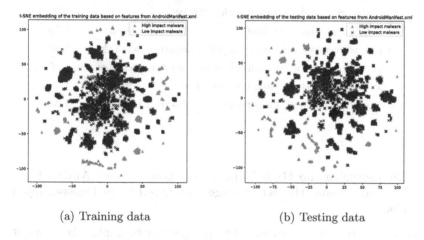

(a) Training data   (b) Testing data

Fig. 2. Visualization of the data using features from *AndroidManifest.xml*

In Table 4(a), we show the confusion matrix of SVM. Totally 7 low impact malware are misclassified as high impact while 17 high impact malware are incorrectly classified as low impact. Table 4(b) is the detailed classification performance of SVM. From the classification results of SVM, we can find that the features from *AndroidManifest.xml* have a better representation for low impact Android malware.

Table 4. The prediction results using *AndroidManifest.xml* features

(a) The confusion matrix of SVM

	LI-malware	HI-malware
LI-malware	6119	7
HI-malware	17	1140

(b) The detection results of SVM

	precision	recall	f1-score
LI-malware	1.00	1.00	1.00
HI-malware	0.99	0.99	0.99
average	1.00	1.00	1.00

We also rank the top 10 features according to their absolute value of importance score in Table 5. From Table 5, among the 10 top features, 5 are permission related and 3 are intent filters related, thus it is reasonable to say that the requested permissions and intent filters are more important in discriminant high impact and low impact malware. Besides, it is easy to find that the common suspicious and sensitive features, e.g., the short message related permission, camera permission or reboot permission are all listed.

Table 5. The top 10 important features from *AndroidManifest.xml*

Feature name	Score
servicelist_com.apperhand.device.android.androidsdkprovider	2.15
requestedpermissionlist_android.permission.receive_sms	1.84
intentfilterlist_android.intent.action.phone_state	1.72
requestedpermissionlist_android.permission.camera	1.65
intentfilterlist_android.intent.action.reboot	1.59
intentfilterlist_android.intent.action.quickboot_poweron	1.50
requestedpermissionlist_android.permission.read_settings	1.40
requestedpermissionlist_android.permission.write_settings	1.40
broadcastreceiverlist_com.ads.smsreceiver	1.36
requestedpermissionlist_android.permission.read_call_log	1.35

4.3 Characterize the High Impact and Low Impact Android Malware Using the Knowledge Extracted from Disassembled Code

For the second research question, in this part, we try to employ the information from the disassembled code to classify the high impact and low impact malware.

Figure 3(a) presents the visualization of the training data, while Fig. 3(b) is the visualization of the testing data. We can see that the high impact and low impact malware is clearly separated, which validate that the semantic knowledge extracted from disassembled code is effective in discriminating high impact and low impact Android malware.

In Table 6, we show the classification results of SVM using the features extracted from disassembled code. The performance is slightly worse than the results using *AndroidManifest.xml* based features but also satisfactory. Another finding is that features from both *AndroidManifest.xml* and disassembled code are weaker in characterizing high impact malware. Thus it is necessary and interesting to explore some more effective features to represent high impact Android malware.

To investigate the representation power of different features, the top 10 features from disassembled code are ranked in Table 7 according to their absolute

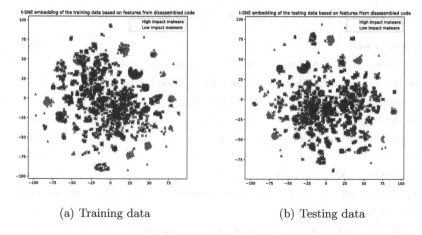

(a) Training data   (b) Testing data

Fig. 3. Visualization of the data using features from disassembled code

value of importance score. Based on the results in Table 7, we can obtain some conclusions. Firstly, the top 3 features are all about used permission, e.g., location access, phone state read and camera request, which are very sensitive and private to mobile users. Secondly, 3 suspicious or sensitive API calls are also listed. To sum up, the permissions and API calls play a dominant role in distinguishing high impact and low impact malware.

Table 6. The detection results employing disassembled code based features

	Precision	Recall	f1-score
LI-malware	0.99	0.99	0.99
HI-malware	0.95	0.96	0.96
Average	0.99	0.99	0.99

4.4 Predict the Impact of Newly Detected Android Malware

As Android malware evolve over time, so do the characteristics of both high impact and low impact applications. It is necessary to take into account such evolution when evaluating our proposed framework, since the accuracy might significantly be affected as newly Android malware may modify their strategies. Evaluating this aspect of our framework constitutes one of our research questions. In this section, we attempt to answer our 3rd research question: can we accurately predict the impact of the newly detected malware?

To address this issue, we train the model using older samples (Both high impact and low impact malware detected between 2011 to 2015) and newer

Table 7. The top 10 important features from disassembled code

Feature name	Score
usedpermissionslist_android.permission.access_fine_location	3.87
usedpermissionslist_android.permission.read_phone_state	3.15
usedpermissionslist_android.permission.camera	2.78
suspiciousapilist_landroid/telephony/telephonymanager.getsubscriberid	2.77
restrictedapilist_android.net.connectivitymanager.getnetworkinfo	2.60
urldomainlist_119.147.23.195	2.16
urldomainlist_52.220.234.108	2.00
restrictedapilist_android.location.locationmanager.getlastknownlocation	1.80
urldomainlist_54.149.205.221	1.79
usedpermissionslist_android.permission.restart_packages	1.77

samples (malware detected in 2016 and 2017) serve as testing set. In the experiments, the training set contains 3409 high impact malware and 15912 low impact malware, while the testing set contains 928 high impact malware and 4537 low impact malware.

We firstly compare the prediction power (here we use F1-score to measure the prediction power) of different feature type using SVM in Fig. 4. Some findings can be obtained from Fig. 4. Firstly, among features in *AndroidManifest.xml*, requested permissions, application components and filtered intents have the optimal prediction power for both high impact malware and low impact malware, while hardware components have the worst prediction power, especially for the high impact malware. Secondly, for features sourced from disassembled code, the prediction power of suspicious API calls and Used permissions is superior to critical API calls and network addresses. To sum up, *AndroidManifest.xml* based features have better prediction power than disassembled code based features.

The prediction results of SVM using all features from disassembled code can be found in Table 8(a), while Table 8(b) is the results using all features from *AndroidManifest.xml*. From these two Tables, we find that the classifier trained on features from *AndroidManifest.xml* has better prediction capability. In addition, the classifier has better prediction performance on low impact malware than on high impact malware. The potential reason is that the number of low impact malware is larger than high impact malware, so the classifier is prone to mis-classify the high impact malware as low impact. The f1-score is 1.00 and 0.99 for low impact and high impact malware, respectively, if we exploit the knowledge not only from *AndroidManifest.xml* but also from disassembled code. However, using more information imply more time- and memory-consuming in feature extraction (especially the code disassembly phase) and model training. In real deployment, it is crucial to select an appropriate feature set to make a balance between accuracy and efficiency.

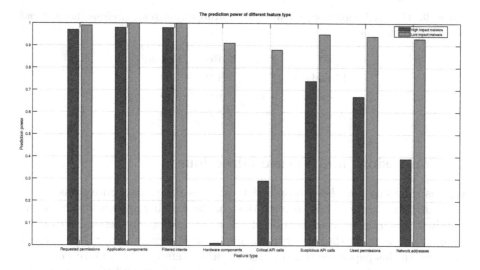

Fig. 4. The prediction power of different feature type

Table 8. The prediction results employing different features

(a) Features in disassembled code

	precision	recall	f1-score
LI-malware	0.99	0.99	0.99
HI-malware	0.96	0.95	0.95
average	0.98	0.98	0.98

(b) Features in *AndroidManifest.xml*

	precision	recall	f1-score
LI-malware	1.00	1.00	1.00
HI-malware	0.98	0.99	0.98
average	1.00	1.00	1.00

5   Limitations Analysis

In this paper, we make an attempt to address a new and challenging research question: How to predict the impact of the detected Android malware. However, our solutions to the research issue have several limitations:

Firstly, the metrics used to evaluate the impact of Android malware are simple and insufficient. Some of the metrics values are not from official sources. Besides, more direct metrics are needed, e.g., the financial loss caused by the malware.

Secondly, in this paper, to simplify the problem, the Android malware are divided into high impact and low impact. It will be more reasonable to introduce a comprehensive impact index considering different factors, e.g., classify the impact of malware into significant, high, medium and low (Table 9).

Thirdly, the created dataset is not large enough, some high impact Android malware samples may be not included in our dataset. What's more, in the paper, we just extract the features from *AndroidManifest.xml* files as well as the disassembled codes. Then these features are embedded into vectors. However, to better characterize the impact of Android malware, more informative and high level features are needed.

Table 9. The prediction results employing features from both *AndroidManifest.xml* and disassembled code

	Precision	Recall	f1-score
LI-malware	1.00	1.00	1.00
HI-malware	0.99	0.99	0.99
Average	1.00	1.00	1.00

6   Conclusions and Future Directions

Various solutions have been proposed to address the great security risk posed by Android malware to our society. Most of the existing approaches, however, mainly focus on binary classification of a new application into being malicious or benign. This is unsatisfactory with the exponential growth of newly discovered malware. In this paper, we raised a new yet important question: Once we detect a new malware application, can we predict its potential impact? We made several concrete steps towards answering this challenging research question. First, we proposed two metrics to separate the high-impact Android malware programs from the low impact ones. Next, based on the metrics, we created a new dataset containing both high impact and low impact Android malware samples. In the end, we employed two feature sets to characterize the Android malware samples, and empirically demonstrated the value of such characterizations in predicting high impact malware.

References

1. Global market share held by the leading smartphone operating systems in sales to end users from 1st quarter 2009 to 1st quarter 2017. https://www.statista.com/statistics/266136/global-market-share-held-by-smartphone-operating-systems/ (2017). Accessed 28 June 2017
2. Aafer, Y., Du, W., Yin, H.: DroidAPIMiner: mining API-level features for robust malware detection in Android. In: Zia, T., Zomaya, A., Varadharajan, V., Mao, M. (eds.) SecureComm 2013. LNICST, vol. 127, pp. 86–103. Springer, Cham (2013). https://doi.org/10.1007/978-3-319-04283-1_6
3. Arp, D., Spreitzenbarth, M., Hubner, M., Gascon, H., Rieck, K., Siemens, C.: DREBIN: effective and explainable detection of android malware in your pocket. In: NDSS (2014)
4. Arzt, S., et al.: Flowdroid: precise context, flow, field, object-sensitive and lifecycle-aware taint analysis for android apps. ACM SIGPLAN Not. **49**(6), 259–269 (2014)
5. Bartel, A., Klein, J., Le Traon, Y., Monperrus, M.: Dexpler: converting android Dalvik Bytecode to Jimple for static analysis with soot. In: Proceedings of the ACM SIGPLAN International Workshop on State of the Art in Java Program analysis, pp. 27–38. ACM (2012)
6. Bläsing, T., Batyuk, L., Schmidt, A.D., Camtepe, S.A., Albayrak, S.: An android application sandbox system for suspicious software detection. In: 2010 5th International Conference on Malicious and Unwanted Software (MALWARE), pp. 55–62. IEEE (2010)

7. Burguera, I., Zurutuza, U., Nadjm-Tehrani, S.: Crowdroid: behavior-based malware detection system for android. In: Proceedings of the 1st ACM Workshop on Security and Privacy in Smartphones and Mobile Devices, pp. 15–26. ACM (2011)

8. Chakradeo, S., Reaves, B., Traynor, P., Enck, W.: MAST: triage for market-scale mobile malware analysis. In: Proceedings of the sixth ACM conference on Security and privacy in wireless and mobile networks, pp. 13–24. ACM (2013)

9. Desnos, A.: Androguard (2011). https://github.com/androguard/androguard

10. Desnos, A., Gueguen, G.: Android: from reversing to decompilation. In: Proceedings of Black Hat Abu Dhabi, pp. 77–101 (2011)

11. Enck, W., et al.: Taintdroid: an information-flow tracking system for realtime privacy monitoring on smartphones. ACM Trans. Comput. Syst. (TOCS) $32(2)$, 5 (2014)

12. Enck, W., Octeau, D., McDaniel, P.D., Chaudhuri, S.: A study of android application security. In: USENIX Security Symposium, vol. 2, p. 2 (2011)

13. Faruki, P., Ganmoor, V., Laxmi, V., Gaur, M.S., Bharmal, A.: AndroSimilar: robust statistical feature signature for Android malware detection. In: Proceedings of the 6th International Conference on Security of Information and Networks, pp. 152–159. ACM (2013)

14. Feng, Y., Anand, S., Dillig, I., Aiken, A.: Apposcopy: semantics-based detection of android malware through static analysis. In: Proceedings of the 22nd ACM SIGSOFT International Symposium on Foundations of Software Engineering, pp. 576–587. ACM (2014)

15. Feng, Y., Bastani, O., Martins, R., Dillig, I., Anand, S.: Automated synthesis of semantic malware signatures using maximum satisfiability. In: NDSS (2017)

16. Gascon, H., Yamaguchi, F., Arp, D., Rieck, K.: Structural detection of android malware using embedded call graphs. In: Proceedings of the 2013 ACM Workshop on Artificial Intelligence and Security, pp. 45–54. ACM (2013)

17. Hearst, M.A., Dumais, S.T., Osuna, E., Platt, J., Scholkopf, B.: Support vector machines. IEEE Intell. Syst. Their Appl. $13(4)$, 18–28 (1998)

18. Hinton, G.E.: Visualizing high-dimensional data using t-SNE. Vigiliae Christianae $9(2)$, 2579–2605 (2008)

19. Hou, S., Ye, Y., Song, Y., Abdulhayoglu, M.: HinDroid: an intelligent android malware detection system based on structured heterogeneous information network (2017)

20. Huang, C.Y., Tsai, Y.T., Hsu, C.H.: Performance evaluation on permission-based detection for Android malware. In: Pan, J.S., Yang, C.N., Lin, C.C. (eds.) Advances in Intelligent Systems and Applications, vol. 2, pp. 111–120. Springer, Heidelberg (2013). https://doi.org/10.1007/978-3-642-35473-1_12

21. Lueg, C.: 8,400 new Android malware samples every day, April 2017. https://www.gdatasoftware.com/blog/2017/04/29712-8-400-new-android-malware-samples-every-day. Accessed 28 June 2017

22. Octeau, D., Jha, S., McDaniel, P.: Retargeting android applications to Java Bytecode. In: Proceedings of the ACM SIGSOFT 20th International Symposium on the Foundations of Software Engineering, p. 6. ACM (2012)

23. Rastogi, V., Chen, Y., Enck, W.: AppsPlayground: automatic security analysis of smartphone applications. In: Proceedings of the third ACM Conference on Data and Application Security and Privacy, pp. 209–220. ACM (2013)

24. Sanz, B., Santos, I., Laorden, C., Ugarte-Pedrero, X., Bringas, P.G., Álvarez, G.: PUMA: permission usage to detect malware in android. In: Herrero, Á., et al. (eds.) International Joint Conference CISIS'12-ICEUTE'12-SOCO'12 Special Sessionspp, pp. 289–298. Springer, Heidelberg (2013). https://doi.org/10.1007/978-3-642-33018-6_30

25. Snell, B.: Mobile threat report: what's on the horizon for 2016. Intel Security and McAfee, 1 March 2016

26. Wognsen, E.R., Karlsen, H.S., Olesen, M.C., Hansen, R.R.: Formalisation and analysis of Dalvik Bytecode. Sci. Comput. Program. **92**, 25–55 (2014)

27. Wu, C., Zhou, Y., Patel, K., Liang, Z., Jiang, X.: AirBag: boosting smartphone resistance to malware infection. In: NDSS (2014)

28. Wu, D.J., Mao, C.H., Wei, T.E., Lee, H.M., Wu, K.P.: DroidMat: android malware detection through manifest and API calls tracing. In: 2012 Seventh Asia Joint Conference on Information Security (Asia JCIS), pp. 62–69. IEEE (2012)

29. Yan, L.K., Yin, H.: DroidScope: seamlessly reconstructing the OS and Dalvik semantic views for dynamic android malware analysis. In: USENIX Security Symposium, pp. 569–584 (2012)

30. Yang, W., Xiao, X., Andow, B., Li, S., Xie, T., Enck, W.: AppContext: differentiating malicious and Benign mobile app behaviors using context. In: 2015 IEEE/ACM 37th IEEE International Conference on Software Engineering (ICSE), vol. 1, pp. 303–313. IEEE (2015)

31. Zhang, Y., et al.: Vetting undesirable behaviors in android apps with permission use analysis. In: Proceedings of the 2013 ACM SIGSAC Conference on Computer & Communications Security, pp. 611–622. ACM (2013)

32. Zheng, M., Sun, M., Lui, J.C.: Droid analytics: a signature based analytic system to collect, extract, analyze and associate android malware. In: 2013 12th IEEE International Conference on Trust, Security and Privacy in Computing and Communications (TrustCom), pp. 163–171. IEEE (2013)

Fault Diagnosis for an Automatic Shell Magazine Using FDA and ELM

Qiangqiang Zhao[✉], Lingfeng Tao, Maosheng Li, and Peng Hong

Jiangsu Jinling Institute of Intelligent Manufacturing Co. Ltd.,
Nanjing 210006, People's Republic of China
zqqlzl@139.com

Abstract. A fault diagnosis method for an automatic shell magazine based on Functional Data Analysis (FDA) and Extreme Learning Machine (ELM) is presented in this paper. A virtual prototype model of the automatic shell magazine includes a mechanical model and control model is built in RecurDyn and Simulink. The failure mechanism of the automatic shell magazine is analyzed, and the corresponding fault factors are selected. Due to an insufficient number of fault samples, a large number of fault samples are generated by the virtual prototype model and the fault samples are analyzed by FDA. Then, the eigenvalues from FDA are used to train ELM to obtain a diagnostic machine. The diagnostic machine is used for the fault diagnosis of the automatic shell magazine and is proved to be very effective.

Keywords: Fault diagnosis · Automatic shell magazine
Functional data analysis · Extreme learning machine

1 Introduction

Since the 1980s, fault diagnosis technology has entered into the intelligent diagnosis stage. Expert systems, neural networks, fuzzy inferences and support vector machines have been introduced in fault diagnosis, and making it capable of logical inference, self-learning, self-diagnosis and self-processing. Many achievements have been made in fault diagnosis on boilers, pressure vessels, nuclear power plants, nuclear reactors and railway vehicles. Currently, fault diagnosis is widely used in most of the developed countries and accounts for as much as 15%–20% of the savings in device maintenance [1].

It is worth noting, a data-driven fault diagnosis method need to analyze large volumes of data. However, the fault signals are processed as time-series signal in the above methods and most other signal processing methods. Thus the continuity and smoothness properties of the data are completely ignored. However, in Functional Data Analysis (FDA) and Functional Principal Component Analysis (FPCA), the observation data are regarded as a whole instead of discrete points, are expressed as smooth curve or continuous function, and are analyzed from a functional perspective. The greatest contributor to the topic of FDA is Ramsay [2–5]. Compared with the traditional methods, FDA rarely depends on modeling and assumed conditions, and it does not matter whether the observation points and observation frequency of the observed object are the same. Another feature of FDA is that it can transform the data from finite

© Springer Nature Switzerland AG 2018
G. Gan et al. (Eds.): ADMA 2018, LNAI 11323, pp. 255–262, 2018.
https://doi.org/10.1007/978-3-030-05090-0_22

dimensional to infinite dimensional; therefore, the data information achieved is more abundant and responsible. By assuming the functions are all derivable, FDA can extract more important information such as exploring the differences among the curves and the dynamic change between the curves by analyzing the first derivative or higher derivatives of the curves. FDA has been used in three-dimensional data analysis of the protein [6], geochemical data analysis [7], vehicle insurance field [8], neuroscience field [9], etc. However, to the best of my knowledge, FDA is a new data processing method, and it has not been used in the fault diagnosis of a mechatronic system.

For this reason, a new fault diagnosis method based on FDA and ELM is proposed in this paper. Considering that in the mechatronics system, most physical quantities change continuously over time with smoothness properties, a new data mining method named FDA is used to analyze the samples from the verified model. Finally, the fault eigenvalues from FDA are used to train the ELM to build a diagnostic machine. And the diagnostic machine is used for the fault diagnosis of an automatic shell magazine to verify the feasibility and availability of this method.

2 Modelling and Simulating

2.1 Mechanical Model of the Automatic Shell Magazine

An automatic shell magazine is a very complex mechatronics system. It consists of a support frame, a driving wheel, a driven wheel, twenty-five magazines, a reduction gearbox, some rollers, an angular velocity sensor (tachometer generator), and two linear potentiometers. The reduction gearbox includes a small gear, a large gear, a worm, and a worm gear. The mechanical model is simplified for reducing the human errors during the modelling and solving processing. The simplified model is shown in Fig. 1. The model consists of only one shell because there is only one shell in the magazine during the testing.

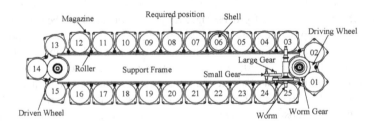

Fig. 1. The mechanical model of the automatic shell magazine

2.2 Control Model of the Automatic Shell Magazine

The automatic shell magazine is driven by two parallel DC series motors, which are controlled by analogue circuits using a position-speed closed-loop system. The rated power and voltage of the motor are 500 Walt and 26 V, respectively. Two linear potentiometers, BQ1 and BQ2, are used as the angle sensors, and the output of the linear potentiometer is voltage. BQ1 is connected with a digital controller to indicate the position of the magazines, and BQ2 is used to provide a position feedback signal to the control system. When the automatic shell magazine rotates once through a circle, the BQ1 also rotates through a circle. Therefore, the digital controller can judge which magazine is on the required position because the output of BQ1 is different for each magazine when it is located in the required position. Different from BQ1, BQ2 rotates once through a circle when the magazine only moves a displacement between two adjacent magazines. The output of BQ2 is 5 V when one magazine is accurately located in the required position.

According to the analysis hereinabove, the control model built in Simulink is shown in Fig. 2.

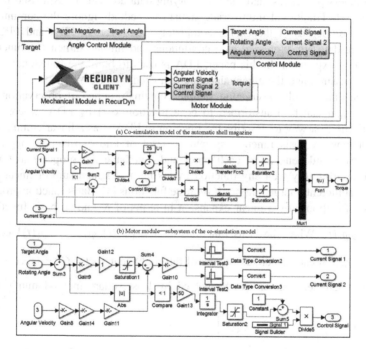

(a) Co-simulation model of the automatic shell magazine

(b) Motor module—subsystem of the co-simulation model

Fig. 2. The control model of the automatic shell magazine

2.3 Sampling and Simulating

The theoretical transmission efficiency of the worm and worm gear is 0.52, the theoretical sensitivity of the tachometer generator is 1, and the rated voltage of the motors is 26 V. The transmission efficiency of the worm and worm gear is in relation to the equivalent friction angle. The equivalent friction angle of the worm and worm gear is 1.72°–6.28°; therefore, the range of the transmission efficiency is 0.471–0.767 [10], obeying the Gumbel distribution [11]. The change of the sensitivity of the tachometer generator directly influences the location accuracy. Here, its range is 0.075 to 0.12, obeying the Weibull distribution [11]. The voltage of the motors is not stable because it is supplied by a battery. Here, its range is 24 V to 28 V, obeying the Gumbel distribution [11]. The three fault factors are sampled by Latin Hypercube Sampling (LHS) for balancing the distribution homogeneity in multi-dimensional space and one-dimensional space, and the sample size for each fault factor is 200.

3   Fault Diagnosis Based on FDA and ELM

3.1   Features Extraction Based on FDA and FPCA

3.1.1   The FDA and FPCA Algorithm

In a mechatronics system, many physical quantities are time-varying with continuity and smoothness properties, and the time-varying data are usually expressed by a discrete point set. Traditional multivariate statistical methods ignore the continuity and smoothness properties of the curves completely. However, in FDA and FPCA, the time-varying data are expressed as a continuous and infinite dimension function by basis function expansion. In this paper, FDA is used to build the functional data and smooth processing, and the eigenvalues are extracted by FPCA.

Assume that the ith observation sample contains a series of observation values $y_{i1}, y_{i2}, \cdots y_{in}$, it can be transformed to be a function $x_i(t)$ by FDA, where t is the argument. The best way to describe the function in FDA is as a linear combination of basis functions. A basis function system is a set of known functions ϕ_k that are mathematically independent of each other and that have the property that we can approximate arbitrarily well any function by taking a weighted sum or linear combination of a sufficiently large number K of these functions [4]. Basis function procedures represent a function x by a linear expansion in terms of K known basis function ϕ_k:

$x(t) = \sum\limits_{k=1}^{K} c_k \phi_k(t)$. When the basis functions $\varphi(t) = \{\varphi_1(t), \varphi_2(t), \ldots \varphi_k(t)\}$ certainty,

only one function can be defined by one of the coefficient vector $C^j = (c_1^j, c_2^j, \ldots c_k^j)^{\mathrm{T}}$.

Generally, coefficient vector can be solved by least square method $\min \sum\limits_{j=1}^{n} [x(t_j) -$

$\hat{x}(t_j)]^2$ or $\min \sum\limits_{j=1}^{n} \left[x(t_j) - \sum\limits_{k=1}^{K} c_k \varphi_k(t) \right]^2$.

FPCA is an expansion of Principal Component Analysis (PCA) to Hilbert space. Assume that $x_i(s)(s \in T), i = 1, 2, \ldots N$, N is a square integrable function in interval T, s is similar to j in multiple principal component analysis, and s is continuous. Let $f_i = \int \beta(s)x_i(s)\mathrm{d}s, i = 1, 2, \ldots N$,   after   standardized   processing, $\mathrm{var}(f) = \frac{1}{N-1}$

$\sum\limits_{i=1}^{N} (\int \beta(s)x_i(s)\mathrm{d}s)^2$, the first principal component meet that if $\int (\beta_1(s))^2 \mathrm{d}s = \beta^2 = 1$,

then $\frac{1}{N-1} \sum\limits_{i=1}^{N} (\int \beta_1 x_i)^2$ meet the maximum. The kth principal component meet that if

$\int (\beta_k(s))^2 \mathrm{d}s = 1, \int \beta_k(s)\beta_m(s)\mathrm{d}s = 0$, when $m = 1, 2, \ldots, k-1$, then $\frac{1}{N-1} \sum\limits_{i=1}^{N} (\int \beta_k x_i)^2$

meet the maximum. Solve the eigenfunction $\int v(s,t)\beta(t)dt = \lambda\beta(s)$, when $v(s,t) = \frac{1}{N-1}\sum_{i=1}^{N} x_i(s)x_i(t)$ is the covariance of the $x(s)$ and $x(t)$. Define that $\int v(s,t)\beta(t)dt = V\beta(s)$, then the eigenvalue and eigenfunctions of $V\beta(s) = \lambda\beta(s)$ can be solved.

For the sample data or the data used in the fault diagnosis system, the eigenvalues of the data can be obtained by calculating the inner products of the principal function and the eigenfunction.

3.1.2   Features Extraction of the Angular Velocity Data of the Arm

In this paper, a 4-order B-spline basis and a 2-order roughness penalty function with smoothing coefficient $\lambda=50000$ are used. Figure 3(a) shows the B-spline basis function system. Each sample datum has a total of 1001 discrete points that are divided into 101 subintervals on the average with 100 knots. Therefore, the number of the basis function K is 104 which is equal to the order of the B-spline basis and the number of knots. Figure 3(b) shows the sample data for the displacement of the shell after this functional procedure. Figure 3(c) shows the first ten principal component functions and Fig. 3(d) shows the proportion of the first ten principal component functions. Here, the proportion of principal component functions denotes the information of the original samples included in each principal component function. As same with PCA, the sum of ratio of the principal component functions need more than 85%; thus, we choose only the first 6 principal component functions. Therefore, the eigenvalues of the 200 sample data are a 200 × 6 matrix, and this eigenvalues matrix is set as the input of ELM to train it.

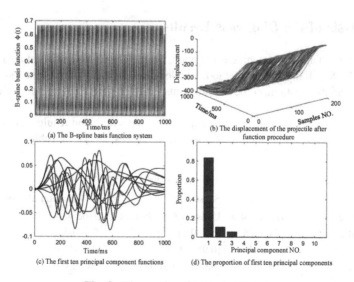

(a) The B-spline basis function system

(b) The displacement of the projectile after function procedure

(c) The first ten principal component functions

(d) The proportion of first ten principal components

Fig. 3. The results of FDA and FPCA

3.2 Fault Diagnosis Based on ELM

ELM is a typical single-hidden-layer feedforward neural network (SLFN). The network consists of an input layer with n nodes corresponding to n input variables, a hidden layer with l nodes, and an output layer with m nodes corresponding to m output variables. Each node in the input layer connects with every node in the hidden layer, and each node in the hidden layer connects with every node in the output layer.

A suitable number of the hidden nodes should be selected before training the ELM. When the number of the hidden nodes is equal to the number of the training samples, ELM with the activation function $g(x)$ can approximate the samples with zero error. However, when the number of the hidden nodes is greater than the number of the training samples, the accuracy for the test set gradually decreases due to the over-fitting phenomena [13]. Therefore, the runtime of the ELM, the average relative error and the coefficient of determination of the training set and test set are compared to select a suitable number of the hidden nodes.

According to Sect. 2.3, we know that the total number of the samples is equal to 200. Here, a training set includes the first 170 samples are used to training the ELM, and a test set includes the last 30 samples are used to verify the performance of the ELM. The ELM runs 10 times with different number of the hidden nodes, and the performance of the ELM is evaluated by the mean values of the runtime of the ELM, the average relative error and coefficient of determination of the training set and test set.

Finally, the parameters of the ELM are determined as follows: the number of the hidden nodes is equal to 44; the activation function is a sigmoid function; 170 samples are used to training the ELM, the input is the eigenvalue matrix obtained from Sect. 3.1, and the output is the sampled values from Sect. 2.3.

4 Analysis of the Diagnosis Results

When the fault diagnosis machine is obtained, 5 simulation data and 5 test data are used to verify the feasibility of this method. The location accuracy of the first test data and first two simulation data meet the requirements, while the location accuracy of the other data are out of tolerance.

According to engineering practice experience and simulation results, the three fault factors are divided into 4 intervals for 4 different states shown in Table 1.

Table 1. The classification of the parameters interval

	Voltage (V)	Sensitivity	Transmission efficiency
Failure	Under 24.5	Under 0.8 or above 1.2	Under 0.5
Mildly abnormal	[24.5–25.5]	[0.8–0.9] [1.1–1.15]	[0.5–0.55]
Normal	[25.5–26.5]	[0.95–1.05]	[0.55–0.6]
Acceptable	[26.5–28]	[0.9–0.95] [1.05–1.1]	Above 0.6

The diagnosis results are shown in Table 2. Here, the numerals 1, 2, 3, and 4 stand for the four different states corresponding to failure, mildly abnormal, normal, and acceptable. The first 5 rows are the diagnosis results of the test data, and the last 5 rows are the diagnosis results of the simulation data. The true states of the transmission efficiency are zero because the true states are unknown.

Table 2. Fault diagnosis results of ELM

	Voltage		Sensitivity		Transmission efficiency	
	True state	Diagnosis state	True state	Diagnosis state	True state	Diagnosis state
Simulation samples	4	4	3	3	3	3
	4	4	4	4	4	4
	3	3	4	4	1	1
	1	1	1	1	4	4
	3	3	2	2	2	2
Test samples	3	3	4	4	NaN	3
	2	3	2	2	NaN	4
	2	2	2	2	NaN	4
	2	2	2	2	NaN	3
	1	1	1	1	NaN	3

The diagnosis results for the three fault factors, the voltage of the motors, the sensitivity of the tachometer generator, and the transmission efficiency of the worm and worm gear, are shown in Table 2. As shown in Table 2, the correct rate of the diagnosis results of ELM for the three fault factors are 90%, 100%, and 100%, respectively; The correct rate of the diagnosis results of BP neural network for the three fault factors are 80%, 90%, and 90%, respectively; And the correct rate of the diagnosis results of RBF neural network for the three fault factors are 80%, 100%, and 100%, respectively. The correct rate of the diagnosis results shows that the fault diagnosis method based on ELM is feasible and effective.

5 Conclusions

In this paper, a novel fault diagnosis method is presented for an automatic shell magazine that includes FDA and ELM. Targeting the fact that it is difficult to acquire enough fault data for complex mechatronics system, a mechanical model of the automatic shell magazine is built in RecurDyn and the corresponding control model is built in Simulink. The virtual prototype model is verified according to the test data. The fault data can be easily acquired from the verified model instead of actual experiment. A new signal processing method named FDA and FPCA are introduced to the data processing of the automatic shell magazine, and the eigenvalues of the displacement of the shell from FPCA are used to train the ELM. The diagnosis results show that the proposed method is feasible and effective.

The method presented in this paper can be used not only for the fault diagnosis of the automatic shell magazine but also for other complex mechatronics systems. Further study will be conducted on the data fusion problem of the displacement of the shell and the rotating angle of the driving wheel.

References

1. Wang, Z.S.: Intelligent fault diagnosis and fault tolerance control. Northwestern Polytechnical University Press, Xi'an (2007)
2. Ramsay, J.O.: when the data are functions. J. Psychometrika **47**, 379–396 (1982)
3. Ramsay, J.O., Dalzell, C.J.: Some tools for functional data analysis. J. Roy. Stat. Soc. Ser. B (Stat. Methodol.) **53**(3), 539–572 (1991)
4. Ramsay, J.O., Silverman, B.W.: Functional Data Analysis, 2nd edn. Springer Science +Business Media, Inc., New York (2005). https://doi.org/10.1007/b98888
5. Ramsay, J.O., Hooker, G., Graves, S.: Functional data analysis with R and MATLAB. Science +Business Media, Inc., New York (2009). https://doi.org/10.1007/978-0-387-98185-7
6. Kayano, M., Konishi, S.: Functional principal component analysis via regularized Gaussian basis expansions and its application to unbalanced data. J. Stat. Plann. Infer. **139**, 2388–2398 (2009)
7. Ordóñez, C., Sierra, C., Albuquerque, T., et al.: Functional data analysis as a tool to correlate textural and geochemical data. J. Appl. Math. Comput. **223**, 476–482 (2013)
8. Segovia-Gonzalez, M.M., Guerrero, F.M., Herranz, P.: Explaining functional principal component analysis to actuarial science with an example on vehicle insurance. J. Insur.: Math. Econ. **45**, 278–285 (2009)
9. Zipunnikov, V., Caffo, B., Yousem, D.M., et al.: Functional principal component model for high-dimensional brain imaging. J. NeuroImage **58**, 772–784 (2011)
10. Chen, L.Y., Gong, Y.P.: Mechanical Design Handbook Volume (2), 5th edn. China Machine Press, Beijing (2010)
11. Gao, X.X., Sun, H.G., Hou, B.L., et al.: Reliability analysis of rotation positioning precision for howitzer shell transfer arm based on virtual experiment. In: Proceedings of the 5th International Conference on Mechanical Engineering and Mechanics, 20–22 August 2014, Yangzhou, China, pp. 428–433. Science Press, USA (2014)
12. Boor, C.D.: A Practical Guide to Splines. Springer-Verlag, New York (2001)
13. Shi, F., Wang, H., Yu, L., et al.: 30 Cases Analysis of Intelligent Algorithm in MATLAB. Beihang University Press, Beijing (2011)

Anomalous Trajectory Detection Using Recurrent Neural Network

Li Song[1], Ruijia Wang[1], Ding Xiao[1], Xiaotian Han[1], Yanan Cai[2], and Chuan Shi[1(✉)]

[1] Beijing University of Posts and Telecommunications, Beijing, China
song200626@gmail.com,
hanxiaotian.h@gmail.com,{Vanrhyga,dxiao,shichuan}@bupt.edu.cn
[2] Big Data Center, PICC Property and Casualty Company Limited, Beijing, China
caiyanan@picc.com.cn

Abstract. Anomalous trajectory detection which plays an important role in taxi fraud detection and trajectory data preprocessing is a crucial task in trajectory mining fields. Traditional anomalous trajectory detection methods which utilize density and isolation approaches mainly focus on the differences of a new trajectory and the historical trajectory dataset. Although these methods can capture the particular characteristics of trajectories, they still suffer from the following two disadvantages. (1) These methods cannot capture the sequential information of the trajectory well. (2) These methods only concentrate on the given source and destination which may lead to data sparsity issues. To overcome above shortcomings, we propose a novel method called **A**nomalous **T**rajectory **D**etection using **R**ecurrent **N**eural **N**etwork (**ATD-RNN**) which characterizes the trajectory by learning the trajectory embedding. The trajectory embedding can capture the sequential information of the trajectory and depict the internal characteristics between anomalous and normal trajectory. To address the potential data sparsity problem, we enlarge the dataset between a source and a destination by taking the relevant trajectories into consideration. Extend experiments on real-world datasets validate the effectiveness of our method.

Keywords: Anomalous trajectory detection · Trajectory embedding · Recurrent neural network

1 Introduction

With the proliferation of global positioning system (GPS) based equipment, massive spatial trajectory data has been generated. The trajectory data represents the mobility of a diversity of moving objects and contains valuable information concerning both Service Providers (SPs) and the customers. A large number of trajectory data mining tasks [25], including map matching, trajectory compression, stay point detection, POIs recommendation, trajectory classification and

© Springer Nature Switzerland AG 2018
G. Gan et al. (Eds.): ADMA 2018, LNAI 11323, pp. 263–277, 2018.
https://doi.org/10.1007/978-3-030-05090-0_23

anomalous trajectory detection, have been widely researched. Anomalous trajectory detection plays an important role in the fields of trajectory data mining. For example, anomalous trajectory detection can enhance the quality of taxi services impressively, since the greedy taxi drivers who overcharge passengers by deliberately taking unnecessary detours could be detected with anomalous trajectory detection technique. Therefore, it allows the taxi companies to monitor the movements of all the taxis and to identify the dishonest drivers who tend to take routes longer than usual. A toy example of anomalous and normal trajectory between a source (S) and destination (D) (SD-Pair) can be seen in Fig. 1(a).

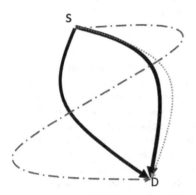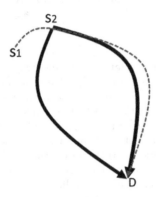

(a) Anomalous and normal trajectory.   (b) Trajectory of two neighbor SD-Pairs.

Fig. 1. The historical trajectories between SD-pairs.

In recent years, automatically anomalous trajectory detection has attracted extensive research attention. Some existing anomalous trajectory detection methods have been proposed to deal with this problem [2,4,8,13,20,23,26] . [13] proposes a partition-and-detect framework for anomalous trajectory detection, which partitions a trajectory into a set of line segments and detects outlying line segments for trajectory outliers based on distance and density. IBAT [23] and the augmented version IBOAT [4] present the isolation-based anomalous trajectory detection methods which detect anomalous by comparing a new trajectory against a large collection of historical trajectories. [26] determines whether an input trajectory is an anomalous by taking spatial and temporal abnormalities into consideration simultaneously. However, most existing methods addressing the anomalous detection problem are only based on density or isolation methods. So although these methods achieve impressive performances, there still remain some limitations to these methods.

1. **Sequential Information.** These methods based on density or isolation cannot characterize the sequential information of a trajectory well. Those methods mainly focus on sub-trajectories or trajectory points and neglect the influence of sequential information of the whole trajectory.

2. **Data Sparsity.** These methods based on density or isolation also suffer from data sparsity problem. If there are N intersections in a city, then the number of SD-pairs will be up to N^2. Even worse, it is only few trajectories in historical dataset for some special intersection. As a result, it is difficult to model these SD-pairs.

3. **Computational Complexity and Space Complexity.** Traditional anomalous trajectory detection methods is time-consuming since these methods need to compute the similarity of every two trajectories in historical datasets. Moreover, for every SD-pair, a corresponding model is required.

Inspired by that word embedding [12,17] models the co-occurrence of words by mapping the words to the low-dimensional vectors, we propose a novel method called **A**nomalous **T**rajectory **D**etection using **R**ecurrent **N**eural **N**etwork (ATD-RNN). ATD-RNN represents the trajectory by the low-dimensional vector using recurrent neural network (RNN) and detects anomalous trajectory in the embedded space. The sequential information is learned through trajectory embedding because RNN can use its internal memory to process time series data. To address the shortcoming of data sparsity, ATD-RNN attempts to train the model through the extended large amount of historical trajectories. For example, as illustrated in Fig. 1(b), there may be few historical trajectories from source S_1 to destination D. But if take the source S_2 into consideration which has sufficient historical trajectories to build a model, the situation for source S_1 could be alleviated. Since different SD-Pairs could be trained together, ATD-RNN has potential to reduce the frequency of accessing to historical trajectories and has little demand for extra space. Therefore, ATD-RNN not only solves the anomalous trajectory detection by capturing the sequential information through trajectory embedding, but also alleviates the data sparsity problem by taking different SD-Pairs into consideration. Experiments on several real-world datasets demonstrate the superior performance of the proposed method. Source code is available at Github[1].

The main contributions of this paper are listed in detail as follows:

- We present a novel anomalous trajectory detection method named ATD-RNN, which captures the sequential information of the trajectory by learning the trajectory embedding through a delicately designed recurrent neural network.
- We take different sources and destinations into consideration to alleviate the data sparsity problem. This can capture the internal characteristics from similar trajectories.
- We evaluate ADT-RNN on real-world datasets collected from 442 taxis in Porto for one year. It achieves remarkable detection results comparing to the famous anomalous trajectory detection baselines.

The rest of this paper is organized as follows. Section 2 reviews the related works. A formal definition of our problem is given in Sect. 3. The details of our

---

[1] https://github.com/LeeSongt/ATD-RNN.

proposed method are introduced in Sect. 4. Experimental results of our method are compared and analyzed in Sect. 5. And Sect. 6 concludes this paper and outlines the future work.

2   Related Work

In this section, we give a brief review of the relevant academic literature on anomalous trajectory detection and trajectory embedding.

Anomalous Trajectory Detection. A considerable amount of researches have been published on anomalous trajectory detection. In general, these studies could be divided into three major categories.

The first major category is clustering, which performs automated grouping based on distance and density. [13] conducts a partition-and-detect framework for trajectory outliers detection. [11] deals with finding outliers in large, multidimensional datasets based on distance. [22] characterizes outliers as moving objects that behaved differently from the majority in trajectory streams by neighbor-based trajectory outliers definitions. However, these distance-based and density-based approaches often need to calculate the measure metrics according to the whole historical database that is time consuming.

The second category is classification, which relies on labeled training data. In order to detect anomalous efficiently and effectively, [14] constructs a multi-dimensional feature space oriented on segmented trajectories and then learns a model to classify trajectories as normal or abnormal. [18] proposes an anomalous behavior detection framework in a video surveillance scenario. However, these classification-based approaches require amount of labeled data which spends huge manpower and material resources.

The third category is based on patterns. [3] proposes an anomalous behavior detection framework using trajectory analysis, which includes a trajectory patterns learning module and an online abnormal detection module. [23] presents an Isolation-Based Anomalous Trajectory (iBAT) detection method which exploits the property that anomalies are susceptible to a mechanism called isolation. IBOAT [4] is an augmented version of iBAT. Specifically, in order to process trajectories online and find the anomalous at early stag, IBOAT builds an inverted index for historical trajectory data. [26] proposes a time-dependent popular routes based algorithm which takes spatial and temporal abnormalities into consideration simultaneously to improve the accuracy of the detection. [20] proposes a probabilistic model-based approach via modeling the driving behavior/preferences from the set of historical trajectories. Generally speaking, the pattern-based approaches investigate the co-occurrence of trajectory points so that the anomalous trajectory which is rarely appeared can be detected.

Trajectory Embedding. Trajectory Embedding is an extended field of word embedding [12, 17] which represents a word as a fixed length numerical vector and the co-occurrence of words can be learned from the embedded vectors. [24] proposes a time-aware trajectory embedding model in next-location recommendation systems to deal with sequential information and data sparsity problem. [19]

predicts next location using a spatial-temporal-semantic neural network algorithm. Concretely, the road networks is divided into significant discrete points in terms of a distance parameter threshold and then a long short-term memory (LSTM) neural network is used to model these sequences. [7] demonstrates that trajectory-user linking problem can be solved by trajectory embedding. After the check-ins in trajectories is embedded into a low-dimensional space, the connection between a particular user and the motion patterns can be learned with a LSTM. In order to learn sequential information, trajectory embedding is applied to model the co-occurrence between the trajectory points.

3   Problem Definition

It is necessary here to clarify exactly what is meant by trajectory. After that, the formal definition of anomalous trajectory detection is introduced.

Definition 1 *(Raw Trajectory). A raw trajectory $T = \{p_1 \rightarrow p_2 \rightarrow \cdots \rightarrow p_n\}$ is a sequence of records, and each recored p_i is represented by (lon_i, lat_i, t_i) where (lon_i, lat_i) is a geographic coordinate and t_i is the timestamp. p_1 and p_n are source and destination of the trajectory, respectively.*

Definition 2 *(Mapped Trajectory). For a given n and m, a map can be split into $n * m$ equal sized grids. Then, a map function $\rho(lon, lat, n, m) = grid_i$ implements a discretization process. A mapped Trajectory corresponding to a raw trajectory $T = \{p_1 \rightarrow p_2 \rightarrow \cdots \rightarrow p_n\}$ is expressed as $tr = \{grid_1 \rightarrow grid_2 \cdots \rightarrow grid_n\}$, where $grid_i = \rho(lon_i, lat_i, n, m)$.*

Definition 3 *(Anomalous Trajectory Detection). Given a set of trajectories $D = \{tr_1, tr_2, \cdots, tr_m\}$, anomalous trajectory detection is to find those trajectory R that are significantly different from the majority in historical datasets.*

Since anomalous trajectory is rarely occurred, we want to learn the normal pattern from historical trajectories. Then, an anomalous can be detected by comparing a trajectory with these normal patterns. However, it is difficult to learn the normal patterns for a given source and destination (SD-pair). Traditional anomalous trajectory detection methods based on density and isolation mainly focus on sub-trajectory and points in the trajectory and cannot capture the sequential information of the trajectory well. In addition, these methods tend to train one model for one SD-Pair that cannot make use of similar characteristics between different SD-Pairs. Even worse, these methods concentrate on an SD-Pair may suffer from data sparsity.

In this paper, anomalous trajectory detection is conducted by learning trajectory embedding through recurrent neural network. Trajectory embedding can capture the internal characteristics between anomalous and normal trajectories.

4   Methodology

4.1   Overview of ATD-RNN

In this paper, we propose a method to detect anomalous trajectories named Anomalous Trajectory Detection using Recurrent Neural Network(**ATD-RNN**). As illustrated in Fig. 2, the proposed method is mainly consists of three steps, trajectory data preprocessing, trajectory embedding and anomalous detection. In trajectory data preprocessing step, we discrete the trajectory points so that the continuous numerical variables could be fed into the trajectory embedding step. After that, a delicately designed neural network, stacked RNN, is applied to learning the trajectory embedding which can capture the sequential information and the internal characteristics of the trajectories. Then, the multilayer perceptron and a softmax layer are used to detect anomaly from the trajectory embeddings. We will describe these steps in detail in the following chapters.

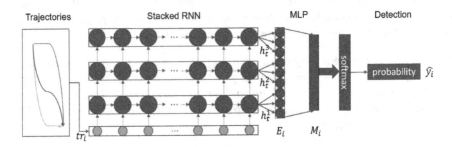

Fig. 2. The architecture of ATD-RNN.

4.2   Trajectory Data Preprocessing

In general, the raw trajectory data is continuous numerical variables. These trajectory points are enormous so that we cannot learn the trajectory embeddings for every points. Even worse, it is difficult to generalize to new points. In trajectory data preprocessing, we need a discretization step. In details, the geographic coordinate space is meshed into equal sized grids according to the hyper parameters n and m. In fact, we adjust n and m so that the size of a grid is about $100m$. Then we label each grid with a unique index. After that, the raw trajectory points located on the grid will have the same index. Although we obtain the mapped trajectories, we have to notice that the length of each trajectory data is not equivalent. Therefore, we utilize some padding operations to align the trajectories. After that, a trajectory tr_i is represented as:

$$tr_i = \{x_1, x_2, \cdots, x_m\}, \tag{1}$$

another critical problem in anomalous detection is that the anomalies are often rarely appear in historical datasets. This may be an obstacle in the process of optimization. To alleviate this problem, we sample some anomalous trajectories and make disturbances to generate new anomalous data. In practice, we randomly select some trajectory points from an anomalous trajectory and replace those points with their neighbors.

4.3   Trajectory Embedding

We use recurrent neural network to learn the sequential information of the trajectory. After the whole trajectory feeds into a stacked recurrent neural network, we can learn the trajectory embedding.

Recurrent Neural Network. Recurrent neural network (RNN) is powerful to process sequential data in many fields including NLP [15] and speech recognition [9]. In details, RNN can change monotonously along with the position in a sequence. A rolled RNN structure can be unfolded as Fig. 3 illustrated. The hidden state h_t and output state o_t are updated as Eq. 2. In order to capture the relationship between trajectory points, we utilize a stacked RNN to learn the trajectory embeddings. And the experiments of visualization of trajectory embeddings illustrate the effectiveness of embedding.

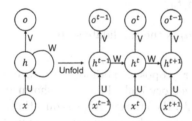
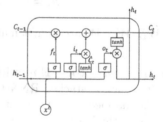

Fig. 3. Recurrent neural network.   **Fig. 4.** The structure of LSTM.

$$h_t = f(W \cdot h_{t-1} + U \cdot x_t + b_h),$$
$$o_t = f(V \cdot h_t + b_o), \tag{2}$$

where W, U, V are weight matrices, and b_h, b_o are biases for hidden state and output state respectively. f is the activation function, and x_t is the embedding for each trajectory points.

Long short-term memory (LSTM) is a special kind of RNN cell, capable of learning long-term dependencies. The structure of LSTM cell is shown in Fig. 4. Specifically, the LSTM cell utilizes three gates to control the cell state. The forget gate f_t decides how much information of the last cell state C_{t-1} will keep to the current state C_t. The input gate i_t and $\hat{C}_t$ decides how much information of the input x_t will keep to the current state C_t. And the output gate o_t decides how much information of the current state C_t will output to h_t. The detail information can be seen in Eq. 3,

$$
\begin{aligned}
f_t &= \sigma(W_f \cdot [h_{t-1}, x_t] + b_f), \\
i_t &= \sigma(W_i \cdot [h_{t-1}, x_t] + b_i), \\
\hat{C}_t &= tanh(W_C \cdot [h_{t-1}, x_t] + b_C), \\
C_t &= f_t * C_{t-1} + i_t * \hat{C}_t, \\
o_t &= \sigma(W_o \cdot [h_{t-1}, x_t] + b_o), \\
h_t &= o_t * tanh(C_t),
\end{aligned}
\tag{3}
$$

where W_f, W_i, W_C, W_o are the weight matrices and b_f, b_i, b_C, b_o are biases of each gate, respectively. σ and $tanh$ refer to the logistic sigmoid and hyperbolic tangent function.

Gated recurrent unit (GRU) is a gating mechanism in recurrent neural network which is similar to LSTM. A GRU has two gate, reset gate and update gate. Formally, the reset gate and update gate are calculated as Eq. 4,

$$
\begin{aligned}
z_t &= \sigma(W_z \cdot [h_{t-1}, x_t]), \\
r_t &= \sigma(W_r \cdot [h_{t-1}, x_t]), \\
\tilde{h}_t &= tanh(W \cdot [r_t * h_{t-1}, x_t]), \\
h_t &= (1 - z_t)h_{t-1} + z_t\tilde{h}_t,
\end{aligned}
\tag{4}
$$

where W_z, W_r, W are weight matrices and $\tilde{h}_t$ is the candidate state.

Learn the Trajectory Embedding In our proposed method, we utilize a delicately designed stacked RNN to learn trajectory embeddings, as shown in Fig. 2. The mapped trajectories are inputed into the stacked RNN sequentially. In every step, the RNN cell can learn a hidden state which summarizes the past sequences by merging the current step input information and the previous hidden state. The stacked RNN not only learns sequential information through the time step, but also memorizes significant information through the RNN layers which use non-linear function to capture trajectory characteristics in high-dimension space. Moreover, to avoid the over-fitting problem, dropout techniques [6] are used to constrain the representation ability of this model. In details, when data information flow pass through different RNN cells, we randomly select some edge and cut off the connection between the stacked layers. After the last step, we can get the trajectory embedding by concatenating the output state of the stacked RNN cells as Eq. 5:

$$
E_i = [h_t^1, h_t^2, \cdots, h_t^l],
\tag{5}
$$

where E_i is the embedding for tr_i, l is the number of stacked RNN layers, and h_t^k is the last output state at layer k $(k = 1..l)$.

4.4   Anomalous Detection

In Sect. 4.3, we obtain the embedding E_i for the i-th trajectory. In order to detect anomalous, we utilize a multilayer perceptron (MLP) to reduce the dimensions of trajectory embedding.,

$$M_i = \sigma(W_i \cdot E_i + b_i) \tag{6}$$

where M_i is output vector, and W_i and b_i are the projection parameters of the MLP layer. After that, as illustrated in Fig. 2, the results are fed into a softmax layer to generate the anomalous probability $\hat{y}_i$ of a trajectory, then given a trajectory datasets $D = \{tr_1, tr_2, \cdots, tr_m\}$, we can train our model by minimize the cross-entropy as following:

$$J = - \sum_{i=1}^{m} [y_i \ log \ \hat{y}_i + (1 - y_i) \ log(1 - \hat{y}_i)]. \tag{7}$$

5   Experiments

5.1   Dataset and Metrics

We conduct all experiments on a real-world taxi trajectory dataset which is collected by 442 taxi in the city of Porto[2], in Portugal from Jan. 07, 2013 to Jun. 30, 2014 and the average sampling rate is $15s/points$. We extract 5 SD-pairs with sufficient historical trajectories. The basic information of those SD-pairs is shown in Table 1. There are about 5% anomalous trajectories.

Table 1. The information of SD-pairs.

	#Trajectories	#Anomalousness(%)	#avgPointsNum
SD-Pair1	1233	54(4.4%)	32
SD-Pair2	765	28(3.7%)	31
SD-Pair3	617	37(6.0%)	61
SD-Pair4	1379	44(3.2%)	51
SD-Pair5	4973	270(5.4%)	67

Instead of labelling the anomalous manually, we follow the solution in [20] which adopts a complete-linkage clustering algorithm to hierarchically cluster the trajectories. Given the prediction and its ground truth, we compute the measure metrics as Eq. 8,

---

[2] https://www.kaggle.com/c/pkdd-15-predict-taxi-service-trajectory-i/data.

$$ACC = \frac{TP + TN}{TP + TN + FP + FN},$$
$$P = \frac{TP}{TP + FP},$$
$$R = \frac{TP}{TP + FN}, \qquad (8)$$
$$F_1 = \frac{2PR}{P + R},$$

where TP, TN, FP and FN are true positive, true negative, false positive and false negative in confusion matrix, respectively.

5.2   Baselines

We compare our method with the baselines based on density and isolation. The baselines are as following:

1. **LCS** [21]: The Longest Common Sub-sequence (LCS) method matches the longest common sub-sequence between two sequences using dynamic programming. LCS is also a widely used representative method for measuring the trajectory similarity. We implement LCS by comparing all trajectories in the training set for every given testing trajectory.
2. **XGBoost** [5]: XGBoost is an ensemble model based on the gradient boosting decision tree (GBDT) and is designed to be highly efficient, flexible, and portable. Those characteristics we extracted from a trajectory contain the distance of a trajectory, the angle between trajectory points and so on.
3. **TOP-EYE** [8]: TOP-EYE uses a decay function to mitigate the influences of the past trajectories on the evolving outlying score, which is defined based on the evolving moving direction and density of the historical trajectories. Since the sampling rate in our dataset is $15s/points$, we conduct this method by counting the density of each grid and compute anomalous score for each test trajectory.
4. **IBOAT** [4]: Anomalous trajectories will be isolated from the majority of routes, while normal trajectories will be supported by a large number of trajectories. IBOAT adopts the inverted index mechanism to fast retrieve the relevant trajectories.

5.3   Settings

We implement the proposed model ATD-RNN with TensorFlow [1] and run the code on an Intel Core i7-4790 with 8-GB RAM. As we introduced in Sect. 4, there are two kinds of RNN cell. In our experiments, ATD-LSTM represents the model using the LSTM cell and ATD-GRU represents the model using GRU cell. There are three key parameters in our model, the embedding dimensions, the number of RNN layers and the dropout probabilities. For generality, we set the dimensions of each point to 64, the size of hidden state of a RNN cell to 64 and the layers of stacked RNN to 5. To alleviate the problem of over-fitting inherent of RNN, we apply dropout technique [6] and the dropout probability is set to 0.5. We utilize Adam [10] to optimize our model.

5.4   Results and Analysis

Performance Evaluation. We evaluate the results on SD-Pair with the above baselines. The results of anomalous trajectory detection in given SD-pair are shown in Table 2. We can observe that in most cases, ATD-LSTM and ATD-GRU achieve the best results. This indicates that recurrent neural network could capture the internal characteristics of anomalous trajectories and normal trajectories. The better performance of ATD-RNN shows that the sequential information of a trajectory is critical to anomalous trajectory detection. In addition, the reason of the undesirable results of LCS and XGBoost may be that those methods only consider the shape of a trajectory and neglect the information of the historical dataset and the sequential information of the trajectory. On the contrary, TOP-EYE and iBOAT achieve considerable performance because those methods utilize the historical trajectories. And the reason of iBOAT outperforms TOP-EYE in most circumstances. The reason is probably that iBOAT not only takes the similarity of historical trajectories into consideration, but also makes use of the local sequential information in the trajectory.

Study on Influences with Multi-SD-Pairs. To explore the influence of anomalous trajectory detection on different SD-Pairs instead of a given SD-pair, we conduct experiments on the union set of two SD-Pairs selected from SD-pair1 to SD-pair5. In details, for SD-Pair1 and SD-Pair2, we add the remaining SD-Pairs into the train process, and then test the performance on SD-Pair1 and SD-Pair2 respectively. For example, we train a model with SD-Pair1 and SD-Pair3, and then test the performance on SD-Pair1. The relative location of these SD-Pairs is shown in Fig. 5(a). We can see that the source and destination of SD-Pair1 are close to SD-Pair2, while SD-Pair3, SD-Pair4 and SD-Pair5 are close to each other. The results are shown in Fig. 5. At the first sight, we can see that if we train a model with different SD-Pairs trajectories dataset, the results are better than a single SD-Pair. This demonstrates that the extra trajectory information is useful in anomalous trajectory detection. Concretely, the merged set of SD-Pair1 and SD-Pair2 gets a significant performance which indicates that the close SD-Pairs could enhance the results of anomalous trajectory detection. The probable reason is that different SD-Pairs which are close to each other can offer valuable information in detection process and trajectory from different SD-Pairs can deliver information through the model. As mentioned in Fig. 1(b), the trajectories from S_2 to D can provide information to the detection process from S_1 to D. In addition, the results indicate that the proposed model ATD-RNN has potential to alleviate data sparsity.

Table 2. Performance evaluation of anomalous trajectory detection on different SD-Pairs. ATD-LSTM is ATD-RNN based on LSTM cell, and ATD-GRU is ATD-RNN based on GRU cell.

		LCS	XGBoost	TOP_EYE	iBOAT	ATD-LSTM	ATD-GRU
SD-Pair1	Acc	0.8434	0.8795	0.9629	0.9506	0.9518	**0.9638**
	P	0.6765	0.9412	0.9583	0.9565	0.9565	0.9231
	R	0.9200	0.6400	0.9200	0.8800	0.8800	0.9600
	F_1	0.7797	0.7619	0.9230	0.9167	0.9167	**0.9412**
SD-Pair2	Acc	0.9444	0.8810	0.9444	**0.9583**	**0.9583**	0.9583
	P	0.9000	1.0000	0.7500	0.9091	0.9091	0.8462
	R	0.7500	0.2857	1.0000	0.8333	0.8333	0.9167
	F_1	0.8182	0.4444	0.8571	0.8696	0.8696	**0.8800**
SD-Pair3	Acc	0.9625	0.8375	0.9625	0.9625	**0.9875**	0.9750
	P	0.9333	0.7143	0.9333	0.8824	0.9412	0.8889
	R	0.8750	0.3125	0.8750	0.9375	1.0000	1.0000
	F_1	0.9032	0.4348	0.9032	0.8276	**0.9697**	0.9412
SD-Pair4	Acc	0.9242	0.9470	0.9470	0.9394	0.9848	**0.9924**
	P	1.0000	0.9412	0.8261	1.0000	0.9545	1.0000
	R	0.5455	0.7273	0.8636	0.6364	0.9545	0.9545
	F_1	0.7059	0.8205	0.8444	0.7778	0.9545	**0.9767**
SD-Pair5	Acc	0.7819	0.7574	0.8431	0.8750	0.9020	**0.9167**
	P	0.9623	0.9756	0.6832	0.8000	0.9537	0.9063
	R	0.3696	0.2898	1.0000	0.8406	0.7463	0.8406
	F_1	0.5340	0.4469	0.8118	0.8198	0.8374	**0.8722**

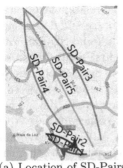
(a) Location of SD-Pairs.

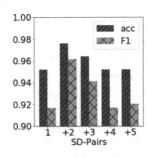
(b) SD-Pair1 performance.

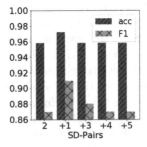
(c) SD-Pair2 performance.

Fig. 5. Anomalous trajectory detection with Multi-SD-pairs. "1" means the model on SD-Pair1, and "+1" means the model of some SD-Pair and SD-Pair1.

Parameter Experiments. In this section, we explore the influence of different parameters. The key parameters include the dimensions of embedding, the dropout probability and the number of stacked RNN layers. All parameter experiments are conducted on SD-Pair4 with LSTM cell. The results are shown in Fig. 6. As shown in Fig. 6(a), we can observe that the accuracy and F_1 get the best score when *embedding dimensions* is 64. As one can see from Fig. 6(b), when *dropout_keep_prob* is 0.5, the model gets the best performance. It is interesting that the change of *dropout_keep_prob* has slightly influence on accuracy, but the influence on F_1 is serious. The reason of this phenomenon may cause by the unbalance of anomalous and normal trajectory proportion. As illustrated in Fig. 6(c), when *num_layers* is 5, we can get a significant performance.

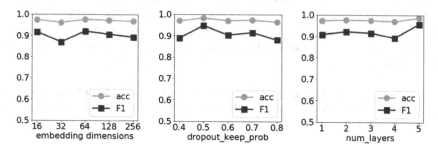

(a) Embedding dimension. (b) Dropout probabilistic. (c) The number of layers.

Fig. 6. Performance of different parameters.

Visualization. To analyze the trajectory embedding, we visualize the vectors of trajectory embedding by t-SNE [16]. The results are shown in Fig. 7. We can observe that the trajectories embeddings can separate the anomalous trajectories from normal historical trajectories dataset. The results indicate that the trajectory embedding learned by recurrent neural network can capture the internal characteristics of the trajectories.

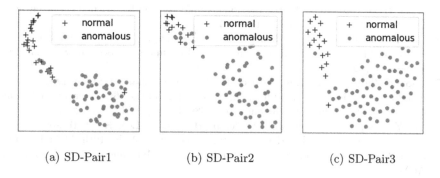

(a) SD-Pair1   (b) SD-Pair2   (c) SD-Pair3

Fig. 7. Visualization of trajectory embeddings.

6   Conclusions and Future Work

In this paper, we propose **A**nomalous **T**rajectory **D**etection using **R**ecurrent Neural Network (**ATD-RNN**). ATD-RNN learns the trajectory embeddings through a delicately designed recurrent neural network. Comparing to traditional methods, ATD-RNN can capture the sequential information of the historical trajectories and the internal characteristics between the anomalous and normal trajectories. Moreover, ATD-RNN is not constrained by the given SD-Pair, but takes different SD-Pairs into consideration. This means that ATD-RNN has potential to alleviate data sparsity. Extensive experiments on real-world datasets demonstrate the effectiveness of ATD-RNN.

In the future, we will extend our model to learn the trajectory embedding with the up-to-date techniques, e.g., attention mechanism and memory augmentation. In addition, we will attempt to solve other trajectory data mining tasks and make use of the additional information to improve the quality of the trajectory embedding.

Acknowledgement. This work is supported in part by the National Natural Science Foundation of China (No. 61772082, 61702296, 61375058), and the Beijing Municipal Natural Science Foundation (4182043).

References

1. Abadi, M., et al.: TensorFlow: a system for large-scale machine learning. In: OSDI, vol. 16, pp. 265–283 (2016)
2. Bu, Y., Chen, L., Fu, A.W.C., Liu, D.: Efficient anomaly monitoring over moving object trajectory streams. In: Proceedings of the 15th ACM SIGKDD International Conference on Knowledge Discovery and Data Mining, pp. 159–168. ACM (2009)
3. Cai, Y., Wang, H., Chen, X., Jiang, H.: Trajectory-based anomalous behaviour detection for intelligent traffic surveillance. IET Intell. Transport Syst. **9**(8), 810–816 (2015)
4. Chen, C., et al.: iBOAT: isolation-based online anomalous trajectory detection. IEEE Trans. Intell. Transport. Syst. **14**(2), 806–818 (2013)
5. Chen, T., Guestrin, C.: XGBoost: A scalable tree boosting system. In: Proceedings of the 22nd ACM SIGKDD International Conference on Knowledge Discovery and Data Mining, pp. 785–794. ACM (2016)
6. Gal, Y., Ghahramani, Z.: A theoretically grounded application of dropout in recurrent neural networks. In: Advances in Neural Information Processing Systems, pp. 1019–1027 (2016)
7. Gao, Q., Zhou, F., Zhang, K., Trajcevski, G., Luo, X., Zhang, F.: Identifying human mobility via trajectory embeddings. In: Proceedings of the 26th International Joint Conference on Artificial Intelligence, pp. 1689–1695. AAAI Press (2017)
8. Ge, Y., Xiong, H., Zhou, Z.h., Ozdemir, H., Yu, J., Lee, K.C.: Top-eye: top-k evolving trajectory outlier detection. In: Proceedings of the 19th ACM International Conference on Information and knowledge Management, pp. 1733–1736. ACM (2010)

9. Graves, A., Mohamed, A.R., Hinton, G.: Speech recognition with deep recurrent neural networks. In: 2013 IEEE International Conference on Acoustics, speech and signal processing (ICASSP), pp. 6645–6649. IEEE (2013)
10. Kingma, D.P., Ba, J.: Adam: a method for stochastic optimization. arXiv preprint arXiv:1412.6980 (2014)
11. Knorr, E.M., Ng, R.T., Tucakov, V.: Distance-based outliers: algorithms and applications. VLDB J. Int. J. Large Data Bases **8**(3–4), 237–253 (2000)
12. Le, Q., Mikolov, T.: Distributed representations of sentences and documents. In: International Conference on Machine Learning, pp. 1188–1196 (2014)
13. Lee, J.G., Han, J., Li, X.: Trajectory outlier detection: a partition-and-detect framework. In: 2008 IEEE 24th International Conference on Data Engineering, 2008 ICDE, pp. 140–149. IEEE (2008)
14. Li, X., Han, J., Kim, S., Gonzalez, H.: ROAM: Rule-and motif-based anomaly detection in massive moving object data sets. In: Proceedings of the 2007 SIAM International Conference on Data Mining, pp. 273–284. SIAM (2007)
15. Luong, M.T., Pham, H., Manning, C.D.: Effective approaches to attention-based neural machine translation. arXiv preprint arXiv:1508.04025 (2015)
16. Maaten, L.V.D., Hinton, G.: Visualizing data using t-SNE. J. Mach. Learn. Res. **9**(Nov), 2579–2605 (2008)
17. Mnih, A., Kavukcuoglu, K.: Learning word embeddings efficiently with noise-contrastive estimation. In: Advances in neural information processing systems, pp. 2265–2273 (2013)
18. Sillito, R.R., Fisher, R.B.: Semi-supervised learning for anomalous trajectory detection. In: BMVC, vol. 1, pp. 1035–1044 (2008)
19. Wu, F., Fu, K., Wang, Y., Xiao, Z., Fu, X.: A spatial-temporal-semantic neural network algorithm for location prediction on moving objects. Algorithms **10**(2), 37 (2017)
20. Wu, H., Sun, W., Zheng, B.: A fast trajectory outlier detection approach via driving behavior modeling. In: Proceedings of the 2017 ACM on Conference on Information and Knowledge Management, pp. 837–846. ACM (2017)
21. Ying, J.J.C., Lee, W.C., Weng, T.C., Tseng, V.S.: Semantic trajectory mining for location prediction. In: Proceedings of the 19th ACM SIGSPATIAL International Conference on Advances in Geographic Information Systems, pp. 34–43. ACM (2011)
22. Yu, Y., Cao, L., Rundensteiner, E.A., Wang, Q.: Detecting moving object outliers in massive-scale trajectory streams. In: Proceedings of the 20th ACM SIGKDD International Conference on Knowledge Discovery and Data Mining, pp. 422–431. ACM (2014)
23. Zhang, D., Li, N., Zhou, Z.H., Chen, C., Sun, L., Li, S.: iBAT: detecting anomalous taxi trajectories from GPS traces. In: Proceedings of the 13th International Conference on Ubiquitous Computing, pp. 99–108. ACM (2011)
24. Zhao, W.X., Zhou, N., Sun, A., Wen, J.R., Han, J., Chang, E.Y.: A time-aware trajectory embedding model for next-location recommendation. Knowl. Inf. Syst. 1–21 (2017)
25. Zheng, Y.: Trajectory data mining: an overview. ACM Trans. Intell. Syst. Technol. (TIST) **6**(3), 29 (2015)
26. Zhu, J., Jiang, W., Liu, A., Liu, G., Zhao, L.: Time-dependent popular routes based trajectory outlier detection. In: Wang, J., Cellary, W., Wang, D., Wang, H., Chen, S.-C., Li, T., Zhang, Y. (eds.) WISE 2015. LNCS, vol. 9418, pp. 16–30. Springer, Cham (2015). https://doi.org/10.1007/978-3-319-26190-4_2

Text and Multimedia Mining

Detecting Spammers with Changing Strategies via a Transfer Distance Learning Method

Hao Chen[1,2(✉)], Jun Liu[2,3], and Yanzhang Lv[1,3]

[1] National Engineering Lab for Big Data Analytics, Xi'an Jiaotong University,
Xi'an 710049, Shaanxi, China
chenhaogs@stu.xjtu.edu.cn
[2] School of Electronic and Information Engineering, Xi'an Jiaotong University,
Xi'an 710049, Shaanxi, China
[3] Shaanxi Province Key Laboratory of Satellite and Terrestrial Network Tech. R&D,
Xi'an, China

Abstract. Social spammers bring plenty of harmful influence to the social networking involving both social network sites and normal users. It is a consensus to detect and filter spammers. Existing social spammer detection approaches mainly focus on discovering discriminative features and organizing these features in a proper way to improve the detection performance, e.g., combining multiple features together. However, spammers are easy to escape being detected by using changing spamming strategies. Various spamming strategies bring differences in data distribution between training and testing data. Thus, previous fixed approaches are difficult to achieve desired performance in real applications. To address this, in this paper, we present a transfer distance learning approach, which combines distance learning and transfer learning to extract informative knowledge underlying training and testing instances in a unified framework. The proposed approach is validated on large real-world data. Empirical experiments results give the evidence that our method is efficient to detect spammers with changing spamming strategies.

Keywords: Transfer distance learning · Social spammer detection
Spamming strategies

1 Introduction

Social networking on-line is a wonderful experience for users to share and enjoy life. However, there are some unsatisfactory phenomena in on-line social network, such as social spamming activities. These social spamming activities are created by spammers, who introduce undesirable or harmful information to normal users. In detail, spammers may spread and promote mass advertisement posts in a discussing topic; mislead readers with fake content; or lurk as predators for some

© Springer Nature Switzerland AG 2018
G. Gan et al. (Eds.): ADMA 2018, LNAI 11323, pp. 281–291, 2018.
https://doi.org/10.1007/978-3-030-05090-0_24

personal gains. These spammers bring harmful to social networking and normal users.

Spammer detection and relative tasks have been studied with different techniques in various social network platforms. A graph based method introduced by Wang et al. [15] captures intricate relationships among different entities to detect spammers in a product review site. Hu et al. [7] designed a uniform framework to detect spammers by combining network and content information together. Marcos Alvarez et al. [11] proposed to detect spammers by comparing the neighborhoods in different views. Yang et al. [19] proposed an innovative framework of spammer detection based on bipartite graph and co-clustering.

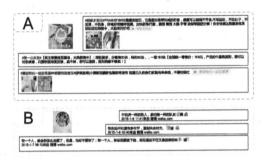

Fig. 1. Examples of spammers with different spamming strategies. User in region A post many advertainment posts, so it is identified as a spammer because of its posted content of its messages. User in region B post many normal contents in its messages just like normal users, however its posting behavior is unnormal (its posting time is regular) just like a bot machine.

Though existing methods [13,18,20] significantly reveal the mechanism of spammer detection, they may suffer a lot sometimes in real applications. In practice, spammers usually use different spamming strategies, e.g., as shown in Fig. 1, the user in region A does spam work mainly through content and the user in region B is identified as a spammer mainly because of its posting behavior. Empirical study indicate that spammers tend to adopt different spamming strategies to escape the detection. This may lead the changing of data distributions of training and testing sets which makes existing fixed machine learning methods difficult to achieve an efficient and stable performance. For example, the detection model learned this month may work well in future half year but be disabled in next year. In short, changing spamming strategies of spammers bring the following three practical challenges and difficulties: (1) the information used as clues to identify a spammer are highly fragmented and inconsistent; (2) lack of enough labeled instances for learning in time; (3) spammers diversity spamming strategies make insightful knowledge for spammers detection is changing and hard to capture thoroughly.

To resolve the above dilemmas and tackle with the challenges, we propose to apply a transfer distance spammer detection (TDSD) method, which uses distance learning to explore more interesting knowledge brought by training instances and leverages importance sampling technique to transfer knowledge from training set to testing set. Our proposed method TDSD is designed with two considerations in mind: (1) Good distance measures may provide insight into the underlying knowledge releasing from real-world data [2,5,16]. (2) Transfer learning can improve the ability of a system to recognize and apply knowledge and skills learned in previous tasks to novel tasks [9,10,12,17]. The novelty and technical contributions of this paper are summarized as follows:

- To the best of our knowledge, ours is the first work to apply a transfer learning under similarity metric learning framework to deal with the spammer detection problem. The proposed unified method learns instances similarity and matches the different data distributions of training and testing instances at the same time.
- Our method learns from both the coming instances and pairwise constraints of predicted instances, which achieves a more stable and better performances.
- We perform comprehensive experiments on the real-world data sets. Experimental results show that TDSD can effectively detect spammers with changing spamming strategies and outperforms the baseline methods overall.

2   Method

In this section, we present our transfer distance spammer detection (TDSD) method for spammer identification, especially for the spammers with changing spamming strategies.

2.1   Problem Formulation

The setup of our spammer detection task is as follows. Denoted the training instances set by $X_{tr} = \{(x_i, y_i) | i = 1, \cdots, n_s\}$, which also called the source-domain labeled data set, where $x_i \in \mathbb{R}^{d \times 1}$ is an input feature vector, and $y_i \in \mathbb{R}$ is the corresponding output. The testing data, which also called target domain, $X_{te} = X_{te}^l \bigcup X_{te}^u = \{(x_i, y_i) | i = n_s + 1, \cdots, n_S + n_T^l\} \bigcup \{x_i | i = n_S + n_T^l, \cdots, n_S + n_T\}$, where $n_T^l \ll n_s$. Our goal is to learn a Mahalanobis distance metric $M \in \mathbb{R}^{d \times d}$ jointly with the importance weights $\beta(x_i, x_j)$, where $\beta(x_i, x_j)$ is to tackle the practical difficulty that the marginal distributions of the source and target domain data are usually different, i.e., $P_{tr}(X) \neq P_{te}(X)$, in real-world data.

2.2   The Overall Framework

In this paper, we propose to improve the spammer detection performance by using a transfer distance learning method, which learns an informative similarity matrix in training set according to the special task and considers the different data distribution between training and testing sets simultaneously. The framework of this unified method is shown in Fig. 2. It has three steps, including *feature extraction* from meta data, *transfer distance* learning and *prediction*. In *feature extraction* step, discriminable features are extracted from meta data of both label and unlabeled instances. Then *transfer distance* learning step will simultaneously learn an importance weights to cope with the different data distributions between training and testing data and learn a similarity matrix to describe the training set. At last, in *prediction* step the unlabeled instances in the testing set will be predicted by a similarity-based classifier based on the output of *transfer distance* learning step. Techniques used in our method is inspired by the previous work [3].

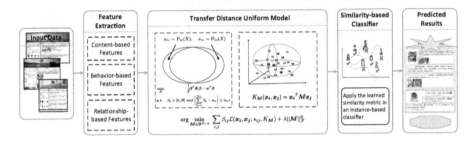

Fig. 2. Framework of the proposed method

2.3   Transfer Distance Learning for Spammer Detection

We first only consider the problem of learning a distance matrix on training set. For training instances, some of them have similar labels while some of them share different labels. Thus, we assume two sets of pairwise similarity constraints as the follow:

$$
\begin{aligned}
\mathcal{S} &= \{(\boldsymbol{x}_i, \boldsymbol{x}_j) : \boldsymbol{x}_i \text{ and } \boldsymbol{x}_j \text{ should be similar}\}, \\
\mathcal{D} &= \{(\boldsymbol{x}_i, \boldsymbol{x}_j) : \boldsymbol{x}_i \text{ and } \boldsymbol{x}_j \text{ should be dissimilar}\},
\end{aligned}
\tag{1}
$$

where users are considered to be similar when they share the same label while users are dissimilar when they have different labels And we define a binary label for each pair of $(\boldsymbol{x}_i, \boldsymbol{x}_j) \in \mathcal{S} \bigcup \mathcal{D}$ as:

$$
s_{ij} =
\begin{cases}
1, & (\boldsymbol{x}_i, \boldsymbol{x}_j) \in \mathcal{S}, \\
-1, & (\boldsymbol{x}_i, \boldsymbol{x}_j) \in \mathcal{D}.
\end{cases}
\tag{2}
$$

Given a set of data samples $\boldsymbol{X} = \{\boldsymbol{x_1}, \boldsymbol{x_2}, \cdots, \boldsymbol{x_n}\} \in \mathbb{R}^d$, the similarity function which measures the similarity between a pair of training instances $\boldsymbol{x}_i$ and $\boldsymbol{x}_j$. is shown as the following form:

$$
K_M(\boldsymbol{x}_i, \boldsymbol{x}_j) = \boldsymbol{x}_i^T M \boldsymbol{x}_j,
\tag{3}
$$

where $M \in \mathbb{R}^{d \times d}$ is a symmetric matrix.

To approximate the underlying semantic similarity, we propose to learn the optimal similarity function parameterized by M such that it best agrees with the constraints in 1. The optimization problem can be formulated solving the following objective function:

$$\arg \min_{M \in \mathbb{R}^{d \times d}} \sum_{(x_i, x_j) \in S \bigcup D} \mathcal{L}(x_i, x_j; s_{ij}, K_M) + \lambda ||M||_{\mathcal{F}}^2, \tag{4}$$

where $||\cdot||_{\mathcal{F}}$ denotes the Frobenius norm and $||M||_{\mathcal{F}}^2$ is a regularizer on the parameters of the learned similarity function. λ is the trade-off parameter to balances the loss function and regularization. $\mathcal{L}(x_i, x_j; s_{ij}, M)$ is the loss function that incurs a plenty when training constraints are violated. Here, log loss is used in our problem empirically, and it is formulated as:

$$\mathcal{L}(x_i, x_j; s_{ij}, K_M) = log(1 + exp(-s_{ij}(K_M(x_i, x_j) - 1))). \tag{5}$$

As we observed that the diplomatic spammers can change their spamming strategies, the similarity function learned based on training set cannot be applied to the testing data directly due to some changes of data. In this work, we also consider that the training instance x_{tr} is drawn from distribution $P_{tr}(X)$ and the testing instance x_{te} is drawn from distribution $P_{te}(X)$. In practice, the above two distributions are usually different. Thus we assume that $P_{tr}(Y|X) = P_{te}(Y|X)$, but $P_{tr}(X) \neq P_{te}(X)$. This phenomenon named *covariate shift*. Then we will investigate how to handle the problem of similarity learning with *covariate shift*.

The *covariate shift* problem can be solved by importance sampling methods [14]. Here we adapt this kind of methods to the similarity learning problem based on Theorem 1 proposed in [3] as follows.

Theorem 1. Suppose that x_i and x_j are drawn independently from $P_{tr}(X)$ and $P_{te}(X)$. $\ell_{te}(X; k_M)$ is the loss on test set. If a similarity learning method is without covariate shift, then the 6 will set up.

$$\ell_{te}(X; k_M) = \sum_{i,j} \beta_{ij} \mathcal{L}(x_i, x_j; s_{ij}, K_M). \tag{6}$$

where $\beta_{ij} = \dfrac{P_{te}(x_i) P_{te}(x_j)}{P_{tr}(x_i) P_{tr}(x_j)}$ is the importance weight.

Thus, the corresponding similarity learning with *covariate shift* problem can be formulated as:

$$\arg \min_{M \in \mathbb{R}^{d \times d}} \sum_{i,j} \beta_{ij} \mathcal{L}(x_i, x_j; s_{ij}, K_M) + \lambda ||M||_{\mathcal{F}}^2, \tag{7}$$

where $x_i, x_j \sim P_{tr}(X)$. In the following, we give a possible way to estimate the importance weights $\beta(x_i, x_j)$ directly and solve the optimization problem 7.

2.4   Optimization

In this paper, we use an efficient algorithm named kernel-mean matching (KMM) proposed by Huang et al. in [8] to estimate the importance weights. Here we briefly introduce KMM and readers can refer to [8] for more details. KMM learns $\beta_p = \dfrac{\mathrm{P}_{te}(\boldsymbol{x}_p)}{\mathrm{P}_{tr}(\boldsymbol{x}_p)}$ directly by matching the means between the training set and test set in a reproducing-kernel Hilbert space by solving the following quadratic programming optimization problem,

$$\min_{\beta} \quad \frac{1}{2}\beta^T K \beta - \kappa^T \beta$$

$$\text{s.t.} \quad \beta_k \in [0, B] \ and \ \left| \sum_{k=1}^{n_{tr}} \beta_k - n_{tr} \right| \leq n_{tr}\epsilon$$

where

$$K = \begin{bmatrix} \boldsymbol{K}_{tr,tr} & \boldsymbol{K}_{tr,te} \\ \boldsymbol{K}_{te,tr} & \boldsymbol{K}_{te,te} \end{bmatrix}$$

and $\boldsymbol{K}_{ij} = k(x_i, x_j)$. $\boldsymbol{K}_{tr,tr}$ and $\boldsymbol{K}_{te,te}$ are kernel matrices for the training set and the test set, respectively. $\kappa_i = \dfrac{n_{tr}}{n_{te}} \sum_{j=1}^{n_{te}} k(x_i, x_{j(te)})$, where $x_i \in \boldsymbol{X}_{tr} \bigcup \boldsymbol{X}_{te}$ and $x_{j(te)} \in \boldsymbol{X}_{te}$. B is a real number and is a trade-off parameter.

Then, the optimization problem 7 can be carried out efficiently by using recent advances in optimization over matrix manifolds [1].

The learned metric can be used in some distance-based classifier. In our experiments we use kNN (k-nearest neighbors) classifier as the basic classifier. The kNN-based method with the transfer distance learning for spammer detection is summarized in Algorithm 1.

Algorithm 1. kNN+TDSD

1: Input: training data $\boldsymbol{X}_{tr}$; testing data $\boldsymbol{X}_{te}$; the learned metric M; parameter k.
2: Output: spammer(+) or not(-) label y_{te} for each instance x_{te}.
3: **for** each $x_{te} \in \boldsymbol{X}_{te}$ **do**
4:    compute *similarity* (x_{te}, x_{tr}) with each instance $(x_{tr}, y_{tr}) \in \boldsymbol{X}_{tr}$;
5:    choose k instance $x_{tr}^s \in \boldsymbol{X}_{tr}$ with the highest *similarity* based on M;
6:    add to $\boldsymbol{X}_{tr}^s$;
7:    $y_{te} \leftarrow \arg\max_{y \in \{+,-\}} \sum_{x_{tr}^s \in \boldsymbol{X}_{tr}} I(y = y_{tr}^s)$
8: **end for**
9: **return** $\{y_{te}\}$.

---

3   Experiments

In this section, we conduct experiments to evaluate our proposed method from different angles. Through the evaluation we mainly want to verify the effectiveness of the idea that transfer distance learning can help improving the detection performance of spammer with changing strategies. We begin by introducing the data and experimental setup and then compare the performances of our proposed method and some the state-of-the art baseline approaches. We conduct the experiments on a quad-core 3.20 GHz PC operating on Windows 7 Professional (64 bit) with 8 GB RAM and 1TB hard disk.

3.1   Data for the Experiments

We now introduce the real-world *Microblog* data used in our experiment.

Data Crawling and Pre-processing. A developed asynchronous parallel crawler for *Microblog* was used from February 2014 to November 2016 to collect data from user profile, user generated content and interaction record. The crawler firstly obtains raw HTML data from web and then extracts metadata based on HTML tags. The crawled real-world data is stored in the relational database MySQL. These raw data contain a mixture of numerical, ordinal and categorical values, as well as Chinese free-text entries. We take the next two steps to pre-process these data. Firstly, we conduct data cleaning to remove extra or unrecognizable symbols. Secondly, we compute statistical values, including *URL* number, *hashtag* number, *mention* number, and *posting* times for each user.

Manually Labelling. We invited five human judges majored in Computer Science or Sociology to identify spammers individually. Judging spammer is a complex task for human. These five judges give their labels according to intuition, background knowledge and searching for additional information. In totally, each judge labeled 16,236 users. The labeling processes of judges are considered dependable if they have high consistency. *Fleiss kappa* [6] was used to evaluate their agreement and in our work the five judges obtained a *kappa* of 0.71, which represents almost substantial agreement. There are 3,371 anomalous users and 10,866 normal users labeled consistently by the five judges.

Real-World Ground-Truth Data. We use the consistently labeled instances as the ground-truth data. And We further divide the real-world data into different subsets according to time periods and topics. Among the subsets, we assumed that data distribution will be different. We totally generated ten sub-sets including Period-I (*April 2014-June 2014*), Period-II (*January 2015-March 2015*), Period-III (*October 2015-December 2015*), Period-IV (*July 2016-September 2016*), Period-IV (*October 2016- December 2016*), Topic-I (*Sports*), Topic-II (*Science*), Topic-III (*Politics*), Topic-IV (*Business*) and Topic-V (*Music*). The statistic of the dataset is presented in Table 1.

Table 1. Description of data sets

	Period-I	Period-II	Period-III	Period-IV	Period-V
#spammer	303	287	366	299	205
#normal user	1226	1599	1738	2002	1365
	Sports	Science	Politics	Business	Music
#spammer	857	776	415	403	628
#normal user	2086	2553	1978	2376	3260

Feature Extraction. *Microblog* is an information sharing platform that allows users post and receive information. This information consists of short texts called *message*, which may contain *URLs* or *hashtags* (#topic#). Relationship links between users in *microblog* are directional, which means that each user has *followers* and *followees*. Users can communicate with each other using *comment* or *repost*. We follow the previous work [4] to extract and organize features in classification methods. These discriminatory features are applicable and enough for our method and the baseline methods.

3.2   Evaluation Methodology and Experimental Setup

To verify the performance of our proposed method TDSD, we consider the following different methods for comparison:

MVSD [13]: a generalized spammer detection framework by jointly modeling multiple view information and a social regularization term into a classification model.

SSDM [7]: a matrix factorization-based model that integrates social-aspect features and content-aspect features as a long vector for social spammer detection. In the experiments, we empirically set parameters $\lambda_1 = 0.1$, $\lambda_2 = 0.1$ and $\lambda_s = 0.1$.

kNN+Euclidean: the typical and widely used classification method kNN with the Euclidean distance measure.

kNN+TDSD: our method. The output of our proposed method TDSD is a similarity matrix. We use k-Nearest-Neighbor (kNN) as the basic classifier in our method. The result of kNN is the values for k ranging from 1 to 5.

For the compared methods, they are trained over several runs of randomly generated 70/30 splits of the training data. (For the purpose of cross-validation, the training sets were further partitioned into training and validation sets.)

We use the area under the curve (AUC) to evaluate TDSD and the baseline methods. AUC is a standard measure used in machine learning to assess the classification model quality and balance the class sizes by considering the trade-off between false positive and true positive rates. Each experiment is repeated ten times independently and the reported numbers represent *mean* and textit-standard deviation.

3.3   Experimental Results and Analysis

The motivation of our work is that considering the diversity and variability of spammers' strategies can benefit spammer detection. The diversity and variability of spammers' strategies results in difference between training instances and testing ones, which brings challenges and difficulties for existing fixed spammer detection methods. The proposed method TDSD connects transfer learning and distance learning in a uniform framework to learn proper instance-based distance for classifier by taking into the data distribution difference between training labeled instances and testing unseen instances into consideration. And we empirically found that spammers usually share different spamming strategies in different time periods or topics. Therefore, we simultaneously conduct experiments on the data which crawled from different time periods and topics. These two scenarios are thoroughly tested with the following settings:

Performance for Different Time Periods. In this experiment, data set Period-V is used as the training set to learn a classifier model. Then the learned model is applied to data sets of Period-I, Period-II, Period-III, Period-IV and Period-V, respectively. The results of all the compared methods are shown in Table 2. From Table 2 we can see that our method outperforms other methods, when the training data and testing data from different time periods. The performance of the proposed method is more stable while baseline methods suffer a performance fluctuation.

Table 2. AUC values for different time periods

Methods	AUC values for different time periods				
	Period-I	Period-II	Period-III	Period-IV	*Period-V*
MVSD	0.811 ± 0.010	0.799 ± 0.012	0.827 ± 0.012	0.830 ± 0.008	*0.857 ± 0.004*
SSDM	0.761 ± 0.007	0.746 ± 0.021	0.758 ± 0.013	0.793 ± 0.008	*0.803 ± 0.007*
kNN+Euclidean	0.731 ± 0.011	0.709 ± 0.011	0.823 ± 0.007	0.788 ± 0.003	*0.826 ± 0.009*
kNN+TDSD	0.871 ± 0.021	0.824 ± 0.013	0.852 ± 0.011	0.899 ± 0.006	*0.903 ± 0.013*

Table 3. AUC values for different topics

Methods	AUC values for different topics				
	Sports	Science	Politics	Business	*Music*
MVSD	0.856 ± 0.003	0.862 ± 0.005	0.827 ± 0.003	0.835 ± 0.007	*0.902 ± 0.005*
SSDM	0.806 ± 0.007	0.792 ± 0.012	0.777 ± 0.004	0.785 ± 0.005	*0.852 ± 0.006*
kNN+Euclidean	0.811 ± 0.008	0.804 ± 0.009	0.762 ± 0.014	0.775 ± 0.005	*0.813 ± 0.008*
kNN+TDSD	0.902 ± 0.003	0.931 ± 0.005	0.914 ± 0.007	0.882 ± 0.012	*0.934 ± 0.005*

Performance for Different Topics. In this experiment, we use data set of Music topic as the training set to learn a classifier and apply it on data sets of

Sports, Science, Politics, Business and Music topics. The reported performance of all methods is shown in Table 3. From Table 3, we observe that, with transfer distance learning in a unified model, the proposed method achieves better performance than baseline ones.

Both results in Tables 2 and 3 show that our method achieves the highest performance. Baseline methods MVSD and SSDM mainly focus on connecting different information from different views as comprehensive as possible, however, it is hard to extract enough information from user's meta data due to data privacy and the worse is that they ignore the changing of the multiple information they pursue. Thus, MVSD and SSDM does not achieve the ideal performance in experiments on real-world data. For baseline kNN+Euclidean, it cannot efficiently obtain the underlying knowledge of the real-world data help classifier either. Our proposed method TDSD can provide more informative distance for instance-based classification methods. Distance learning can more sufficiently capture the task's own semantic notion of similarity and transfer learning setting can relieve the potential failure caused by data differences. Our method which applies transfer learning under metric learning framework is more efficient than many existing methods to detect the spammers in real-world social networking site *microblog*.

4 Conclusion

In this paper, we propose to apply a transfer distance learning method to detect social spammers with changing spamming strategies. The transfer distance learning method is designed to address the problem that knowledge may be not sufficiently extracted from training examples and the data distributions may be different between training and testing instances. Our proposed method TDSD combines similarity learning and instance-based transfer learning in a unified framework. Evaluations on real-world data show that TDSD is capable to improve the spammer detection performance. For future work, we plan to make our method more adaptive for non-linear problem by introducing kernel trick.

References

1. Absil, P.A., Mahony, R., Sepulchre, R.: Optimization Algorithms on Matrix Manifolds. Princeton University Press, Princeton (2009)
2. Bellet, A., Habrard, A., Sebban, M.: A survey on metric learning for feature vectors and structured data. arXiv preprint arXiv:1306.6709 (2013)
3. Cao, B., Ni, X., Sun, J.T., Wang, G., Yang, Q.: Distance metric learning under covariate shift. In: proceedings of the Twenty-Second International Joint Conference on Artificial Intelligence, Barcelona, Catalonia, Spain, p. 1204 (2011)
4. Chen, H., Liu, J., Lv, Y., Li, M.H., Liu, M., Zheng, Q.: Semi-supervised clue fusion for spammer detection in Sina Weibo. Inf. Fusion **44**, 22–32 (2018)

5. Davis, J.V., Kulis, B., Jain, P., Sra, S., Dhillon, I.S.: Information-theoretic metric learning. In: Proceedings of the 24th International Conference on Machine Learning, pp. 209–216. ACM (2007)
6. Fleiss, J.L., Cohen, J.: The equivalence of weighted kappa and the intraclass correlation coefficient as measures of reliability. Educ. Psychol. Measur. **33**(3), 613–619 (1973)
7. Hu, X., Tang, J., Zhang, Y., Liu, H.: Social spammer detection in microblogging. In: IJCAI, vol. 13, pp. 2633–2639 (2013)
8. Huang, J., Gretton, A., Borgwardt, K.M., Schölkopf, B., Smola, A.J.: Correcting sample selection bias by unlabeled data. In: Advances in Neural Information Processing Systems, pp. 601–608 (2007)
9. Jiang, J., Zhai, C.: Instance weighting for domain adaptation in NLP. In: Proceedings of the 45th Annual Meeting of the Association of Computational Linguistics, pp. 264–271 (2007)
10. Liao, X., Xue, Y., Carin, L.: Logistic regression with an auxiliary data source. In: Proceedings of the 22nd International Conference on Machine Learning, pp. 505–512. ACM (2005)
11. Marcos Alvarez, A., Yamada, M., Kimura, A., Iwata, T.: Clustering-based anomaly detection in multi-view data. In: Proceedings of the 22nd ACM International Conference on Information & Knowledge Management, pp. 1545–1548. ACM (2013)
12. Pan, S.J., Yang, Q.: A survey on transfer learning. IEEE Trans. Knowl. Data Eng. **22**(10), 1345–1359 (2010)
13. Shen, H., Ma, F., Zhang, X., Zong, L., Liu, X., Liang, W.: Discovering social spammers from multiple views. Neurocomputing **225**, 49–57 (2017)
14. Shimodaira, H.: Improving predictive inference under covariate shift by weighting the log-likelihood function. J. Stat. Plann. Infer. **90**(2), 227–244 (2000)
15. Wang, G., Xie, S., Liu, B., Philip, S.Y.: Review graph based online store review spammer detection. In: 2011 IEEE 11th International Conference on Data Mining (ICDM), pp. 1242–1247. IEEE (2011)
16. Weinberger, K.Q., Saul, L.K.: Distance metric learning for large margin nearest neighbor classification. J. Mach. Learn. Res. **10**(Feb), 207–244 (2009)
17. Wu, P., Dietterich, T.G.: Improving SVM accuracy by training on auxiliary data sources. In: Proceedings of the Twenty-First International Conference on Machine Learning, p. 110. ACM (2004)
18. Xu, Z., Zhang, Y., Wu, Y., Yang, Q.: Modeling user posting behavior on social media. In: Proceedings of the 35th International ACM SIGIR Conference on Research and Development in Information Retrieval, pp. 545–554. ACM (2012)
19. Yang, W., Shen, G.W., Wang, W., Gong, L.Y., Yu, M., Dong, G.Z.: Anomaly detection in microblogging via co-clustering. J. Comput. Sci. Technol. **30**(5), 1097–1108 (2015)
20. Zhu, Y., Wang, X., Zhong, E., Liu, N.N., Li, H., Yang, Q.: Discovering spammers in social networks. In: AAAI (2012)

Adversarial Learning for Topic Models

Tomonari Masada[1]($\boxtimes$)(iD) and Atsuhiro Takasu[2]

[1] Nagasaki University, 1-14 Bunkyo-machi, Nagasaki-shi, Nagasaki, Japan
masada@nagasaki-u.ac.jp
[2] National Institute of Informatics, 2-1-2 Hitotsubashi, Chiyoda-ku, Tokyo, Japan
takasu@nii.ac.jp

Abstract. This paper proposes adversarial learning for topic models. Adversarial learning we consider here is a method of density ratio estimation using a neural network called discriminator. In generative adversarial networks (GANs) we train discriminator for estimating the density ratio between the true data distribution and the generator distribution. Also in variational inference (VI) for Bayesian probabilistic models we can train discriminator for estimating the density ratio between the approximate posterior distribution and the prior distribution. With the adversarial learning in VI we can adopt implicit distribution as an approximate posterior. This paper proposes adversarial learning for latent Dirichlet allocation (LDA) to improve the expressiveness of the approximate posterior. Our experimental results showed that the quality of extracted topics was improved in terms of test perplexity.

Keywords: Topic models · Adversarial learning · Variational inference

1 Introduction

This paper proposes adversarial learning for topic models. Topic modeling [2, 3] is a widely used text analysis method for extracting diverse topics from a given document set. Topic modeling represents each latent topic as a probability distribution defined over vocabulary words. By visualizing such word probability distributions as word clouds (cf. Fig. 2), we can intuitively grasp what kind of things are discussed in the given document set by going through the word clouds. Topic modeling can also provide topic proportions for each document [2, Fig. 1]. Therefore, after finding favorite topics by going through the word clouds, we may retrieve the documents where the proportions of our favorite topics are large and read only those documents carefully. In this way, topic modeling mitigates the burden imposed on us by the extreme diversity of contents a massive set of documents delivers. It can be said that topic modeling provides us a bird's-eye view over diverse document contents.

Since topic modeling considered in this paper is Bayesian modeling, we need to infer posterior distribution. However, the true posterior distribution is intractable. We can approximate the true posterior either by drawing many

© Springer Nature Switzerland AG 2018
G. Gan et al. (Eds.): ADMA 2018, LNAI 11323, pp. 292–302, 2018.
https://doi.org/10.1007/978-3-030-05090-0_25

samples with MCMC or by seeking a tractable approximate distribution as a surrogate of the true posterior. In this paper, we consider the latter way of posterior inference and propose GAN-like [7] adversarial learning for effectively approximating the true posterior in variational inference (VI) for topic models.

The adversarial learning we consider here is a density ratio estimation using a neural network called discriminator [7]. VI approximates the true posterior by maximizing the evidence lower bound (ELBO). This maximization requires an estimation of the KL-divergence from the approximate posterior distribution to the prior distribution.[1] When we choose an approximate posterior whose density function is explicitly given, the KL-divergence can be easily estimated. However, the expressiveness of the approximate posterior is then limited. We thus propose to use implicit distribution, i.e., the probability distribution whose log-likelihood function is not explicitly given [10, 13], for approximating the true posterior. Even when we adopt implicit distribution, we can estimate the KL-divergence by using adversarial learning.

This paper provides adversarial learning for latent Dirichlet allocation (LDA) [3]. We aim to improve the expressiveness of the approximate posterior for LDA. In VI for LDA we consider the log evidence for each document: $\log p(\boldsymbol{x}_d; \boldsymbol{\Phi}) = \int p(\boldsymbol{x}_d|\boldsymbol{\theta}_d; \boldsymbol{\Phi})p(\boldsymbol{\theta}_d)d\boldsymbol{\theta}_d$, where $\boldsymbol{x}_d$ is the observed word count vector of document d, $\boldsymbol{\theta}_d$ is the document-wise parameter vector of categorical distribution over K latent topics, and $\boldsymbol{\Phi}$ denotes the set of free parameters for modeling topic-wise probabilities of V vocabulary words. The elements $(\theta_{d,1}, \ldots, \theta_{d,K})$ of $\boldsymbol{\theta}_d$ represent the probabilities of K latent topics in document d. Our main task is to approximate the true posterior $p(\boldsymbol{\theta}_d|\boldsymbol{x}_d)$ with a surrogate distribution $q(\boldsymbol{\theta}_d|\boldsymbol{x}_d)$. This approximation is equivalent to the maximization of the following evidence lower bound (ELBO):

$$\mathcal{L}(\boldsymbol{\theta}_d, \boldsymbol{\Phi}) = \mathbb{E}_{q(\boldsymbol{\theta}_d|\boldsymbol{x}_d)}[\log p(\boldsymbol{x}_d|\boldsymbol{\theta}_d; \boldsymbol{\Phi})] - \mathrm{KL}(q(\boldsymbol{\theta}_d|\boldsymbol{x}_d)\|p(\boldsymbol{\theta}_d)) \leq \log p(\boldsymbol{x}_d; \boldsymbol{\Phi}) \quad (1)$$

The estimation of $\boldsymbol{\Phi}$ can be performed by maximizing the log-likelihood term $\mathbb{E}_{q(\boldsymbol{\theta}_d|\boldsymbol{x}_d)}[\log p(\boldsymbol{x}_d|\boldsymbol{\theta}_d; \boldsymbol{\Phi})]$ in Eq. (1) when $\boldsymbol{\theta}_d$ is given. Our proposal concerns the approximation of the KL-divergence in Eq. (1). We propose to use implicit distribution as $q(\boldsymbol{\theta}_d|\boldsymbol{x}_d)$, which is represented by a neural network called *encoder*, and to make $q(\boldsymbol{\theta}_d|\boldsymbol{x}_d)$ more expressive than when we use explicit distribution. However, this makes the estimation of the KL-divergence in Eq. (1) intractable. Therefore, we provide adversarial learning. The proposed adversarial learning employs a neural network called *discriminator*, with which we can obtain an approximation of the KL-divergence from $q(\boldsymbol{\theta}_d|\boldsymbol{x}_d)$ to $p(\boldsymbol{\theta}_d)$ in Eq. (1). The evaluation experiment showed that the latent topics extracted by the proposed method were better than those extracted by collapsed Gibbs sampling (CGS) [8] for LDA in terms of test perplexity.

---

[1] We do not consider the joint contrastive form of ELBO [10] in this paper.

2   Method

2.1   Adversarial Learning in VI for LDA

Adversarial learning we consider here is a method of density ratio estimation using a discriminator network. The density ratio we estimate is $\frac{q(\boldsymbol{\theta}_d|\boldsymbol{x}_d)}{p(\boldsymbol{\theta}_d)}$, which is used for approximating the KL-divergence $\mathrm{KL}(q(\boldsymbol{\theta}_d|\boldsymbol{x}_d)\|p(\boldsymbol{\theta}_d))$ in Eq. (1). We propose to represent $q(\boldsymbol{\theta}_d|\boldsymbol{x}_d)$ with an encoder network $\boldsymbol{\theta}_d = g(\boldsymbol{x}_d, \boldsymbol{\epsilon})$, whose input is the concatenation of a word count vector $\boldsymbol{x}_d$ and a noise vector $\boldsymbol{\epsilon}$ drawn from the standard multivariate Gaussian distribution, i.e., $\boldsymbol{\epsilon} \sim \mathcal{N}(\mathbf{0}, \boldsymbol{I}_e)$. The output of the encoder then has randomness thanks to $\boldsymbol{\epsilon}$. However, the density function cannot be explicitly given for the distribution of the encoder output. Therefore, we provide adversarial learning, where we use a discriminator network $r(\boldsymbol{\theta}_d, \boldsymbol{x}_d)$ to estimate the density ratio between $q(\boldsymbol{\theta}_d|\boldsymbol{x}_d)$ and $p(\boldsymbol{\theta}_d)$ [10,13]. By maximizing the following objective function with respect to the discriminator parameters, we obtain an approximation of the logarithmic density ratio $\log \frac{q(\boldsymbol{\theta}_d|\boldsymbol{x}_d)}{p(\boldsymbol{\theta}_d)}$ as $r(\boldsymbol{\theta}_d, \boldsymbol{x}_d)$:

$$\ell(r) = \sum_{d=1}^{D} \Big[\mathbb{E}_{p(\boldsymbol{\theta}_d)} \log \big(1 - \sigma(r(\boldsymbol{x}_d, \boldsymbol{\theta}_d)) \big) + \mathbb{E}_{q(\boldsymbol{\theta}_d|\boldsymbol{x}_d)} \log \big(\sigma(r(\boldsymbol{x}_d, \boldsymbol{\theta}_d)) \big) \Big] \quad (2)$$

where $\sigma(t) = \frac{1}{1+e^{-t}}$ is the standard sigmoid function. We assume that the prior $p(\boldsymbol{\theta}_d)$ is an isotropic Gaussian distribution, whose mean and standard deviation parameters are estimated by empirical Bayes approach.

It should be noted that, for the optimal discriminator r^* obtained by maximizing $\ell(r)$ in Eq. (2), it holds that $\sigma(r^*) = \frac{1}{1+\exp(-\log(q/p))} = \frac{q}{p+q}$, which corresponds to the optimal discriminator D^* in GANs [7]. Therefore, precisely speaking, it is an abuse of terminology to call r discriminator, because r is not equal to the discriminator D in GANs. However, we think that it is harmless to call r discriminator, because r corresponds D through the mapping $D = \sigma(r)$.

By using $r(\boldsymbol{\theta}_d, \boldsymbol{x}_d)$, the ELBO in Eq. (1) can be rewritten as

$$\mathcal{L}(g, \boldsymbol{\Phi}) = \mathbb{E}_{\boldsymbol{\epsilon} \sim \mathcal{N}(\mathbf{0}, \boldsymbol{I}_e)}[\log p(\boldsymbol{x}_d|g(\boldsymbol{x}_d, \boldsymbol{\epsilon}); \boldsymbol{\Phi}) - r(g(\boldsymbol{x}_d, \boldsymbol{\epsilon}), \boldsymbol{x}_d)] \quad (3)$$

We update the parameters of the encoder network $g(\boldsymbol{x}_d, \boldsymbol{\epsilon})$ and the model parameters $\boldsymbol{\Phi}$ by maximizing $\mathcal{L}(g, \boldsymbol{\Phi})$ in Eq. (3) for a fixed $r(\boldsymbol{\theta}_d, \boldsymbol{x}_d)$. The expectation with respect to $\boldsymbol{\epsilon} \sim \mathcal{N}(\mathbf{0}, \boldsymbol{I}_e)$ is approximated by Monte Carlo integration.

We call this inference *adversarial variational Bayes (AVB)* by following [13]. Our AVB for LDA approximates the true posterior with two multilayer perceptrons (MLPs), i.e., one for encoder and another for discriminator (cf. Fig. 1). The number of hidden layers of each MLP was set differently for each data set. The experiment showed that two hidden layers were enough to achieve good results and that the results were not improved by increasing the number of hidden layers. We next discuss how the topic-wise word probabilities $\boldsymbol{\Phi}$ are modeled.

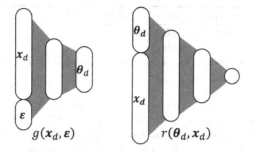

Fig. 1. Schematic depiction of the encoder network (left) and the discriminator network (right) in AVB-LDA. The softmaxed encoder output $\boldsymbol{\theta}_d$ is the parameter vector of document-wise categorical distribution over topics. The discriminator output is an approximation of the logarithmic density ratio between $q(\boldsymbol{\theta}_d|\boldsymbol{x}_d)$ and $p(\boldsymbol{\theta}_d)$. See Eq. (3).

2.2 Topic-Wise Word Probabilities

In the generative story of the original LDA [3] we draw a topic for each word token from the document-wise categorical distribution over latent topics, i.e., Categorical($\boldsymbol{\theta}_d$). Let the topic drawn for the i-th word token in document d be denoted by $z_{d,i} \in \{1, \ldots, K\}$. We then draw a word from the topic-wise categorical distribution over vocabulary words Categorical($\boldsymbol{\phi}_{z_{d,i}}$), which corresponds to the drawn topic $z_{d,i}$. This generative story assigns each word token to a topic $z_{d,i}$ randomly chosen based on the document-wise topic proportions $\boldsymbol{\theta}_d$. Our AVB for LDA collapses the discrete latent variables $z_{d,i}$ in the same manner as [5,18] to reduce computational burden in VI. We thus obtain the log-likelihood $\log p(\boldsymbol{x}_d|\boldsymbol{\theta}_d; \boldsymbol{\Phi})$, which is expressed by only using continuous parameters $\boldsymbol{\Phi} = \{\boldsymbol{B}, \boldsymbol{b}_0\}$, as follows:

$$\log p(\boldsymbol{x}_d|\boldsymbol{\theta}_d; \boldsymbol{\Phi}) = \boldsymbol{x}_d^\top \text{LogSoftmax}(\boldsymbol{B}\boldsymbol{\theta}_d + \boldsymbol{b}_0) \tag{4}$$

The parameter vector $\boldsymbol{\theta}_d$ is affine-transformed with the $V \times K$ matrix $\boldsymbol{B}$ and the V dimensional vector $\boldsymbol{b}_0$. The computation $\boldsymbol{B}\boldsymbol{\theta}_d + \boldsymbol{b}_0$ in Eq. (4) can be regarded as forward pass of a single-layer neural network. We call this network *decoder*. By applying the log-softmax function to the decoder output, we obtain the logarithmic word probabilities in document d. The log-likelihood of document d is then obtained as the inner product of the word count vector $\boldsymbol{x}_d$ with the logarithmic word probability vector as in Eq. (4). We use no prior distribution for word probabilities in our LDA. However, the V-dimensional bias vector $\boldsymbol{b}_0$ plays a similar role to smoothing parameter [4], because $\boldsymbol{b}_0$ depends on no particular topics. We train the decoder also by maximizing the ELBO in Eq. (3). While we tested decoders having hidden layers accompanied with nonlinear activation function in the evaluation experiment, the results were not improved. Therefore, word probabilities are modeled in this simple way.

The pseudo code of the proposed adversarial learning for LDA, abbreviated as AVB-LDA, is given in Algorithm 1, where C is the mini-batch size, and M is

---

Algorithm 1. Adversarial variational Bayes for LDA

1: **procedure** AVB-LDA($\mathcal{D}, r, g, \boldsymbol{\Phi}, C, M$)
2:   **repeat**
3:     Sample C items $\boldsymbol{x}_{(1)}, \ldots, \boldsymbol{x}_{(C)}$ from the training set $\mathcal{D}$ to make a mini-batch.
4:     **for** $m = 1$ to M **do**
5:   ▷ discriminator update
6:       Sample C noise vectors $\boldsymbol{\epsilon}_{(1)}, \ldots, \boldsymbol{\epsilon}_{(C)}$ from $\mathcal{N}(\boldsymbol{0}, \boldsymbol{I}_e)$.
7:       Compute $g(\boldsymbol{x}_{(c)}, \boldsymbol{\epsilon}_{(c)})$ for each $c \in \{1, \ldots, C\}$.
8:       Sample C document-wise topic proportions $\boldsymbol{\theta}_{(1)}, \ldots, \boldsymbol{\theta}_{(C)}$ from $p(\boldsymbol{\theta})$.
9:       Maximize ℓ in Eq. (2) with respect to the parameters of r.
10:   ▷ encoder update
11:     Sample C noise vectors $\boldsymbol{\epsilon}_{(1)}, \ldots, \boldsymbol{\epsilon}_{(C)}$ from $\mathcal{N}(\boldsymbol{0}, \boldsymbol{I}_e)$.
12:     Compute $g(\boldsymbol{x}_{(c)}, \boldsymbol{\epsilon}_{(c)})$ for each $c \in \{1, \ldots, C\}$.
13:     Maximize $\mathcal{L}$ in Eq. (3) with respect to the parameters of g.
14:   ▷ decoder update
15:     Sample C noise vectors $\boldsymbol{\epsilon}_{(1)}, \ldots, \boldsymbol{\epsilon}_{(C)}$ from $\mathcal{N}(\boldsymbol{0}, \boldsymbol{I}_e)$.
16:     Compute $g(\boldsymbol{x}_{(c)}, \boldsymbol{\epsilon}_{(c)})$ for each $c \in \{1, \ldots, C\}$.
17:     Maximize $\mathcal{L}$ in Eq. (3) with respect to $\boldsymbol{\Phi}$.
18:   ▷ prior update
19:     Sample C noise vectors $\boldsymbol{\epsilon}_{(1)}, \ldots, \boldsymbol{\epsilon}_{(C)}$ from $\mathcal{N}(\boldsymbol{0}, \boldsymbol{I}_e)$.
20:     Compute $g(\boldsymbol{x}_{(c)}, \boldsymbol{\epsilon}_{(c)})$ for each $c \in \{1, \ldots, C\}$.
21:     Maximize $\mathcal{L}$ in Eq. (3) with respect to the parameters of the prior.
22:   ▷ postprocessing
23:     Adjust learning rate.
24:   **until** change in parameters is negligible.

---

the number of iterations for the training of discriminator. In our experiment, the initialization method of the parameters of the encoder g and the discriminator r was chosen from among Xavier uniform, Xavier normal, Kaiming uniform, and Kaiming normal [6,9]. The activation function was chosen from among ReLU and LeakyReLU. For the parameters of the decoder, $\boldsymbol{B}$ was initialized by standard normal random numbers and $\boldsymbol{b}_0$ by zeros.

3   Experiment

3.1   Document Sets for Evaluation

In the evaluation experiment we used three data sets, whose specifications are given in Table 1, where D_{train} and D_{test} are the numbers of documents in training and test sets, respectively, and V is the vocabulary size. The first data set, denoted by NIPS, is the set of NIPS full papers obtained from UCI Bag of Words Data Set.[2] The second data set is a subset of the questions from the StackOverflow data set available at Kaggle.[3] We denote this document set by

---

[2] https://archive.ics.uci.edu/ml/datasets/bag+of+words.
[3] https://www.kaggle.com/stackoverflow/rquestions.

STOF. The third one is a set of New York Times articles also obtained from UCI Bag of Words Data Set. We refer to this by NYT. Each document set is split into training and test sets as given in Table 1. Based on the perplexity computed over training set, we tuned free parameters for each compared method. The final evaluation is performed in terms of the perplexity computed over test set. The perplexity is defined as the exponential of the negative log likelihood of the corpus, where the log likelihood is normalized by the number of word tokens.

Table 1. Specifications of data sets used in comparison experiment

NIPS			STOF			NYT		
D_{train}	D_{test}	V	D_{train}	D_{test}	V	D_{train}	D_{test}	V
1,050	225	12,419	20,486	2,561	13,184	239,785	29,968	102,660

3.2 Evaluation Results

We compared the proposed adversarial learning for LDA, denoted by AVB-LDA, to the following three methods.

The first compared method is collapsed Gibbs sampling (CGS) for LDA [8]. This choice aims to compare our proposal to the vanilla topic modeling. CGS is a sampling-based posterior inference and is thus time-consuming. However, it is known that CGS often gives better test perplexity than VI [1]. In CGS we have two free parameters, i.e., the hyperparameter α of the symmetric Dirichlet prior distribution for document-wise topic categorical distributions and the hyperparameter β of the symmetric Dirichlet prior distribution for topic-wise word categorical distributions. We tuned these two hyperparameters by grid search [1] based on training set perplexity. We call this method CGS-LDA.

The second compared method is the adversarial learning for document modeling, not for topic modeling. This choice aims to compare the effectiveness of adversarial learning for topic modeling to that for plain document modeling. By plain document modeling, we mean modeling of documents by merely mapping them into some lower-dimensional space. We refer to this method as AVB-DM. AVB-DM uses the following three MLPs. The first MLP is encoder with a single hidden layer for mapping document vectors to their lower-dimensional representation. The second one is decoder with a single hidden layer for mapping the encoded representation to the reconstructed document vector. The dimension of encoded representations is set to K, i.e., equal to the number of topics in CGS-LDA and AVB-LDA. The encoder input is a concatenation of a word count vector $\boldsymbol{x}_d$ and a noise vector $\epsilon \sim \mathcal{N}(\boldsymbol{0}, \boldsymbol{I}_e)$ as described in [13]. The dimension e of the noise vector was tuned based on training set perplexity. Since the encoder represents an implicit distribution, we adopt adversarial learning. AVB-DM uses the third MLP as discriminator in a similar manner to AVB-LDA. The number of the hidden layers of discriminator was set to two only for NIPS data set and one for the other data sets. Since AVB-DM is not a topic model, it does not provide latent topics as categorical distributions over words. AVB-DM provides no straightforward ways to obtain diverse document contents as word lists, each possibly visualized as word clouds. Therefore, it can be said that AVB-DM is inferior to AVB-LDA with respect to the interpretability of analysis results.

The third compared method is the variational autoencoder (VAE) [11] for document modeling. This method, denoted by VAE-DM, is more straightforward than AVB-DM, because no implicit distribution is used for approximating the true posterior. This choice aims to clarify what we lose by discarding adversarial learning. In VAE-DM we consider here the encoder MLP maps each document vector x_d to a concatenation of the mean parameter vector μ_d and the log standard deviation parameter vector $\log \sigma_d$ of diagonal Gaussian distribution $\mathcal{N}(\mu_d, \mathrm{diag}(\sigma_d^2))$, where $\mathrm{diag}(\sigma_d^2)$ is the diagonal matrix whose diagonal elements are σ_d^2. The dimension of μ_d and σ_d^2 is K. We can draw samples from this diagonal Gaussian by using reparameterization trick [16, 19]. That is, the samples from $q(z_d|x_d)$ are obtained as $\epsilon \odot \sigma_d + \mu_d$, where $\odot$ is element-wise product and ϵ is drawn from the standard multivariate Gaussian distribution $\mathcal{N}(0, I_K)$. The number of the encoder hidden layers was set to two only for STOF data set and one for the other data sets. There is no difference between AVB-DM and VAE-DM with respect to the decoder MLP, which maps the encoded representation to a reconstruction of the input document vector. The number of the decoder hidden layers was set to one for all data sets. We made the mini-batch size for VAE-DM larger than AVB-DM and AVB-LDA for obtaining better training set perplexity. While the VAE-DM we consider here is almost the same with neural variational document model (NVDM) [14], we applied dropout to the input vectors in every mini-batch, because perplexity was improved dramatically.

The compared methods were implemented in PyTorch[4] except CGS-LDA, which uses no neural networks. Our implementation of AVB-LDA is available at github[5]. For AVB-LDA, the number of the encoder hidden layers was set to two only for STOF data set and one for the other data sets. Further, the number of the discriminator hidden layers was set to one only for NYT data set and two for the other data sets. The number of the hidden layers were determined based on training set perplexity for all cases.

Table 2 contains all evaluation results in terms of test set perplexity, where we also present the optimal settings obtained based on training perplexity. For example, $h_{\mathrm{Enc},1}$ means the size of the first hidden layer of the encoder, and η_{Dis} the initial learning rate of the discriminator. As Table 2 shows, AVB-LDA achieved the best test perplexity among the compared methods. However, AVB-DM gave the second best results for all data sets. Therefore, we can say that adversarial learning works both for topic modeling and for plain document modeling. VAE-DM led to a comparable result only for NIPS data set, and the perplexity of CGS-LDA exhibits a tendency similar to that of VAE-DM. It can be concluded that AVB-LDA is the best choice if our aim is to achieve excellent test perplexity.

---

[4] https://pytorch.org/.

[5] https://github.com/tomonari-masada/adversarial-learning-for-topic-models.

Table 2. Evaluations in terms of test perplexity with the optimal hyperparameters. B is the mini-batch size. e is the dimension of noise vector for encoder. h represents the hidden layer size. For example, $h_{Enc,2}$ is the size of the second hidden layer of encoder. η denotes the learning rate. For example, η_{Dis} is the learning rate of discriminator. α and β are Dirichlet hyperparameters in LDA.

	NIPS	STOF	NYT
VAE-DM	1494.57 ± 2.90	884.66 ± 2.00	2609.80 ± 0.95
	$B = 4000$	$B = 10000$	$B = 1,000$
	$h_{Enc} = 1200$	$h_{Enc,1} = 1600$	$h_{Enc} = 1000$
		$h_{Enc,2} = 800$	
	$h_{Dec} = 1200$	$h_{Dec} = 800$	$h_{Dec} = 1000$
	$\eta_{Enc} = 0.001$	$\eta_{Enc} = 0.0002$	$\eta_{Enc} = 0.003$
	$\eta_{Dec} = 0.001$	$\eta_{Dec} = 0.0002$	$\eta_{Dec} = 0.01$
AVB-DM	1473.32 ± 0.12	630.22 ± 0.11	1794.88 ± 0.01
	$B = 200, e = 300$	$B = 200, e = 300$	$B = 200, e = 200$
	$h_{Enc} = 600$	$h_{Enc} = 1600$	$h_{Enc} = 800$
	$h_{Dis,1} = 600$	$h_{Dis} = 1600$	$h_{Dis} = 800$
	$h_{Dis,2} = 300$		
	$h_{Dec} = 600$	$h_{Dec} = 1600$	$h_{Dec} = 800$
	$\eta_{Enc} = 0.00002$	$\eta_{Enc} = 0.0001$	$\eta_{Enc} = 0.001$
	$\eta_{Dis} = 0.003$	$\eta_{Dis} = 0.0001$	$\eta_{Dis} = 0.001$
	$\eta_{Dec} = 0.0004$	$\eta_{Dec} = 0.003$	$\eta_{Dec} = 0.003$
AVB-LDA	$\mathbf{1443.45 \pm 0.04}$	$\mathbf{613.35 \pm 0.09}$	$\mathbf{1747.42 \pm 0.01}$
	$B = 200, e = 300$	$B = 200, e = 400$	$B = 200, e = 400$
	$h_{Enc} = 800$	$h_{Enc,1} = 1600$	$h_{Enc} = 1000$
		$h_{Enc,2} = 800$	
	$h_{Dis,1} = 800$	$h_{Dis,1} = 1600$	$h_{Dis} = 1000$
	$h_{Dis,2} = 400$	$h_{Dis,1} = 800$	
	$\eta_{Enc} = 0.04$	$\eta_{Enc} = 0.001$	$\eta_{Enc} = 0.001$
	$\eta_{Dis} = 0.01$	$\eta_{Dis} = 0.001$	$\eta_{Dis} = 0.001$
	$\eta_{Dec} = 0.001$	$\eta_{Dec} = 0.05$	$\eta_{Dec} = 0.1$
CGS-LDA	1524.47 ± 0.30	843.57 ± 0.32	3001.17 ± 0.58
	$\alpha = 0.2, \beta = 0.025$	$\alpha = 0.07, \beta = 0.005$	$\alpha = 0.01, \beta = 0.01$

We depict an example of high probability words in several topics obtained by AVB-LDA as word cloud in Fig. 2.

4   Previous Work

The variational autoencoder (VAE) proposed by Kingma and Welling [11] has initiated a creative interaction between deep learning and Bayesian probabilistic modeling. The VAE was firstly applied to plain document modeling by Miao et al. [14] and then firstly applied to LDA by Srivastava et al. [18]. The VI presented in the original LDA paper [3] prepares the approximate posterior of document-wise topic categorical distributions separately for each document. When compared to this, VAE for LDA has a special feature that it considers not only the document-wise structure but also the global structure of how the input vector x_d provides the corresponding approximate posterior $q(\theta_d|x_d)$. The inference having this feature is called *amortized* inference [17]. Adversarial learning in VI [10,13] shares this special feature with VAE and has an additional feature that the expressiveness of the approximate posterior is improved. Our work is not just an application of the already proposed adversarial learning to VI for LDA, because our experiment showed that relatively simple MLPs, i.e., MLPs having at most two hidden layers, achieved better test perplexity than other methods. These are highly practical results, which could not be obtained only by considering a theoretical possibility of the application of adversarial learning to VI for LDA.

Fig. 2. High probability topic words extracted by AVB-LDA from NIPS data set.

5 Conclusions

In this paper, we proposed adversarial learning for LDA, where we used discriminator multilayer perceptron for estimating the log density ratio between the approximate posterior distribution and the prior distribution. The experimental results showed that our proposal AVB-LDA could achieve better test perplexity than the three compared methods, i.e., CGS-LDA, AVB-DM, and VAE-DM. It should be noted that the proposed method is not the only way to use implicit distribution for approximating the true posterior in VI [15,20]. Further, it is an interesting research direction to use other types of neural network, e.g., RNN [5] and CNN [12], as encoder and to provide adversarial learning for such encoders.

References

1. Asuncion, A., Welling, M., Smyth, P., Teh, Y.W.: On smoothing and inference for topic models. In: Proceedings of the Twenty-Fifth Conference on Uncertainty in Artificial Intelligence, UAI 2009, pp. 27–34 (2009)
2. Blei, D.M.: Probabilistic topic models. Commun. ACM **55**(4), 77–84 (2012)
3. Blei, D.M., Ng, A.Y., Jordan, M.I.: Latent Dirichlet allocation. J. Mach. Learn. Res. **3**, 993–1022 (2003)
4. Chen, S.F., Goodman, J.: An empirical study of smoothing techniques for language modeling. In: Proceedings of the 34th Annual Meeting on Association for Computational Linguistics, ACL 1996, pp. 310–318 (1996)
5. Dieng, A.B., Wang, C., Gao, J., Paisley, J.W.: TopicRNN: a recurrent neural network with long-range semantic dependency. CoRR abs/1611.01702 (2016). http://arxiv.org/abs/1611.01702
6. Glorot, X., Bengio, Y.: Understanding the difficulty of training deep feedforward neural networks. In: Proceedings of the Thirteenth International Conference on Artificial Intelligence and Statistics, AISTATS 2010, pp. 249–256 (2010)
7. Goodfellow, I.J., et al.: Generative adversarial nets. In: Proceedings of the 27th International Conference on Neural Information Processing Systems, NIPS 2014, vol. 2, pp. 2672–2680 (2014)
8. Griffiths, T.L., Steyvers, M.: Finding scientific topics. Proc. Natl. Acad. Sci. **101**(Suppl. 1), 5228–5235 (2004)
9. He, K., Zhang, X., Ren, S., Sun, J.: Delving deep into rectifiers: surpassing human-level performance on ImageNet classification. In: Proceedings of the 2015 IEEE International Conference on Computer Vision, ICCV 2015, pp. 1026–1034 (2015)
10. Huszár, F.: Variational inference using implicit distributions. CoRR abs/1702.08235 (2017). http://arxiv.org/abs/1702.08235
11. Kingma, D.P., Welling, M.: Auto-encoding variational Bayes. CoRR abs/1312.6114 (2013). http://arxiv.org/abs/1312.6114
12. Lea, C., Vidal, R., Reiter, A., Hager, G.D.: Temporal convolutional networks: a unified approach to action segmentation. In: Hua, G., Jégou, H. (eds.) ECCV 2016. LNCS, vol. 9915, pp. 47–54. Springer, Cham (2016). https://doi.org/10.1007/978-3-319-49409-8_7
13. Mescheder, L.M., Nowozin, S., Geiger, A.: Adversarial variational Bayes: unifying variational autoencoders and generative adversarial networks. In: Proceedings of the 34th International Conference on Machine Learning, ICML 2017, pp. 2391–2400 (2017)
14. Miao, Y., Yu, L., Blunsom, P.: Neural variational inference for text processing. In: Proceedings of the 33rd International Conference on International Conference on Machine Learning, ICML 2016, vol. 48, pp. 1727–1736 (2016)
15. Mohamed, S., Lakshminarayanan, B.: Learning in implicit generative models. CoRR abs/1610.03483 (2016). http://arxiv.org/abs/1610.03483
16. Rezende, D.J., Mohamed, S., Wierstra, D.: Stochastic backpropagation and approximate inference in deep generative models. In: Proceedings of the 31st International Conference on International Conference on Machine Learning, ICML 2014, vol. 32, pp. II-1278–II-1286 (2014)
17. Shu, R., Bui, H.H., Zhao, S., Kochenderfer, M.J., Ermon, S.: Amortized inference regularization. CoRR abs/1805.08913 (2018). http://arxiv.org/abs/1805.08913
18. Srivastava, A., Sutton, C.: Autoencoding variational inference for topic models. CoRR abs/1703.01488 (2017). http://arxiv.org/abs/1703.01488

19. Titsias, M.K., Lázaro-Gredilla, M.: Doubly stochastic variational Bayes for non-conjugate inference. In: Proceedings of the 31st International Conference on International Conference on Machine Learning, ICML 2014, vol. 32, pp. II-1971–II-1980 (2014)
20. Uehara, M., Sato, I., Suzuki, M., Nakayama, K., Matsuo, Y.: Generative adversarial nets from a density ratio estimation perspective. CoRR abs/1610.02920 (2016). http://arxiv.org/abs/1610.02920

Learning Concise Relax NG Schemas Supporting Interleaving from XML Documents

Yeting Li[1,2], Xiaoying Mou[1,2], and Haiming Chen[1(✉)]

[1] State Key Laboratory of Computer Science, Institute of Software,
Chinese Academy of Sciences, Beijing 100190, China
{liyt,mouxy,chm}@ios.ac.cn
[2] University of Chinese Academy of Sciences, Beijing, China

Abstract. Relax NG is a popular and powerful schema language for XML, which concerns the relative order among the elements. Since many XML documents in practice either miss schemas or lack valid schemas, we focus on inferring a concise Relax NG from some XML documents. The fundamental task of Relax NG inference is learning regular expressions. Previous work in this direction lacks support of all operators allowed in Relax NG especially for interleaving. In this paper, by analysis of large-scale real-world Relax NG, we propose a restricted subclass of regular expressions called *chain regular expressions with interleaving (ICREs)*. Meanwhile, we develop a learning algorithm to infer a *descriptive generalized* ICRE from XML samples, based on *single occurrence automata* and the *maximum clique*. We conduct experiments on real benchmark from *DBLP*. Experimental results show that ICREs are expressive enough to cover the vast majority of practical Relax NG. Our algorithm can effectively learn from small and large dataset, and our results are concise and more precise than other popular methods.

Keywords: Relax NG · Regular expressions · Interleaving
XML documents · Schema inference

1 Introduction

EXtensible Markup Language (XML), as a standard format for data representation and exchange, is widely used in various applications on the Web [1]. In many XML documents, the relative order of elements is an important part of their structural properties. For example, to describe an element *date*, there may be many possible structures listed in Fig. 1. According to the convention, *year* usually presents in the last position, and the order of *month* and *day* is arbitrary. That is, for elements *month* (*day*) and *year*, they have a sequential order, and

H. Chen—Work supported by the National Natural Science Foundation of China under Grant No. 61472405.

G. Gan et al. (Eds.): ADMA 2018, LNAI 11323, pp. 303–317, 2018.
https://doi.org/10.1007/978-3-030-05090-0_26

for *month* and *day*, they are shuffled. Sometimes, *day*, or even *month* may be missed. If we use above separate descriptions of the *date*, it is too clumsy to manage data with a waste of space and time. Thus, it is necessary to design a schema for such XML documents, which can precisely regulate legal elements and the structure of elements in the *date*.

```
<date>                          <date>
<day> 1 </day>                  <month> January </month>
<month> May </month>           <year> 2018 </year>
<year> 2018 </year>            </date>
</date>
                               
<date>                          <date>
<month> May </month>           <year> 2017 </year>
<day> 2 </day>                 </date>
<year> 2017 </year>
</date>
```

Fig. 1. Fragments of XML documents describing the *date*

A schema for XML can describe the structural properties and constrains of admissible XML documents, and is useful at data processing, integration, transformations, validation and so on [10,16,27–30]. However, in practice, most documents are either not accompanied by a schema or the schema is not valid [2,5,23]. Nowadays, the most popular XML schema languages include Document Type Definition (DTD), XML Schema Definition (XSD) and Relax NG. We focus on the Relax NG schema, since it is more powerful than the first two schemas [25] and has unrestricted supports for unordered content and mixed content. That is, a Relax NG schema allows the interleaving operator and support the interleaving to be mixed with other operators. But DTD does not support interleaving, and XSD only permits a strongly limited interleaving formed like: $a_1^{[m_1,n_1]} \& a_2^{[m_2,n_2]} \& \cdots$ [1]. This causes the Relax NG schema to be feasible and concrete in many cases. For example, an XSD schema for the above *date* is $(month, day^?, year)|((day, month)^?, year)^2$; and the corresponding Relax NG schema of the *date* is shown in Fig. 2, which can be represented as $(day^? \& month)^?, year$. Besides, Relax NG schemas can follow the syntax of XML. Since a schema actually uses regular expressions to define allowed content models of XML documents, the essential task in schema inference is learning regular expressions from samples [4,6,7,20]. Hence, the problem of inferring a Relax NG schema from XML documents can be reduced to learning regular expressions from samples.

In the study of language learning, Gold [22] proposed a classical language learning model (*learning in the limit*) and pointed out that the class of regular expressions cannot be identifiable from positive examples only. Hence, we concentrate on studying the restricted subclasses of regular expressions with

---

[1] Where $m_i \in \mathbb{N}, n_i \in \mathbb{N} \setminus \{0\} \cup \{\infty\}$ and $i = \{1, 2 \cdots \}$..

[2] The content model of XSD must be deterministic expressions formally defined in [9]. So nondeterministic forms like $((day^?, month)|(month, day^?))^?, year$ are illegal.

```
<element name="date" xmlns="http://relaxng.org/ns/structure/1.0">
    <optional>
        <interleave>
            <optional>
                <element name="day"> <text/> </element>
            </optional>
            <element name="month"> <text/> </element>
        </interleave>
    </optional>
    <element name="year"> <text/> </element>
</element>
```

Fig. 2. A simplified fragment of the Relax NG document for describing the *date*

interleaving from positive samples. Previous researches have discussed various subclasses of standard deterministic regular expressions [4,6,15,17], and most of them have corresponding learning algorithms. Bex et al. [6] proposed learning algorithms for single occurrence regular expressions (SORE) and chain regular expressions (CHARE). Freydenberger and Kötzing improved these algorithms to make sure the learning results to meet the descriptive generalization [17]. Bex et al. [4] learned a subclass k-ORE of deterministic expressions. In [15], a simple form of deterministic expressions is directly inferred from positive data. There are also some subclasses of deterministic regular expressions with interleaving or unordered concatenation. The latter can be viewed as a weaker form of interleaving. Two subclasses DME and ME were given in [8,11]. They support to describe complete unordered data without the sequential order and ME even uses no union operator. The inference algorithm of DME is discussed in [11]. Peng et al. [31] developed an approximate method to learn the restricted subclass (called SIRE), without concerning the union operator. Besides, Kim et al. [26] focused on inferring a Relax NG schema using hedge grammars. Their learning algorithm can deal with both positive and negative samples, however their learning results excluded the interleaving.

We address the problem of learning restricted regular expressions with interleaving from positive samples. Our main contributions are listed as follows:

- We propose a restricted subclass of regular expressions with interleaving, called chain regular expressions with interleaving (ICREs): it is a suitable formalism to specify XML with both ordered and unordered data.
- We design a polynomial time algorithm *GenICRE* to learn a descriptive generalized ICRE, including two core steps: constructing the single occurrence automaton and extracting the maximum cliques. We obtain the learning result with the help of an approximate method for the clique problem [14].
- We provide a case study to help understanding our learning algorithm. Meanwhile, we show how to construct a Relax NG documents by means of the learning result, which includes some details about converting the ICRE into

the content model of Relax NG and adding some datatypes referring to a repository of practical Relax NG.

- We introduce the way to crawling large-scale real-world Relax NG documents. By analysis of these schema documents, ICREs cover the vast majority of practical Relax NG. Then we crawl small and large sets of XML documents from $DBLP^3$ for the learning test. By comparing $GenICRE$ with other learning algorithms (DME [8,11], SIRE [31]) and the tool (InstanceToSchema [13]), experimental results show that our learning results are concise and more precise.

The rest of the paper is organized as follows. Section 2 contains basic definitions. Section 3 describes the learning algorithm, and the techniques it implements. In Sect. 4, we show the process on inferring a Relax NG documents with the help of our learning algorithm. Section 5 reports experimental results. Section 6 concludes.

2   Preminilaries

2.1   Regular Expressions with Interleaving (RE(&))

Let Σ be an alphabet of symbols. The set of all finite strings over Σ is denoted by Σ^*. The empty string is denoted by ε. The interleaving of two strings in Σ^* is defined as: $u\&\varepsilon = \varepsilon\&u = \{u\}$, $au\&bv = \{aw \mid w \in u\&bv\} \cup \{bw \mid w \in au\&v\}$ with $a, b \in \Sigma$ and $u, v \in \Sigma^*$. For two languages L_1, L_2, $L_1\&L_2 = \{u\&v \mid u \in L_1, v \in L_2\}$ [3].

The class of standard regular expressions over Σ, denoted by RE, is defined in the standard way: $\emptyset, \varepsilon$ or $a \in \Sigma$ is a regular expression, the concatenation $E_1 \cdot E_2$, the union $E_1|E_2$, or the Kleene star E_1^* is a regular expression for regular expressions E_1 and E_2. Notice that the optional $E^?$ and the plus E^+ are also used in the rest of the paper, which are the abbreviation of $E|\varepsilon$ and $E \cdot E^*$, respectively, and parentheses can be added if necessary. The class of regular expressions with interleaving, denoted by $RE(\&)$, is extended from RE by using the interleaving operator: $E_1\&E_2$ is a regular expression for regular expressions E_1 and E_2. For a regular expression E, the language specified by E is denoted by $L(E)$. The language of $E_1\&E_2$ is defined as $L(E_1\&E_2) = L(E_1)\&L(E_2)$, for the others, please refer to [24]. The set of symbols that occur in an expression E is denoted by $sym(E)$, this notation is extended for strings in the obvious way.

2.2   Definitions

Single Occurrence Regular Expressions (SORE) [6]. A standard regular expression E belongs to the class $SORE$, if every symbol in E occurs at most once.

We apply this definition to expressions in RE(&), that is, for an expression $E \in$ RE(&), if every symbol occurs at most once, then E belongs to a single

---

[3] http://dblp.uni-trier.de/xml/.

occurrence regular expression with interleaving (*ISORE*). Our new restricted subclass is designed based on this popular subclass.

Definition 1. *A chain regular expression with interleaving (ICRE) E is an ISORE over Σ, defined as follows:*

$$
\begin{aligned}
E &:= F_1^{p_1} \cdot \ldots \cdot F_n^{p_n}, & (n \geq 0, p_i \in \{?, 1\}) \\
F_i &:= D_1 \& \ldots \& D_k & (i \in [1, n], k \geq 1), \\
D_j &:= a_1^{mul_1} | \ldots | a_m^{mul_m}, & (j \in [1, k], m \geq 1)
\end{aligned}
$$

where $mul_o \in \{1, ?, *, +\}$ and $a_o \in \Sigma$ for $o \in [1, m]$.

We also introduce some basic definitions for developing our learning algorithm. There will be many operations on the single occurrence automata (SOA).

Definition 2 ([6,17]). *An SOA over Σ is a finite directed graph $G = (V, D)$ which satisfies the following two conditions:*

1. *$src, sink \in V$ and all nodes in $V \setminus \{src, sink\}$ are distinct symbols.*
2. *src has only outgoing edges; $sink$ has only incoming edges, and every $v \in V$ lies on a path from src to $sink$.*

A generalized single-occurrence automata (generalized SOA) over Σ is a finite directed graph in which each node $v \in V \setminus \{src, sink\}$ is an *ISORE* and all nodes are disjoint *ISOREs*. For a generalized SOA G, the set of terminal symbols in G is $V \setminus \{src, sink\}$. We define the relation $\rightarrow$ on V which equals D. The accepted language by G denoted as $L(G)$ is the set of strings $w = a_1 \ldots a_n (n \geq 0)$ such that $src \rightarrow a_1 \rightarrow \cdots \rightarrow a_n \rightarrow sink$.

Level Number (ln) and Skip Level [17]. For a generalized SOA $G = (V, D)$ where G is a directed acyclic graph, there is $ln(src) = 0$, and for each node $v \in V$, $ln(v)$ is the longest path from src. For a node $v \in V$, if there exist two nodes v_1 and v_2 with $ln(v_1) < ln(v) < ln(v_2)$, such that $v_1 \rightarrow v_2$, then $ln(v)$ is a skip level.

Definition 3. *For a set S of symbols and a string $s = a_1 \ldots a_n$ over Σ, the projection function $fun(S, s)$ returns a new string $s' = a'_1 \ldots a'_n$ over Σ, such that if $a_1 \in S$ then $a'_1 = a_1$; otherwise $a'_1 = \varepsilon$. Clearly, s' can be reduced by $a\varepsilon = \varepsilon a = a$, for $a \in \Sigma$.*

Example 1. Let $S = \{b, c, r\}$ and $s = ebbdfc$, we get $s' = fun(S, s) = bbc$.

3   The Learning Algorithm *GenICRE*

In the following, we focus on the learning problem, that is, given a set of positive samples S, we can learn an expression $E \in$ ICREs such that $S \subseteq L(E)$, and it meets descriptive generalization. We first introduce some definitions which will be used in this section.

Positive Samples. A sample is actually a string served as an example in learning. When the string is accepted by the targeting language, then it is positive, otherwise it is negative. In our learning process, a positive sample must belong to $L(E)$ that E is an ICRE.

Definition 4 ([11]). *For a set of strings S, the Conflict Set (CS) contains pairs of symbols occurring in S such that:*
$$CS(S) = \{(a, b) \mid \nexists s \in S, a \in sym(s) \wedge b \in sym(s)\}.$$

3.1 Algorithm Implementation

The main procedures of the algorithm *GenICRE* are illustrated as follows:

- *Step 1:* For the set of samples S, we construct an SOA $G = (V, D)$ using the method *2T-INF* [18]. Then each $v \in V$ is labelled with a terminal symbol.
- *Step 2:* For each node $v \in V$ with a self-loop, label it with "v^+", and remove the self-loop. Then a generalized SOA is obtained.
- *Step 3:* Find all non-trival strongly connected components (NSCC) in the generalized SOA (the node with v^+ in each NSCC is restored to the label v). For each $ns \in$ NSCC, carry out the following operations. (i) construct the new sample $S' = fun(ns, s \in S)$. (ii) compute the conflict set $CS(S')$ and construct an undirected graph $A = (V_A, D_A)$ in which $V_A = ns$ and $D = CS(S')$. (iii) find all *maximum cliques* in A recursively and carry out the step (iv) to get a new node from each clique. Then combine all new nodes $v_i'(i \in [1, n])$ labelled with $v_1'\&v_2'\&...\&v_n'$. To be mentioned, after finding one *maximum clique*, remove all nodes in the *maximum clique* and corresponding edges from A. (iv) For each *maximum clique*, combine each node $v_i(i \in [1, n])$ in the clique to a new node v' labelled with $v_1|v_2...|v_n$. (v) Each ns is replaced with the node labelled with the expression obtained in step (iii). All incoming edges for any node in ns will build the relation to the new node. All outgoing edges from any node in ns will keep the relation from the new node.
- *Step 4:* Compute the level number for each node v of the directed acyclic graph and find all skip levels. The labels of nodes of each level number are turned to one or more chain factors of the result expression. At last, we reduce the expression by the rule that $E^{+?} = E^*$, and get an expression $E \in$ ICREs such that $S \subseteq L(E)$.

The algorithm *GenICRE* is mainly based on *SOA* and the *maximum clique*. The computation of the *maximum clique* is a well-known NP-hard problem [19]. Hence, we present an approximate solution to this learning problem [14].

In detail, we present the pseudocode of *GenICRE* in Algorithm 1. The function $G.setln()$ is to set the level number for each node of G. $AIT_U(V)$ and $AIT_I(V)$ are to combine all elements in V with the union and the interleaving, respectively. $reduce()$ reduces the expression by the rule that $E^{+?} = E^*$.

Algorithm 1. *GenICRE*

   input : A set of positive samples S
   output: A regular expression $E \in$ ICREs

1 construct an SOA $G(V, D)$ for S and initialize a set S';
2 remove the self-loop for $v \in V$ and label it with v^+; compute NSCCs C;
3 **for** *each $c_i \in C$* **do**
4     $S'.add(fun(c_i, s \in S))$; $F_i = cliqueTore(S')$;
5 update the graph G
6 $G.setln()$; $E \leftarrow \varepsilon$; $level = 1$
7 **while** $level \leq (ln(G.sink)) - 1$ **do**
8     $V_T \leftarrow \forall v \in V$ with $ln(v) = level$ and $|sym(v)| \geq 2$;
9     $V_S \leftarrow \forall v \in V$ with $ln(v) = level$ and $|sym(v)| = 1$;
10   **for** *each $v_T \in V_T$* **do**
11   **if** *level is a skip level or* $(|V_T| + |V_S|) > 1$ **then** $E \leftarrow E \cdot v_T^?$;
12   **else** $E \leftarrow E \cdot v_T$;
13   **if** $|V_S| > 0$ **then**
14   **if** *level is a skip level or* $|V_T| > 0$ **then** $E \leftarrow E \cdot AIT_U(V_S)^?$;
15   **else** $E \leftarrow E \cdot AIT_U(V_S)$;
16     $level = level + 1$
17 reduce(E)

Algorithm 2. *cliqueTore(S)*

   input : A set of positive samples S
   output: A regular expression $E \in$ ICREs

1 scan S to find the conflict set M, and construct an undirected graph $G(V, D)$;
2 initialize two lists *all_clique* and r
3 **while** $V \neq \emptyset$ **do**
4     $clique = find\_clique(G)$ [14]; $all\_clique.append(clique)$; $V = V \backslash clique$;
5 **for** *each $clique \in all\_clique$* **do**
6     $r.append$ $(AIT_U(clique))$;
7 $E \leftarrow AIT_I(r)$;
8 **for** *each symbol a in E and each set $s \in S$* **do**
9   **if** $\exists s$, *a occurs more than once* **then**
10   **if** $\forall s$, *a occurs* **then** $a \leftarrow a^+$;
11   **else** $a \leftarrow a^*$;
12   **else if** $\exists s$, *a does not occur* **then** $a \leftarrow a^?$;

3.2 Time Complexity on *GenICRE*

It needs polynomial time to construct an SOA for the example set S, such that
$G(V, D) = SOA(S)$. Let $n = |V|$ and $m = |D|$, it costs $O(n)$ to find all nodes
with loops and $O(m + n)$ to find all NSCCs using depth-first search algorithm
[12]. The function $find\_clique()$ can be finished in polynomial time [14]. The
function $setln()$ can be finished in time of $O(m + n)$ using breadth-first search

[12]. All nodes will be converted into specific chain factors of ICRE in $O(n)$. Therefore, the time complexity of *GenICRE* is polynomial time.

3.3   Descriptive Generalization

Definition 5 ([17]). *Let $\mathcal{D}$ be a kind of regular expressions over Σ. For a given nonempty language $S \in \Sigma^*$ and a regular expression $r \in \mathcal{D}$, if $S \subseteq L(r)$ and there does not exist another regular expression $\delta \in \mathcal{D}$ that satisfies the relation $S \subseteq L(\delta) \subset L(r)$, then r is a $\mathcal{D}$-descriptive generalized regular expression or automata for S over $\mathcal{D}$.*

For the set S, *GenICRE* can learn a regular expression E in ICREs such that $S \subseteq L(E)$. In this section, we aim to prove that E meets the definition of descriptive generalization. That is,

Theorem 1. *For a set of given string sample S, $\alpha = GenICRE(S)$. There does not exist an expression β such that $S \subseteq L(\beta) \subset L(\alpha)$.*

Proof. We prove this claim by supposing that there exists the expression β satisfying that $S \subseteq L(\beta) \subset L(\alpha)$, and by induction we show that $L(\beta) = L(\alpha)$ holds. For space reason, details are available online https://github.com/yetingli/ICREs/.

4   A Case Study

Learning an Expression in ICREs. Given a set of positive examples $S = \{ab, cada, addc, eeba, caca, acda\}$, we show how to learn an ICRE by Algorithm 1. We present some results of the main steps during the learning process in Fig. 3.

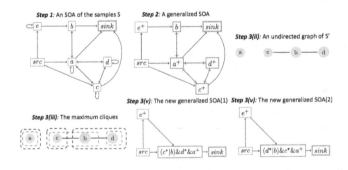

Fig. 3. Results of the main steps in the learning process

We first construct an SOA of the sample S. Then we deal with nodes with self-loops and get a generalized SOA at step 2. At the same time, we mark the NSCC= $\{\{b, a, c, d\}\}$ in red. Based on the component $\{b, a, c, d\}$, we obtain a new sample $S' = \{ab, cada, addc, ba, caca, acda\}$, and $CS(S') = \{(b, c), (b, d)\}$. At step 3(ii), the undirected graph G is constructed according to the set $CS(S')$. We can compute the set of maximum clique partitions $\{\{c, b\}, \{a\}, \{d\}\}$ or $\{\{b, d\}, \{a\}, \{c\}\}$, which is not unique. For G, both of them are optimal. Then $cliqueTore(\ S')$ returns the expression $(c^*|b)\&d^*\&a^+$ or $(d^*|b)\&c^*\&a^+$.

We replace each NSCC with a new node labelled with the returned expression. Followed by this step, we compute the level number and skip level of each node, then expressions are generated. At last, we reduce the expression and get $E = e^* \cdot ((c^*|b)\&d^*\&a^+)$ or $E = e^* \cdot ((d^*|b)\&c^*\&a^+)$.[4]

Constructing a Relax NG Document. We convert a regular expression in ICREs into the content model of Relax NG by the syntax tree. Take one result $E = e^* \cdot ((c^*|b)\&d^*\&a^+)$ as an example, the converting process shows in Fig. 4. As we can see, the syntax tree can clearly represent the structure of an expression, which corresponds to the structure of the content model of the Relax NG document. We use the *interleave* indicator in Relax NG to replace $\&$, *choice* for $|$, *optional* for ?, *zeroOrMore* for *, *oneOrMore* for + and sequential representation of elements for $\cdot$.

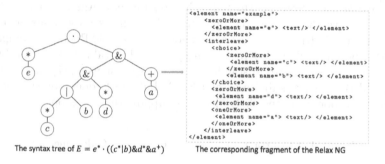

The syntax tree of $E = e^* \cdot ((c^*|b)\&d^*\&a^+)$ The corresponding fragment of the Relax NG

Fig. 4. Converting E into a fragment of the Relax NG

Adding Datatypes. In practice, the element name is usually meaningful like elements *year, month, day* in the *date* introduced in Sect. 1, instead of $a, b...$ in the above example. Thus, the element may have some constrains on datatype according to its real-world semantics. For instance, the datatype of *year* can be <data type="integer"/>. For *telephone-number*, the datatype can be designed more complex like <data type="string"> <param name="maxLength"> 11 </param> </data>, which requires the length of telephone number is not more than 11. It seems hard to teach machines learning the semantics of each element and giving each one a datatype. Here, we use a shortcut for adding datatypes for the fragment of the Relax NG. We crawl large-scale real-world Relax NG schemas (see Sect. 5), and extract the most frequent definition of each element

---

[4] The learning results are not unique, but *GenICRE* only returns one of the solutions.

to form a repository. Then when elements and the schema structure are learned, we can efficiently add most reasonable datatype for each element. The critical shortcoming is obvious that we cannot give datatypes for elements not appearing in the repository.

5   Experiments

In this section, we validate our algorithm by crawling large-scale real-world Relax NG schemas. First, we analyze the practicality of ICREs compared with other popular subclasses. Then we compare *GenICRE* with other learning algorithms of the subclasses (DME [8,11], SIRE [31])) and the XML tool (Instance-ToSchema [13]). All experiments are conducted on a machine with Intel Core i5-5200U@2.20GHz, 4G memory. All codes are written in python 3.

5.1   Analysis on Practicality

We collect Relax NG schemas by mainly three ways: Google by querying *filetype: rng*, data sets from Github and jars from Maven. Finally, we extract 137,286 regular expressions from 13,946 Relax NG documents. Among them, 38.45% expressions have the interleaving operator, which shows its widespread use of unordered data. ICREs take a vast proportion on real-world Relax NG comparing with other popular subclasses. Results are presented in Table 1. Note that SORE and CHARE are subclasses of standard regular expressions [6,17], and DME [8, 11] and SIRE [31] are subclasses of RE(&), but DME does not support the concatenation operator, and SIRE does not support the union operator.

Table 1. Proportions of subclasses on Relax NG documents

Subclass	SORE	CHARE	DME	SIRE	ICRE
Percentage score	58.30%	56.78%	50.62%	76.83%	85.55%

5.2   The Comparison Among Learning Algorithms

We conduct the experiments based on two kinds of data sets: small and large sets of XML documents, which are both extracted from *DBLP*. *DBLP* is data-centric database of information on major computer science journals and proceedings. We download the file *dblp-2015-03-02.xml.gz*. The elements chosen from *DBLP* are *book* and *article* with $14,845$ (small) and $1,737,265$ (large) samples, respectively.

We compare *GenICRE* with the XML tool (*InstanceToSchema*), which is a tool to generate a Relax NG schema from XML instances, and two learning algorithms (*learner$_{DME}^{+}$* for DME [11], *conMiner* for SIRE [31]). All of these support learning a regular expression with interleaving from XML data. To be mentioned, our learning algorithm *GenICRE* applies an approximate method to solve the maximum clique problem. Since the data set for testing is not too large to compute the maximum clique, we also give a result by using an exact solution replacing the approximate one of the maximum clique problem in *GenICRE*, which is denoted by *exact-GenICRE*.

Conciseness. Referring to the Definition 1, ICREs are composed of chain subexpressions, whose form is readable for both human and machine. However, we still need a measure to show the conciseness of expressions. In [21], they apply one part of the Minimum Description Length (MDL [32]) principle to calculate the length of expressions of XML schemas in bits as a direct measure. We apply this definition to an expression $E \in$ ICREs, the length of E in bits is

$$Len = n * \lceil \log_2(|\Sigma| + |\mathcal{M}|) \rceil$$

where $|\Sigma|$ is the number of distinct symbols occurring in E, $\mathcal{M}$ is the set of metacharacters $\{|, \cdot, \&, ?, *, +, (,)\}$ and n is the length of E including symbols and metacharacters. An expression with a smaller value of *Len* is more concise.

Preciseness. Referring to [4,21], it is common to use another part of the MDL principle called *datacost* to measure the preciseness of expressions. However, the computation of *datacost* for an expression E needs tedious work to compute the number of subset words of $L(E)$, and the length of words ranges from 1 to $2m+1$[5]. Here, we introduce a simple definition *Combinatorial Cardinality(CC)* to value the preciseness of expressions. *CC* is limited to measure the preciseness of expressions satisfying that different expressions have the same symbols with the same unary operators. The smaller value of *CC*, the more precise an expression. It is defined as follows:

Definition 6. *For an expression $E \in RE(\&)$, the Combinatorial Cardinality (CC) of E can be computed inductively:*
$CC(a) = CC(\epsilon) = 1$, *where* $a \in \Sigma$;
$CC(E^?) = CC(E^*) = CC(E) + 1;$ $\qquad CC(E^+) = CC(E);$
$CC(E_1 | E_2) = CC(E_1) + CC(E_2);$ $\qquad CC(E_1 \cdot E_2) = CC(E_1) \times CC(E_2);$
$CC(E_1 \& E_2) = \sum_{m=1}^{m=CC(E_1)} \sum_{n=1}^{m=CC(E_2)} \frac{(|E_1'| + |E_2'|)!}{|E_1'|! \times |E_2'|!},$
where E_1' and E_2' mean the m-ith string generated by E_1 and the n-ith string generated by E_2, and $|E_1'|$ $(|E_2'|)$ is the length of strings E_1' (E_2').

Experimental results are shown in Table 2 (*book*) and Table 3 (*article*), perspectively. The original schemas for *book* and *article* in *DBLP* are also represented in tables. To save space, we use the short names of words and the list of abbreviations is shown in Table 4.

We analyze the experimental results in Tables 2 and 3. Our learning results almost are the same concision with expressions learned by other methods, since all of them are in the same order of magnitude. However, only our results can support all operators allowed in Relax NG. Referring to the value of *CC*, expressions learned by our algorithms are more precise than the others, since all results

---

[5] m is the length of the expression excluding operators, $\emptyset$ and ϵ [4].

Table 2. Results of inference using different methods on **book**

Sample size	From	Element name	Len	CC
14845	DBLP	Book		
Method		Regular Expression E		
Original Schema		$(a_1\|a_2\|a_3\|\cdots\|a_{21}\|a_{22}\|a_{23})^*$	240	–
InstanceToSchema		$a_7^*\&a_1^*\&a_{10}^?\&a_{17}^+\&a_2^*\&a_{19}^*\&a_{14}\&a_4^?\&a_{15}^*\&a_{16}^?\&$ $a_{12}^*\&a_9^?\&a_{21}^*\&a_{20}^*\&a_{13}^?\&a_5^*\&a_3^?$	245	$3.56*10^{14}$
conMiner		$a_1^* \cdot a_2^* \cdot a_4^? \cdot a_9^? \cdot a_{19}^* \cdot a_{10}^?\&a_{14}^* \cdot a_7^* \cdot a_5^*\&$ $a_{21}^* \cdot a_3^? \cdot a_{15}^*\&a_{17}^?\&a_{12}^*\&a_{13}^?\&a_{16}^*\&a_{20}^*$	250	$1.37*10^{10}$
$learner^+_{DME}$[a]		$(a_5^*\|a_{16}^?\|a_9^?)\&(a_{19}^*\|a_4^?)\&(a_2^*\|a_1^*)\&(a_{21}^*\|a_3^?)\&$ $(a_{15}^*\|a_{10}^?)\&a_7^*\&a_{20}^*\&a_{12}^*\&a_{13}^?\&a_{14}^?\&a_{17}^?$	300	$1.92*10^9$
exact-GenICRE		$(a_1^+\|a_2^+)^? \cdot ((a_9^?\|a_5^*\|a_{21}^*\|a_3^?)\&a_{17}^?\&a_{16}^?\&$ $a_{15}^*\&a_{20}^*\&a_{12}^*\&a_{13}^?\&a_7^*\&a_{14}^?\&a_{19}^?) \cdot a_4^? \cdot a_{10}^?$	275	$2.90*10^7$
GenICRE		$(a_1^+\|a_2^+)^? \cdot ((a_{12}^*\|a_5^*\|a_9^?)\&(a_3^?\|a_{21}^*)\&a_7^*\&a_{20}^*\&$ $a_{16}^?\&a_{15}^*\&a_{14}^?\&a_{17}^?\&a_{19}^?\&a_{13}^?) \cdot a_4^? \cdot a_{10}^?$	295	$4.4*10^7$

[a] We replace the unordered concatenation operator ($\|$) with interleaving ($\&$) in DME [11], since they have the same semantics when operating on single symbol.

consist with the given samples and our results include a minimum number of strings. This means *GenICRE* has avoided some generalization during the learning process. What's more, the results of *GenICRE* are similar with that of exact-*GenICRE*. They are both more precise than others and their *CC* values are on the same order of magnitude. Hence, when the alphabet is large and the maximum clique problem costs heavily, we can efficiently use *GenICRE* to get a good result.

From the semantic point of view, take *article* as an example in Table 2. About the *Original Schema*, it only uses union operators to accept all possible strings. The results learned by *InstanceToSchema* are rigid, which are only symbols combined with the interleaving. Clearly, these kinds of learning results are over-generalization. *conMiner* and $learner^+_{DME}$ try to avoid over-generalization but

Table 3. Results of inference using different methods on **article**

Sample size	From	Element name	Len	CC
1737265	DBLP	Article		
Method		Regular Expression E		
Original Schema		$(a_1\|a_2\|a_3\|\cdots\|a_{21}\|a_{22}\|a_{23})^*$	240	–
InstanceToSchema		$a_1^*\&a_2^*\&a_3^?\&a_4^?\&a_5^*\&a_6^?\&a_7^?\&a_8^*\&a_9^?\&$ $a_{10}^*\&a_{11}^?\&a_{12}^?\&a_{13}^?\&a_{14}\&a_{15}^?\&a_{16}^?\&a_{17}^?$	245	$3.56*10^{14}$
conMiner		$a_{14}^+ \cdot a_6^? \cdot a_{11}^? \cdot a_{13}^? \cdot a_5^*\&a_3^? \cdot a_9^?$ $a_4^? \cdot a_{10}^?\&a_1^* \cdot a_7^?\&a_2^* \cdot a_{12}^?\&a_8^?\&a_{15}^?\&a_{16}^?\&a_{17}^?$	250	$3.09*10^{10}$
$learner^+_{DME}$		$(a_1^*\|a_5^*\|a_6^?\|a_9^?)\&(a_3^?\|a_8^?)\&(a_{11}^?\|a_{13}^?)\&$ $a_2^*\&a_4^?\&a_7^*\&a_{10}^*\&a_{12}^?\&a_{14}^?\&a_{15}^?\&a_{16}^?\&a_{17}^?$	280	$7.66*10^9$
exact-GenICRE		$a_1^* \cdot ((a_5^*\|a_3^?\|a_9^?)\&(a_4^?\|a_6^?\|a_{13}^?)\&$ $a_{10}^?\&a_8^?\&a_{12}^?\&a_{15}^?\&a_{11}^?\&a_{17}^?\&a_{14}^?\&a_7^?\&a_{16}^?\&a_2^*)$	280	$4.31*10^9$
GenICRE		$a_1^* \cdot ((a_5^*\|a_6^?\|a_9^?)\&(a_3^?\|a_8^?)\&(a_{11}^?\|a_{13}^?)\&$ $a_2^*\&a_4^?\&a_7^?\&a_{10}^?\&a_{12}^?\&a_{14}^+\&a_{15}^?\&a_{16}^?\&a_{17}^?)$	290	$5.75*10^9$

they still cannot learn more detailed information as we do. This is because the form of their expressions is simple and exclude some operator. In detail, for the symbol $a_1(editor)$ in Table 3, the result learned by $learner^+_{DME}$ or $conMiner$ cannot reflect that it always appears in the first position, but ours can. Similar cases can be seen from Table 1. Since $learner^+_{DME}$ cannot learn expressions with the concatenation operator and the expression learned by $conMiner$ usually is connected by interleaving in the outermost layer, their learning results preform poorly at describing data with sequence order. ICREs focus on describing the relative order of data, so our learning results are semantically precise.

Table 4. The list of abbreviations for words in **DBLP**

Word	editor	author	booktitle	cdrom	cite	crossref	ee	journal	month	note	number	pages
Abbr.	a_1	a_2	a_3	a_4	a_5	a_6	a_7	a_8	a_9	a_{10}	a_{11}	a_{12}
Word	publisher	title	url	volume	year	address	isbn	series	school	chapter	publnr	
Abbr.	a_{13}	a_{14}	a_{15}	a_{16}	a_{17}	a_{18}	a_{19}	a_{20}	a_{21}	a_{22}	a_{23}	

6 Conclusion

This paper focused on learning a Relax NG schema for a collection of XML documents supporting to describe both ordered and unordered data. We solved this problem by learning a restricted subclass of regular expressions with interleaving (called ICREs), converting expressions into the content model of Relax NG and adding datatypes. We highlighted the part of learning an expression in ICREs from positive samples. A polynomial time learning algorithm $GenICRE$ was proposed to learn an ICRE based on the SOA and the maximum clique partition. Experiments showed that ICREs cover the vast majority of real-world Relax NG. Dataset of practical XML documents from $DBLP$ were collected to test the effectiveness of $GenICRE$. The results showed that $GenICRE$ can learn concise and more precise results than other popular methods.

References

1. Abiteboul, S., Buneman, P., Suciu, D.: Data on the Web: from Relations to Semistructured Data and XML. Morgan Kaufmann, Burlington (2000)
2. Barbosa, D., Mignet, L., Veltri, P.: Studying the XML web: gathering statistics from an XML sample. World Wide Web-Internet Web Inf. Syst. **9**(2), 187–212 (2006)
3. Beek, M.H.T., Kleijn, J.: Infinite unfair shuffles and associativity. Theor. Comput. Sci. **380**(3), 401–410 (2007)
4. Bex, G.J., Gelade, W., Neven, F., Vansummeren, S.: Learning deterministic regular expressions for the inference of schemas from XML data. ACM Trans. Web **4**(4), 1–32 (2010)

5. Bex, G.J., Neven, F., Bussche, J.V.D.: DTDs versus XML schema: a practical study. In: International Workshop on the Web and Databases, pp. 79–84 (2004)
6. Bex, G.J., Neven, F., Schwentick, T., Vansummeren, S.: Inference of concise regular expressions and DTDs. ACM Trans. Database Syst. **35**(2), 1–47 (2010)
7. Bex, G.J., Neven, F., Vansummeren, S.: Inferring XML schema definitions from XML data. In: International Conference on Very Large Data Bases, University of Vienna, Austria, September, pp. 998–1009 (2007)
8. Boneva, I., Ciucanu, R., Staworko, S.: Simple schemas for unordered XML. In: International Workshop on the Web and Databases (2015)
9. Brüggemann-Klein, A.: Unambiguity of extended regular expressions in SGML document grammars. In: Lengauer, T. (ed.) ESA 1993. LNCS, vol. 726, pp. 73–84. Springer, Heidelberg (1993). https://doi.org/10.1007/3-540-57273-2_45
10. Che, D., Aberer, K., Özsu, M.T.: Query optimization in XML structured-document databases. VLDB J. **15**(3), 263–289 (2006)
11. Ciucanu, R., Staworko, S.: Learning schemas for unordered XML. Computer Science (2013)
12. Cormen, T.H., Leiserson, C.E., Rivest, R.L., Stein, C.: Introduction to Algorithms, 2nd edn, p. 1297C1305 (2001)
13. Demany, D.: InstanceToSchema: a RELAX NG schema generator from XML instances (2003). http://www.xmloperator.net/i2s/
14. Feige, U.: Approximating maximum clique by removing subgraphs. SIAM J. Discret. Math. **18**(2), 219–225 (2006)
15. Fernau, H.: Algorithms for learning regular expressions. Inf. Comput. **207**(4), 521–541 (2009)
16. Florescu, D.: Managing semi-structured data. ACM Queue **3**(8), 18–24 (2005)
17. Freydenberger, D.D., Kötzing, T.: Fast learning of restricted regular expressions and DTDs. Theory Comput. Syst. **57**(4), 1114–1158 (2015)
18. Garcia, P., Vidal, E.: Inference of k-testable languages in the strict sense and application to syntactic pattern recognition. IEEE Trans. Pattern Anal. Mach. Intell. **12**(9), 920–925 (2002)
19. Garey, M.R., Johnson, D.S.: Computers and Intractability: A Guide to the Theory of NP-Completeness. W.H. Freeman and Company, New York (1979)
20. Garofalakis, M., Gionis, A., Shim, K., Shim, K., Shim, K.: XTRACT: learning document type descriptors from XML document collections. Data Mining Knowl. Discov. **7**(1), 23–56 (2003)
21. Garofalakis, M.N., Gionis, A., Rastogi, R., Seshadri, S., Shim, K.: XTRACT: a system for extracting document type descriptors from XML documents. In: Proceedings of the 2000 ACM SIGMOD International Conference on Management of Data, Dallas, Texas, USA, 16–18 May 2000, pp. 165–176 (2000)
22. Gold, E.M.: Language identification in the limit. Inf. Control **10**(5), 447–474 (1967)
23. Grijzenhout, S., Marx, M.: The quality of the XML web. Web Semant.: Sci. Serv. Agents World Wide Web **19**, 59–68 (2013)
24. Hopcroft, J.E., Motwani, R., Ullman, J.D.: Introduction to Automata Theory, Languages, and Computation. Addison-Wesley Series in Computer Science, 2nd edn. Addison-Wesley-Longman, Boston (2001). ISBN: 978-0-201-44124-6
25. Clark, J., Murata, M.: Organization for the Advancement of Structured Information Standards (OASIS). Relax NG specification (2001)
26. Kim, G.-H., Ko, S.-K., Han, Y.-S.: Inferring a relax NG schema from XML documents. In: Dediu, A.-H., Janoušek, J., Martín-Vide, C., Truthe, B. (eds.) LATA 2016. LNCS, vol. 9618, pp. 400–411. Springer, Cham (2016). https://doi.org/10.1007/978-3-319-30000-9_31

27. Koch, C., Scherzinger, S., Schweikardt, N., Stegmaier, B.: Schema-based scheduling of event processors and buffer minimization for queries on structured data streams. In: Thirtieth International Conference on Very Large Data Bases, pp. 228–239 (2004)

28. Manolescu, I., Florescu, D., Kossmann, D.: Answering XML queries on heterogeneous data sources. In: International Conference on Very Large Data Bases, pp. 241–250 (2001)

29. Martens, W., Neven, F.: Typechecking top-down uniform unranked tree transducers. In: International Conference on Database Theory, pp. 64–78 (2003)

30. Martens, W., Neven, F.: Frontiers of tractability for typechecking simple XML transformations. In: ACM SIGMOD-SIGACT-SIGART Symposium on Principles of Database Systems, pp. 23–34 (2004)

31. Peng, F., Chen, H.: Discovering restricted regular expressions with interleaving. In: Cheng, R., Cui, B., Zhang, Z., Cai, R., Xu, J. (eds.) APWeb 2015. LNCS, vol. 9313, pp. 104–115. Springer, Cham (2015). https://doi.org/10.1007/978-3-319-25255-1_9

32. Quinlan, J.R., Rivest, R.L.: Inferring decision trees using the minimum description length principle. Inf. Comput. 80(3), 227–248 (1989)

Deep Group Residual Convolutional CTC Networks for Speech Recognition

Kai Wang[1], Donghai Guan[1,2], and Bohan Li[1,2,3(✉)]

[1] College of Computer Science and Technology,
Nanjing University of Aeronautics and Astronautics, Nanjing, China
`mitsuha1994@gmail.com`, {`dhguan,bhli`}`@nuaa.edu.cn`
[2] Collaborative Innovation Center of Novel Software Technology
and Industrialization, Nanjing, China
[3] Jiangsu Easymap Geographic Information Technology Corp., Ltd., Yangzhou,
China

Abstract. End-to-end deep neural networks have been widely used in the literature to model 2D correlations in the audio signal. Both Convolutional Neural Networks (CNNs) and Long Short-Term Memory (LSTM) have shown improvements across a wide variety of speech recognition tasks. Especially, CNNs effectively exploit temporal and spectral local correlations to gain translation invariance. However, all CNNs used in existing work assume each channel's feature map is independent of each other, which may not fully utilize and combine information about input features. Meanwhile, most CNNs in literature use shallow layers may not be deep enough to capture all human speech signal information. In this paper, we propose a novel neural network, denoted as GRCNN-CTC, which integrates group residual convloutional blocks and recurrent layers paired with Connectionist Temporal Classification (CTC) loss. Experimental results show that our proposed GRCNN-CTC achieve 1.11% Word Error Rate (WER) and 0.48% Character Error Rate (CER) improvements on a subset of the LibriSpeech dataset compared to the baseline automatic speech recognition (ASR) system. In addition, our model greatly reduces computational overhead and converges faster, leading to scale up to deeper architecture.

Keywords: Residual neural network · Group convolution
Gated recurrent unit · Connectionist temporal classification
Speech recognition

1 Introduction

In the past few years, there has been significant progress in automatic speech recognition (ASR) [1,2]. Deep Neural Networks (DNNs) play an indispensable role in large vocabulary continuous speech recognition (LVCSR) tasks and have led to significant improvements in various components of speech recognizers. The current state-of-art automatic speech recognition systems are mainly composed

© Springer Nature Switzerland AG 2018
G. Gan et al. (Eds.): ADMA 2018, LNAI 11323, pp. 318–328, 2018.
https://doi.org/10.1007/978-3-030-05090-0_27

of three sub-modules: acoustic, pronunciation and language modeling. However, the pipeline of such hybrid systems requires dictionaries of hand-crafted pronunciation, phoneme lexicons and a multi-stage training procedure to make the components work together. Moreover, the optimization goals of separately-trained acoustic, pronunciation and language model components vary, leading to the final recognition accuracy that is not necessarily good.

Recently, more researchers focused on the end-to-end paradigm [3–5], which directly integrates all components into one neural network and joinly optimizes without intermediate process. CNNs, RNNs [20], LSTMs [6] and their variants are the cornerstone of these end-to-end ASR systems. In ASR, CNNs operate on both time and frequency axis, which can model temporal as well as spectral local correlations and gain translation invariance in speech signals, while LSTMs and their variants, GRUs [8], make full use of hidden state from previous time moments, which are suitable for temporal modeling.

As [14] points out, one disadvantage of gated recurrent layer is it needs to store multiple gating neural responses at each time-step and unfold the time steps during training and test stages, which results in a computational bottleneck for long sequences in ASR. From the perspective of computational overhead, CNNs are an alternative type of neural network that reduce spectral variations and model spectral correlations in signals at feature preprocessing stage.

Inspired by the ResNet [15] and the concept of channel-wise group convolutions explored in computer vision recently [18], we follow the framework of CLDNN and propose a deep group residual convolutional CTC network, denoted as GRCNN-CTC. The model first feed input features (i.e, log-mel filterbank features) into two 2D-invariant convolutional layers with large fitler size to achieve semanteme-level local information and help to disentangle underlying factors of variation within both spectral and temporal axis. The output of the 2D-invariant convolutions is then fed into the group residual convolutional blocks, which effectively reorganize and combine channel features, to plenarily capture the high-level features. Then, the output of last group residual convolutional blocks is fed into a few bidirectional Gated Recurrnet Unit [8] layers, paired with Connectionist Temporal Classification (CTC) [16] loss, to model sequence-to-sequence dependency. In summary, our contributions in this paper are threefolds:

1. We propose a group residual convolutional CTC network, which combines 2D-invariant convolutional layers, group residual convolutional blocks and bidirectional GRU layers, paired with CTC loss for ASR task. It is a deep and wide architecture that fully captures the human speech signal information and obtains better generalization capacity.
2. We incorporate the group residual convolutional blocks [18], in our proposed framework. It is a *splitting, transforming, aggregating* architecture that breaks down the assumption of independence between each channel and divides the input channels in the form of groups. The channel feature maps within the group share the same hyper-parameters. To the best of our knowledge, such blocks has not been applied to ASR before.

3. We evaluate our method on a subset of the LibriSpeech dataset. Emperically, our proposed GRCNN-CTC network can achieve lower WER and CER compared with baseline automatic speech recognition system. Meanwhile, our proposed network, while doubling the model depth but reducing the training time by half, gives the model more room to improve performance.

The reminder of this paper is orginized as follows. We begin with a review of releated work in deep learning, end-to-end speech recognition in Sect. 2. Section 3 describes the overall architecture of our proposed neural network and details each component of the model. Experimental results and analyses are reported in Sect. 4. Finally, we summarize the whole paper in Sect. 5.

2   Related Work

In the past few years, the DNN/HMM hybrid model dominates the field of ASR in industry, the DNN is trained to estimate the posterior probability of continuous density HMM's state given the acoustic observations. The drawback of this feed-forward architecture is it only uses fixed-length sliding window of frames without considering long range correlations in speech utterence.

As a sequnence-to-sequence task, the RNNs have natural advantages in sequential modeling. The RNN takes advantage of encoding sequence history information into their internal states and predicts the current frame based on the encoded state. LSTMs [6], Bidirectional LSTM [7] and GRU [8] have been successfully applied to ASR and shown superiority over DNNs on a variety of ASR tasks [9,10]. Although stacking multiple layers of LSTMs or GRUs can bring about an increase in performance, the gradient vanishing issue still exists if the network goes too deep. On the other hand, it needs to store multiple gating neural responses at each time-step, which results in a great deal of computational overhead. In view of the above reasons and the ability of CNNs to deal with variability along the frequency and time-axis, CNN has an increasing proportion in the network design. In early work [11], the models only use one to two CNN layers, stacked with additional fully-connected DNN layers. Later, addition RNN layers, e.g., LSTMs, were integrated into the model to form CLDNN [17]. DeepSpeech2 [4] follows the architecture of CLDNN. It uses two large kernel convolutional layers as the feature preprocessing layers, combined with multiple bidirectional GRU layers and paired with CTC loss.

More recently, primarily motivated by the successes in image recognition, various architectures of deep CNNs [12,14] have been proposed and evaluated for ASR, which attribute to ResNet [15] with its identity mapping as the skip connection in residual blocks. In [12], it proposed a deep convolutional network with batch normalization (BN), residual connections and convolutional LSTM structure. Convolutional LSTM uses convolutions to replace the inner products within LSTM units. Thanks to the residual connections, the network can achieve compelling convergence and high accuracy to train deeper network. Another network architecture was proposed in [14], it uses a large 41×11 large filter with 32 features maps to deal with variability in spectral and temporal local region, followed by 4 groups of residual blocks. The residual block takes the architecture in [13], which widens the convolutional layers by adding more feature maps in each residual block. To date, all CNNs used in existing work assume each channel's feature map is independent of each other, which may not effectively combine the channel information.

3   Method

In this section, we will detail the techniques used in our proposed GRCNN-CTC network. The complete architecture is shown in Table 1.

Table 1. Architecture of deep group residual convolutional CTC network

Stage	[input channel, kernel, output channel]	Stride
Conv1	$[1, 41 \times 11, 32]$	(2, 2)
Conv2	$[32, 21 \times 11, 32]$	(2, 1)
Group ResBlock1	$\begin{bmatrix} 32, 1 \times 1, 32 \\ 32, 3 \times 3, 64, \text{Group} = 16 \\ 64, 1 \times 1, 64 \end{bmatrix} \times N$	1
Group ResBlock2	$\begin{bmatrix} 64, 1 \times 1, 64 \\ 64, 3 \times 3, 128, \text{Group} = 16 \\ 128, 1 \times 1, 128 \end{bmatrix} \times N$	1
Recurrent Layers	[bidirectional GRU with hidden size 1536] $\times$ 3	–
Fully-connected	–	–
CTC	–	–

3.1   Channel-Wise Group Convolutions

CNN is a special feedforward neural network, which can effectively exploit variable-length contextual information. The filters in the convolution operation are called kernels, which typically are 4-dimensional tensors (kernel height, kernel width, input channel, output channel). Given the above definition, the formula of conventional convolution operation is defined as follows:

$$z_{ijl}(\boldsymbol{K}, \boldsymbol{X}) = \sum_{c=1}^{C} \boldsymbol{K}_{cl} \cdot \boldsymbol{X}_{ijc} \qquad (1)$$

where $z_{ijl}(\boldsymbol{K}, \boldsymbol{X})$ denotes the convolution result of output channel l at ith step along the vertical direction and jth step along the horizontal direction, $\boldsymbol{K}_{cl}$ of size (H_k, W_k) denotes the kernel matrix associated with input channel c and output channel l, the same size as the patch $\boldsymbol{X}_{ijc}$ in channel c.

The conventional convolution operation can also be viewed as a combination of *splitting, transforming,* and *aggregating* [18] as shown in Fig. 1. The procedure can be depicted as: first, the receptive fields x of size (H_k, W_k, C) split along the channel axis into C low-dimensional embedding of size (H_k, W_k). Then, the low-dimensional embedding is transformed, weighted by the corresponding kernel w of size (H_k, W_k). Finally, the transformations in all embeddings are aggregated by $\sum_{i=1}^{C}$. In this setting, each channel's low-embedding is independent of each other.

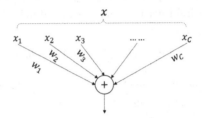

Fig. 1. Schematic diagram of splitting, transforming and aggregating procedure

However, in computer vision, the channels of visual representations are not entirely independent. Classical features of SIFT, HOG and GIST are group-wise representations by design, where each group of channels is constructed by some kind of histogram. In [18], it proposes an aggregated transformations as:

$$\mathcal{F}(x) = \sum_{i=1}^{G} T_i(x) \tag{2}$$

where T_i is an arbitrary function that projects x into embedding and then transform it. In its setting, all T_i have the same topology and be set as bottleneck-shaped architecture, as shown in Fig. 2 (left). In Eq. (2), G is the size of the set of transformations to be aggregated, also refers to as *cardinality*. The dimension of cardinality controls the number of more complex transformations.

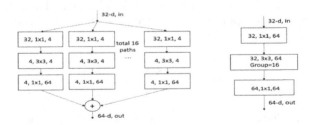

Fig. 2. The structure of group convolutional block and its equivalent form

The schematic diagram of aggregated transformation is shown in Fig. 2 (left), its bottleneck-shaped architecture is the basic unit of residual neural network (show in next subsection) and the reformulation of grouped convolution can be proven equivalence as follow. In the first 1×1 layers, all the low-dimensional embeddings can be replaced by a wider layer (e.g., $1 \times 1, 64$-d in Fig. 2 (right)). At the *splitting* stage, it's essentially equivalent to be done by the grouped convolutional layer when it divides its input channels into groups. In Fig. 2 (right), the 64-d input channels performs 16 groups of convolutions whose input and output channels are 4-dimensional. At the *transforming* and *aggregating* stage, it's equivalent to concatenate the outputs of all the grouped convolutional layers and then transform to the final results.

To the best of my knowledge, group structure hasn't been used in ASR. Due to its success in image clssification task, we take the blocks of aggregated transformation as encoder in our system. We explore different cardinality and width to improve the accuracy, the results will show in Sect. 4.

3.2   Residual-Based Neural Networks

The depth of representations is of central importance for many visual recognition tasks. The depth demonstrates the superiority of integrating multi-level features and classifiers into an end-to-end pattern, leading to state-of-art results in specified tasks.

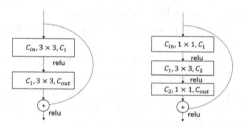

Fig. 3. Architecture of two residual blocks

However, training deep neural networks faces with many difficulties, especially degradation. The degradation of deep neural networks has been comprehensively explored in [15], it proposes a *deep residual learning* framework, motivated by the fact that the neural networks with multiple nolinear layers have difficulties in approximating the identity mappings. Suppose the desired underlying mapping, to be fit by a few stack layers, is denoted as $\mathcal{H}(x)$. Instead of letting the stacked nonlinear layers directly approximate the desired underlying mapping $\mathcal{H}(x)$, it attempts to approximate the residual function: $\mathcal{F}(x) = \mathcal{H}(x) - x$. Formally speaking, a building block is defined as:

$$y = \mathcal{F}(x, \{W_i\}) + x \tag{3}$$

where x and y are the input and output vectors of the layer considered. The residual function $\mathcal{F}(x, \{W_i\})$ represents the residual mapping to be learned. More generally, the Eq. (3) can be modified if considering the dimension of vector x and y mismatch:

$$y = \mathcal{F}(x, \{W_i\}) + W_s x \qquad (4)$$

The residual blocks have two types of structure, as shown in Fig. 3. We choose the bottleneck-shaped architecture (Fig. 3 (right)) as the first 1×1 layers can explore the cross-channel correlations and reduce dimension to compress redundant information. Stacked residual blocks alleviate the degradation problem and reduce time complexity and model capacities simultaneously. In our experiments, it reduces the training time by nearly half compared to the baseline network.

3.3   CTC

Traditional acoustic model is based on frame-level labels with cross-entropy criterion (CE), which requires pre-segmented training sequence data and postprocessing to transform network outputs to label sequence. The drawback of such label alignment procedure are (1) the manual segmentation requires a significant amount of task specific knowledge (2) the alignment compulsively assign label to each frame, which is apt to result in misalignment at the segment boundaries. (3) the frame-by-frame labeling that optimized by CE criterion is not completely equivalent to the optimization of ultimate target.

Since the speech recognition task is a sequence-to-sequence translation task, we take CTC [16] loss as the object function, which let neural network automatically to learn the alignment mappings from speech frames to corresponding label sequences.

Denote x as the input speech sequence, l as the original label sequence, and a one-to-many map $\mathbf{B}^{-1}(l)$, which maps label sequence to all the corresponding CTC paths. By incorporating the blank label, which denotes the state of mute or boundary at different label, the likelihood of l can be evaluated as the sum of the probabilities of corresponding CTC paths:

$$P(l|x) = \sum_{z \in \mathbf{B}^{-1}(l)} P(z|x) \qquad (5)$$

Given this distribution, we can backpropagate the errors from the negative logarithm of $P(l|x)$, i.e, $\ln P(l|x)$ and further update the network parameter.

4   Experiments

In this section, we present and analyze the experimental results of our proposed group channel residual convolutional CTC network. Our experiments are conducted on subsets of a benchmark dataset, LibriSpeech [19], which is a corpus of approximately 1000 h of 16 kHz read English speech.

4.1 Experimental Setup

All experiments use 40 dimensional log-mel filterbanks features, computed with a 25 ms window and shifted every 10 ms, with delta and delta-delta configurations. Then, the features are normalized via mean subtraction and variance normalization. Data augmentation was performed on all utterances of training set by using Sox Tookit to randomize the tempo and gain perturbations.

We establish the DeepSpeech2 architecture, detailed in [4], as our baseline ASR system. We refer to the architecture with 5 bidirectional GRU layers (DS2-5GRU) as baseline A, the architecture with 3 bidirectional GRU layers (DS2-3GRU) as baseline B. Both baseline architectures take two 2D-invariant convolutional layers with the cliped rectified-linear (ReLU) function as nonlinearity, followed by three bidirectional GRU layers (768 nodes per direction).

The details of our proposed GRCNN-CTC network architecture is shown in Table 1. In particular, we take the same configuration of 2D-invariant convolutional layers in baseline ASR system. In order to get a trade-off between computational complexity and performance, each group of Group Residual Blocks use 2 basic bottleneck architecture units ($N = 2$).

Stochastic Gradient Descent (SGD) with Nestrov was used for training all models. A learning rate of $3e-4$ was used with a 1.1 learning rate annealing per epoch with early stopping strategy.

4.2 Results and Analysis

In this part, we will show the results of our experiments on the LibriSpeech [19] from the three perspectives of error rate, convergence and average training time per epoch.

Table 2. The comparison results on the dataset in [19]

Model	WER	CER	Average epoch time(s)
Baseline A	25.66	7.54	2885.5
Baseline B	27.51	8.01	1283.3
Ours	**24.55**	**7.06**	1495.8

As shown in Table 2, our proposed GRCNN-CTC network achieves 24.55% WER and 7.06% CER, which is a greater performance improvement compared to the other two baseline model.

The experiments shows that the increasement of number of GRU layers brings the WER and CER a performance improvement. However, this improvement is no longer evident when the GRU has 5 or more layers. Instead, it imposes a computational burden, as average epoch training time shows in Table 2.

From Table 2, we can infer that the Group Residual Blocks can replace or even exceed the effectiveness of the recurrent layers as an encoder to a certain extent through multi-channel information fusion without introducing much computational overhead. Meanwhile, the deeper architecture brings our model more capacity to learn representative features. Not only that, the parameter-free, identity shortcuts in Group Residual Blocks helps the network converge faster, as shown in Fig. 4.

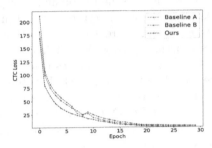

Fig. 4. Comparison of model convergence

4.3 Influence of Cardinality

Cardinality, the size of the set of transformations to be aggregated in Group Residual Blocks, is the important hyper-parameter in architecture design. In this part, we explore the influence of cardinality to the WER and CER.

Table 3. The influence of cardinality on the dataset in [19]

Cardinality	1	2	4	8	16	32
WER	26.86	26.02	25.31	24.89	**24.55**	24.78
CER	7.83	7.55	7.34	7.17	**7.06**	7.12

As shown in Table 3, with cardinality increasing from 1 to 32, both WER and CER keep reducing except the condition that cardinality equals 32. It suggests that the performances gradually saturate, but it is not worthwhile to keep increasing cardinality because group statistics tend to randomize as the number of participating channels decreases, it is no longer statistically significant, ie, it is no longer an abstract feature. Hence, the trade-off between cardinality and bottleneck width should be taken into consideration.

5　Conclusions

In this paper, we introduce group residual convolutional blocks to the CLDNN architecture and achieve great improvements on both WER and CER. The residual blocks rationally divide and reorganize the input channel information, which produce feature representations with more group statistics. Meanwhile, the parameter-free, identity shortcuts extend the network to a deeper structure without extra computational overhead, leading to more model capacity and faster convergence. The researches in this paper show the promise of replacing the recurrent layers to group convolutional residual blocks. In the future, we intend to investigate from this viewpoint, which is a challenge problem.

Acknowledgements. This work was supported by the Fundamental Research Funds for the Central Universities NS2018057, NJ20160028.

References

1. Yu, D., Li, J.: Recent progresses in deep learning based acoustic models. IEEE/CAA J. Autom. Sinica **4**(3), 396–409 (2017)
2. Rao, K., Sak, H., Prabhavalkar, R.: Exploring architectures, data and units for streaming end-to-end speech recognition with RNN-transducer (2018)
3. Hannun, A., Case, C., Casper, J., et al.: Deep speech: scaling up end-to-end speech recognition. Computer Science (2014)
4. Amodei, D., Anubhai, R., Battenberg, E., et al.: Deep speech 2: end-to-end speech recognition in english and mandarin. Computer Science (2015)
5. Chan, W., Jaitly, N., Le, Q.V., et al.: Listen, attend and spell. Computer Science (2015)
6. Hochreiter, S., Schmidhuber, J.: Long short-term memory. Neural Comput. **9**(8), 1735–1780 (1997)
7. Graves, A., Schmidhuber, J.: Framewise phoneme classification with bidirectional LSTM and other neural network architectures. Neural Netw. **18**(5–6), 602–610 (2005)
8. Jozefowicz, R., Zaremba, W., Sutskever, I.: An empirical exploration of recurrent network architectures. In: International Conference on International Conference on Machine Learning, pp. 2342–2350. JMLR.org (2015)
9. Sak, H., Senior, A., Beaufays, F.: Long short-term memory recurrent neural network architectures for large scale acoustic modeling. In: Fifteenth Annual Conference of the International Speech Communication Association (2014)
10. Graves, A., Mohamed, A., Hinton, G., Speech recognition with deep recurrent neural networks. In: 2013 IEEE International Conference on Acoustics, Speech and Signal Processing (ICASSP), pp. 6645–6649. IEEE (2013)
11. Abdel-Hamid, O., Mohamed, A., Jiang, H., et al.: Convolutional neural networks for speech recognition. IEEE/ACM Trans. Audio, Speech Lang. Process. **22**(10), 1533–1545 (2014)
12. Zhang, Y., Chan, W., Jaitly, N.: Very deep convolutional networks for end-to-end speech recognition. In: 2017 IEEE International Conference on Acoustics, Speech and Signal Processing (ICASSP), pp. 4845–4849. IEEE (2017)
13. Zagoruyko, S., Komodakis, N.: Wide residual networks. arXiv preprint arXiv:1605.07146 (2016)
14. Wang, Y., Deng, X., Pu, S., et al.: Residual convolutional CTC networks for automatic speech recognition (2017)

15. He, K., Zhang, X., Ren, S., et al.: Deep residual learning for image recognition. In: Proceedings of the IEEE Conference on Computer Vision and Pattern Recognition, pp. 770–778 (2016)
16. Graves, A., Fernández, S., Gomez, F., et al.: Connectionist temporal classification: labelling unsegmented sequence data with recurrent neural networks. In: Proceedings of the 23rd International Conference on Machine Learning, pp. 369–376. ACM (2006)
17. Sainath, T.N., Vinyals, O., Senior, A., et al.: Convolutional, long short-term memory, fully connected deep neural networks. In: IEEE International Conference on Acoustics, Speech and Signal Processing, pp. 4580–4584. IEEE (2015)
18. Xie, S., Girshick, R., Dollár, P., et al.: Aggregated residual transformations for deep neural networks. In: 2017 IEEE Conference on Computer Vision and Pattern Recognition (CVPR), pp. 5987–5995. IEEE (2017)
19. Panayotov, V., Chen, G., Povey, D., et al.: LibriSpeech: an ASR corpus based on public domain audio books. In: 2015 IEEE International Conference on Acoustics, Speech and Signal Processing (ICASSP), pp. 5206–5210. IEEE (2015)
20. Chen, W., Wang, S., Zhang, X., et al.: EEG-based motion intention recognition via multi-task RNNs. In: Proceedings of the 2018 SIAM International Conference on Data Mining (2018)

Short Text Understanding Based on Conceptual and Semantic Enrichment

Qiuyan Shi[1], Yongli Wang[1(✉)], Jianhong Sun[2], and Anmin Fu[1]

[1] School of Computer Science and Engineering,
Nanjing University of Science and Technology Library, Nanjing 210094, China
yongliwang@mail.njust.edu.cn
[2] School of Electronic Engineering and Optical Engineering,
Nanjing University of Science and Technology Library, Nanjing 210094, China

Abstract. Due to the limited length and freely constructed sentence structures, short text is different from normal text, which makes traditional algorithm of text representation does not work well on it. This paper proposes a model called Conceptual and Semantic Enrichment with Topic Model (CSET) by combining Biterm Topic Model (BTM), a widely used probabilistic topic model which is designed for short text with Probase, a large-scale probabilistic knowledge base. CSET is able to capture semantic relations between words to enrich short text. Our model enables large amount of applications that rely on semantic understanding of short text, including short text classification and word similarity measurement in context.

Keywords: Short text · Text enrichment · Similarity

1 Introduction

Pieces of information in the social web are usually short, such as tweets, commodity title and news headlines. Algorithm of text representation on short text is needed for further information extraction and data mining. However, weak description ability, sparse feature and discrete semantics of short text hamper efforts to apply traditional text represent models.

Many short text enrichment methods have been proposed to enhance performance of short text analysis. Some methods use web data like HowNet [6] or Wikipedia [1] to expand features while some others treat short text as a search query and use answers to enrich original text [9]. Although these methods reduce impact of sparse feature in short text in some degree, they still have limits. Efficiency of methods that based on other data resources highly depend on resources quality. If a word from short text is not in the data resource or the meaning of this word in data resource is different from what in text, a resource-based method would not give an ideal result. Meanwhile, for those queries that are not popular, the answer from a search-based methods usually is unrelated noise.

© Springer Nature Switzerland AG 2018
G. Gan et al. (Eds.): ADMA 2018, LNAI 11323, pp. 329–338, 2018.
https://doi.org/10.1007/978-3-030-05090-0_28

Our innovative ideal is to combine the advantage of topic model and concept graph to enrich short text with conceptualization and semantic relationship. A Conceptual and Semantic Enrichment with Topic Model (CSET) is built, by combining Biterm Topic Model (BTM) with Probase to capture semantic relations between words to enrich short text.

The following are the major contributions of this paper: (i) a Conceptual and Semantic Enrichment Model is built and utilized for short text enrichment, which enhances the quality of short text in both way; (ii) we provide a way to combine BTM topic model with conception knowledge constructor to captures semantic relations which is especially suitable fro short text.

The rest of this paper is organized as follows. Section 2 reviews related works. Section 3 introduces the knowledge bases and the topic models framework adopted from the work. Section 4 describes the model and methods of CSET. Section 5 presents the experimental results. Section 6 concludes this paper and describes future work.

2   Related Works

Traditional text processing models and analysis algorithms usually don't work well on short text considering data sparsity. Many researches have been done on this particular area. In recent years, the research focuses on two main aspects in order to gain more semantic information from short text. One aspect focus on decreasing data sparsity in short text with extra information. The other one derives latent topics as additional features to expand short text.

The method of feature extension is mainly to expand the feature space of short text with search engine or external corpus. In [9], the authors use titles and snippets to expand the Web queries. And their model achieves better classification accuracy compared to using the queries alone. However, such search-based method has potential problems in efficiency and the reliability issues, especially when the set of short text under consideration is large. To address these issues, some researchers [4] use Wikipedia to build a concept thesaurus to enhance traditional content similarity measurement. Due to the difficulty of finding a suitable external corpus when a few short texts are very specialized or with special language, using predefined taxonomy has one possible shortcoming: lacking of adaptability.

To overcome the shortcoming mentioned above, researchers turn to derive latent topics to expand short text features. Some [3,7] use Latent Dirichlet Allocation (LDA)[2] model to do latent topics mining from a set of documents from Wikipedia and then use the topics as additional features to expand the short text. LDA adds Dirichlet priors on topic distributions, resulting in a more complete generative model. However, traditional topic model like LDA has limits in measuring short text similarity, they utilize word co-occurrences as structure priors for topic-word distribution, rather than directly modeling their generation process.

3   Preliminary

3.1   Probase

Probase [11] is a web-scale taxonomy that contains 10 millions of concepts/terms and uses IS-A relationship to describe connection between concepts and terms. It is an automatically constructed knowledge base which is built with linguistic patterns scaned from billions of web pages. It contains numbers of entities and concepts and formalized relationships between one anthor with their co-occurrence frequency. Probase data are publicly available. Figure 1 shows the relationship of entity "iphone" and its concepts.

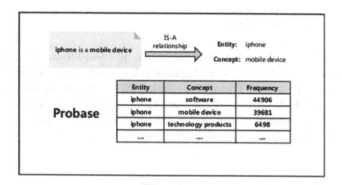

Fig. 1. Structure of Probase.

Using concepts can enrich short text and disambiguate terms in certain degree, but there still are many terms have strong semantic relationship while been conceptualized in separate groups. For example, given a short text "mouse from apple", which has two entities {mouse, apple}, these two terms have no common concepts ({device, animal} and {company, fruit}) and all the concepts are independent. If we only use IS-A relationship from Probase, it does not explicitly capture the specific semantic relationships between these two terms. So we also need to use a topic model—BTM [12] to model the co-occurrence relationship, as it can estimate words semantic relationship based on their co-occurrence statistics.

3.2   Biterm Topic Model

We choose Biterm Topic Model (BTM)[12] to model co-occurrence relationship which was proposed in 2013 specific for short text. BTM models word co-occurrence patterns to enhance the topic learning and uses aggregated patterns in the whole corpus for learning topics to solve the problem of sparse word co-occurrence patterns at document-level. As shown in Fig. 2, it represents the topic distribution of whole corpus in BTM, is the topic-specific word distribution, is a topic assignment, represent two words in a biterm, and is the number of biterms in whole corpus.

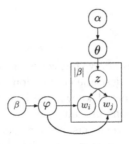

Fig. 2. Graphical representation of BTM.

4   Models and Methods

Although words in Probase are delivered in two different kinds: term vocabulary and concept vocabulary, topic model cannot tell the difference between them. In this work, we propose CSET model which enrich short text in two aspects: (i) enrich short text with terms and concepts through Probase relationship knowledge; (ii) Ppredict semantic relationship between terms and concepts via a topic model to find out semantic related words that has few conceptually relationship.

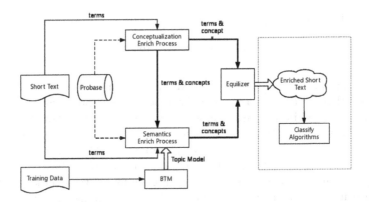

Fig. 3. Enrichment and classification Process of CSET.

4.1   Conceptual Enrichment

The processes of preprocessing are (1) breaking each short text S into terms set $\{t_i\}$, where t_i means a term that can be recognized by Probase; (2) finding out related conceptions and co-occurring terms T through IS-A relationship in Probase.

4.1.1   Related Conceptions

Finding conceptions of each term t_i is a querying process in Probase. For each term t_i, Probase outputs a list of concepts $C = \{c_i\}$ with its probability $P(c_i|t_i)$ which measures the typicality of c_i given t_i. Then we use the Naive Bayesian mechanism [10] to measure the probability of relationship between concept c and terms set $T = \{t_i\}$:

$$P(c|T) = \frac{p(T|c)}{p(T)} \propto p(c) \prod_i^n p(t_i|c) \tag{1}$$

Then we can get a new concept set C_{new} which has the most related concepts with T.

4.1.2   Co-occurring Terms

After finding out related conceptions, short text S has been enriched into $\{T, C_{new}\}$, then we use C_{new} to find co-occurring terms which shares similar conceptions with S:

$$P(t_{co}|S) = p(t_{co}|C_{new}) = \sum_{c_i} p(c_i, C_{new})p(t_{co}|c_i) \tag{2}$$

After the whole pre-processing, short text S is enriched into $\{T, C_{new,T_{co}}\}$.

4.2   Semantic Enrichment

4.2.1   Topic Model Distribution

Given a text corpus, let B be its whole biterm set. We train BTM with part of the corpus and get the trained model M. With the definition of BTM and inspiration of conceptualization method in [5], we calculate the conditional probability of topic z for each biterm $b = (w_i, w_j)$ of whole corpus, where Z_{-b} denotes the topic assignments for all biterms except b:

$$P(z|Z_{-b}, B, M) \propto (n_z + \alpha)\frac{(n_{(w_i|z)} + \beta)(n_{w_j|Z} + \beta)}{(\sum_w n_{(w|Z)} + N\beta)^2} \tag{3}$$

where n_z is the number of times that biterm b is assigned to topic z, $n_{(w|z)}$ is the times that word w is assigned to topic z, α and β are the symmetric Dirichlet priors as defined in BTM, N is the number of texts in corpus. In this way, we can get the topic distribution $\overrightarrow{Z}$.

4.2.2   Estimating Concept Distribution with

M In this step, given the term biterm $b_t = (t_i, t_j)$, we use topic distribution $\overrightarrow{Z}$ to calculate the possibility of concept c as followed:

$$P(c|b_t, \overrightarrow{Z}) \propto p(c|b_t) \sum_z p_{bz} p_{cz} \tag{4}$$

$$p(c|b_t) = p(c|t_i)p(c|t_j)$$

$$p_{bz} = P(z|Z_{-b}, B, M)$$

$$p_{cz} = \frac{n_{(c|z)} + \beta}{\sum_w n_{(c|z)} + N\beta'}$$

where $P(c|b_t)$ is probability of concept c given the term b_t, which can be calculated from Probase, p_{bz} is the inferred probability through Eq. 3, p_{cz} is the possibility of concept given the topic z.

4.2.3   Estimating Terms Distribution with

M Similar to step 2, we can calculate possibility of term t given $b_c = (c_i, c_j)$ and $\overrightarrow{Z}$ as:

$$P(t|b_t, \overrightarrow{Z}) \propto p(t|b_c) \sum_z p_{bz} p_{tz} \tag{5}$$

4.2.4   Balance Between Conception and Semantics

With former two steps, we dig up semantic relation between terms and concepts, and calculate a new set of concepts C_{sem} and a new set of terms T_{sem} for each text. Now the origin short text S has been enriched into $\{T, C_{new}, T_{co}, C_{sem}, T_{sem}\}$. In order to balance the impaction between conception and semantics, we define the degree of importance as follow:

$$D_t = \gamma D_{t\_co} + (1 - \gamma)D_{t\_sem} \tag{6}$$

$$D_c = \delta D_{c\_new} + (1 - \delta)D_{c\_sem} \tag{7}$$

where D_{t_co} and D_{c_new} are the possibility of term/concept which are defined separately in Eqs. 1 and 2, D_{t_sem} and D_{c_sem} are the possibility of term and concept under semantics which are defined separately in Eqs. 4 and 5. γ and δ are meta parameters that explore the trade-off between the two component

---

Algorithm 1. CSET short text enrich algorithm based on BTM algorithm

---

Input: a set of original short text S, topic number k
Output: a set of enriched short text S_{result}
1: $M = BTM(S, k);$ // BTM modeling
2: **for** each text $s_i \in S$ **do**
3:     $s_i = s_i \cup s_c;$ //enrich s_i with conceptualization terms s_c
4: **end for**
5: **for** each conceptualized text $s_i \in S$ **do**
6:     $s_i = s_i \cup s_s;$ //enrich s_i with Semantic terms s_s
7: **end for**
8: **for** each enriched text $s_i \in S$ **do**
9:     $s_{result} = BalanceImpact(s_c, s_s)$ //Balance adjustment
10: **end for**
11: **return** enriched text S_{result}

---

scores in each equation. With this measurement, we re-rank all additional terms and concepts to get final enriched text as $\{T, C_{cset}, T_{cset}\}$.

Algorithm 1 shows the process of CSET. According to [11], time complexity of BTM is $O(k|B|\bar{l}(\bar{l}-1)/2)$, $|B|$ is the number of biterms in the corpus and represents the average length of a document. Time complexity of conceptualization enrichment and semantic enrichment are $O(|B|\bar{l}(\bar{l}-1)/2)$ and $O(k\bar{l}(\bar{l}-1)/2)$. So the whole complexity of CSET is $O((k|B| + |B| + k)\bar{l}(\bar{l}-1)/2)$.

5   Experiment

5.1   Data Sources

We choose Google Snippets [7] data set to do the experiment which is widely used in short text classification. This data set consists of two subsets. One is Corpus which contains 10060 labeled short text data; the other is short text with 2280 data. We use Corpus training BTM to support the enrichment and classification on short text.

5.2   Experiment Settings

For learning topic assignments with BTM, we set $\alpha = 1/k$ and $\beta = 0.01$ where k is the number of topics, and Gibbs sampling was run for 1000 iterations. For the sake of efficiency, we set meta parameters $\gamma = \delta = 0.5$. For enrichment, we select top 30 related terms/concepts from Probase in each iteration. We use MaxEnt and LibSVM as classifiers to evaluate the qualities of our methods for short text enrichment. MaxEnt and SVM have been successfully applied in many text mining tasks [8] which proves that MaxEnt is much faster in both training and inference while SVM is more robust.

5.3   Experiment Result and Comparison

5.3.1   Enrichment Performance

In order to explore the performance of CSET, we employ LDA as the usage in [7] as baseline. In baseline, we first learn LDA model based on Wikipedia corpus as [4], then use the latent topics drawn by LDA to enrich the original short text. The topics distribution corresponding to each snippet is fed to LibSVM classifier, and the enriched snippets with latent topics are treated as new features for MaxEnt classifier. Meanwhile, BTM is trained directly on the search snippets.

Then, we compute classification accuracy when topic number k changes, and the result are shown in Fig. 4. We can find that our method outperforms the baselines obviously, and obtain the highest accuracy of 0.8461 when $k = 10$, which reduce classification error by 8.01% compared to [3] and by 22.95% compared to [7].

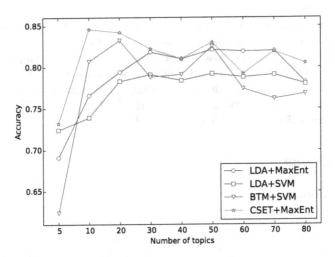

Fig. 4. Accuracy vary with topic numbers of 4 methods

5.3.2 Importance Comparison Between Conceptual Enrichment and Semantic Enrichment

In order to do full test on the validity of CSET model in the aspect of enrichment for short texts. We denote the approach CE which only use Conceptual Enrichment as CE-MaxEnt, and SE which only use Semantics Enrichment as SE-MaxEnt. In this experiment, we set topic number $k = 10$ according formal experiments. We compare the average accuracy CE-MaxEnt, SE-MaxEnt and CSET-MaxEnt under three different size of training data.

Table 1. Perfomance of 3 methods in different size of training data.

Training size	Method		
	CE	SE	CSET
30%	73.27%	70.54%	**74.22%**
60%	73.27%	76.83%	**79.34%**
100%	73.27%	80.41%	**84.61%**

As we can see from the Table 1 above, since CE-based method simply enriches text with concepts from Probase which is got from open APIS, the result is not sensitive to the training size which is used for topic modeling. When training data size is small, CE-based method works better than another two which are using topic model because the lack of data decreases the impact of topic model, and the pure relationship between terms and concepts is more efficient. With the increase of training data, the benefit of topic model embodies its superiority. SE-based method shows better performance than CE-based method because the meaning of a term is able to explicitly express its true meaning under different topics. CSET outperforms in each size of training data. This implies that our proposed method CSET enriching short data in both conceptualization and semantics because it grasps the semantic relationship between entities and concerns in Porbase.

6   Conclusion and Future Work

In order to solve the shortcoming of short text with high dimension and semantics sparsity, this paper proposes a short text enrich method CSET which combines the conception knowledge in Probase with latent semantic relationship in the short text. Experiments show that CSET is able to enhance the performance of text classification. Optimization for combination of conception knowledge and semantics and customization of similarity measurement in CSET model is the direction of our future research efforts.

Acknowledgement. This work is supported in part by the National Natural Science Foundation of China under Grant 61170035, 61272420 and 81674099, Six talent peaks project in Jiangsu Province (Grant No. 2014 WLW-004), the Fundamental Research Funds for the Central Universities (Grant No. 30916011328, 30918015103), Jiangsu Province special funds for transformation of science and technology achievement (Grant No. BA2013047).

References

1. Banerjee, S., Ramanathan, K., Gupta, A.: Clustering short texts using wikipedia. In: International ACM SIGIR Conference on Research and Development in Information Retrieval, pp. 787–788 (2007)
2. Blei, D.M., Ng, A.Y., Jordan, M.I.: Latent dirichlet allocation. J Mach. Learn. Res. Arch. **3**, 993–1022 (2003)
3. Chen, M., Shen, D., Shen, D.: Short text classification improved by learning multi-granularity topics. In: International Joint Conference on Artificial Intelligence, pp. 1776–1781 (2011)
4. Hu, J., et al.: Enhancing text clustering by leveraging Wikipedia semantics, pp. 179–186 (2008)
5. Kim, D., Wang, H., Oh, A.: Context-dependent conceptualization. In: International Joint Conference on Artificial Intelligence, pp. 2654–2661 (2013)
6. Ning, Y.H., Zhang, L., Ju, Y.R., Wang, W.J., Li, S.Q.: Using semantic correlation of hownet for short text classification. Appl. Mech. Mater. **513–517**, 1931–1934 (2014)
7. Phan, X.H., Nguyen, L.M., Horiguchi, S.: Learning to classify short and sparse text & web with hidden topics from large-scale data collections. In: WWW, pp. 91–100 (2015)
8. Pietra, S.A.D., Pietra, S.A.D., Pietra, S.A.D.: A maximum entropy approach to natural language processing. Comput. Linguist. **22**, 39–71 (1996)
9. Shen, D., et al.: Query enrichment for web-query classification. ACM Trans. Inf. Syst. **24**(3), 320–352 (2006)

10. Song, Y., Wang, H., Wang, Z., Li, H., Chen, W.: Short text conceptualization using a probabilistic knowledgebase. In: International Joint Conference on Artificial Intelligence, pp. 2330–2336 (2011)
11. Wu, W., Li, H., Wang, H., Zhu, K.Q.: Probase: a probabilistic taxonomy for text understanding, pp. 481–492 (2012)
12. Yan, X., Guo, J., Lan, Y., Cheng, X.: A biterm topic model for short texts, pp. 1445–1456 (2013)

LDA-PSTR: A Topic Modeling Method for Short Text

Kai Zhou and Qun Yang[(✉)]

College of Computer Science and Technology,
Nanjing University of Aeronautics and Astronautics,
Nanjing 210016, Jiangsu, China
1131966336@qq.com, qun.yang@nuaa.edu.cn

Abstract. Topic detection in short text has become an important task for applications of content analysis. Topic modeling is an effective way for discovering topics by finding document-level word co-occurrence patterns. Generally, most of conventional topic models are based on bag-of-words representation in which context information of words are ignored. Moreover, when directly applied to short text, it will arise the lack of co-occurrence patterns problem due to the sparseness of unigrams representations. Existing work either performs data expansion by utilizing external knowledge resource, or simply aggregates these semantically related short texts. These methods generally produce low-quality topic representation or suffer from poor semantically correlation between different data resource. In this paper, we propose a different method that is computationally efficient and effective. Our method applies frequent pattern mining to uncover statistically significant patterns which can explicitly capture semantic association and co-occurrences among corpus-level words. We use these frequent patterns as feature units to represent texts, referred as pattern set-based text representation (PSTR). Besides that, in order to represent text more precisely, we propose a new probabilistic topic model called LDA-PSTR. And an improved Gibbs algorithm has been developed for LDA-PSTR. Experiments on different corpus show that such an approach can discover more prominent and coherent topics, and achieve significant performance improvement on several evaluation metrics.

Keywords: Topic modeling · Short text · Text representation
Frequent pattern LDA

1 Introduction

1.1 Background

Topic models for traditional normal documents (e.g. online news and academic articles) have been extensively studied and achieve broad success in the past years. Statistical topic models like Probabilistic Latent Semantic Analysis (PLSA) [1, 22, 23] and Latent Dirichlet Allocation (LDA) [2] are generative models which implicitly discover co-occurrence information of words to uncover latent topics. They assume that a document is generated with a mixed topic distribution and a topic samples words with

© Springer Nature Switzerland AG 2018
G. Gan et al. (Eds.): ADMA 2018, LNAI 11323, pp. 339–352, 2018.
https://doi.org/10.1007/978-3-030-05090-0_29

multinomial distribution over the vocabulary. As the mixture of topics can be treated as high-level semantic representation for a document, topic models are usually regarded as a dimension reduction method [3].

Conventional topic models achieve success for discovering latent topics of normal documents by capturing document level co-occurrence information of words. However, applying these models directly on short texts often suffer from the lack of co-occurrence of words and the semantic gaps because of unigram representations caused by sparsity. More specially, the occurrences of words in short text are less compared to long texts in which the model can get enough words to know how words are related. Besides, compared with long texts, the meaning of word is more refined in short texts. Limited context information offered by single word makes it more difficult to determine polysemy and synonymy. How to enhance semantic accuracy becomes more important in short text.

1.2   Related Work

Several attempts have been made to address the sparsity problem in short texts, these methods attempt to aggregate short texts into an artificial lengthy documents. In [18], tweets are aggregated into one document before training LDA. In [4], Hong et al. aggregated the tweets containing same word and got better modeling performance. The work in [19] proposed various tweets pooling schemes. Some other methods utilize external knowledge bases to enrich the original data. In [21], Somnath et al. proposed a method to enrich short text representation with additional features from Wikipedia. In [14], Jin et al. learned topics on short texts via transfer learning. However, additional context information and knowledge bases such as user information or hashtags for such heuristic data aggregation and expansion methods [5] is not always available, and aggregated pseudo-documents may not be semantically consistent with original documents due to intrinsic noisy operation. Moreover, topic models based on bag-of-words representation drop semantic accuracy because it treats one word as a unit. In view of this problem, Wallach et al. proposed Bigram LDA model which treats bigram as a unit to model texts [6]. Wang et al. proposed topical N-grams(TNG) to discover phrases and topics within topical context simultaneously [7]. These methods can partly break the bag-of-words assumption limitation, yet they are still specific to a few models and suffered from high complexity.

Success of frequent pattern mining in transaction data has inspired its application in topical modeling. Therefore, we proposed a pattern set-based text representation (PSTR) method by introducing frequent pattern mining into topic modeling. Because pattern mining can fully make use of information of words contained in original text data, we propose to mine frequent patterns among the corpus-level words and use these patterns as feature units to represent texts. Then we use this enriched text representation to perform topic modeling.

Compared with unigram representation of text, our approach has several advantages: By using patterns explicitly it can capture abundant co-occurrence of words from the whole text corpus, and further capture more semantic associations by treating several words as a feature unit. Our method fully takes advantage of data itself to enhance the feature representation of input data without utilizing external data resource. It is purely data-driven. Topic representations discovered by the method are more explicable, we incorporate constraint for inferring topics of constituent terms within a pattern, and this doesn't need additional latent variables and increase too much complexity of model.

The following steps should be followed when applying the method proposed in this paper. Firstly, we need to process different type records into a corpus. Then the corpus will be preprocessed, which contains removing stop words, filtering non-target language, and extracting stems. Next step, we use FP-growth algorithm to mine frequent patterns from the corpus. Finally, Gibbs sample algorithm is applied to estimate all parameters.

1.3   Overview

The rest of this paper is organized as follows. In Sect. 2, we briefly review the LDA framework and frequent pattern mining. Then we introduce our proposed approach LDA-PSTR detailed. Experiments and analysis are presented in Sect. 3, and we conclude our work in Sect. 4.

2   LDA-PSTR

2.1   Latent Dirichlet Allocation

Latent Dirichlet Allocation (LDA) is a Bayesian probability graphic mode. It uses word bag method, which treats each document as a word frequency vector, to transform the text information into easy-to-model digital information. However, the word bag method does not consider the order between words, which simplifies the complexity of the problem and also provides opportunities for the improvement of the model. Each document represents a probability distribution of topics, and each topic represents a probability distribution of words.

LDA has been used successfully in the topic modeling of regular texts, while it is not so effective when applied to topic modeling of short texts. Because different short texts contain the same word less, and it prevents us from effectively assigning documents that share the same topic but different words to the same topic. So we need to find something that appears frequently in short texts to replace words in short texts representation.

2.2   Pattern Set-Based Text Representation

The successful application of pattern mining in transaction data mining inspired us to introduce the concept of pattern into short text representation. In this paper, we mainly use frequent patterns, which are some sets of words that co-occur frequently among short texts, to represent short texts. Frequent patterns abandon order information between words to simplify calculations, and the loss of order does not affect the accuracy of our modeling of short text topics. We map the feature space of short text modeling from frequent word space to frequent pattern space, which solves the sparseness of word frequency matrix. Next we will introduce how to mine frequent patterns from short text corpus.

Above all, we need to introduce the concept of support which specifies the frequency of patterns appear in the corpus. It defined as follow:

$$\sup(c_i) = \frac{N_{c_i}}{N} \tag{1}$$

Where N_{c_i} denotes the number of short texts which attain pattern c_i, and N is the number of all short texts in the corpus.

And we define the minimum support of frequent pattern as min_sup, which is set by ourselves. Mining frequent patterns in short text corpus is divided into three steps. The first step is preprocessing, which includes removing stop words and stemming. The second step is to set the parameters min_sup. And the last step is to use FP-growth algorithm to find out all frequent patterns, which can be represented as follows:

$$C = \left\{ c_1, c_2, \ldots, c_{|C|} \right\} \tag{2}$$

$$c_i = \left\{ w_1, w_2, \ldots, w_{l_{p_i}} \right\} \tag{3}$$

Where C denotes the set of all the frequent patterns, w_t $(t = 1, 2, \ldots, l_{p_i})$ denotes the t-th word in the vocabulary, l_{p_i} denotes the number of words in pattern p_i.

According to unigram text representation, a document d_j can be represented as a set of words:

$$d_j = \left\{ w_1, w_2, \ldots, w_{l_{d_j}} \right\} \tag{4}$$

Then, document d_j can be represented as set of patterns:

$$d_j = \left\{ c_i | \forall w_t \in c_i, w_t \in d_j \right\} \tag{5}$$

We call the above equation pattern set, and this is pattern set-based text representation (PSTR).

2.3   LDA-PSTR

In the previous subsection, we represent the short texts in the patterns space by mapping the word bag to pattern bag. Each pattern is a set of words which appear simultaneously and frequently. In order to express short texts more accurately, we propose a new probabilistic topic model called LDA-PSTR. And an improved Gibbs algorithm has been developed for LDA-PSTR.

Given a corpus of short texts, we firstly mine all the frequent patterns with the method we have introduced in Subsect. 2.2. Then we create a frequent pattern library C, which contains all the frequent patterns. So, we can use a pattern frequency vector to represent short text:

$$\vec{d_j} = (n_1, n_2, \ldots, n_{|C|}) \tag{6}$$

Where n_1 denotes the occur frequency that the first frequent pattern in the C occurs in the short text d_j. And we define n_i as the number of occurrences of the word with the least number of occurrences in pattern c_i. $|C|$ represents the number of all patterns in the C.

In general, our goal is to find the distribution of topics in each document (short text) and the distribution of frequent patterns in each topic. In the LDA-PSTR model, we need to first assume a number of topics K, so that all distributions are based on K topics. As shown in Fig. 1.

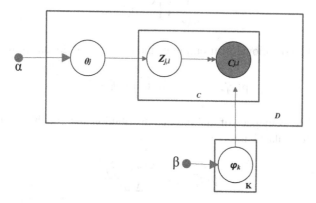

Fig. 1. Graphical model for LDA-PSTR

The LDA-PSTR assumes that the document's prior distribution on the topics and the topic's prior distribution on the frequent patterns both are Dirichlet distributions, as shown below:

$$\theta_j = \text{Dirichlet}(\vec{\alpha}) \tag{7}$$

$$\varphi_k = \text{Dirichlet}\left(\vec{\beta}\right) \tag{8}$$

Where, $\vec{\alpha}$ and $\vec{\beta}$ are the hyper-parameter of the distribution, which is a $|C|$ dimension vector.

Our goal is to find the two distributions above. In this paper, we improved the Gibbs algorithm to solve this problem. Its inputs are the number of topics K and two hyper-parameter of Dirichlet distributions $\vec{\alpha}$, $\vec{\beta}$. θ_j and φ_k are outputs. The main idea of our algorithm is that if we can first find the joint distribution $p(\vec{c}, \vec{z})$ of $\vec{c}$ and $\vec{z}$, and then we can find the conditional probability distribution $p(z_i = k|\vec{c}, \vec{z}_{\neg i})$ of a pattern c_i corresponding to the topic number z_i. Among them, $\vec{z}_{\neg i}$ denotes the topic distribution after removing the pattern c_i. With the conditional probability distribution $p(z_i = k|\vec{c}, \vec{z}_{\neg i})$, we can perform Gibbs sampling. Finally, we can get the topic of pattern c_i after Gibbs sampling convergence. If we get the topic of all patterns by sampling, then by counting the topics, we can get the pattern distribution of each topic. In the same way, by counting the topics of the corresponding patterns in each document, the topic distribution of each document can be obtained.

Here's how to derive expressions $p(z_i = k|\vec{c}, \vec{z}_{\neg i})$:

Firstly, simplify the expression of Dirichlet distribution, $\Delta(\vec{\alpha})$ is a normalized parameter:

$$\text{Dirichlet}(\vec{p}|\vec{\alpha}) = \frac{\Gamma\left(\sum_{k=1}^{K}\alpha_k\right)}{\prod_{k=1}^{K}\Gamma(\alpha_k)} \prod_{k=1}^{K} p_k^{\alpha_k-1} \tag{9}$$

$$= \frac{1}{\Delta(\vec{\alpha})} \prod_{k=1}^{K} p_k^{\alpha_k-1} \tag{10}$$

Calculate the conditional distribution of the topic of the document d_j:

$$p\left(\vec{z_j}, \vec{\alpha}\right) = \int p\left(\vec{z_j}|\vec{\theta_j}\right) p(\theta_j|\vec{\alpha})d\vec{\theta_j} \tag{11}$$

Because of the conjugacy of multinomial distribution and Dirichlet distribution, we can transform formula 11 into:

$$p\left(\vec{z_j}, \vec{\alpha}\right) = \frac{\Delta\left(\vec{\alpha}+\vec{n_j}\right)}{\Delta(\vec{\alpha})} \tag{12}$$

where the number of patterns in topic z_k of document d_j is defined as $n_j^{(k)}$, so $\vec{n_j}$ can be expressed as:

$$\vec{n_j} = \left\{n_j^{(1)}, n_j^{(2)}, \ldots, n_j^{(k)}\right\} \tag{13}$$

According to Eq. 12, we can get all of the condition distribution:

$$p(\vec{z_j}|\vec{\alpha}) = \prod_{j=1}^{M} \frac{\Delta\left(\vec{\alpha}+\vec{n_j}\right)}{\Delta(\vec{\alpha})} \tag{14}$$

In the same way, we can get condition distribution of patterns corresponding to the topic z_j:

$$p(\vec{c_i}|\vec{z}, \vec{\beta}) = \prod_{k=1}^{K} \frac{\Delta\left(\vec{\beta}+\vec{n_j}\right)}{\Delta\left(\vec{\beta}\right)} \tag{15}$$

Finally, we can get joint distribution of topics and patterns:

$$p(\vec{c}, \vec{z}) \propto p(\vec{c}, \vec{z}|\vec{\alpha}, \vec{\beta}) = \prod_{j=1}^{M} \frac{\Delta(\vec{\alpha} + \overrightarrow{n_j})}{\Delta(\vec{\alpha})} \prod_{k=1}^{K} \frac{\Delta(\vec{\beta} + \overrightarrow{n_j})}{\Delta(\vec{\beta})} \tag{16}$$

Now start calculating the conditional probability for Gibbs sampling:

$$p(z_i = k|\vec{c}, \overrightarrow{z_{\neg i}}) = \frac{n_{d,\neg i}^{k} + \alpha_k}{\sum_{s=1}^{K} n_{d,\neg i}^{s} + \alpha_k} \frac{n_{k,\neg i}^{t} + \beta_k}{\sum_{f=1}^{V} n_{d,\neg i}^{f} + \beta_k} \tag{17}$$

With this formula, we can use Gibbs sampling to sample the themes of all patterns. When Gibbs samples converge, we get the sampling topics for all patterns.

3 Experiments

3.1 Data Sets

We use three real-world data sets for our experiments. We processed the raw content by the following normalization steps: (1) convert letters into lower case; (2) remove non-Latin characters and stop words; (3) remove words with document frequency less than 5.

TWEETS. We select 30 hashtags and sample a collection of 223,159 tweets from the TREC2011 microblog track. After text preprocessing we got 11,234 unique words. In the Tweets2011 collection we use their hashtags as label information, since each hashtag denotes a specific topic labeled by its author, we organized documents with the same hashtag into a cluster. These hashtags are content label and they could be treated as cluster labels for every tweet.

DBLP Titles. We also collect titles of scientific papers from DBLP as another short text corpus. We collect titles of all conference papers from the DBLP database. This data set contains 40190 short documents and 9393 unique words with labels of 22 different conferences.

20NG. This data set contains 18,774 newsgroup documents labeled in 20 categories, with a vocabulary of 30451 unique words. The statistics of dataset is shown in Table 1.

Table 1. Statistics of data set

Data set	#docs	#vocabulary	Average length
Tweets	223,159	11,234	6.5
DBLP	40,190	9,393	5.7
20NG	18,774	4,561	104.2

3.2   Parameters Setting

Pattern Mining. In pattern mining, there is one parameter we can tune: min_-sup. Min_sup decides the threshold of frequentness. For example, with min_-sup = 10%, the target pattern should occur more than or equal to 10 times in 100 cases. In our experiment, we try various minimum supports (0.01%, 0.05%, 0.1%, 1%, and 3%) and choose the best performance via cross validation.

Topic Modeling. For LDA, the number of topics varies from 20 to 150 in different dataset. We run Gibbs sampling with 1,000 iterations, we ran 10 times and calculate average value. Generally, let $\alpha = 50/K$, $\beta = 0.01$.

3.3   Comparison Setting

We compare our method (denoted as LDA-PSTR) with the following models.

LDA-U. Standard topic model based on unigram representation.

TNG. This method uses a hierarchical dirichlet to share the topic across each word within a bigram. TNG is a state-of-the-art approach to n-gram topic modeling that uses additional latent variables and word-specific multinomial to model bigrams.

Mixture of Unigrams (Mix-U). Mixture of unigrams assumes that each document is generated by only one topic. This simple model forces the topic representation of a document to adopt the largest sparsity. It may be a reasonable assumption that each document only contains one topic in short text.

3.4   Evaluation Metrics and Results

Some commonly used metrics such as the perplexity or the likelihood of held-out data cannot be used directly to measure the semantic coherence of the learned topics. The likelihood of the held-out data is not always a good indicator of topic coherence. We adopt the following metrics to compare the topic models.

Classification Accuracy. One of the most important products of topic modeling is the topic proportion of each document, which provides a latent semantic representation of that document. As an alternative to using all words as features, we can use a topic-based representation as a low-dimensional representation in external text mining tasks such as text classification. Our evaluation is aims at how accurate and discriminative of the topical representation of documents. More specifically, a document d_k could be treated as k-dimension topics feature:

$$\overrightarrow{d_k} = [p(z_1), \cdots, p(z_k)] \tag{18}$$

where K is the number of topics.

By using the linear SVM classifier LIBLINEAR, we trained and tested data set and recorded average accuracy on 5-fold cross validation. The accuracy of the whole document collection D is calculated by the following equation:

$$\text{ACC}(D) = \frac{1}{|D|} \sum_{d \in D} I(C_d = P_d) \tag{19}$$

Where $I(C_d = P_d)$ is indicator function, C_d and P_d are respectively the true label and the predicted label of document d in a text classification task.

We first perform text classification tasks on 20NG and DBLP to evaluate the effectiveness of the topic representation of documents. The topic distribution is used as the feature representation instead of the conventional bag-of-words representation. We randomly split data set and use 70% of the data for training and 30% for testing. The parameters are set to the same value in all comparison candidates. In order to find the optimal parameters of SVM, we conducted a 5-fold cross validation. Besides the topic models, we also include a base representation which represents documents using the conventional term frequencies (TF).

Figures 2 and 3 show the classification accuracy on 20NG and DBLP under different topic numbers. We can observe that LDA-PSTR consistently outperforms model LDA using unigram as feature representation (LDA-U) and Mix-U, TNG on both data sets. On the 20NG data set, the sample TF representation outperforms the other latent topic model. On 20NG data set, because the data is normal text which carries sufficient feature information and covers multiple topics so that TF representation can fit the classifier well. At the same time, Mix-U is unable to capture multiple specific topics so it underperforms other model.

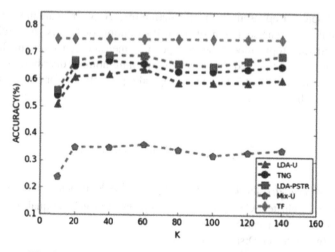

Fig. 2. Accuracy on 20NG over different topic number K

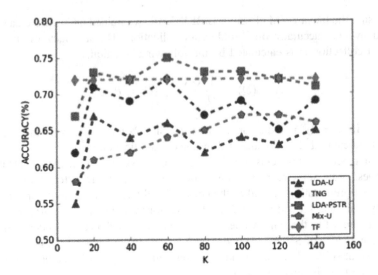

Fig. 3. Accuracy on DBLP over topic number K

On the DBLP data set, short text can't capture more feature unit, thus our method always outperforms all other model include the TF representation. The Mix-U outperforms LDA-U suggests that LDA using unigram as feature representation is not a good choice for short texts due to the data sparsity problem. When keyword features are rare, the experimental results indicate that LDA-PSTR is potentially very useful in text classification tasks with short documents. This demonstrates that our method can successfully model short text and the improvement is more prominent compared with the improvement on normal text.

Quality of Topical Representation of Documents. Another evaluation is to investigate the quality of representation of topic distribution of tweets. According to the Eq. 14, considering topic models as a type of dimension reduction methods, each document can be represented by a vector of multinomial distribution of topics. Therefore, we can compare the distance of topical representation for two documents by Jensen-Shannon divergence:

$$JS(d_i, d_j) = \frac{1}{2}(D_{KL}(d_i||m) + D_{KL})(m||d_j) \tag{20}$$

Where m $= \frac{1}{2}(d_i, d_j)$ and $D_{KL}(p||q) = \sum_i p_i \ln \frac{p_i}{q_i}$ is Kullback-Leibler divergence.

Given a set of clusters C $= \{C_1, C_2, \ldots, C_k\}$, we can evaluate quality of the set of clusters by using two distance scores:

Average Intra-Cluster Distance in [15]:

$$\text{IntraDis}(C) = \frac{1}{k} \sum_k \left[\sum_{d_i, d_j \in C} \frac{2dis(d_i, d_j)}{|C_k||C_{k-1}|} \right] \tag{21}$$

Average Inter-Cluster Distance in [16]:

$$\text{InterDis}(C) = \frac{1}{k(k-1)} \sum_{C_l, C_m \in C} \left[\sum_{d_i \in C_j} \sum_{d_j \in C_m} \frac{dis(d_i, d_j)}{|C_l||C_m|} \right] \tag{22}$$

We use these two distance metrics coupled with label of each document to evaluate the quality of topic. If the average inter-cluster distance is small and the average intra-cluster distance is big, the topical representation of documents matches well with labeled clusters (via hashtag). We calculate the ratio of IntraDis and InterDis to evaluate the quality of topical representation of documents in [17]:

$$H = \frac{\text{IntraDis}(C)}{\text{InterDis}(C)} \tag{23}$$

Good topical representation for a document is the one which has the minimum H score.

Table 2 shows the H score for all comparisons under different topic numbers. From the results, we can see that LDA-PSTR significantly outperforms other three methods. TNG perform better than conventional baseline method LDA-U implying that TNG captures more important keyword or phrase feature in short text. When the number of topic K is 50, the LDA-PSTR achieves the best performance (H score is smallest, 0.317). At the same time, the Mix-U outperforms LDA-U and TNG in this short text. It suggests that data sparsity problem affects LDA with unigrams features, However the H score of Mix-U is still worse than LDA-PSTR.

Table 2. H Score for tweets with different topic number K

Methods	Topic numbers		
	20	50	70
LDA-U	0.496 ± 0.004	0.427 ± 0.004	0.432 ± 0.011
TNG	0.421 ± 0.007	0.412 ± 0.002	0.425 ± 0.005
LDA-PSTR	0.331 ± 0.012	0.317 ± 0.006	0.310 ± 0.014
Mix-U	0.342 ± 0.005	0.336 ± 0.012	0.324 ± 0.001

In order to further understand the quality of topics found by the models, we perform another evaluation task. Here, we use Normalized mutual Information (NMI) to measure the quality by how likely the topics agree with the true category (hashtag). The NMI can be defined as follows:

$$\text{NMI}(\Omega, C) = \frac{I(\Omega, C)}{[H(\Omega) + H(C)]/2} \qquad (24)$$

Where $I(\Omega, C)$ is mutual information between set Ω and C and H(A) is the entropy. NMI is always a number between 0 and 1. NMI may achieve 1 if the clustering results can exactly match category labels, while 0 if two sets are independent. The topical representation of tweets is achieved as the same as above task. For each tweet we use the maximum value of topic multinomial to determine its cluster, which leads to a hard clustering result. From the Fig. 4 we can see that NMI is low in general. But again the LDA-PSTR significantly outperforms the other models. Meanwhile, the NMI values by the LDA-U are very low, indicating that LDA-U do not model good topical representation for short text tweets. As discussed before, LDA-U and TNG have poor effect when there are rare feature unit. Therefore, the quality of topical representation learned is comparatively poor.

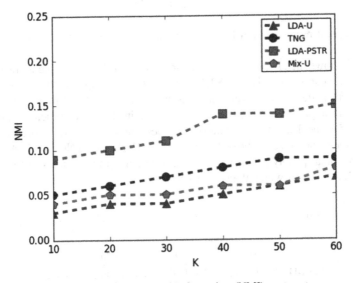

Fig. 4. Normalized Mutual Information (NMI) on tweets.

In summary, the experiments result shows that our method can achieve an effective topic representation of documents, especially for sparse short text. Multiple tasks evaluate the effectiveness of the proposed method LDA-PSTR. The performance outperforms classical topic models (i.e. LDA-U) and the state-of-the-art n-gram topic modeling (TNG). The effectiveness is especially significant on short text.

4 Conclusions

In this paper, we proposed a novel method to improve topic modeling performance for short text. Our LDA-PSTR method can use frequent patterns to represent short text. The topical representation modeled by LDA-PSTR provides a nice low-dimensional latent semantic representation of documents, which is useful in many applications such as text classification. Experimental results demonstrate LDA-PSTR can achieve an effective topic representation of documents, the classification accuracy and H scores outperform the classical topic model using unigram representation and the state-of-the-art approach n-gram topic modeling.

Though the advantage of our method is obvious, we can further improve it by exploiting sophisticated strategy to represent text, for example we can take the pattern support into account to get weights of a pattern in the pattern-based representation. What's more, we can try more evaluation and comparison approach to help to understand the proposed method.

References

1. Hofmann, T.: Probabilistic latent semantic indexing. In: Proceedings of the 22nd Annual International ACM SIGIR Conference on Research and Development in Information Retrieval, pp. 50–57. ACM (1999)
2. Blei, D.M., Ng, A.Y., Jordan, M.I.: Latent Dirichlet allocation. J. Mach. Learn. Res. 3(Jan), 993–1022 (2003)
3. Griffiths, T.L., Steyvers, M., Tenenbaum, J.B.: Topics in semantic representation. Psychol. Rev. 114(2), 211 (2007)
4. Hong, L., Davison, B.D.: Empirical study of topic modeling in Twitter. In: Proceedings of the First Workshop on Social Media Analytics, pp. 80–88. ACM (2010)
5. Tang, J., Zhang, M., Mei, Q.: One theme in all views: modeling consensus topics in multiple contexts. In: Proceedings of the 19th ACM SIGKDD International Conference on Knowledge Discovery and Data Mining, pp. 5–13. ACM (2013)
6. Wallach, H.M.: Topic modeling: beyond bag-of-words. In: Proceedings of the 23rd International Conference on Machine Learning, pp. 977–984. ACM (2006)
7. Wang, X., McCallum, A., Wei, X.: Topical n-grams: phrase and topic discovery, with an application to information retrieval. In: Seventh IEEE International Conference on Data Mining, pp. 697–702. IEEE (2007)
8. Blei, D., Lafferty, J.: Correlated topic models. Adv. Neural Inf. Process. Syst. 18, 147 (2006)
9. Teh, Y.W., Jordan, M.I., Beal, M.J.: Hierarchical Dirichlet processes. J. Am. Stat. Assoc. (2012)
10. Mcauliffe, J.D., Blei, D.M.: Supervised topic models. In: Advances in Neural Information Processing Systems, pp. 121–128 (2008)
11. Kim, H.D., Park, D.H., Lu, Y.: Enriching text representation with frequent pattern mining for probabilistic topic modeling. Proc. Am. Soc. Inf. Sci. Technol. 49(1), 1–10 (2012)
12. Mihalcea, R., Corley, C., Strapparava, C.: Corpus-based and knowledge-based measures of text semantic similarity. In: AAAI, vol. 6, pp. 775–780 (2006)
13. Phan, X.H., Nguyen, L.M., Horiguchi, S.: Learning to classify short and sparse text & web with hidden topics from large-scale data collections. In: Proceedings of the 17th International Conference on World Wide Web, pp. 91–100. ACM (2008)

14. Jin, O., Liu, N.N., Zhao, K.: Transferring topical knowledge from auxiliary long texts for short text clustering. In: Proceedings of the 20th ACM International Conference on Information and Knowledge Management, pp. 775–784. ACM (2011)

15. Bordino, I., Castillo, C., Donato, D.: Query similarity by projecting the query-flow graph. In: Proceedings of the 33rd International ACM SIGIR Conference on Research and Development in Information Retrieval, pp. 515–522. ACM (2010)

16. Yan, X., Guo, J., Lan, Y.: A biterm topic model for short texts. In: Proceedings of the 22nd International Conference on World Wide Web, pp. 1445–1456. ACM (2013)

17. Guo, J., Cheng, X., Xu, G.: Intent-aware query similarity. In: Proceedings of the 20th ACM International Conference on Information and Knowledge Management, pp. 259–268. ACM (2011)

18. Weng, J., Lim, E.P., Jiang, J.: TwitterRank: finding topic-sensitive influential Twitterers. In: Proceedings of the Third ACM International Conference on Web Search and Data Mining, pp. 261–270. ACM (2010)

19. Mehrotra, R., Sanner, S., Buntine, W.: Improving LDA topic models for microblogs via tweet pooling and automatic labeling. In: Proceedings of the 36th International ACM SIGIR Conference on Research and Development in Information Retrieval, pp. 889–892. ACM (2013)

20. Lin, T., Tian, W., Mei, Q.: The dual-sparse topic model: mining focused topics and focused terms in short text. In: Proceedings of the 23rd International Conference on World Wide Web, pp. 539–550. ACM (2014)

21. Banerjee, K.: Clustering short texts using Wikipedia. In: Proceedings of the 30th Annual International ACM SIGIR Conference on Research and Development in Information Retrieval, pp. 787–788. ACM (2007)

22. Chen, W., et al.: EEG-based motion intention recognition via multi-task RNNs. In: Proceedings of the 2018 SIAM International Conference on Data Mining, pp. 279–287. Society for Industrial and Applied Mathematics (2018)

23. Yue, L., Chen, W., Li, X., Zuo, W., Yin, M.: A survey of sentiment analysis in social media. Knowl. Inf. Syst. 1–47 (2018)

Vertical and Sequential Sentiment Analysis of Micro-blog Topic

Shuo Wan[1], Bohan Li[1,2,3(✉)], Anman Zhang[1], Kai Wang[1],
and Xue Li[1,4]

[1] College of Computer Science and Technology,
Nanjing University of Aeronautics and Astronautics, Nanjing, China
{shuowan, bhli}@nuaa.edu.cn
[2] Collaborative Innovation Center of Novel Software Technology and
Industrialization, Nanjing, China
[3] Jiangsu Easymap Geographic Information Technology Corp., Ltd.,
Nanjing, China
[4] School of Information Technology and Electrical Engineering,
University of Queensland, Brisbane, Australia

Abstract. Sentiment analysis of micro-blog topic aims to explore people's attitudes towards a topic or event on social networks. Most existing research analyzed the micro-blog sentiment by traditional algorithms such as Naive Bayes and SVM based on the manually labelled data. They do not consider timeliness of data and inwardness of the topics. Meanwhile, few Chinese micro-blog sentiment analysis based on large-scale corpus is investigated. This paper focuses on the analysis of sequential sentiment based on a million-level Chinese micro-blog corpora to mine the features of sequential sentiment precisely. Distant supervised learning method based on micro-blog expressions and sentiment lexicon is proposed and fastText is used to train word vectors and classification model. The timeliness of analysis is guaranteed on the premise of ensuring the accuracy of classifier. The experiment shows that the accuracy of the classifier reaches 92.2%, and the sequential sentiment analysis based on this classifier can accurately reflect the emotional trend of micro-blog topics.

Keywords: Vertical sentiment analysis · fastText · Distant supervision
Sequential analysis

1 Introduction

Sina Weibo, the largest social network platform in China, has more than 150 million daily active users and an average of two hundred million micro-blogs published daily. People can push their life and opinion to Weibo, or comment on hot events. These

This work is supported by National Natural Science Foundation of China (61672284, 41301407), Funding of Security Ability Construction of Civil Aviation Administration of China (AS-SA2 015/21), Fundamental Research Funds for the Central Universities (NJ20160028, NT201 8028, NS2018057).

© Springer Nature Switzerland AG 2018
G. Gan et al. (Eds.): ADMA 2018, LNAI 11323, pp. 353–363, 2018.
https://doi.org/10.1007/978-3-030-05090-0_30

subjective data bring great convenience to the study of sentiment analysis. The real-time sentiment information mining of micro-blog can accurately reflect the trend of micro-blog topics and provide early warnings, which has positive significance for individuals, businesses and governments.

At present, most work about sentiment analysis focus on improving the performance of classification algorithms [1] based on existing datasets, especially the sentiment analysis of Chinese micro-blog, which lacks sound sentiment lexicons and large-scale training corpus. This paper mainly discusses the sequential sentiment analysis based on large-scale Chinese micro-blog corpus. The popular micro-blog topic of "ZTE Denial Order" was selected for detailed sequential analysis, which fully excavated the timeliness of micro-blog data. Main work of this paper are as follows:

(1) We designed a micro-blog general crawler and collected 35 million Sina Weibo data. A training set and test set for a Chinese micro-blog sentiment analysis was constructed by using a method of distant supervision based on Emoji. Meanwhile, we use the Sentiment Lexicon to further filter the training set.
(2) fastText [2] is used to train word vector and generate classification models. Experiment shows that based on the classification model trained by the training set, the accuracy reaches 92.2% on the test set.
(3) We designed a micro-blog specialized crawler, which collects data on a vertical topic on Weibo and performs dynamic sentiment analysis on timeline. The experiments have proved that the sequential sentiment analysis can accurately reflect the micro-blog user's attitude toward a topic or event and the trend of the incident.

The rest of the paper is organized as follows: Sect. 2 introduces the related work of sentiment analysis and Chinese micro-blog sentiment analysis. Section 3 introduces the collection and cleaning of data and proves the validity of data. Section 4 introduces the algorithm of the construction of corpus and the training of classifier. Section 5 introduces the vertical and sequential sentiment analysis of micro-blog topics. Section 6 concludes the whole paper.

2   Related Works

Research on the sentiment classification of micro-blog can be divided into two categories: sentiment classification based on sentiment dictionary and sentiment classification based on feature selection and corpus.

The research based on sentiment dictionary focuses on the creation and expansion of sentiment dictionaries. Ku et al. [3] conducted emotional mining of news and micro-blogs and generated the NTUSD sentiment dictionary. Zhu et al. [4] calculated the similarity of words based on HowNet. Sentiment analysis based on feature selection and corpus mostly use machine learning methods to train classifiers. The most representative research is Pang et al. [5] using Naive Bayes, SVM and Maximum Entropy classifier to classify sentiment in English movie reviews. The earliest research on sentiment analysis based on micro-blog short text is distant supervision method proposed by Go [6] in 2009. The method of distant supervision, which applied the method proposed by Read [7] to construct classified datasets using Emoji, constructed a sentiment dataset containing 1.6 million tweets and achieved a classification accuracy of 83%. Based on this, Pak et al. [8] proposed a method for building a sentiment analysis corpus automatically. Iosifidis et al. [9] built a larger dataset on this basis and applied self-learning and collaborative training methods to expand the dataset.

In recent years, the machine learning method has gradually become the mainstream sentiment classification method. The proposed word2vec [10] for distributed word vector training implements automatic extraction of text features. Especially in the study of micro-blog short text with an average length of only 70, the distributed word vector has the edge over the traditional methods. fastText [11], open-sourced by facebook, can implement unsupervised word vector training and supervised classifier training to train word vector and classifier comparable to deep learning methods in time. As far as we know, there is no precedent for applying fastText to large-scale Chinese micro-blog datasets for training and classification.

The dataflow of micro-blogs is of real-time and timeliness, only by grasping the timeliness of micro-blog information and analyzing the latest topic data can we make even greater use of the value of the data. Culotta et al. [12] monitored the influenza epidemic by analyzing real-time Twitter data. Nahar et al. [13] realized real-time monitoring of cyber bullying based on the Twitter platform. Paul [14] exploited Twitter to construct a spatio-temporal sentiment analysis system to analyze the sentiment trends of voters during the US election. However, most of the researches on analysis of micro-blog focus on the deep learning method to improve the classification performance and apply the typical Stanford Twitter sentiment analysis datasets[1] to train, and there is no vertical and sequential analysis for a particular topic or field in micro-blog. This paper focuses on the sequential study of micro-blog data, seize the timeliness of the data to maximize the value of micro-blog.

3 Data Acquisition and Cleaning

This section will introduce the acquisition and cleaning of Sina Weibo data, concentrate on the design of the general crawler and specialized crawler, the process of data cleaning, and the validity of the micro-blog dataset.

3.1 Data Acquisition

The first way to request micro-blog data is the public interface[2] provided by Sina Weibo. The interface can only obtain the latest 200 micro-blogs of related topics, in this paper, the data of Sina weibo are collected by web crawler.

We designed two different crawler programs. The general crawler, which applies multi-thread and proxy to achieve the concurrent crawl of 580 thousand micro-blogs per day, is used to collect a large amount of weibo data. The specialized crawler

---

[1] https://www.kaggle.com/kazanova/sentiment140.

[2] http://open.weibo.com/wiki/2/search/topics.

focuses on a certain topic and achieves the micro-blog text within a specific period on the specific topic. From December 1, 2017 to January 31, 2018, the general crawler crawled 35 million micro-blogs in two months, with a total size of 6.34 GB. At the same time, specialized crawler collected micro-blog topic texts from several famous companies from January 2018 to May 2018 for dynamic and sequential sentiment analysis as shown in Table 1.

Table 1. Number of microblogs collected on each topic

ZTE	MI	MEIZU	HUAWEI	LeEco	Apple
38,000	34,000	59,000	58,000	73,000	47,000

3.2   Data Cleaning

This section conducts micro-blog cleaning of datasets, including the cleaning of unique attributes of Weibo, the cleaning of links, mailboxes, special characters and short micro-blogs. The size of the dataset after each step of cleaning is shown in Fig. 1.

First, special symbols like @ and ## are filtered from dataset. Second, statistics show that 0.67 million texts in the micro-blog dataset contain url links or email address, that is, for every 100 micro-blogs, there are two micro-blogs that contain links and email address. We use regular expressions to match and filter url links and email address. Thirdly, special and meaningless characters are filtered in the preprocessing stage. The last step is the filter of short micro-blogs. We stipulate that the micro-blog with length of less than 5 is an invalid micro-blog, in which one Chinese character is counted as one character. After the cleaning of short micro-blogs, 2.28 million invalid short micro-blogs were filtered out, and finally there were 33.48 million valid micro-blogs left in the dataset.

3.3   Validity of Data

After the data cleaning, the jieba[3] word segmentation tool was used to segment each micro-blog and generate a dictionary with the size of 649,334. The distribution of word frequency in the dictionary is shown in Fig. 2. We can see from the picture that the number of words that appear once are more than that of twice, and the number of words that appear twice are more than that of third, and so on. This indicates that the distribution of word frequency follows Zipf's Law, which proves the validity of the dataset from the perspective of word frequency distribution.

We also test the validity of the data collected by specialized crawler according to chronological order. Taking *"ZTE Topic"* as an example, the daily micro-blog statistics from March 21, 2018 to April 27, 2018 are shown in Fig. 3. On April 16th, the number of micro-blogs on ZTE's topic has soared, which corresponds to the incident that ZTE sanctioned by the US Department of Commerce. The validity of the hourly data has also been tested, as shown in Fig. 4. We selected ZTE topic's micro-blogs on March

---

[3] https://github.com/fxsjy/jieba.

13, 15 and April 16 for statistics. The number of micro-blogs on March 13, 15 is normal, however, the curves at noon, afternoon and evening with three peaks, which correspond to the three peak periods of people's landing of Sina weibo. The number of micro-blogs in April 16 was not unusual at noon and afternoon, however, after 9:00 pm, the news that ZTE was sanctioned became a hot spot and the number of micro-blogs increased rapidly. In summary, the data collected by specialized crawler is time-sensitive, which can reflect the public's attitude towards a topic and can be used for sequential analysis.

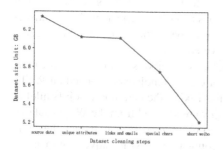

Fig. 1. Cleaning steps of dataset

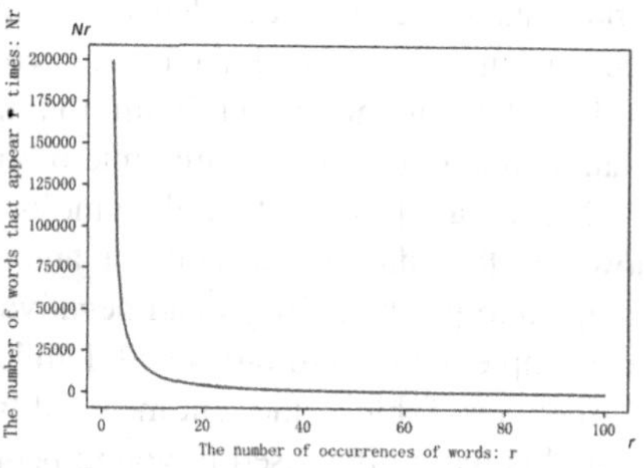

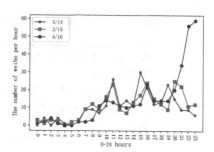

Fig. 2. Word frequency distribution

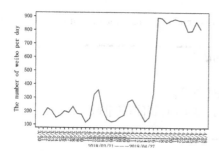

Fig. 3. Number of ZTE weibo from March 21 to April 27

Fig. 4. The 24-hour weibo number of ZTE

4 Corpus Construction and Classifier Training

Section 3 introduces data collection and cleaning and verifies the validity of the data in the end. This section will mainly introduce the construction of corpus and the training of classifiers.

4.1 Corpus Construction

In this paper, we divide people's sentiments into positive sentiment and negative sentiment. Correspondingly, micro-blogs are divided into positive and negative micro-blogs, and sentiment-neutral micro-blogs will not be discussed. Chinese characters are extensive and profound, in many cases, it is difficult to tell whether a micro-blog's sentiment is neutral or not. The third law of Weibo Arguing Laws [15] also points out that people only have positive and negative parties when they quarrel.

The accuracy of manual classification is difficult to control, especially in the condition of big data of 33.48 million micro-blogs. Therefore, we adopt the method of distant supervision learning, using micro-blog Emoji and Sentiment Lexicon to extract positive and negative micro-blogs.

Emoji can be used to express the user's emotion in micro-blog, to a large extent, it represents the sentiment tendencies of micro-blog text. If a micro-blog contains emojis such as [angry] and [sad], then this micro-blog has a negative tendency. After statistics of dataset, we found that 11.79 million micro-blogs have emojis, accounting for 33.5%. That is, one out of every three micro-blogs have emojis.

Xie et al. [16] first apply emojis for Chinese weibo sentiment classification, however, they did not consider the ambiguity of certain emojis and mapped all Weibo emojis into positive emojis and negative emojis. Taking the common emoji [smile] as an example, two micro-blogs with [smile] emoji were selected from the Weibo dataset. As shown in Table 2, the sentiments of two micro-blogs are converse. In fact, the emoji [smile] is more often used in young people's groups to express negative sentiment, so this emoji is ambiguous and cannot be used as a positive flag.

Table 2. Examples of Chinese micro-blog containing ambiguous emoji

表情词：*[微笑] (in English [smile])*
微博消息：
1. 不想和你理论，你开心就好[微笑]
2. 今天终于见到了我家爱豆，手动笔芯[微笑][微笑][微笑]

We manually selected feature emojis with strong sentiment, including 18 typical negative emojis and 37 typical positive emojis. Table 3 lists part of selected emojis. The NTUSD sentiment dictionary is also used as a double filter for Weibo. The detailed filtering process of sentiment micro-blog is shown in Algorithm 1.

Table 3. Typical examples of feature emojis

Algorithm 1. Emotional Weibo Corpus Construction Algorithm

---

input: Emotion dictionary NTUSD, Weibo data set *weibos*, Weibo expression dictionary *emoji_dict*

output: Positive weibo collection *pos_set*, Negative weibo collection *neg_set*

1. *pos_set* ← ∅

2. *neg_set* ← ∅ // init set

3. *pos_emotions, neg_emotions* ← load(NTUSD)

4. *pos_emojis, neg_emojis* ← load(emoji_dict)

5. **for** *weibo* **in** *weibos* **do**

6.   *words* ← set(*weibo*)

7.   **if** len(*words* & *pos_emojis*) > 0 **and** len(*words* & *neg_emotions*) == 0 **and** len(*words* & *neg_emojis*) == 0 **then** // & intersection

8.   add *weibo* to *pos_set*

9.   **end if**

10. **if** len(*words* & *neg_emojis*) > 0 **and** len(*words* & *pos_emotions*) == 0 **and** len(*words* & *pos_emojis*) == 0 **then**

11.   add *weibo* to *neg_set*

12.   **end if**

13. **end for**

14. **return** *pos_set, neg_set*

The corpus generation algorithm extracts 4.2 million positive micro-blogs and 680,000 negative micro-blogs from the dataset. To prevent the classifier from assigning a bigger priori probability of positive sentiment to the micro-blog in the process of training, we randomly select the same number of micro-blogs as the negative micro-blogs from the positive micro-blogs set, which constitute a corpus of Chinese micro-blog for sentiment analysis.

4.2 Sentiment Classifier Training

In this section, we will apply fastText to train distributed word vectors and sentiment classifiers and introduce some tricks for training sentiment classifier.

Statistics show that the average length of Sina Weibo text is 70, and the short text Weibo with a length of less than 30 accounts for 26%. Short text has the characteristics of large amount of information and close context. fastText considers context and N-gram features at the input layer and can quickly train and classify short texts. fastText has two efficient word vector training models: CBOW and Skip-gram. Both are three-layer shallow neural network models. The difference is that the CBOW model predicts the current word w_t on the premise of context w_{t-2}, w_{t-1}, w_{t+1}, w_{t+2}, Skip-gram model predicts its context w_{t-2}, w_{t-1}, w_{t+1}, w_{t+2} on the premise of the current word w_t, and its maximum likelihood function is shown in formula (1). This paper adopts a better Skip-gram model to train word vectors and classifiers.

$$L_{Skip-gram} = \sum_{w \in C} log \, p(Context(w)|w) \tag{1}$$

Where C is the corpus and *Context* is the set of context words for word w.

Considering that stopwords will affect the training of word vectors, we trained the word vectors in two kinds of training sets: training set with stopwords and training set without stopwords. The training of the distributed word vector model considers the context of the words, so we speculate that there is no need to eliminate stopwords in the Chinese text word vector training. We also stripped off the emojis from each micro-blog to prevent fastText from using emojis as a feature in the process of training the classifier.

The experiment results are shown in Fig. 5. When training datasets with 34 million vocabularies and a dictionary size of 360,000, fastText took less than 100 s to implement the training of each word's 100-dimension word vector. The classifier achieved more than 90% accuracy on the test set, and the classifier based on the training set with stopwords was 0.4% higher than the classifier trained without stopwords, which proved that the stopwords play an important role in the training of distributed word vectors. We also conducted 200-dimension and 300-dimension word vector training, as the training time doubled, the accuracy of classifier only increased by 0.1%. The classifier with the highest accuracy is the 300-dimension classifier trained with stopwords, which achieves an accuracy of 92.2% compared to the 83% achieved by Alec Go's maximum entropy method.

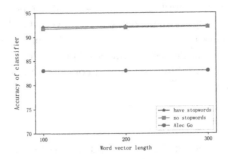

Fig. 5. Accuracy of word vector classifier

5 Experiments

In this section, we apply sentiment classifier trained in Sect. 4 to perform dynamic and sequential sentiment analysis of micro-blog topic data collected in Sect. 3.

Considering that ZTE has a hot topic in the first four months of 2018, this section focuses on a detailed analysis of ZTE's topic. In Sect. 3, we collected 38,000 micro-blogs of ZTE, with an average of 310 micro-blogs per day and 13 micro-blogs per hour. Micro-blog with a probability greater than 0.7 is marked as corresponding positive or negative micro-blog, which can be used to further filter the noise micro-blogs. Finally, the sentiment classification results of each micro-blog are obtained in chronological order. As shown in Fig. 6, a sequential sentiment analysis of ZTE Topics from March 21 to April 27, 2018 was conducted in units of days. From the diagram, we can intuitively see that the topic sentiment from March 21 to April 15 are all positive. Until the incident in which ZTE was sanctioned by the US Department of Commerce on April 16, negative micro-blogs increased sharply and exceeded positive micro-blogs, reflecting the huge crisis faced by ZTE. We also conducted the more vertical analysis of the 24 h on April 16th. It can be seen from Fig. 7 that the number of positive micro-blogs exceeds the negative micro-blogs in most of the time. After 21 o'clock, negative micro-blogs rose suddenly and continued to increase, which corresponds exactly to the *"ZTE Denial Order"* at 9:00 a.m. in US.

Fig. 6. Number of positive weibo and negative weibo from 3/21 to 4/16 about ZTE

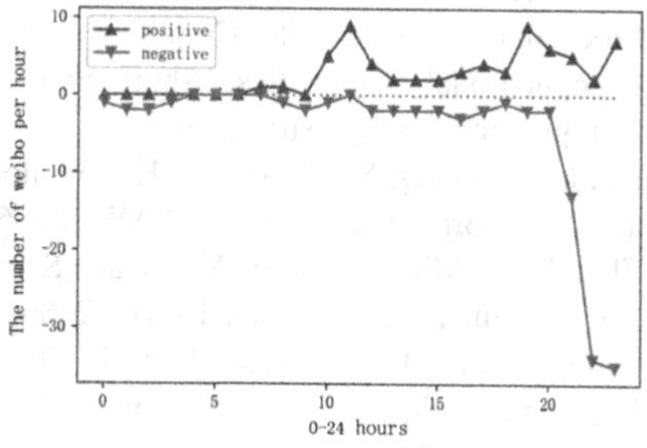

Fig. 7. Number of positive weibo and negative weibo in 4/16 about ZTE

The analysis above proves the validity of the vertical and sequential analysis of Sina Weibo topics. Different from the degree of heat provided by Baidu index and Weibo micro index, the analysis method realized in this paper not only considers the popularity of topic, but also realizes the dynamic sequential analysis of positive emotions and negative emotions, which can quickly and intuitively reflect the sentiment changes of the topic.

6 Conclusion

The vertical and sequential sentiment analysis of micro-blog has broad application prospects. This paper collects data from Sina Weibo and proposes a distant supervision learning method based on Emojis and Sentiment Lexicon. A large-scale Chinese micro-blog sentiment analysis corpus is constructed, and fastText is used to train word vectors and sentiment classifiers. Experiments show that the classifier surpasses traditional classifier in performance and can accurately reflect the sentiment trend of a topic in practical applications. Limited by the length of the article, the results of other Weibo topics, manually selected emojis and ZTE topics' Weibo data will be shared on Github[4]. Next, we will focus on the classification of neutral emotion micro-blog and apply the research results to the public opinion analysis and early warning of public sentiment.

References

1. Wang, S., Manning, C.D.: Baselines and bigrams: simple, good sentiment and topic classification. In: Meeting of the Association for Computational Linguistics: Short Papers, 08–14 July 2012, Jeju, Island, Korea, pp. 90–94. Association for Computational Linguistics (2012)
2. Joulin, A., Grave, E., Bojanowski, P., et al.: Bag of tricks for efficient text classification. In: Proceedings of the 15th Conference of the European Chapter of the Association for Computational Linguistics: Short Papers, April 2017, pp. 427–431. Association for Computational Linguistics (2017)
3. Ku, L.W., Liang, Y.T., Chen, H.H.: Opinion extraction, summarization and tracking in news and blog corpora. In: AAAI-CAAW (2006)
4. Zhu, Y.L., Min, J., Zhou, Y., et al.: Semantic orientation computing based on HowNet. J. Chin. Inf. Process. 20(1), 14–20 (2006)
5. Pang, B., Lee, L., Vaithyanathan, S.: Thumbs up?: Sentiment classification using machine learning techniques. In: Proceedings of EMNLP, pp. 79–86. Association for Computational Linguistics (2002)
6. Go, A., Bhayani, R., Huang, L.: Twitter sentiment classification using distant supervision. Cs224n Project Report (2009)
7. Yue, L., Chen, W., Li, X., Zuo, W., Yin, M.: A survey of sentiment analysis in social media. Knowl. Inf. Syst. 1–47 (2018)
8. Park, A., Paroubek, P.: Twitter as a corpus for sentiment analysis and opinion mining. In: International Conference on Language Resources and Evaluation, LREC 2010, 17–23 May 2010, Valletta, Malta. DBLP (2010)
9. Iosifidis, V., Ntoutsi, E.: Large scale sentiment learning with limited labels. In: ACM SIGKDD International Conference on Knowledge Discovery and Data Mining, pp. 1823–1832. ACM (2017)
10. Mikolov, T., Chen, K., Corrado, G., et al.: Efficient estimation of word representations in vector space. Comput. Sci. (2013)

---

[4] https://github.com/wansho/Weibo_Sentiment_Analyze.

11. Bojanowski, P., Grave, E., Joulin, A., et al.: Enriching word vectors with subword information (2016)
12. Culotta, A.: Towards detecting influenza epidemics by analyzing Twitter messages. In: Proceedings of the First Workshop on Social Media Analytics, Washington, DC, Columbia, 25–28 July 2010, pp. 115–122. ACM (2010)
13. Nahar, V., Al-Maskari, S., Li, X., Pang, C.: Semi-supervised learning for cyberbullying detection in social networks. In: Wang, H., Sharaf, M.A. (eds.) ADC 2014. LNCS, vol. 8506, pp. 160–171. Springer, Cham (2014). https://doi.org/10.1007/978-3-319-08608-8_14
14. Paul, D., Li, F., Teja, M.K., et al.: Spatio temporal sentiment analysis of US Election what Twitter says!. In: ACM SIGKDD International Conference on Knowledge Discovery and Data Mining, pp. 1585–1594. ACM (2017)
15. Du, J.F.: Weibo arguing laws: a sketch of social psychology on the Internet. News Writ. (06), 65 (2017)
16. Xie, L., Zhou, M., Sun, M.-S.: Hierarchical structure based hybrid approach to sentiment analysis of Chinese micro blog and its feature extraction. J. Chin. Inf. Process. **26**(1), 73–83 (2012)

Abstractive Document Summarization
via Bidirectional Decoder

Xin Wan, Chen Li, Ruijia Wang, Ding Xiao, and Chuan Shi$^{(\boxtimes)}$

Beijing University of Posts and Telecommunications, Beijing, China
{wanxin,shichuan}@bupt.edu.cn, http://www.shichuan.org

Abstract. Sequence-to-sequence architecture with attention mechanism is widely used in abstractive text summarization, and has achieved a series of remarkable results. However, this method may suffer from error accumulation. That is to say, at the testing stage, the input of decoder is the word generated at the previous time, so that decoder-side error will be continuously amplified. This paper proposes a **Sum**marization model using a **Bi**directional decoder (**BiSum**), in which the backward decoder provides a reference for the forward decoder. We use attention mechanism at both encoder and backward decoder sides to ensure that the summary generated by backward decoder can be understood. Also, pointer mechanism is added in both the backward decoder and the forward decoder to solve the out-of-vocabulary problem. We remove the word segmentation step in regular Chinese preprocessing, which greatly improves the quality of summary. Experimental results show that our work can produce higher-quality summary on Chinese datasets TTNews and English datasets CNN/Daily Mail.

Keywords: Abstractive summarization · Bidirectional decoder
Attention mechanism · Sequence-to-sequence architecture

1 Introduction

In the era of information explosion, summarization that can help people quickly extract knowledge is of great significance. Abstractive summarization is a technology of generating summary from source documents using deep learning methods, and hopes to keep the original meaning of the documents to the utmost extent.

Sequence-to-sequence architecture with attention mechanism (Seq2Seq-Attn) is a general solution for abstractive summarization in academic circles. In existing Seq2Seq-Attn summarization model, decoder-side input at the next time step is up to the referred summary while training. But at the testing stage, decoder-side input depends on its output at the previous time step. Hence, there will be a problem of error accumulation during testing. Once the decoder generates a wrong word, it will have a negative impact on the following predictions, which may result in the subsequent incorrect summary.

© Springer Nature Switzerland AG 2018
G. Gan et al. (Eds.): ADMA 2018, LNAI 11323, pp. 364–377, 2018.
https://doi.org/10.1007/978-3-030-05090-0_31

To solve this problem, we proposed an abstractive **Summ**arization model based on **Bi**directional decoder (**BiSum**), and Fig. 1 is a diagram of it. A backward decoder is added in Seq2Seq-Attn model to generate the summary from right to left. The result of the backward decoder can provide a reference for the final summary through attention mechanism, thereby avoiding the error of the latter part of summary. Our model follows these steps: (1) generate the summary by the backward decoder from right to left in a similar manner to the Seq2Seq-Attn model; (2) apply attention mechanism to both encoder and backward decoder, so that the forward decoder can generate the summary from left to right. At the same time, the pointer mechanism [19] is also embedded in the forward decoder and backward decoder to address the out-of-vocabulary (OOV) problem. This problem is due to the limitation of the vocabulary size, that is, the vocabulary cannot cover all the words in the source documents and the referred summary. Both the Chinese summarization datasets TTNews and the English summarization datasets CNN/Daily Mail are conducted in our model. Since that the scale of Chinese summarization datasets is generally not large, the source documents are not segmented by words, but trained character by character. We find that this trick can significantly improve the quality of generated summary of BiSum and other models. Experimental results demonstrate that BiSum achieves excellent results on both datasets.

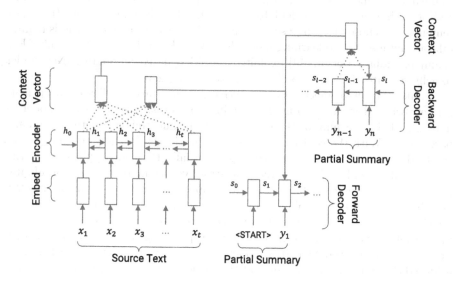

Fig. 1. BiSum is an Seq2Seq-Attn model with bidirectional decoder.

To sum up, the main contributions of our work include: (1) apply the bidirectional decoder into the abstractive summarization for the first time to avoid the accumulated errors; (2) integrate the pointer mechanism into the forward decoder and the backward decoder to solve the OOV problem; (3) propose not to segment the words in the source documents, which improves the quality of generated summary a lot; (4) verify the effectiveness and universality of our model on both Chinese and English datasets.

We organize this paper in the following sections. Section 2 summarizes the related work of text summarization and bidirectional decoder, and compares it with our work. Section 3 introduces the basic Seq2Seq-Attn model and defines BiSum on this basis. Experimental results, sample summary and the analysis of them are in the Sect. 4. We conclude our work in the last section, and propose possible directions for future improvement.

2   Related Work

2.1   Text Summarization

Text summarization problem studies how to automatically obtain a summary from the source texts. At present, two methods are mainly used for the problem: extractive and abstractive.

Extractive summarization is to select some sentences from the original text to summarize it. It guarantees the readability of the summary, but cannot maintain the logic between sentences, and there may be confusion in referential relation. The early extractive works [2,5] utilized the position of the sentence, cue word and other information to calculate the weight of each sentence, and the top-n sentences are chosen to compose a summary. Modern extractive methods mainly take advantage of deep learning methods [3,14,16] to achieve better results. For example, Nallapati et al. [14] exploited the GRU-based SummaRuNNer model to achieve state-of-the-art results in extractive summarization.

Unlike extractive methods, abstractive summarization relies on deep neural network and is closer to human thinking patterns. It requires the model to understand the meaning of the entire document and thus generate a summary based on it. With the continuous development of deep learning technologies in the field of natural language processing, especially the wide application of the Seq2Seq-Attn model in recent years, abstractive summarization has gradually become a new research hotspot. Since the input and output lengths often differ greatly, summarization problem is subdivided into the sentence-level summarization (generate headline based on the sentence) and the paragraph-level summarization (generate sentences based on the paragraph). Rush et al. [18] first proposed applying Seq2Seq-Attn model in abstractive summarization field and achieved the state-of-the-art results at that time on the DUC-2004 and Gigaword (two sentence-level summarization datasets).

In recent years, the optimization of this model is springing up. Zhou et al. [23] presented a sentence-level summarization method to copy the phrases in the source document from the model output in bundles. See et al. [19] introduced a paragraph-level summarization method to copy the input sequence into the output sequence by considering the attention distribution of the input sequence. Both of the above tasks are trying to solve the OOV problem, and they have achieved good results in the field. See et al. [19] proposed a coverage mechanism to reduce the weight of words that has been taken care of in the attention mechanism. Salesforce [17] put forward a self-attention mechanism to reduce the probability of generating duplicate words. Tan et al. [20] raised a hierarchical encoder and used a graph-based attention mechanism to decrease the probability of focusing on the same part of the source document. All three mentioned above attempted to prone the duplicate words in paragraph-level summarization. (There are a few duplicates in sentence-level summarization, because the generated headlines are short enough.)

In general, the current sentence-level summarization has been basically readable and has outperformed the extractive methods in some datasets. However, readability of existing paragraph-level summarization models is still not satisfying. There is still much room for improvement. Meanwhile, all of the above paragraph-level summarization works are based on the English datasets CNN/Daily Mail. Hu et al. [9] crawled news data from Sina Weibo to build a large-scale Chinese sentence-level summarization datasets, LCSTS. Chinese paragraph-level summarization work and datasets are rare nowadays [8], which does limit the development of it.

2.2 Bidirectional Decoder

In this work, we mainly use a bidirectional decoder to perform the summarization. Bidirectional decoder has been a pop topic in the field of machine translation and has been widely used in Neural Machine Translation (NMT) system. Watanabe et al. [21] first referred to the translation results generated in the forward and backward directions in the NMT. In recent years, the idea of bidirectional decoder has been continuously improved. Liu et al. [11] trained two bi-LSTM models at the same time, and then directly added the predict distributions of two models together to generate translations. Hoang et al. [7] proposed an approximate inference framework based on continuous optimization to decode the bidirectional model. Zhang et al. [22] used two independent attention distribution to let the forward decoder pay attention to the backward decoder.

To the best of our knowledge, our work is the first one to use the bidirectional decoder in the field of abstractive summarization. The bidirectional decoding mechanism in this paper is similar to the work of Zhang et al. [22]. The main difference between two works: (1) We add a pointer mechanism to both forward and backward decoders to avoid the OOV problem; (2) Different quality requirements for the summary generated at the backward decoder side. Duplicate output is common in abstractive summarization, but is rare in machine translation. Therefore, the output of backward decoder is stressed in BiSum.

3   Our Model

This section introduces the basic Seq2Seq-Attn model, then the definition of BiSum[1] and other tricks in it.

3.1   Seq2Seq-Attn

Seq2Seq-Attn is a regular solution to the abstractive summarization problem [12, 15,18] and the method here is similar to [15]. Figure 2 describes this process.

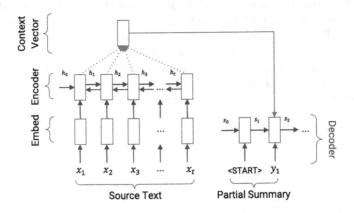

Fig. 2. The Seq2Seq-Attn model.

Seq2Seq architecture can be divided into the encoder and decoder. Encoder is usually a bidirectional, recurrent neural network (single-layer bi-LSTM in BiSum) that encodes the source text into a semantically hidden state vector h_i (i in h_i represents the ith word in the sequence). The decoder is usually a recurrent neural network (single-layer LSTM in BiSum). It obtains the decoder's current state s_t based on the decoder's hidden state vector s_{t-1} at the previous time step. It should be noted that the decoder input y will be the words of referred summary during training but the words of generated summary during testing. This is the reason why error accumulation occurs when testing mentioned in Sect. 1.

We use the same attention mechanism as Bahdanau et al. [1], so that each encoder's hidden state vector h_i has an effect on every word generated by the decoder, but the attention distribution is different from each other. It can be derived as Eq. 1.

$$e_i^t = v^T \tanh(W_h h_i + W_s s_{t-1} + b_{attn})$$
$$a_t = softmax(e_i^t) \qquad (1)$$
$$c = \sum_i a_i^t h_i$$

---

[1] https://github.com/AnnieAldo/BiSum.

where v, W_h, W_s, b_{attn} are the learnable variables, a_t is the attention distribution of h_i, and c is the context vector. c is a weighted sum of h_i and can be viewed as a collection of source text hidden state vectors of interest.

According to the context vector c and the decoder current state s_t, we can calculate $P_{vocab}(w)$ (the probability distribution of our predicted word w in the vocabulary), which also can be $P(w)$ (the probability distribution of the decoder output word).

$$P_{vocab} = softmax(V'(V[s_t, c] + b) + b')$$
$$P(w) = P_{vocab}(w) \tag{2}$$

In Eq. 2, V', V, b, b' are the learnable variables, and P_{vocab} is the probability distribution of all vocabulary words, which is calculated through two linear layers. Assuming there is a training datasets $D = (x, y)$, where x is the source document and y is the reference summary, the objective function is defined as Eq. 3.

$$J(D) = \frac{1}{|D|} argmax \sum_{(x,y) \in D} \log P(w) \tag{3}$$

3.2 BiSum

The bidirectional decoder was implemented in the field of machine translation for the past two years [7,11,22] and can effectively solve the problem of error accumulation on the decoder-side. However, as far as we know, this article is the first work to apply it to abstractive summarization. The bidirectional decoder in this paper mainly refers to the work of Zhang et al. [22] in machine translation.

Encoder. BiSum has exactly the same encoder as Seq2Seq-Attn model. Attention distribution of the encoder is also consistent with the Eq. 1.

Backward Decoder. Compared to the Seq2Seq-Attn model, BiSum has one more backward decoder. That is, before the summary is generated from left to right, a reverse summary is generated from right to left. Without considering the next steps, the encoder and the backward decoder can be understood as a Seq2Seq-Attn model, and only the direction of the generated summary is different from it. Therefore, the probability distribution of a vocabulary word generated by forward decoder is defined as Eq. 4.

$$\hat{P}(w) = softmax(V'(V[s_t, c] + b) + b') \tag{4}$$

where hat symbol $\hat{}$ represents operations related to the backward decoder, such as $\hat{P}(w)$ here.

Forward Decoder. Combining the results of the encoder and the backward decoder, we can further calculate the results of the forward decoder. At this point, we apply the attention mechanism to both the encoder and the backward decoder. For the encoder, the attention distribution a_t can be calculated as Eq. 5.

$$e_i^t = v^T \tanh(W_h h_i + W_s s_{t-1} + b_{attn})$$
$$a_t = softmax(e_i^t) \tag{5}$$
$$c = \sum_i a_i^t h_i$$

where v, W_h, W_s, b_{attn} are the learnable variables, and s_{t-1} is the hidden state vector of the forward decoder at the previous time step. a_t is the attention distribution of h_i, and c is the context vector.

For the backward decoder, the attention distribution $\hat{a}_t$ is:

$$\hat{e}_{t'}^t = \hat{v}^T \tanh(W_{\hat{s}} \hat{s}_{t'} + W_s' s_{t-1} + \hat{b}_{attn})$$
$$\hat{a}_t = softmax(\hat{e}_{t'}^t) \tag{6}$$
$$\hat{c} = \sum_{t'} \hat{a}_{t'}^t \hat{s}_{t'}$$

where $\hat{v}, W_{\hat{s}}, W_s', \hat{b}_{attn}$ are the learnable variables. $\hat{s}_{t'}$ is the hidden state vector of backward decoder at t'th time step, and s_{t-1} is the hidden state vector of forward decoder at the previous time step. It should be noted here that when we utilize the attention mechanism to the backward decoder, the attention distribution is based on its hidden state vector $\hat{s}_{t'}$. Otherwise, if the mechanism is applied to the generated reverse summary, the error will affect the forward decoder more easily, which we do not wish to.

Founded on two attention mechanisms, we can derive the final probability distribution of the predicted word w like Seq2Seq-Attn model.

$$P(w) = softmax(V'(V[s_t, c, \hat{c}] + b) + b') \tag{7}$$

where V', V, b, b' are the learnable variables.

Finally, in order to guarantee the quality of the summary produced by the backward decoder, we reconstructed the objective function. The maximum likelihood of the backward decoder is added to the objective function and is balanced with the hyperparameter λ.

$$J(D) = \frac{1}{|D|} argmax \sum_{(x,y) \in D} \left[\lambda \cdot \log P(w) + (1 - \lambda) \cdot \log \hat{P}(\hat{w}) \right] \tag{8}$$

In general, when we train a model incorporating a bidirectional decoder, we will first encode the input sequence, and then use a backward decoder to generate a reverse summary, and finally, employ a forward decoder to produce summary using the information provided by the former encoder and the backward decoder. Since the time complexity of beam search is too large, we use a smaller beam size in backward decoder to generate the reverse summary, which speeds up the model to some extent.

Pointer Mechanism. Pointer mechanism is used in the forward and backward decoder. It uses a soft attention distribution mechanism to make the output sequence map the input sequence, and can be very helpful to solve the OOV problem. The pointer mechanism used in this article is the same as that of See et al. [19], that is, controls the model's generating and pointing by p_{gen}. $p_{gen} \in [0,1]$ is defined as a probability of generating a word from the vocabulary. According to this definition, $(1 - p_{gen})$ represents the probability of copying a word from the input. The p_{gen} at time step t is calculated as shown in Eq. 9.

$$p_{gen} = \sigma(w_c^T c + w_s^T s_t + w_y^T y_t + b_{ptr}) \tag{9}$$

where w_c, w_s, w_y, b_{ptr} are learnable parameters, and y_t is the input of decoder. σ is the sigmoid function. Thus, we have got an extended vocabulary that contains all words in the vocabulary and all words that can be copied from the original texts. Considering p_{gen} as a soft switch, the probability distribution of this extended vocabulary is shown in Eq. 10.

$$P(w) = p_{gen} P_{vocab}(w) + (1 - p_{gen}) \sum_{i:w_i=w} a_i^t \tag{10}$$

No Word Segmentation. Intuitively speaking, when doing Chinese processing, we segmented both the source text and the referred summary word by word. However, in the following experiments, we find that if we remove the word segmentation part and train the text sequence character by character, the results of the model are significantly better regardless of the model we used. This is due to the fact that when we segment the corpora, we naturally get a very large vocabulary, which is detrimental to smaller datasets and can lead to sparse data. But when we generate the summary character by character, the vocabulary size is reduced, and the relevance between characters is strengthened. Considering that there are no large-scale Chinese paragraph-level summarization datasets in academic circles and industrial circles, it is a good way to improve the summary quality.

4 Experiments

4.1 Datasets

We evaluated our model on two paragraph-level summarization datasets. They are NLPCC 2017 TTNews in Chinese and CNN/Daily Mail in English.

TTNews. It contains a large number of news articles from Toutiao.com and corresponding manual summaries for news feeds. In addition, the training set also contains another set of news articles without summaries (perhaps provided for semi-supervised methods, not used in this paper). As far as we know, TTNews is the largest single-document paragraph-level summarization corpus in Chinese, with 50,000 news articles containing summaries and 52,000 news articles without summaries.

CNN/Daily Mail. This corpus was recently widely used in the paragraph-level summrization field. [6,15,17,19] It contains news articles in CNN/Daily Mail and manual summaries for them. We used the data preprocessing scripts provided by Nallapati et al. [15] to obtain 312,084 article and summary pairs. We did not use Named Entity Recognition [13] (NER) technology to replace proper nouns, because of the pointer mechanism in the forward and backward decoder.

The specific statistics of the two corpus are shown in Table 1.

Table 1. Information of the two datasets TTNews and CNN/daily mail.

Datasets		The number of documents	The average length of documents	The average length of summaries
TTNews	Training set (with summary)	50,000	1,036	45
	Training set (without summary)	50,000	1,526	/
	Test set	2,000	1,037	45
CNN/ daily mail	Training set	287,226	781	56
	Test set	13,368	781	/
	Validation set	11,490	781	56

For the CNN/Daily Mail datasets, the file format is .story, and for the TTNews datasets is .txt. We format them and further convert them into binary files. Also, for the convenience of training, every 1000 samples are integrated as a chunk.

4.2 Baseline

Seq2Seq-Attn. [15] The Seq2Seq-Attn model mentioned in Sect. 3.

Seq2Seq-Attn + Pointer. [19] The Seq2Seq-Attn model with the pointer mechanism mentioned in Sect. 3. It's worth mentioning that we generated the summary from left to right and from right to left, respectively, to ensure the effectiveness of the bidirectional decoder.

4.3 Setup

For the TTNews datasets, due to the absence of the referred summary test set, we first separate 5000 samples from the training set into the test set and 5000 samples into the validation set. Meanwhile, because the word segmentation does

Table 2. Experimental results of the baseline and BiSum.

Datasets	Model/evaluation	ROUGE		
		1	2	L
CNN/ daily mail	Seq2Seq-Attn	30.49	11.17	28.08
	Seq2Seq-Attn + Pointer (left to right)	36.44	15.66	33.42
	Seq2Seq-Attn + Pointer (right to left)	35.46	15.30	33.28
	BiSum	**37.01**	**15.95**	**33.66**
TTNews (segmented)	Seq2Seq-Attn	32.71	15.42	29.26
	Seq2Seq-Attn + Pointer (left to right)	35.03	**18.03**	30.38
	Seq2Seq-Attn + Pointer (right to left)	34.59	17.51	30.14
	BiSum	**35.18**	17.70	**30.61**
TTNews (non-segmented)	Seq2Seq-Attn	36.43	21.17	30.41
	Seq2Seq-Attn + Pointer (left to right)	39.85	23.69	32.52
	Seq2Seq-Attn + Pointer (right to left)	39.14	23.62	32.38
	BiSum	**40.89**	**25.04**	**34.97**

not take place, the vocabulary size is greatly reduced. For the data with word segmentation, vocabulary size is set to be 60k. And for the data without word segmentation, vocabulary size is 6k.

For the CNN/Daily Mail datasets, Seq2Seq-Attn model uses a vocabulary size of 150k (because it encounters the OOV problem), and other models have 50k words in the vocabulary.

In all the experiments in this paper, the dimension of the hidden state vector is 256, and the dimension of the word embedded vector is 128. Instead of using the pre-trained word vectors, we learn them with other parameters during training. Adagrad [4] is used for training with a learning rate of 0.15 and an initial accumulator value of 0.1. The maximum gradient norm is configured to be 5. To prevent over-fitting, we achieved early stop by observing the losses on the validation set.

Since limiting the length of documents and summaries speeds up the training and testing and improves the performance of the model [19], we limit document length to 400 words and the summary length to 100 words in training, 120 words in testing. The model is trained on a single GTX 1080 Ti GPU, and the batch size is 16. Beam search is used to generate the summary with width 2 in backward decoder and width 4 in forward decoder.

For the Seq2Seq-Attn model, we trained about 236 K iterations on the CNN/Daily Mail dataset which spent 1 day 17 h. Segmented TTNews is trained about 125k iterations which took 14 h 40 min, and 85k iterations took 8 h 40 min for non-segmented datasets. For BiSum, English, segmented and non-segmented Chinese datasets trained about 103K, 160k, and 90k iterations and took 21 h 7 min, 22 h 40 min, and 18 h 58 min, respectively.

4.4 Effectiveness Experiments

The experimental results are shown in Table 2. We use the standard ROUGE metric [10] to evaluate BiSum, and give the F1 scores for ROUGE-1, ROUGE-2,

Source Document
-lrb- cnn -rrb- anthony ray hinton is thankful to be free after nearly 30 years on alabama's death row for murders he says he didn't commit. and incredulous that it took so long. hinton, 58, looked up, took in the sunshine and thanked god and his lawyers friday morning outside the county jail in birmingham, minutes after taking his first steps as a free man since 1985. he spoke of unjustly losing three decades of his life, under fear of execution, for something he didn't do. "all they had to do was to test the gun, but when you think you're high and mighty and you're above the law, you don't have to answer to nobody," hinton told reporters. "but i've got news for you -- everybody that played a part in sending me to death row, you will answer to god." jefferson county circuit court judge laura petro had ordered hinton released after granting the state's motion to dismiss charges against him. hinton was convicted of murder in the 1985 deaths of two birmingham-area, fast-food restaurant managers, john davidson and thomas wayne vason. but a new trial was ordered in 2014 after firearms experts testified 12 years earlier that the revolver hinton was said to have used in the crimes could not be matched to evidence in either case, and the two killings couldn't be linked to each other. (...)

Referred Summary
anthony ray hinton goes free friday, decades after conviction for two murders. court ordered new trial in 2014, years after gun experts testified on his behalf. prosecution moved to dismiss charges this year.

Seq2Seq-Attn
"i can't believe that i can't believe," he says. new: "we don't have to do so," he says. new: "i can't believe that i can't do anything," he says.

Seq2Seq-Attn + Pointer (Left to Right)
anthony ray hinton is thankful to be free after nearly 30 years on alabama's death row for murders. the state race, poverty, he didn't commit. declined "race, poverty, he didn't commit. everybody "race, poverty, he didn't commit. hinton "race, poverty, i didn't commit.

Seq2Seq-Attn + Pointer (Right to Left)
he says he didn't commit. you don't have to answer to nobody," hinton told reporters. you don't have to answer to nobody," hinton told reporters. but a new trial was ordered in 2014 after firearms experts testified 12 years earlier that the two killings couldn't be linked to each other.

BiSum
anthony ray hinton is thankful to be free after nearly 30 years on alabama 's death row for murders he says he didn't commit. hinton was convicted of murder in the 1985 deaths of two birmingham-area, fast-food restaurant managers, john davidson and thomas wayne vason.

Fig. 3. Sample English summaries of the models. Texts in blue show the correct referred summary. Texts in green indicate that although they are inconsistent with the referred summary, they can be subjectively interpreted as a correct summary. (Color figure online)

and ROUGE-L (character overlap, 2-grams overlap, and longest common subsequence overlap for the generated summary and referred summary, respectively).

Our work outperforms the baseline in most cases, which shows that bidirectional decoders do play a positive role in the summarization. It can be also found that the performance of the two Seq2Seq-Attn + Pointer models are always similar, but the ROUGE scores of left-to-right model are often higher. We speculate that the model is more likely to focus on the first sentences of the source document when summary is generated from left to right, so that characteristic of the news datasets (the first few sentences are likely to be a good summary) leads to this phenomenon. In all experiments, the performance of non-segmented models is better than segmented models.

4.5 Case Study

We showed some sample summaries in the Figs. 3 and 4. Obviously, reverse summaries generated by the right-to-left pointer model tend to have better performance at the tail of the summaries, which are in line with our expectations and form a complementary with the left-to-right summaries. However, the summaries are sometimes likely to repeat themselves, and this problem also appears in BiSum. We will try to address this problem in further work.

Source Document
楚天都市报讯<Paragraph>本报记者余皓<Paragraph>湖北日报大学生记者团杨帆新洲区一精神病医院发生惨案: 一精神病人竟趁病友熟睡时, 用湿纸巾捂其口鼻, 又用被褥毛巾捂头脸致其窒息身亡。昨日, 涉嫌故意杀人的黄某在武汉中院受审, 他为自己辩护称, "这是看电影看书学来的, 只想玩玩他没想到他就死了"。精神病院凌晨杀人案今年27岁的黄某是新洲人, 患精神分裂症长达9年, 父母于去年底将他送至郏城街刘集精神病医院。黄某与病友方某同住在三楼大号病房, 病房住有30多个病人。按病人反映, 黄某初贝病房就惹事, 因有病友不给烟抽就罚跪地, 还对病友掌掴。检方指控, 案发前黄某向方某要烟抽未果, 怀恨在心。去年12月17日凌晨6时许, 黄某趁方某熟睡之机, 用被褥毛巾等物长时间捂住方的头、面部, 压扼其颈部。方某经医院抢救无效死亡, 经鉴定方某系被他人捂住口鼻致机械性窒息死亡。(...)

Referred Summary
武汉: 一精神病人"玩"死病友, 涉嫌故意杀人受审, 辩称"这是看电影看书学来的, 只想玩玩他没想到他就死了"。

Seq2Seq-Attn (Segmented)
武汉一精神病患者欲举报物业[UNK]硫酸[UNK]当场身亡; 死前为自己好友[UNK]因[UNK][UNK][UNK][UNK][UNK]精神疾病[UNK]留遗书称男子患有精神分裂症[UNK]已索赔(图)

Seq2Seq-Attn + Pointer (Segmented, Left to Right)
徐州: 精神病发生惨案: 一精神病人竟趁病友熟睡病友熟睡病友熟睡病友熟睡病友, 窒息身亡。窒息身亡。身亡; 证词病友。因有男子群殴

Seq2Seq-Attn + Pointer (Segmented, Right to Left)
青岛一精神病患精神病患精神分裂症长达9年[UNK]父母于去年底将其送至医院[UNK]医院抢救无效死亡; 医院抢救无效死亡。

BiSum (Segmented)
武汉: 精神病人杨帆新洲区一精神病医院受审, 疑因趁病友熟睡时, 用湿纸巾鼻[UNK]脸致其窒息身亡, 近日被检方指控其窒息身亡。详细

Seq2Seq-Attn (Non-segmented)
楚皓院凌晨杀人用湿纸巾贴其口鼻, 用湿纸巾贴其口鼻, 用湿纸巾贴其口鼻, 用湿纸巾贴其口鼻, 机械性窒息死亡。

Seq2Seq-Attn + Pointer (Non-segmented, Left to Right)
团记鉴团杨帆新洲区一精神病人竟趁病友熟睡时, 用湿纸巾贴其口鼻, 又用被褥毛巾捂头脸致其窒息身亡。

Seq2Seq-Attn + Pointer (Non-segmented, Right to Left)
武汉涉长黄生记者团杨帆新洲区一精神病人竟趁病友熟睡时, 用湿纸巾贴其口鼻, 又用被褥毛巾捂头脸致其窒息身亡。

BiSum (Non-segmented)
武汉一精神病人趁病友熟睡时, 用湿纸巾鼻被捂头脸致其窒息身亡, 目前已被警方刑事拘留。

Fig. 4. Sample Chinese summaries of the models. Texts in red point to the source of error. (Color figure online)

In CNN/Daily Mail datasets, we can find that although the summary generated by BiSum is not exactly the same as the referred summary, the sentences they attended to are adjacent. Since summarization is a subjective work, it can be considered that BiSum produces a meaningful summary here. In TTNews datasets, the results of the segmented model are significantly more prone to bias. That is because the vocabulary is too large but the datasets are small: when the two documents have a same word, the decoder may generate the words in another document.

5   Conclusion and Further Work

Based on the traditional Seq2Seq-Attn model, this work introduces a bidirectional decoder for the problem of error accumulation when generating summaries. We added the pointer mechanism and remove the word segmentation for the datasets (only for TTNews), which effectively reducing the vocabulary size and improving the model's performance. The experimental results indicate that our model can obtain remarkable results in Chinese and English summarization tasks. Also, we find that there are still repeated words appearing in the generated abstract, and we hope to deal with this problem in the follow-up work.

Acknowledgement. This work is supported in part by the National Natural Science Foundation of China (No. 61772082, 61806020, 61375058), and the Beijing Municipal Natural Science Foundation (4182043).

References

1. Bahdanau, D., Cho, K., Bengio, Y.: Neural machine translation by jointly learning to align and translate. arXiv preprint arXiv:1409.0473 (2014)
2. Baxendale, P.B.: Machine-made index for technical literature-an experiment. IBM J. Res. Dev. **2**(4), 354–361 (1958)
3. Cao, Z., Wei, F., Dong, L., Li, S., Zhou, M.: Ranking with recursive neural networks and its application to multi-document summarization. In: AAAI, pp. 2153–2159 (2015)
4. Duchi, J., Hazan, E., Singer, Y.: Adaptive subgradient methods for online learning and stochastic optimization. J. Mach. Learn. Res. **12**(Jul), 2121–2159 (2011)
5. Edmundson, H.P.: New methods in automatic extracting. J. ACM (JACM) **16**(2), 264–285 (1969)
6. Hermann, K.M., Kocisky, T., Grefenstette, E., Espeholt, L., Kay, W., Suleyman, M., Blunsom, P.: Teaching machines to read and comprehend. In: NIPS, pp. 1693–1701 (2015)
7. Hoang, C.D.V., Haffari, G., Cohn, T.: Towards decoding as continuous optimisation in neural machine translation. In: EMNLP, pp. 146–156 (2017)
8. Hou, L., Hu, P., Bei, C.: Abstractive document summarization via neural model with joint attention. In: Huang, X., Jiang, J., Zhao, D., Feng, Y., Hong, Y. (eds.) NLPCC 2017. LNCS (LNAI), vol. 10619, pp. 329–338. Springer, Cham (2018). https://doi.org/10.1007/978-3-319-73618-1_28
9. Hu, B., Chen, Q., Zhu, F.: LCSTS: a large scale Chinese short text summarization dataset. arXiv preprint arXiv:1506.05865 (2015)
10. Lin, C.Y.: Rouge: a package for automatic evaluation of summaries. ACL workshop, Text Summarization Branches Out (2004)

11. Liu, L., Utiyama, M., Finch, A., Sumita, E.: Agreement on target-bidirectional neural machine translation. In: Proceedings of the 2016 Conference of the North American Chapter of the Association for Computational Linguistics: Human Language Technologies, pp. 411–416 (2016)
12. Lopyrev, K.: Generating news headlines with recurrent neural networks. arXiv preprint arXiv:1512.01712 (2015)
13. Nadeau, D., Sekine, S.: A survey of named entity recognition and classification. Lingvisticae Investig. **30**(1), 3–26 (2007)
14. Nallapati, R., Zhai, F., Zhou, B.: SummaRuNNer: a recurrent neural network based sequence model for extractive summarization of documents. In: AAAI, pp. 3075–3081 (2017)
15. Nallapati, R., Zhou, B., Gulcehre, C., Xiang, B., et al.: Abstractive text summarization using sequence-to-sequence RNNs and beyond. arXiv preprint arXiv:1602.06023 (2016)
16. Narayan, S., Papasarantopoulos, N., Cohen, S.B., Lapata, M.: Neural extractive summarization with side information. arXiv preprint arXiv:1704.04530 (2017)
17. Paulus, R., Xiong, C., Socher, R.: A deep reinforced model for abstractive summarization. arXiv preprint arXiv:1705.04304 (2017)
18. Rush, A.M., Chopra, S., Weston, J.: A neural attention model for abstractive sentence summarization. arXiv preprint arXiv:1509.00685 (2015)
19. See, A., Liu, P.J., Manning, C.D.: Get to the point: summarization with pointer-generator networks. In: Proceedings of the 55th Annual Meeting of the Association for Computational Linguistics, vol. 1, pp. 1073–1083 (2017)
20. Tan, J., Wan, X., Xiao, J.: Abstractive document summarization with a graph-based attentional neural model. ACL **1**, 1171–1181 (2017)
21. Watanabe, T., Sumita, E.: Bidirectional decoding for statistical machine translation. In: Proceedings of the 19th International Conference on Computational Linguistics-Volume 1, pp. 1–7. Association for Computational Linguistics (2002)
22. Zhang, X., Su, J., Qin, Y., Liu, Y., Ji, R., Wang, H.: Asynchronous bidirectional decoding for neural machine translation. arXiv preprint arXiv:1801.05122 (2018)
23. Zhou, Q., Yang, N., Wei, F., Zhou, M.: Sequential copying networks. In: AAAI (2018)

Miscellaneous Topics

CBPF: Leveraging Context and Content Information for Better Recommendations

Zahra Vahidi Ferdousi$^{(\boxtimes)}$, Dario Colazzo, and Elsa Negre

Paris-Dauphine University, PSL Research University, CNRS UMR 7243, LAMSADE, Paris, France
{zahra.vahidiferdousi,dario.colazzo,elsa.negre}@dauhphine.fr

Abstract. Recommender systems (RS) help users to find their appropriate items among large volumes of information. Among the different types of RS, context-aware recommender systems aim at personalizing as much as possible the recommendations based on the context situation in which the user is. In this paper we present an approach integrating contextual information into the recommendation process by modeling either item-based or user-based influence of the context on ratings, using the Pearson Correlation Coefficient. The proposed solution aims at taking advantage of content and contextual information in the recommendation process. We evaluate and show effectiveness of our approach on three different contextual datasets and analyze the performances of the variants of our approach based on the characteristics of these datasets, especially the sparsity level of the input data and amount of available information.

Keywords: Context-aware recommender system
Contextual information integration · Pre-filtering recommender system

1 Introduction

Nowadays, we are faced with a rise of the amount of data on the web, provided by different sources. As a consequence, a user can quickly be overwhelmed by the huge volume of information. *Recommender systems (RS)* [9] aim to help the user to find her appropriate information among all others. Traditional recommender systems have proved their effectiveness in different areas [12], but they have the limitation of not considering the contextual situation in which the user is, at the moment she wants to use the item. In fact this information can roughly influence her preferences for items [2]. As an example, when choosing a movie to watch, the user will have different preferences depending on whether she wants to watch the movie with a kid or with her partner. In this case, a context-aware recommender system (CARS), integrating such contextual information about the user in the recommendation process, can provide more relevant recommendations [1]. A particular class of CARSs are based on *pre-filtering*, based on the idea of pre-procesing contextual data so as to tune the input of a given (traditional) RS in order to increase its effectiveness. Along the lines of a preliminary investigation

© Springer Nature Switzerland AG 2018
G. Gan et al. (Eds.): ADMA 2018, LNAI 11323, pp. 381–391, 2018.
https://doi.org/10.1007/978-3-030-05090-0_32

presented in the workshop paper [15], where we proposed a pre-filtering CARS that integrates contextual information about users by modeling them with item-based influence of context on ratings, in this paper we propose the user-based version of this approach, and we present here results on a much more extensive experimental analysis. With respect to CARS state of the art (discussed later on) our approach, named Correlation-Based Pre-Filtering is, in a sense, more user-centric, as we propose to model item/user-based influence based on the item- or user-based Pearson Correlation Coefficient (PCC) [5] between context and ratings. The distinctive feature of using PCC allows us to catch more precisely the influence of context on ratings, and so to compute more accurate similarities between contexts, which is a crucial point in our pre-filtering process. In addition, we use content information about items/users to improve our model, like, for instance, the category of a film or the age or gender of users. Our experimental analysis on three typically used datasets show improvements over state of the art approaches.

With respect to our preliminary investigation presented in the workshop paper [15], in this paper our new contributions are the followings: we propose to model the context by relying on the user-based influence of contexts on ratings; we compare the item- and user-based approaches, and study the cases where each one of these versions can perform the best. In our experimental analysis we use three different datasets from three different domains to highlight previously unseen properties. And we demonstrate that our approach can deal well with either sparse or dense data.

The remainder of the paper is organized as follows: in the next section we present a state of the art of the subject. In Sect. 3 we describe our approach. In Sects. 4 and 5 we respectively describe the setup and results of our experimental analysis. Finally we discuss our results and make conclusive remarks.

2   Related Work

CARSs aim to take into account the users' contextual information, in the most efficient way, in order to propose more relevant and personalized recommendations [2]. So instead of the 2D rating function of traditional RSs ($R : user \times item \rightarrow rating$), in CARSs we have the multidimensional function, $R : user \times item \times context \rightarrow rating$ [1]. The context of a user is composed of a number of context factors like *time, location, weather, companion, etc.* To each one of these context factors some values can be associated, called context conditions (e.g. possible context conditions for *time* could be *morning, afternoon, evening and night*).

The integration of contextual information in CARSs can be done by relying on either *pre-filtering*, *post-filtering* or *contextual modeling* [2]. Several approaches have been proposed for each one of these categories. Among them, we are especially interested into the following approaches: *DSPF, deviation-based and similarity based CAMF and DCM*, which are the most similar to our approach. In fact, they try to model the influence of the context on their model, but based on different points of view: *DSPF* [7] models the influence of context on ratings based on the difference between context-free rating and the rating given in the specific context. But, differently from our technique, this influence computation is not user-centric enough because of the way the context-free rating is estimated (for more explanations please refer to the extended version of the paper [14]).

CAMF [4] is an extension of matrix factorization [10]. The *deviation-based* version tries to take into account the context situation of users by integrating additional model parameters in the matrix factorization equation. And the *similarity-based* version integrates a similarity function that estimates the similarity between a contextual situation and a non-contextual situation. These CAMF approaches proved their effectiveness to improve recommendation performance in comparison to context-free recommendation and some of the context-aware recommendation approaches, but like other contextual modeling approaches, and differently from ours, they have the disadvantage of needing to be implemented from scratch, with no possibility of re-using recommendation techniques already in production.

DCM [16] is based on the user-based collaborative filtering algorithm [13]. The authors propose to separate the algorithm into different functional components, and apply differential context constraints to each component, in order to maximize the performance of the whole algorithm. Differently, our approach try to model the context from a different point of view.

3   Methodology

In this paper we propose a new pre-filtering approach, based on the influence of context on ratings, by modeling it based on the user-based correlation between context and ratings, computed by the Pearson Correlation Coefficient.

A recommendation problem is often viewed as a matrix/tensor completion problem. A recommender system will firstly estimate missing ratings, and then it will recommend to each user her corresponding items with higher estimated rates. In the case of pre-filtering CARSs, we want to integrate the contextual information into the estimation phase of missing ratings. Our correlation-based pre-filtering approach, like the reduction-based pre-filtering approach [1], makes the hypothesis that a user will rate an item similarly in two similar contexts. Based on this hypothesis, to recommend an item to a user in a specific context, we can identify ratings given in similar contexts of this specific context, and apply a traditional 2D recommendation technique on this selection. The whole recommendation process can be decomposed in five steps:

Step 1: To be able to find similar contexts, we need a strong representation of context. In [15] we proposed to represent contexts based on their item-based influence on ratings. In fact, the context can influence the ratings differently, according to items. For example in the case of points of interest recommendation, a snowy weather will have a positive influence on some winter sport centers, but a negative influence on natural parks. This is why it is important to compute this influence according to items.

In this paper we propose to represent contexts based on their user-based influence on ratings. Indeed, we can say that the influence of context on ratings also depends on users, and will differ from one user to another. For example, a "family person" could like to practise activities with her family, whereas another person may not like this and prefer to practise activities with her friends. So the social context will influence differently these two persons.

We can compute this influence by calculating the Pearson Correlation Coefficient (PCC) of the rating variable r, and each context condition variable c_j, with $j \in [1, n]$, where n is the total number of context condition variables. We choose this correlation measure because in statistics, PCC is widely used to measure the strength of linear association between two variables, and this corresponds to what we want, since we want to catch the influence of context conditions on ratings.

In a context-aware environment, an observation will be the cross-tabulation of the variables of user, item, rating and m different context factors (e.g. *daytype, season, location, social, etc*). To apply PCC, we transform context factors into binary variables. So let us denote with $X_t = (u_t, i_t, r_t, c_{1t}, c_{2t}, ..., c_{nt})$ the tth observation, which represents the evaluation r_t of the user u_t for the item i_t in the context situation $c_{1t}, c_{2t}, ..., c_{nt}$, where as said before, n is the total number of context conditions, and $c_{mt} = 1$ means that the m-th context condition is present in the context of the user, and $c_{mt} = 0$ means that it is not present. For instance, in a movie RS with a notation from 1 to 5 stars, $X_1 = $ *(John, Star Wars, 4, weekend = 1, workingday = 0, holiday = 0, summer = 1, winter = 0, spring = 0, autumn = 0, home = 1, public_place = 0, friend's home = 0, alone = 1, partner = 0, friends = 0)* means that *John* had evaluated the movie *Star Wars* by *4* stars, when he watched the movie *alone*, at *home* in a *weekend* of *summer*.

So the user-based correlation between the rating r and a context condition c_j is calculated as follows in Eq. 1.

$$w_{c_j u} = PCC_u(r, c_j) = \frac{\sum_{k \in K}(r_k - \overline{r_u})(c_{jk} - \overline{c_u})}{\sqrt{\sum_{k \in K}(r_k - \overline{r_u})^2}\sqrt{\sum_{k \in K}(c_{jk} - \overline{c_u})^2}} \tag{1}$$

where K is the set of observations $X_k = (u, i_k, r_k, c_{1k}, c_{2k}, ..., c_{nk})$ with user u, $\overline{r_u}$ is the mean of the ratings given by the user u, while $\overline{c_u}$ is the mean value of the context condition c over observations for user u.

Based on the above explanations, we can build a vector representation for each context condition. The size of this vector will be the total number of users, and the values (between -1 and 1) of this vector are equal to the user-based PCC between the rating vector and the binary context condition vector.

In real world recommendation problems, the total number of users is often very large, and the correlation calculation would be computationally consuming. To overcome this computational cost we propose to cluster users into a limited number of groups, and to compute the influence based on clusters of users. This clustering could be done based on the available static information about users' characteristics (e.g. age, sex, etc), or directly based on the ratings.

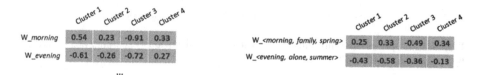

Fig. 1. Examples of representation of cluster-based context condition (Step 1)

Fig. 2. Examples of representation of context situation by aggregation (Step 2)

Figure 1 illustrates some examples of the resulting context condition representations.

Step 2: We can now represent each context situation based on its composing context conditions. This representation can be obtained in two ways:

- *Aggregation:* Each context situation can be represented by a vector with values equal to the mean aggregation of the values of its corresponding composed context condition, as illustrated in Fig. 2 (also used by [7]).
- *Concatenation:* In order to limit the risk of neutralizing the influence of each context condition by the aggregation [15], we can represent a context situation with a larger vector built from the concatenation of its composing context conditions (Fig. 3).

Step 3: Now, we can find the contexts most similar to the target context situation $s*$ by computing the similarity between every context situation s and the target context situation $s*$, based on the cosine similarity between their vector representations (Eq. 2, where d is the dimension of the context representation vector $\overrightarrow{w_s}$).

$$sim(s, s^*) = cosine(\overrightarrow{w_s}, \overrightarrow{w_{s*}}) = \frac{w_s^T w_{s*}}{\sqrt{\sum_{i=0}^d w_{s,i}^2} \sqrt{\sum_{i=0}^d w_{s*,i}^2}} \quad (2)$$

Step 4: We select the ratings given in the similar context situations, and make a *local dataset*.

Step 5: We then apply a traditional 2D recommendation technique on this selection of ratings (local dataset), to obtain recommendations (*local model*).

4   Experimental Analysis

4.1   Datasets

We evaluated our approach on three real world datasets, which are well-known among the CARS community: (a) *CoMoDa*, a contextual dataset for movie recommendation, collected from surveys [11], (b) *STS* [6], a tourism dataset, containing contextual ratings for places of interest, collected using a mobile tourist

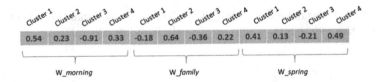

Fig. 3. Example of the representation of the context situation $W_{(morning, family, spring)}$ by concatenation (Step 2)

application, and (c) the *Music* dataset, contains ratings for contextual music recommendation, collected by an in-car music recommender developed by [3] (this dataset has the specificity that for each context situation, the value of only one context factor is known).

Table 1 illustrates some descriptive statistics about these datasets. Note that we calculated the sparsity by means of the following formula: $1 - \frac{\#ratings}{\#users \times \#items}$.

Table 1. Datasets' descriptive statistics

Characteristics	CoMoDa	STS	Music
#ratings	2296	2534	4012
#users	121	325	42
#items	1197	249	139
Rating scale	1–5	1–5	1–5
Rating's mean	3.83	3.47	2.37
Rating's median	4	4	2
Rating's standard deviation	1.05	1.29	1.48
Sparsity	98.41%	96.86%	31.27%
#context factors	12	14	8
#context conditions	49	59	26
#items characteristics	7	1	2
#users characteristics	2	7	0

4.2   Modeling and Evaluation Parameters

We cluster items/users based on their available characteristics (content information), by applying a Hierarchical Clustering. For a sake of space, you can find more details about some pre-treatments and modeling parameters in the extended version of the paper [14]. The traditional (context-free) recommendation technique used in the last step of our approach is the Biased Matrix Factorization model [10], which is one of the best-performing techniques reported in the state of the art [7] (from LibRec Java API [8]).

For the evaluation of our approach, we avoided to exclude items or users with low counts, in order to match as closely as possible the conditions of real recommendation applications. Due to the relatively small size of our dataset, we evaluated our approach based on 5-fold cross-validation. As many researches in the domain, we used MAE and RMSE metrics to evaluate the rating estimation.

5   Results and Discussion

We evaluated our approach in 3 steps: (a) we compared the performances of the derived versions of our approach with each other, (b) we compared our context-aware recommendation approach with a context-free recommendation approach and two baselines, and (c) we compared our approach with similar state of the art CARS approaches to ours.

Table 2. MAE/RMSE of the derived techniques of our CBPF approach

Models	CoMoDa		STS		Music	
	MAE	RMSE	MAE	RMSE	MAE	RMSE
CBPF-IB	0.84	1.055	0.95	1.19	1.25	1.50
CBPF-CIB-AG	0.82	1.03	0.84	1.08	1.05	1.29
CBPF-CIB-CN	**0.73**	**0.93**	0.82	1.03	**0.72**	**0.87**
CBPF-UB	0.85	1.06	0.96	1.20	1.06	1.30
CBPF-CUB-AG	0.83	1.04	0.85	1.08	—	—
CBPF-CUB-CN	0.81	1.02	**0.80**	**1.02**	—	—

Table 2 illustrates the performances of the derived techniques of our approach, in terms of rating estimation. *CBPF-IB* and *CBPF-UB* refer to the item- and user-based correlation model, *CBPF-CIB-AG* and *CBPF-CUB-AG* refer to the correlation models based on the cluster of items or users, with the aggregation technique, and finally *CBPF-CIB-CN* and *CBPF-CUB-CN* refer to the same model, but with the concatenation technique.

As expected, the last version, which is the cluster-based approach with the concatenation technique for the context representation (*CBPF-CIB-CN* and *CBPF-CUB-CN*), has the best performances. In fact by clustering items/users we not only gain in term of computation cost but also in term of performance. We can explain this gained performance by the fact that in general, a correlation computation gives more precise results when calculated on a larger number of data. And the clustering allows to gather more data together, and so results on a better correlation computation and global performance. Moreover, the concatenation technique allows to preserve the real influence of each one of the context conditions, and so to gain in performance.

Another interesting point is that there is not a single winner between the item-based influence model and the user-based. As we can see in the table, contrary to *STS*, where *CBPF-CUB-CN* has the best performance, in *CoModa*, the item-based model gives better results (with *CBPF-CIB-CN*). So we can say that the choice between the item- or user-based model depends on the data, and in particular on the amount of available information about items'/users' characteristics. In fact, as Table 1 shows, for *CoMoDa*, we have more characteristics about items than about users, while the opposite holds for *STS*. So we can say that having more items' or users' characteristics implies better clusters. And as a result, we will be able to compute more relevant correlations and create a better model to do the recommendations. Note that we couldn't compare this effect on the *Music* dataset, because we did not have information about users' characteristics.

Figure 4 illustrates the MAE improvement that our approach makes over the context-free recommendation and the baselines. The context-free recommendation technique used in this experimentation is a Matrix Factorization (MF) technique named BiasedMF [10]. The comparison of our context-aware recommendation and this context-free matrix factorization confirms that users' contextual information can help the recommender to improve its performance.

As baselines, we used the *exact pre-filtering* approach [1] and a second baseline (*binary pre-filtering*) that we built as follow: we represented the context by means of a binary vector with a size equal to the total number of context conditions, where the value of each cell is equal to 1 if the corresponding context condition is present in the context situation, or equal to 0 if it is not present. We did a pre-filtering recommendation using this binary context representation. As the Fig. 4 shows, our approach outperforms these two baselines. The improvement over the *exact pre-filtering* shows that the idea of filtering the ratings based on the ones done in similar contexts is effective. And the improvement over the *binary pre-filtering* shows the positive effect of representing the context based on the influence of context on ratings.

Finally, we compared our approach with four state of the art approaches, which are more closer to our approach: (a) *DSPF* (Distributional Semantic Pre-Filtering) [7], (b) *Deviation-based CAMF* (Context-Aware Matrix Factorization) [4], (c) *Similarity-based CAMF* [18] and (d) *DCM* (Differential Context Modeling) [16]. In fact *DSPF* and *deviation-based CAMF* approaches try to model the context based on the influence of contexts on ratings, and *similarity-based CAMF*, *DCM* and *DSPF* uses the similarities among contexts in their approaches. Each one of these approaches have different versions (cited in Sect. 2). We tested all the possible versions, and the performances of the best version of each approach, in terms of rating estimation are illustrated in Table 3 for each dataset. We tested the state of the art algorithms by relying on the CARSkit Java API [17]. So for the *CoMoDa* dataset, we report the performances of CBPF-CI-CN, DSPF-IB, CAMF-CU, CAMF-ICS and DCW. For *STS*: CBPF-CU-CN, DSPF-IB, CAMF-CU, CAMF-ICS and DCW, and for *Music*: CBPF-CI-CN, DSPF-UB, CAMF-CU, CAMF-ICS and DCW.

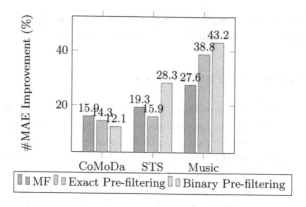

Fig. 4. MAE improvement (%) of our approach with respect to context-free MF and baselines

Table 3. Comparison with state of the art

Models	CoMoDa		STS		Music	
	MAE	RMSE	MAE	RMSE	MAE	RMSE
CBPF	**0.73**	**0.93**	**0.80**	**1.02**	**0.72**	**0.87**
DSPF	0.86	1.08	1.26	1.62	1.76	2.49
Deviation-based CAMF	0.76	1.02	1.03	1.37	0.82	1.06
Similarity-based CAMF	**0.73**	**0.92**	0.94	1.17	**0.72**	1.09
DCM	0.79	1.04	0.96	1.24	1.11	1.41

The reported results are average of multiple executions based on 5-fold cross-validation. For each dataset, the values in bold are statistically significant better (95% confidence level) than other approaches. The statistical significance has been calculated using the Wilcoxon rank test.

The illustrated performances in Table 3 show that, for all the three datasets, our approach can have better or comparable performances in comparison to state of the art (lower values of these metrics show better performances). Note that when comparing to CAMF, we obtain better, even if comparable, performances. However, our pre-filtering approach has the important advantage of being easily pluggable into any existing recommender system which is already in production, in order to improve it, while adopting CAMF would imply to re-implement the whole process.

Finally, we can state that our proposed modeling approach is able to produce good performances on roughly different datasets from various domains. In fact, the three datasets used in our experimentations are from three different domains (movie, tourism and music), with different characteristics in terms of density, rating distribution and the number of available context and content information. Moreover, clustering items or users has shown to have beneficial effect. In fact, our results indicate that grouping items or users can roughly help the model to catch more significantly the influence of context on ratings.

6   Conclusions and Further Work

In this paper, we present a correlation-based pre-filtering (CBPF) approach, that integrates contextual information into the recommendation process by modeling the influence of the context on ratings. This influence is computed by the item- or user-based Pearson Correlation Coefficient (PCC) between the context and the ratings. CPBF tries to take advantage of content and context information in its recommendation process to improve its performances. In our experimental analysis, we evaluate our approach on three different contextual datasets (*Comoda*, *STS* and *Music*), and analyze performances based on the characteristics of these datasets. Experiments validate the positive effect of taking into account contextual information about the user in the recommendation process, and show that our approach outperforms state of the art techniques in most cases. Furthermore, experimental results show that the PCC can efficiently catch the influence of context on ratings. Also, due to the large number of items/users, grouping them not only reduces computational cost, but also increases performances.

As just mentioned, by clustering items/users, we gain in terms of computation cost. We can gain even more by clustering context situations. So in the future, we would like to apply a clustering on context situations (like [7]) to limit the local model building computations.

In our approach, we compute the influence of context on ratings based on the PCC. In the future we plan to test other statistical models (e.g., ANOVA) for the correlation computation.

In our experiments we cluster items/users based on their available characteristics information. But this kind of information is not always available, so it would be interesting to test other clustering strategies, like clustering items/users based on ratings.

In real world applications, not all context factors have the same importance and impact on ratings. Depending on the application, some context factors can play a more important role than others. For example, in the case of recipe recommendation, factors like season, available tools around the user, and her cooking competence would be more important. While in music recommendation, activity and psychological context would be more influencing. So in future work we plan to take this fact into consideration, and rely on weighted context factors, based on their importance.

References

1. Adomavicius, G., Sankaranarayanan, R., Sen, S., Tuzhilin, A.: Incorporating contextual information in recommender systems using a multidimensional approach. ACM Trans. Inf. Syst. (TOIS) **23**, 103–145 (2005)
2. Adomavicius, G., Tuzhilin, A.: Context-aware recommender systems. In: Ricci, F., Rokach, L., Shapira, B., Kantor, P.B. (eds.) Recommender Systems Handbook, pp. 217–253. Springer, Boston (2011). https://doi.org/10.1007/978-0-387-85820-3_7
3. Baltrunas, L., et al.: InCarMusic: context-aware music recommendations in a car. In: Huemer, C., Setzer, T. (eds.) EC-Web 2011. LNBIP, vol. 85, pp. 89–100. Springer, Heidelberg (2011). https://doi.org/10.1007/978-3-642-23014-1_8
4. Baltrunas, L., Ludwig, B., Ricci, F.: Matrix factorization techniques for context aware recommendation. In: RecSys, pp. 301–304. ACM (2011)
5. Benesty, J., Chen, J., Huang, Y., Cohen, I.: Pearson correlation coefficient. In: Cohen, I., Huang, Y., Chen, J., Benesty, J. (eds.) Noise Reduction in Speech Processing. STSP, vol. 2, pp. 1–4. Springer, Berlin (2009). https://doi.org/10.1007/978-3-642-00296-0_5
6. Braunhofer, M., Elahi, M., Ricci, F., Schievenin, T.: Context-aware points of interest suggestion with dynamic weather data management. In: Xiang, Z., Tussyadiah, I. (eds.) Information and Communication Technologies in Tourism 2014, pp. 87–100. Springer, Cham (2013). https://doi.org/10.1007/978-3-319-03973-2_7
7. Codina, V., Ricci, F., Ceccaroni, L.: Distributional semantic pre-filtering in context-aware recommender systems. User Model. User-Adap. Inter. **26**, 1–32 (2016)
8. Guo, G., Zhang, J., Sun, Z., Yorke-Smith, N.: LibRec: a Java library for recommender systems. In: UMAP Workshops (2015)
9. Jannach, D., Zanker, M., Felfernig, A., Friedrich, G.: Recommender Systems: An Introduction. Cambridge University Press, Cambridge (2010)
10. Koren, Y., Bell, R., Volinsky, C.: Matrix factorization techniques for recommender systems. Computer **42**(8), 30–37 (2009)
11. Košir, A., Odic, A., Kunaver, M., Tkalcic, M., Tasic, J.F.: Database for contextual personalization. Elektrotehniški vestnik **78**, 270–274 (2011)
12. Ricci, F., Rokach, L., Shapira, B.: Introduction to recommender systems handbook. In: Ricci, F., Rokach, L., Shapira, B., Kantor, P.B. (eds.) Recommender Systems Handbook, pp. 1–35. Springer, Boston (2011). https://doi.org/10.1007/978-0-387-85820-3_1
13. Su, X., Khoshgoftaar, T.M.: A survey of collaborative filtering techniques. In: Advances in artificial intelligence (2009)
14. Vahidi Ferdousi, Z., Colazzo, D., Negre, E.: CBPF: leveraging context and content information for better recommendations. arXiv preprint arXiv:1810.00751 (2018)
15. Vahidi Ferdousi, Z., Colazzo, D., Negre, E.: Correlation-based pre-filtering for context-aware recommendation. In: PerCom Workshops (CoMoRea). IEEE (2018)
16. Zheng, Y., Burke, R., Mobasher, B.: Optimal feature selection for context-aware recommendation using differential relaxation. In: ACM RecSys 2012 (2012)
17. Zheng, Y., Mobasher, B., Burke, R.: CARSKIT: a Java-based context-aware recommendation engine. In: Proceedings of the 15th IEEE ICDM Workshops (2015)
18. Zheng, Y., Mobasher, B., Burke, R.: Similarity-based context-aware recommendation. In: Wang, J., et al. (eds.) WISE 2015. LNCS, vol. 9418, pp. 431–447. Springer, Cham (2015). https://doi.org/10.1007/978-3-319-26190-4_29

Discovering High Utility Change Points in Customer Transaction Data

Philippe Fournier-Viger[1]([envelope]), Yimin Zhang[2], Jerry Chun-Wei Lin[3],
and Yun Sing Koh[4]

[1] School of Natural Sciences and Humanities,
Harbin Institute of Technology (Shenzhen), Shenzhen, China
philfv@hit.edu.cn
[2] School of Computer Sciences and Technology,
Harbin Institute of Technology (Shenzhen), Shenzhen, China
mrzhangym@126.com
[3] Department of Computing, Mathematics and Physics,
Western Norway University of Applied Sciences (HVL), Bergen, Norway
jerrylin@ieee.org
[4] Department of Computer Sciences,
University of Auckland, Auckland, New Zealand
ykoh@cs.auckland.ac.nz

Abstract. High Utility Itemset Mining (HUIM) consists of identifying
all sets of items (itemsets) that have a high utility (e.g. have a high
profit) in a database of customer transactions. Important limitations
of traditional HUIM algorithms is that they do not consider that the
utility of itemsets varies as time passes and that itemsets may not have
a high utility in the whole database, but in some specific time periods.
To overcome these drawbacks, this paper defines the novel problem of
discovering change points of high utility itemsets, that is to find the
time points where the utility of an itemset is changing considerably.
An efficient algorithms named HUCP-Miner is proposed to mine these
change points. Experimental results show that the proposed algorithm
has excellent performance and can discover interesting patterns that are
not identified by traditional HUIM algorithms.

Keywords: Pattern mining · High-utility itemsets · Change points

1 Introduction

Frequent Itemset Mining (FIM) [1] is a well-studied task in data mining. Given
a database of customer transactions, FIM consists of enumerating all sets of
items (itemsets) that frequently appear together. *High-Utility Itemset Mining*
(HUIM) [2,4–7] is a generalization of FIM. It consists of finding all itemsets
having a utility (e.g. importance or profit) that is equal or greater than a user-
defined threshold in a customer transaction database. HUIM is more challenging

© Springer Nature Switzerland AG 2018
G. Gan et al. (Eds.): ADMA 2018, LNAI 11323, pp. 392–402, 2018.
https://doi.org/10.1007/978-3-030-05090-0_33

than FIM because the utility measure used in HUIM is not monotonic nor anti-monotonic, that is an itemset may have a utility that is smaller, equal or greater than that of its supersets or subsets. Because of this, search space reduction techniques of traditional FIM algorithms [1] cannot be used to reduce the search space of HUIM. To design HUIM algorithms that efficiently enumerate all high utility itemsets, upper-bounds on the utility measure have been proposed, which have anti-monotonic properties, to reduce the search space [5]. To mine high utility itemsets, several algorithms were designed such as UP-Growth [6], HUI-Miner [4] and FHM [2]. They use different search strategies and data structures.

HUIM has several applications such as biomedical applications, customer behavior analysis, and click stream analysis [4, 6, 7]. But a drawback of traditional HUIM is that the timestamps of transactions are not taken into account, although they provide valuable information that can be used to understand how the utility of itemsets changes over time. A few studies about FIM and HUIM have considered time. For instance, *periodic high-utility itemset mining* [8] aims at finding itemsets that periodically appear and yield a high profit in transactional data. Although this work consider the sequential ordering of transactions, it ignores timestamps. Moreover, the algorithm presented in that study tends to discover patterns having a utility that is stable over time. An algorithm was proposed for detecting time points where the frequency of itemsets changes in data streams [9]. Although it considers the sequential ordering of transactions, it ignores timestamps. Hence, it assumes that all consecutive transactions have the same time gap, which is unrealistic.

As time passes, the utility of itemsets may change. It is desirable to find the time points where the utility of itemsets changes dramatically. Discovering this information allow understanding trends in the data to support decision-making. For example, discovering that an upward/downward trend in the utility of some items is happening at a time point allows to improve inventory management by either increasing retail stocks or implementing strategies to reduce stocks.

To fulfill this needs, this paper defines the problem of identifying change points of high utility itemsets, that is time points indicating a considerable increase or decrease of utility. To be able to discover change points that represents significant trends in the utility distributions of itemsets, this paper adapts the concept of moving average crossover, which is used in time series analysis. A change point occurs for an itemset when short-term and long-term moving averages of its utility cross. However, it is impractical to mine all the change points of all the itemsets in a database, and it is also unnecessary since some itemsets seldomly appear and yield a low utility. The solution of this paper is to only find the change points of local high utility itemsets [11]. For example, this can be used to discover that an itemset {*sunglasses, suncream, beach_towel*} has a change point at the begining of the summer representing an increase and another change point at the end of the summer representing a decrease, and this itemset may not be a HUI in the whole database.

The rest of this paper is organized as follows. Section 2 presents preliminaries and defines the problem of identifying change points of HUIs. Section 3 describes the designed algorithm. Section 4 reports results from an experimental evaluation. Finally, Sect. 5 concludes the paper.

2   Preliminaries and Problem Statement

Consider a set of products (items) sold in a retail store, denoted as $I = \{i_1, i_2, \ldots, i_n\}$. A customer transaction T is a set of items $T \subseteq I$. A *transaction database* contains multiple customer transactions $D = \{T_1, T_2, \ldots, T_m\}$, In addition, each transaction T has a timestamp denoted as $t(T)$, which may not be unique. Besides, a positive number $p(i)$ called external utility is defined for each item $i \in I$ to indicate its relative importance. In the context of customer transaction analysis, $p(i)$ can be defined as the unit profit of i. In each transaction T, a positive number $q(i, T)$ called internal utility specifies the purchase quantity of each item i. For instance, a small transaction database is provided in Table 1. This database contains eight transactions, denoted as $T_1, T_2, \ldots, T_8$, and five items, denoted as a, b, c, d, e, where purchase quantities are shown as integers beside items. This database will be used as running example for the rest of this paper. In this database, timestamps of transactions $T_1, T_2 \ldots T_8$ are denoted as $d_1, d_3, \ldots d_{10}$. These values can represent days ($d_i = i$-th day) or other time units such as seconds, or milliseconds. Consider transaction T_2. It indicates that 4, 3, 2 and 1 units of items b, c, d and e were purchased, respectively. According to Table 2, which provides the external utility values of items, the unit profits of these items are 2, 1, 2 and 3, respectively.

Table 1. A transaction database

Transaction	Items $(item, purchase\_quantity)$	Time
T_1	$(b, 2), (c, 2), (e, 1)$	d_1
T_2	$(b, 4), (c, 3), (d, 2), (e, 1)$	d_3
T_3	$(b, 2), (c, 2), (e, 1)$	d_3
T_4	$(a, 2), (b, 10), (c, 2), (d, 10), (e, 2)$	d_5
T_5	$(a, 2), (c, 6), (e, 2)$	d_6
T_6	$(b, 4), (c, 3), (e, 1)$	d_7
T_7	$(a, 2), (c, 2), (d, 2)$	d_9
T_8	$(a, 2), (c, 6), (e, 2)$	d_{10}

Table 2. External utilities of items

Item	a	b	c	d	e
Unit profit	5	2	1	2	3

Definition 1 (Utility of an item/itemset). Let there be an item i and a transaction T from a transaction database. The utility of i in T is defined and calculated as $u(i, T) = p(i) \times q(i, T)$. Let X be an *itemset* (a set of items), i.e. $X \subseteq I$. The utility of X in T is defined as $u(X, T) = \sum_{i \in X \wedge X \subseteq T} u(i, T)$. The utility of X in the database is defined as $u(X) = \sum_{T \in D \wedge X \subseteq T} u(X, T)$.

For instance, the utility of item b in T_1 is $u(b, T_1) = 2 \times 2 = 4$. The utility of itemset $\{b, c\}$ in T_1 is $u(\{b, c\}, T_1) = u(b, T_1) + u(c, T_1) = 2 \times 2 + 1 \times 2 = 6$. The utility of itemset $\{b, c\}$ in the database is $u(\{b, c\}) = u(\{b, c\}, T_1) + u(\{b, c\}, T_2) + u(\{b, c\}, T_4) + u(\{b, c\}, T_6) = 6 + 11 + 6 + 11 = 34$.

An itemset X is called *high utility itemset* (HUI) if its utility $u(X)$ is no less than a user-specified positive threshold *minutil* [5]. HUIM is the task of finding all HUIs in a transaction database. The utility measure is not anti-monotonic, that is a superset of a HUI may not be a HUI. Thus, pruning strategies used in FIM cannot be directly used in HUIM. To reduce the search space in HUIM, the Transaction Weighted Utilization (TWU) upper-bound was introduced [5].

Definition 2 (Transaction weighted utilization). The utility of a transaction T is defined as $tu(T) = \sum_{i \in T} u(i, T)$. The TWU of an itemset X is the sum of the utilities of transactions containing X, i.e. $TWU(X) = \sum_{X \subseteq T \wedge T \in D} tu(T)$.

For itemsets $X \subseteq X'$, $u(X') \leq TWU(X') \leq TWU(X)$ [5]. Thus, for an itemset X, $TWU(X)$ is an upper-bound on $u(X)$, and the TWU is anti-monotonic. Hence, if $TWU(X) < minutil$, all supersets of X can be pruned.

Limitations of HUIM are that transaction timestamps are ignored and that an itemset's utility may vary over time. To identify time periods where an itemset has a high utility, a concept of time window is used. It will then be used to identify change points in the utility of HUIs indicating upward/downward trends.

Definition 3 (Window). A window denoted as $W_{i,j}$ is the set of transactions from time i to j, i.e. $W_{i,j} = \{T | i \leq t(T) \leq j \wedge T \in D\}$, where i, j are integers. The length of a window $W_{i,j}$ is defined as $length(W_{i,j}) = j - i + 1$. The length of a database D containing m transactions is $W_D = t(T_m) - t(T_1) + 1$. A window $W_{k,l}$ is said to **subsume** another window $W_{i,j}$ iff $W_{i,j} \subsetneq W_{k,l}$. The utility of an itemset X in a window $W_{i,j}$ is defined as $u_{i,j}(X) = \sum_{T \in W_{i,j} \wedge X \subseteq T} u(X, T)$.

For example, $W_{d_1,d_3} = \{T_1, T_2, T_3\}$, $length(W_{d_1,d_3}) = 3 - 1 + 1 = 3$, and W_{d_1,d_3} subsumes W_{d_3,d_3}. But W_{d_2,d_3} does not subsume W_{d_3,d_3} because $W_{d_2,d_3} = \{T_2, T_3\} = W_{d_3,d_3}$. The utility of $\{b, c\}$ in the window W_{d_1,d_3} is $u_{d_1,d_3}(\{b, c\}) = u(\{b, c\}, T_1) + u(\{b, c\}, T_2) + u(\{b, c\}, T_3) = 6 + 11 + 6 = 23$.

A concept of change point is proposed by adapting the *moving average crossover* technique. In time series analysis, the *moving average* of a time-series is the mean of the previous n data points. It is used to smooth out short-term fluctuations. The larger n is, the more smoothing is obtained. The *moving average crossover* is the time points where two moving averages based on different degrees of smoothing cross each other. Moving average crossover is widely used by stock trading systems [10], where crossovers indicate points for buying/selling stocks. The crossovers can be interpreted as follows. If the short moving average crosses above the long moving average, it indicates an upward trend, while if it crosses below, it indicates a downward trend.

To find change points where an itemset's utility has upward/downward trends, this paper utilizes the concept of moving crossover. This allows to consider time points that are not equally spaced in time. The moving average utility of an itemset is defined as follows. Let there be a smoothing parameter γ representing a time length. The *moving average* utility of an itemset X for a timestamp t is defined as $mau_\gamma(X,t) = \dfrac{u_{t-\frac{\gamma-1}{2},t+\frac{\gamma-1}{2}}(X)}{\gamma}$, that is the average utility of X before and after t in a window of length γ. For example, let $\gamma = 3$, the moving average utility of itemset $\{a,c\}$ at timestamp d_6 is

$$mau_3(\{a,c\}, d_6) = \frac{u_{d_6-\frac{3-1}{2},d_6+\frac{3-1}{2}}(\{a,c\})}{3} = \frac{u_{d_5,d_7}(\{a,c\})}{3} = 28.$$

To apply the concept of moving average crossover, it is necessary to define two moving averages, respectively having a short and a long window. For an itemset X and a timestamp t, the short term moving average utility $mau_\gamma(X,t)$ is calculated over the window $W_{t-\frac{\gamma-1}{2},t+\frac{\gamma-1}{2}}$. This window has a length of γ. The long term moving average utility $mau_{\lambda\times\gamma}(X,t)$ is calculated using a window $W_{t-\frac{\lambda\times\gamma-1}{2},t+\frac{\lambda\times\gamma-1}{2}}$ having length of $\lambda \times \gamma$, where $\lambda \geq 1$ is a user-defined parameter called the *moving average crossover coefficient*. Based on these windows, the concept *change point* is proposed, indicating time points where the utility of an itemset follows an upward or downward trend.

Definition 5 (Change point). For an itemset X, a time point t is a change point if $mau_{winLength}(X,t) \geq (\leq)mau_{\lambda\times winLength}(X,t) \wedge mau_{winLength}(X,t-1) < (>)mau_{\lambda\times winLength}(X,t-1)$, where $winLength$ and λ are two user-defined parameters ($winLength > 1$, $\lambda \geq 1$), which are the short term window length and the scaling factor to obtain the long term window. If $\lambda = 1$, the short and long moving average utility are equal and there is no change point. If $\lambda = \infty$, the change points of X are the time points where its short moving average utility crosses the average utility in the whole database.

For example, if $winLength = 3$ and $\lambda = 1.67$, d_5 is a change points of itemset $\{a,c\}$ because $mau_3(\{a,c\}, d_5 - 1) = u_{d_3,d_5}(\{a,c\})/3 = 4 < mau_{3\times1.67}(\{a,c\}, d_5 - 1) = u_{d_2,d_6}(\{a,c\})/5 = 5.6$, and $mau_3(\{a,c\}, d_5) = 7 > mau_{3\times1.67}(\{a,c\}, d_5 - 1) = 5.6$. Similarly, we can find that d_7 is also a change point of $\{a,c\}$. And it can be seen that d_5 is where itemset $\{a,c\}$ begins to appear and d_7 is where itemset $\{a,c\}$ begins to disappear in the database.

It is not necessary to find change points of every itemsets, since some seldomly appear or generate little utility. Instead, we only find change points of local high utility itemset [11], a concept defined as follows.

Definition 6 (Local High Utility Itemset). An itemset X is said to be a Local High Utility Itemset (LHUI) if there exists a time point t such that $mau_{winLength}(X,t) > minMau$, where $minMau$ and $winLength$ are two user-defined parameters [11].

Definition 7 (Mining High utility Change Points). The problem of Mining High utility Change Points is to find the change points of all LHUIs, given the parameters $winLength$, $minMau$ and λ, set by the user.

For example, given the database of Table 1, $winLength = 3$, $minMau = 10$ and $lambda = 1.67$, 24 Local High Utility Itemsets with their change points are found, including $\{d\}:\{d_3, d_6\}$, $\{b,d\}:\{d_3, d_6\}$ and $\{a,c\}:\{d_5, d_7, d_9, d_{10}\}$.

3   Proposed Algorithm

This paper proposes an algorithm to efficiently mine change points of LHUIs, named HUCP-Miner. It extends the LHUI-Miner [11] algorithm for LHUI mining. The HUCP-Miner algorithm explores the search space of itemsets by following a total order $\succ$ on items in I. It is said that an itemset Y is an *extension* of an itemset X if $Y = X \cup \{j\} \wedge \forall i \in X, j \succ i$. HUCP-Miner relies on a data structure called *Local Utility-list* (LU-list) [11] to store information about each itemset. This structure extends the utility-list [4] structure by storing information about time periods. HUCP-Miner initially scans the database to build a LU-list for each item. Then, it performs a depth-first search to explore the search space, while joining pairs of itemsets to generate their extensions and corresponding LU-lists. Checking if an itemset is a LHUI and calculating its moving average utility is done using is LU-list (without scanning the database). The LU-list structure is defined as follows. Let there be an itemset X. Its *LU-list* contains a tuple for each transaction where X appears. A *tuple* has the form $(tid, iutil, rutil)$, where tid is the identifier of a transaction T_{tid} containing X, $iutil$ is the utility of X in T_{tid}. i.e. $u(X, T_{tid})$, and $rutil$ is defined as $\sum_{i \in T_{tid} \wedge \forall j \in X, i \succ j} u(i, T_{tid})$ [4]. Moreover, the LU-list contains two sets named $iutilPeriods$ and $utilPeriods$, which stores the maximum *LHUI periods* and *PLHUI periods* of X, respectively. The concepts of LHUI periods and PLHUI periods of X were defined in the LHUI-Miner algorithm [11].

A LU-list allow to derive key information about an itemset. First, if $iutilPeriods$ of an itemset X is not empty, then there exists at least a time point t, where $mau_{winLenth}(X, t) > minMau$. In other words, X is a LHUI. On the other hand, if $utilPeriods$ is empty, all transitive extensions of X are not LHUIs and can be pruned from the search space. The proof of this property is not presented due to the page limitation. The LU-list of an itemset also allows to calculate its moving average utility without scanning the database.

The HUCP-Miner Algorithm. The input of HUCP-Miner is a transaction database, $minMau$, $winLength$ and λ. HUCP-Miner outputs all LHUIs with their change points. The algorithm first reads the database once to obtain each item's TWU, and construct an array, called $tid2time$. The i-th position of $tid2time$ contains the timestamp of transaction $t(T_i)$. Thereafter, HUCP-Miner only considers items having a TWU no less than $minMau \times winLength$, denoted as I^*. Based on the calculated TWU values of items, the total order $\succ$ on I^* is then set as the ascending order of TWU values [2]. The database is then read again and items in each transaction are reordered according to $\succ$ to create the LU-list of each item $i \in I^*$. Then, HUCP-Miner perform a depth-first search of itemsets by calling the recursive $HUCP\text{-}Search$ procedure with $\emptyset$, the LU-lists of 1-itemsets, $minMau$, $winLength$ and λ.

The input of the $HUCP\text{-}Search$ (Algorithm 1) procedure is (1) an itemset P, (2) extensions of P, (3) $minMau$, (4) $winLength$, and (5) λ. $HUCP\text{-}Search$ first verifies if $iutilPeriods$ is empty for each extension Px of P according to their respective LU-lists. If empty, then Px is a LHUI and it is output with its change points. Besides, if $utilPeriods$ is not empty, extensions of Px will be explored. This is done by combining Px with each extension Py of P such that $y \succ x$ to form an extension of the form Pxy containing $|Px| + 1$ items. The LU-list of Pxy is then constructed using the $Construct$ procedure of $HUI\text{-}Miner$, which join the tuples in the LU-lists of P, Px and Py. Thereafter, $iutilPeriods$, $utilPeriods$ and change points of Pxy are constructed by calling the $generateCP$ procedure. Then, $HUCP\text{-}Search$ is called with Pxy to calculate its utility and explore its extension(s) using a depth-first search. The $HUCP\text{-}Miner$ procedure starts from single items, it recursively explores the search space of itemsets by appending single items and it only prunes the search space based on the properties of LU-list. It can be easily seen that this procedure is correct and complete to discover all LHUIs and their change points.

Algorithm 1. HUCP-Search

---

    input : P: an itemset, $ExtensionsOfP$: extensions of P, $minMau$: a user-specified
                 threshold, $winLength$: a window length threshold, λ: a user-specified parameter
    output: the set of LHUIs and their change points

1   **foreach** itemset $Px \in ExtensionsOfP$ **do**
2   **if** $Px.LUList.iutilPeriods \neq \emptyset$ **then** output Px with $Px.LUList.iutilPeriods$;
3   **if** $Px.LUList.utilPeriods \neq \emptyset$ **then**
4           $ExtensionsOfPx \leftarrow \emptyset$;
5   **foreach** itemset $Py \in ExtensionsOfP$ such that $y \succ x$ **do**
6               $Pxy.LUList \leftarrow$ Construct (P, Px, Py);
7   generateCP $(Pxy, minMau, winLength)$;
8               $ExtensionsOfPx \leftarrow ExtensionsOfPx \cup Pxy$;
9   **end**
10   HUCP-Miner $(Px, ExtensionsOfPx, minutil, winLength, \lambda)$;
11   **end**
12 **end**

The $generateCP$ procedure (Algorithm 2) takes as input (1) a LU-list lUl, (2) $minMau$, (3) $winLength$, and λ. The procedure slides a window over lUl using two variable $winStart$ (initialized to 0; the first element of lUl), and $winEnd$. The procedure first scan lUl to find $winEnd$ (the end index of the first window), $iutils$ (sum of $iutil$ values in the first window) and $rutils$ (sum of $rutil$ values in the first window). Then, it repeats the following steps until the end index $winEnd$ reaches the last tuple of the LU-list: (1) increase the start index $winStart$ until the timestamp changes, and at the same time decrease $iutils$ ($rutils$) by the $iutil$ ($rutil$) values of tuples that exit the current window, (2) increase the end index until the window length is no less than $winLength$, and at same time increase $iutils$ ($rutils$) by the $iutil$ ($rutil$) values of tuples that enter the current window, (3) compare the resulting $iutils$ and $iutils + rutil$ values with $minMau$ to determine if the current period should be merged with the previous period or added to $iutilPeriods$ and $utilPeriods$ (line 14 to 15). Merging is performed to obtain the maximum LHUI and PLHUI periods, (4) the moving average utility of the center of a window can be calculated as $mau_\gamma = iutils/winLength$. Similarly, the long term moving average utility can be calculated using a larger window. Then, the moving average are compared to determine if the current window center is a change point. If so it is stored.

Algorithm 2. The generatePeriods procedure

input : lUl: a LU-list, $minMau$: a user-specified utility threshold, $winLength$: a user-specified window length threshold, λ: a user-specified parameter

1 $winStart = 0$;
2 Find $winEnd$ (the end index of the first window in ul), $iutils$ (sum of $iutil$ values of the first window), $rutils$ (sum of $rutils$ values of the first window);
3 **while** $winEnd < lUl.size$ **do**
4   **while** $ul.get(winStart).time$ *is same as previous index* **do**
5          $iutils = iutils - lUl.get(winStart).iutil$;
6          $rutils = rutils - lUl.get(winStart).rutil$;
7          $winStart = winStart + 1$;
8   **end**
9   **while** $ul.get(winEnd).time \leq ul.get(winStart).time + winLength$ **do**
10         $iutils = iutils + lUl.get(winEnd).iutil$;
11         $rutils = rutils + lUl.get(winEnd).rutil$;
12         $winEnd = winEnd + 1$;
13   **end**
14   merge the $[winStart, winEnd]$ period with the previous period if $iutils \geq minMau \times winLength$. Otherwise, add it to $lul.iutilPeriods$;
15   merge the $[winStart, winEnd]$ period with the previous period if $iutils + rutils \geq minMau \times winLength$. Otherwise, add it to $lul.utilPeriods$;
16     $mau_\gamma = iutils/winLength$, use similar strategy to calculate $mau_{\gamma \times \lambda}$, compare to determine if current center of window is a change point, if so, store it.
17 **end**

Optimizations. As HUCP-Miner is based on LHUI-Miner [11], all its optimizations can be used to improve the performance of HUCP-Miner.

4 Experimental Evaluation

An experiment was done to compare the performance of HUCP-Miner with a non-optimized version and the HUI-Miner algorithm for mining HUIs. Three datasets were used, which are commonly utilized to benchmark HUIM algorithms: *retail*, *kosarak* and *e-commerce*, representing the main types of data (long transactions, dense and sparse). Let $|I|$, $|D|$ and A represents the number of distinct items, transactions and average transaction length. *kosarak* is a dataset that contains many long transactions ($|I| = 41{,}270$, $|D| = 990{,}000$, $A = 8.09$). *retail* is a sparse dataset with many different items ($|I| = 16{,}470$, $|D| = 88{,}162$, $A = 10{,}30$). *e-commerce* is a real-world dataset ($|I| = 3{,}803$, $|D| = 17{,}535$, $A = 15.4$), containing customer transactions from 01/12/2010 to 09/12/2011 of an online store. For *retail* and *kosarak*, external utilities of items were generated between 1 and 1,000 using a log-normal distribution and quantities of items are generated randomly between 1 and 5, as in [4,6]. Moreover, transactions timestamps for the three databases were generated using the same distribution as the e-commerce database. The source code of algorithms and datasets can be downloaded as part of the SPMF [3] open source data mining library at http://www.philippe-fournier-viger.com/spmf/.

HUCP-Miner was run with $winLength = 90\ days$ for $e\text{-}commerce$ and 30 days for the other datasets, and $\lambda = 2$. In the following, $hucp\text{-}op$ denotes HUCP-Miner with optimizations and $hucp\text{-}non\text{-}op$ denotes HUCP-Miner without optimization. Algorithms were run on each dataset, while decreasing $minMau$ (for HUI-Miner $minutil = minMau \times \lceil \frac{W_D}{winLength} \rceil$) until they became too long to execute, ran out of memory or a clear trend was observed. Figure 1 compares the execution times of HUCP-Miner with and without optimization. Figure 2 compares the numbers of HUCPs and HUIs, respectively generated by these algorithms.

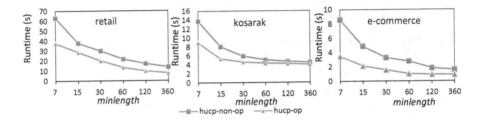

Fig. 1. Execution times

It can be observed that optimizations generally reduce the runtime. In some cases, HUCP-Miner can be one time faster than the non-optimized version. We also compared the execution time of HUI-Miner ($minutil = minMau \times \lceil \frac{W_D}{winLength} \rceil \times W_D$) with these algorithms. HUI-Miner is often much faster and produces much less patterns. However, when the number of patterns found by HUI-Miner is similar to the proposed algorithms, their runtimes are similar. Thus, because HUI-Miner is defined for a different problem its results are not shown in Fig. 1.

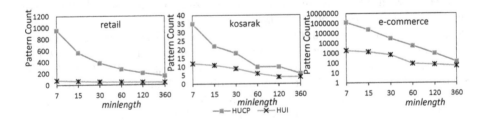

Fig. 2. Number of patterns found

A second observation is that the number of LHUIs is much more than the number of HUIs in most cases. This is reasonable since an itemset is much more likely to be high utility in at least one window than in the whole database.

For example, on mushroom ($W_D = 180$ $days$), $minutil = 500,000$, $minMau = 83,333$, $winLength = 30$ $days$, there are 168 HUIs and 549,479 LHUIs. Lastly, another observation is that for $kosarak$, the difference between the number of LHUIs and HUIs is very small compared to other datasets. The reason is that the utilities of patterns do not vary much over time in $kosarak$.

Among all patterns found, some interesting high utility change points are found in $e\text{-}commerce$. For instance, for $minMau = 16,000$ and $winLength = 90$ $days$, the itemset {white hanging heart T-light holder} has a positive trend on 2011/3/6 and 2011/9/1, and a negative trend on 2011/6/13 and 2011/12/8, while the itemset is not a HUI in the whole database for $minutil = minMau \times 373 = 5,968,000$. This information is important for retail store management since it indicates that this item has a sale increase and decrease at the end of July and in early November, respectively. This can be useful to replenish stocks and offer promotions on this set of products at these time points for the following year.

5   Conclusion

To find itemsets that yield a high utility in non-predefined time periods and consider timestamps of transactions, this paper defined the problem of mining high utility change points. Those are time points where an upward or downward trend is observed in the utility of local high utility itemsets. An experimental evaluation has shown that the proposed HUCP-Miner algorithm can discover useful patterns that traditional HUIM could not find and that optimizations reduced the runtime and memory consumption. For future work, we will adapt the concept of change points developed in this paper to other pattern mining problems such as sequential pattern mining and episode mining.

References

1. Agrawal, R., Srikant, R.: Fast algorithms for mining association rules in large databases. In: Proceedings of the 20th International Conference on Very Large Databases, pp. 487–499. Morgan Kaufmann, Santiago (1994)
2. Fournier-Viger, P., Wu, C.-W., Zida, S., Tseng, V.S.: FHM: faster high-utility itemset mining using estimated utility co-occurrence pruning. In: Andreasen, T., Christiansen, H., Cubero, J.-C., Raś, Z.W. (eds.) ISMIS 2014. LNCS (LNAI), vol. 8502, pp. 83–92. Springer, Cham (2014). https://doi.org/10.1007/978-3-319-08326-1_9
3. Fournier-Viger, P., Gomariz, A., Gueniche, T., Soltani, A., Wu, C., Tseng, V.S.: SPMF: a Java open-source pattern mining library. J. Mach. Learn. Res. (JMLR) **15**, 3389–3393 (2014)
4. Liu, M., Qu, J.: Mining high utility itemsets without candidate generation. In: 22nd ACM International Conference on Information and Knowledge Management, pp. 55–64. ACM, Maui (2012)
5. Liu, Y., Liao, W., Choudhary, A.: A two-phase algorithm for fast discovery of high utility itemsets. In: Ho, T.B., Cheung, D., Liu, H. (eds.) PAKDD 2005. LNCS (LNAI), vol. 3518, pp. 689–695. Springer, Heidelberg (2005). https://doi.org/10.1007/11430919_79

6. Tseng, V.S., Shie, B.-E., Wu, C.-W., Yu, P.S.: Efficient algorithms for mining high utility itemsets from transactional databases. IEEE Trans. Knowl. Data Eng. **25**(8), 1772–1786 (2013)
7. Peng, A.Y., Koh, Y.S., Riddle, P.: mHUIMiner: a fast high utility itemset mining algorithm for sparse datasets. In: Proceedings of 22nd Pacific-Asia Conference on Knowledge Discovery and Data Mining, pp. 196–207. ACM (2017)
8. Lin, J.C.W., Zhang, J., Fournier-Viger, P., Hong, T.P., Zhang, J.: A two-phase approach to mine short-period high-utility itemsets in transactional databases. Adv. Eng. Inform. **33**, 29–43 (2017)
9. Wan, Q., An, A.: Discovering transitional patterns and their significant milestones in transaction databases. IEEE Trans. Knowl. Data Eng. **21**(12), 1692–1707 (2009)
10. Ni, Y., Liao, Y.C., Huang, P.: MA trading rules, herding behaviors, and stock market overreaction. Int. Rev. Econ. Finan. **39**, 253–265 (2015)
11. Fournier-Viger, P., Zhang, Y., Lin, J.C.-W., Fujita, H., Koh, Y.S.: Mining local high utility itemsets. In: Hartmann, S., Ma, H., Hameurlain, A., Pernul, G., Wagner, R.R. (eds.) DEXA 2018. LNCS, vol. 11030, pp. 450–460. Springer, Cham (2018). https://doi.org/10.1007/978-3-319-98812-2_41

Estimating Interactions of Functional Brain Connectivity by Hidden Markov Models

Xingjuan Li[1(✉)], Yu Li[1], and Jiangtao Cui[2]

[1] School of Information Technology and Electrical Engineering,
University of Queensland, Brisbane 4067, Australia
x.li4@uq.edu.au
[2] School of Computer Science and Technology, Xidian University,
Xi'an 710071, Shaanxi, China

Abstract. The brain activity reflected by functional magnetic resonance imaging (fMRI) is temporally organized as a combination of sensory inputs from environment and its own spontaneous activity. However, temporal patterns of brain activity in a large number of subjects remain unclear. We propose a regularized hidden Markov model (HMM) to estimate dynamic functional connectivity among distributed brain regions and discover repeating connectivity patterns from resting-state functional connectivity across a group of subjects. We found that functional brain connectivity are hierarchically organized and exhibit three repeated patterns across subjects with attention deficit hyperactivity disorder (ADHD). We have examined the temporal characteristics of functional connectivity by its occupancy. And we validated our method by comparing the classification performance with state-of-the-art methods using the same dataset. Experimental results show that our method can improve the classification performance compared to other functional connectivity modelling methods.

Keywords: Functional brain connectivity · Dynamical modelling
Hidden Markov models

1 Introduction

Understanding dynamical neural activity across many spatiotemporal scales is an important challenge in clinical and basic neuroscience. For resting-state functional magnetic resonance imaging (rs-fMRI) studies, the aim is to discover neural substrates for understanding principles and mechanisms related to high-level brain functions [1–3]. At resting-state, the brain is not involved in a specific task, but it is still dynamically processing information, adapting its ability to the current environment and producing its own spontaneous activity [5, 7, 9]. Because of that, it is important to model relationships between time series fMRI signals and behavior changes as comprehensive as possible. Particularly, identifying repeating patterns (also refer to metastates) of rs-fMRI data plays a critical role to characterize and interpret differences of brain activity for understanding of cognition [3, 5, 8, 12, 19]. In this paper, we focus on characterizing the spontaneous dynamical brain activity.

© Springer Nature Switzerland AG 2018
G. Gan et al. (Eds.): ADMA 2018, LNAI 11323, pp. 403–412, 2018.
https://doi.org/10.1007/978-3-030-05090-0_34

The most commonly used techniques to describe dynamic brain activity at resting-state is sliding window analysis. The sliding window approach investigates dynamic functional brain activity by segmenting the time series from spatial locations into a set of temporal windows and characterize fluctuations in functional brain connectivity [13, 14]. However, sliding window analysis is challenged by the choice of window size and parameters, which would undermine the conclusions of analysis. It has been suggested that dynamical functional connectivity derived by sliding window correlation cannot directly reveals the presence of fluctuations in fMRI signals [11].

An alternative approach to produce less biased network connectivity is by time-frequency analysis, in which complex interplays of dynamical functional connectivity can be captured by a range of frequency bands. It has been shown that brain activity related to fMRI signals is limited to the analysis of low-frequency fluctuations (0.01 – 0.1 Hz) due to haemodynamic response function [16]. Higher frequency fluctuations are more likely to be related with physiological noise, such as the influence of respiratory and cardiac pulsations [16]. For example, wavelet transform has been used to investigate amplitude and phase characteristics of functional brain connectivity, revealing previously unreported within-network anti-correlation and across-network correlation [6, 15]. A critical limitation of this approach is that only one global functional connectivity has been estimated at a given time, whereas the functional brain connectivity has shown hierarchically organized temporal nature of interacting brain regions [19, 20].

In this paper, we propose to implement a hidden Markov model (HMM) to estimate interactions of function brain connectivity at a group level and discover connectivity patterns repeating over time. Within HMM framework, dynamic functional connectivity states are estimated by a multivariate Gaussian observation model, in which each state is characterized by the mean μ_k and covariance Σ_k. Most HMM uses a standard variational inference which requires loading the entire dataset and sequential forward pass followed by a backward pass on multiple time samples. This strategy is hampered by the high computational load problem. To solve these problems, we propose to implement a regularization-based HMM framework, in which we constrain the probability of state transitions and penalize frequently appeared samples to be selected again in current iteration. It should be noted that, compared to previous work based on sliding window techniques, our method offers a probabilistic model through the process of Bayesian inference, in which time series data are modelled in a self-contained approach.

In the following sections, we describe and justify the proposed method, including how we constrain the state transition probability by parameterized regularization, and how subjects are selected at each iteration. We then show our experimental results on functional brain connectivity derived from ADHD-200 dataset. Lastly, we compare our method with other state-of-the-art approaches.

2   Method

2.1   Problem Statement

Let $X = x_1, \ldots, x_T$ denote a d-dimensional multivariate fMRI time series, where $x_t \in \mathbb{R}^d$. We seek to model fMRI time series in terms of a sequence of hidden states, where each segment corresponds to a subsequence $X_{i...i+m} = x_i, \ldots, x_{i+m}$ and maps to a predictive (latent) state z, represented as a one-of-K vector, where $|z| = K$ and $\sum_{i=1}^{K} z_{t,i} = 1$. For simplicity of notation, let $z_t = k$ denote $z_{t,k} = 1$ and let $Z = z_1, \ldots, z_T$ denote the sequence of latent states. Then for all x_t mapping to state k, it satisfies the following condition.

$$
\begin{aligned}
\Pr(x_{t+1} | X_{1,\ldots,t}, z_t = k) &= \Pr(x_{t+1} | z_t = k) \\
&= \Pr(x_{t'+1} | z_{t'} = k) \\
&= \Pr(x_{t'+1} | X_{1,\ldots,t'}, z_{t'} = k)
\end{aligned}
\tag{1}
$$

To model transitions between states, we use HMM. HMM is a family of models that describe time series using a sequence of states. All states have the same probabilistic distribution but different distribution parameters.

2.2   Proposed Method

Total brain connectivity activation is formulated as a regularized denoising problem. Let $X = \{x_1, x_2, \ldots, x_m\}$ denote multivariate time series, where $x_m \in R^{V \times T}$, V and T denote the number of voxels of input and the number of time points. Given such a time series, we look for the output $Y \in R^{V \times T}$ such that

$$
\begin{aligned}
\hat{Y} = \underset{Y}{\mathrm{argmin}} \frac{1}{2} \|X - Y\|_F^2 + \sum_{v=1}^{V} \lambda_T(v) \sum_{t=1}^{T} |U_s(v,t)| \\
+ \sum_{t=1}^{T} \lambda_s(t) \sum_{m=1}^{M} \sqrt{\sum_{v \in M_m} \Delta_{lap}\{X(v,t)\}^2}.
\end{aligned}
\tag{2}
$$

Where $\lambda_T(v)$ denotes the temporal regularization term for voxel v, $U(v, \cdot) = \Delta_L\{C(v, \cdot)\}$ is the deconvolved BOLD signals, λ_s is the spatial regularization term at time point t, M is brain regions defined by AAL atlas.

Here, we hypothesize that the transition probabilities of each state at each time point for a given functional brain connectivity evolve dynamically as a function of the activity levels of others. Specifically, we use a Gaussian observation model to estimate likelihood of observing a specific observation.

For each time point t, the corresponding state transition probability represents the probability of a state to be active in that time point (see Fig. 1). More explicitly, the transition probability of a connectivity x_k from state s_i at time point t to state s_j at time point $t+1$, which refers to the actual transitions $Pa(s_t = k)$, thus depends on the intrinsic transition probability of connectivity network x_k itself and the influences of other connectivity networks (with $x \neq l$).

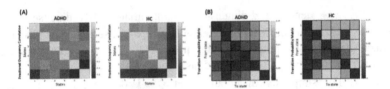

Fig. 1. Transitions between functional brain connectivity of ADHD and HC. (A) is fractional occupancy correlations across each pair of connectivity networks, indicating the total time spent in each state per subject. It exhibits strong correlations between states across subjects. (B) shows the probability of transitioning from one state to another, showing some transitions are more likely to happen.

$$Pa(s_t = k) = \sum_l \theta_{l,k} Pa(s_{t-1} = l). \tag{3}$$

Where $\theta_{l,k}$ denotes the transition probabilities. Based on this formulation, the observed data at each time point can be modelled as a mixture of Gaussian distributions, with weights given by $w_{tk} = Pa(s_t = k)$.

A standard inference for HMM is based on a dynamic programming algorithm (the Viterbi algorithms) with forward and backward variables, which requires a complete sequential forward pass through the data followed by a complete backward pass to estimate state time series. Although the entire process can be improved by parallelization formulation, it is still very time consuming when the time series are long or the number of samples is large. To solve this problem we propose to implement a stochastic variational inference with regularization strategy. We seek to minimize the free energy, which is a quantity to measure the divergence between real and factorized distributions and the entropy of factorized distribution. Whereas standard variational inference deploys a completely random process, the proposed method picks up samples in a decay order at each iteration. This will discourage frequently selected subjects to be selected in the current iteration, leading to improved computation efficiency. This is performed by the following equation

$$w_i = \gamma^{f_i}. \tag{4}$$

Where w_i is the unnormalised probability of selecting subject i, f_i is the appearance frequency of subject i in previous iterations, $\gamma \leq 1$ is the parameter controlling the confidence of subjects selected in current iteration.

To control the computational demanding for long time series and a large dataset, we adopt regularization strategy to constrain the stochastic variational inference. Let $\theta^k_{l,t \rightarrow t+1}$ denotes the transition probability of connectivity k evolve from time point t to time point $t+1$, it satisfies

$$\sum_{l \neq k} \theta^k_{l,t \to t+1} < \rho_{k,t}.$$ (5)

Where $\rho_{k,t}$ denotes the regularization level of connectivity k at time point t. We describe the HMM steps using following pseudocode:

Algorithm: Dynamical modelling of fMRI time series

Input: N |* the length of observation sequence *|, L |* state set of length *|
Output: Q |* a posteriori most probable state sequence *|
begin initialize Q ← { }, t ← 0
for t ← t + 1, j ← j + 1, **until** t = T **do**
    for j ← j + 1, **until** j = c **do**
      $_j$ (t) ← $b_{j,k}\upsilon(t)\sum_{i=1}^{c} \alpha_i (t-1)\alpha_{i,j}$
      j' ← arg $max_j \alpha_j(t)$
      Append ω'_j to Q
    end
end
Return Q |* a posteriori most probable state sequence *|

---

3 Experiments and Results

3.1 Dataset

Data analyzed in this report is collected from the ADHD-200 dataset (http://fcon_1000. projects.nitrc.org/indi/adhd200/). We include all individuals meeting the diagnostic and statistical manual of mental disorders, fourth edition, revised (DSM-IV-TR) criteria of combined ADHD, hyperactive ADHD, or inattentive ADHD, collectively referred to as ADHD group. We also include healthy controls (HC) as HC group for the purpose of comparison. Demographic information for all samples is summarized in Table 1.

Table 1. Demographic information of participants.

Variables	HC	ADHD
Number	116	78
Age	11.71	12.37
Gender(Female/Male)	45/71	7/71
Verbal IQ	49.55	51.03
Performance IQ	50.03	52.85

IQ = intelligence quotient.

3.2   Estimation and Analysis of Dynamic Functional Brain Connectivity

In the following, we analyzed our experimental results on functional brain connectivity obtained with 6 hidden states for ADHD and HC. We first visualized inferred states estimated at the group level. Each state represented distinct patterns of functional brain connectivity. Figure 1 shows an illustrative inferred states estimated by proposed HMM model and the result of correlating fractional occupancy (FO) of each pair of states across samples. FO is a metric to measure temporal characteristics of the states [20]. The functional connectivity matrix was projected to a circle graph. For unclut-tered, we showed the top 15% weights in each state. These maps reflect functional brain connectivity in rs-fMRI and functional states spontaneously change over the scanning time.

We used the temporal structure of the inferred functional brain states to reflect the temporal organization of brain activity. Particularly, we examined the transition probability matrix estimated by the proposed method, in which the probability of transitioning form one state to another is specified. Here, we also examined the FO correlations for each pair of states across subjects, which can reflect temporal char-acteristics of states. Figure 2 shows the results of FO correlation of each pair of states across subjects and the estimated transition probabilities.

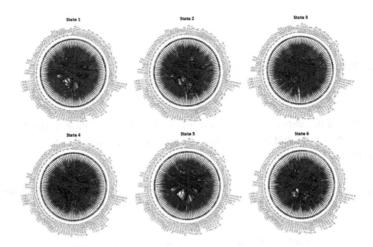

Fig. 2. Inferred functional states of ADHD samples. Note that only the top 15% of weights are presented and there is no correspondence between states of ADHD, for examples, the i-th state is not necessarily related with the j-th state.

In order to revel functional characteristics involved sensory and motor system [15], attention system [10], visual system [5] and default-mode system [4], we decomposed estimated functional connectivity into brain regions related to these areas. Figure 3 shows one functional brain state among regions on above mentioned systems.

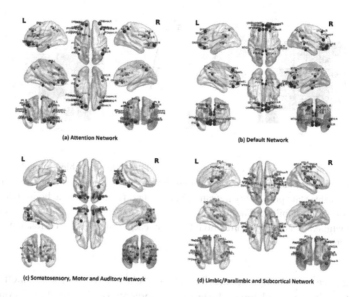

Fig. 3. Visualization of one functional brain state, which decompose ROIs into regions into different type of functional connectivity networks. Red color nodes represent attention connectivity network. Blue color nodes represent default-mode connectivity network. Purple color nodes represent somatosensory, motor and auditory connectivity network. Green color nodes represent limbic/paralimbic and subcortical connectivity network. (Color figure online)

3.3 Identifying Repeating Connectivity Patterns

In order to find repeating connectivity patterns in whole-brain connectivity across a group of subjects, we performed Louvain hierarchical clustering algorithm to find communities on resulting FO correlation matrix. The repeating patterns (metastates) are associated to a set of brain states whose FOs are strongly correlated across multiple subjects. Experimental results indicate a clear hierarchical organization of temporal functional brain connectivity with three metastates (Fig. 4B, C). Specifically, state 5 is strongly correlated with state 6 while less uncorrelated with other states. Figure 5 illustrates the result of transition probability matrix in ADHD and performs hierarchical clustering on the resulting correlation matrix.

3.4 Assessing the Relation Between States and Clinical Outcomes

To validate the effectiveness of proposed method, we compared our method for ADHD diagnosis with resting-state functional brain connectivity with Pearson correlation (baseline model), sliding window proposed by Liegeois *et al.* (code can be found in: https://github.com/CyclotronResearchCentre/SlidingWindowFC), and standard HMM proposed by Diego *et al.* (code can be found in: https://github.com/vidaurre/HMM-fMRI) on the same ADHD dataset.

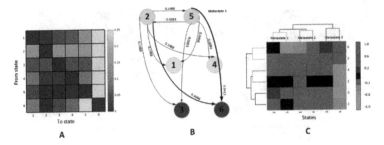

Fig. 4. Transition between brain connectivity in ADHD displays hierarchical organization. A shows transition probability matrix revealing the probability of transiting from one state to another. B indicates state probability transition in metastate 1. Nodes denote brain states and links denote transition probability. The thickness of links represent state transition probability. C shows fractional occupancy correlation between each state. These correlations display a hierarchical structure.

- Baseline: we first computed the Pearson correlation for regional functional connectivity. We extracted local clustering coefficients of the functional connectivity using Brain Connectivity Toolbox (https://sites.google.com/site/bctnet).
- Sliding window: This method uses a sliding window to process time series BOLD signals and estimate dynamic functional connectivity. We used a window size of 50 time points on 116 regional time series of ADHD data. The length of each time series is 232 in our case. For each sample, the process results a volume of size $183 \times 116 \times 116$, representing the temporal functional brain connectivity networks. We extracted local clustering coefficients as features for each sample.
- Standard HMM: This method uses a standard HMM model with Bayesian inference to estimate the temporal structure of brain states. Temporal brain states are reflected by transition probability matrix. We selected 6 hidden states, same as our method. The data has been preprocessed to have a mean of 0 and standard deviation of 1.

We used the proposed model to estimate the dynamical functional connectivity on all subjects collected from ADHD-200. Then, we deployed a linear support vector machine (SVM) as classifier to evaluate the performance of models. And we performed a leave-one-out cross-valuation technique. We quantitatively evaluated accuracy, precision and recall and compared the classification performance with other methods (see Table 2). In terms of the classification accuracy, the proposed method achieved the highest accuracy of 71.58%. Compared to the baseline method, our method improved by 11.29%. In the mean-time, sliding window and standard HMM achieved 67.71% and 65.32% respectively. In terms of the precision and recall, our method achieved 69.59% and 72.33% respectively.

Table 2. A summary of performances of the competing methods on ADHD-200 dataset.

Methods	Accuracy	Precision	Recall
Baseline	59.30%	60.90%	58.00%
Sliding window [13]	67.71%	62.32%	70.53%
Standard HMM [18]	65.32%	61.57%	63.07%
Proposed method	**71.58%**	**69.59%**	**72.33%**

4 Conclusion

In this paper, we developed a regularized HMM model to estimate spatiotemporal nature of functional brain connectivity of interacting brain regions, characterizing its properties by repeating patterns identified across a group of subjects. Specifically, the hierarchical organization of functional brain connectivity revealed three metastates. It suggests transitions between functional brain connectivity are nonrandom, with certain connectivity patterns are more likely to appear than others. By building such a framework, we can estimate the likelihood of input resting-state fMRI features belonging to a disease status. We validated our method with other state-of-the-art methods on the same dataset and showed improved performance in identifying clinical labels of ADHD and HC on test data.

References

1. Bassett, D.S., Wymbs, N.F., Porter, M.A., Mucha, P.J., Carlson, J.M., Grafton, S.T.: Dynamic reconfiguration of human brain networks during learning. Proc. Nat. Acad. Sci. **108**(18), 7641–7646 (2011)
2. Betzel, R.F., Fukushima, M., He, Y., Zuo, X.N., Sporns, O.: Dynamic fluctuations coincide with periods of high and low modularity in resting-state functional brain networks. NeuroImage **127**, 287–297 (2016)
3. Binnewijzend, M.A.A., et al.: Brain network alterations in Alzheimer's disease measured by Eigenvector centrality in fMRI are related to cognition and CSF biomarkers. Hum. Brain Map. **35**(5), 2383–2393 (2014)
4. Buckner, R.L., Andrews-Hanna, J.R., Schacter, D.L.: The brain's default network. Ann. N. Y. Acad. Sci. **1124**(1), 1–38 (2008)
5. Calhoun, V.D., Adali, T., McGinty, V.B., Pekar, J.J., Watson, T.D., Pearlson, G.D.: fMRI activation in a visual-perception task: network of areas detected using the general linear model and independent components analysis. NeuroImage **14**(5), 1080–1088 (2001)
6. Chai, X.J., Castañón, A.N., Öngür, D., Whitfield-Gabrieli, S.: Anticorrelations in resting state networks without global signal regression. Neuroimage **59**(2), 1420–1428 (2012)
7. Chang, C., Glover, G.H.: Time–frequency dynamics of resting-state brain connectivity measured with fMRI. Neuroimage **50**(1), 81–98 (2010)
8. de Haan, W., van der Wiesje, M., Flier, T.K., Smits, L.L., Scheltens, P., Stam, C.J.: Disrupted modular brain dynamics reflect cognitive dysfunction in Alzheimer's disease. Neuroimage **59**(4), 3085–3093 (2012)
9. De Pasquale, F., et al.: Temporal dynamics of spontaneous MEG activity in brain networks. Proc. Nat. Acad. Sci. **107**(13), 6040–6045 (2010)

10. Fan, J., McCandliss, B.D., Fossella, J., Flombaum, J.I., Posner, M.I.: The activation of attentional networks. Neuroimage **26**(2), 471–479 (2005)
11. Hindriks, R., et al.: Can sliding-window correlations reveal dynamic functional connectivity in resting-state fMRI? Neuroimage **127**, 242–256 (2016)
12. Hoekzema, E., et al.: An independent components and functional connectivity analysis of resting state fMRI data points to neural network dysregulation in adult ADHD. Hum. Brain Map. **35**(4), 1261–1272 (2014)
13. Kiviniemi, V., et al.: A sliding time-window ICA reveals spatial variability of the default mode network in time. Brain Connect. **1**(4), 339–347 (2011)
14. Liégeois, R., et al.: Cerebral functional connectivity periodically (de) synchronizes with anatomical constraints. Brain Struct. Funct. **221**(6), 2985–2997 (2016)
15. Pa, J., Hickok, G.: A parietal–temporal sensory–motor integration area for the human vocal tract: Evidence from an fMRI study of skilled musicians. Neuropsychologia **46**(1), 362–368 (2008)
16. Preti, M., Bolton, T.A.W., Van De Ville, D.: The dynamic functional connectome: state-of-the-art and perspectives. Neuroimage **160**, 41–54 (2017)
17. Rolls, E.T., Joliot, M., Tzourio-Mazoyer, N.: Implementation of a new parcellation of the orbitofrontal cortex in the automated anatomical labeling atlas. Neuroimage **122**, 1–5 (2015)
18. Song, J., Nair, V.A., Gaggl, W., Prabhakaran, V.: Disrupted brain functional organization in epilepsy revealed by graph theory analysis. Brain Connect. **5**(5), 276–283 (2015)
19. Tang, W., et al.: Dynamic connectivity modulates local activity in the core regions of the default-mode network. Proc. Nat. Acad. Sci. **114**(36), 9713–9718 (2017)
20. Vidaurre, D., Smith, S.M., Woolrich, M.W.: Brain network dynamics are hierarchically organized in time. Proc. Nat. Acad. Sci. **114**(48), 12827–12832 (2017)

From Complex Network to Skeleton: m_j-Modified Topology Potential for Node Importance Identification

Hanning Yuan[1], Kanokwan Malang[1(✉)], Yuanyuan Lv[1],
and Aniwat Phaphuangwittayakul[2]

[1] School of Computer Science and Technology, Beijing Institute of Technology,
No. 5 South Zhong Guan Cun Street, Beijing 100081,
People's Republic of China
kanokwan.malang@yahoo.com
[2] International College of Digital Innovation, Chiang Mai University,
No. 239 Nimmanhaemin Road, Suthep, Muang, Chiang Mai 50200, Thailand

Abstract. Node importance identification is a crucial content in studying the substantial information and the inherent behaviors of complex network. On the basis of topological characteristics of nodes in complex network, we introduce the idea of topology potential from data field theory to capture the important nodes and view it as the skeleton nodes. Inspired by an assumption that different mass of node (m_j parameter) reflects different quality and interaction reliability over the network space. We propose TP-KS method that is an improved topology potential algorithm whose m_j is identified by k-shell centrality. The important nodes identified by TP-KS is ranked and verified by SIR epidemic spreading model. Through the theoretical and experimental analysis, it is proved that TP-KS can effectively extract the importance of nodes in complex network. The better results from TP-KS are also confirmed in both real-world networks and artificial random scale-free networks.

Keywords: Complex network · Skeleton network · Topology potential
Node importance evaluation

1 Introduction

There exist the rapid growth of data from social-network in term of amount, size, transmission and complexity [1]. Analyzing these kinds of data from real-world and complex network is brought about some challenges since network contains nodes and edges heterogeneity. Those are not only critical for network data management, but also difficult to understand universal principles and dynamical process on complex networks. With this reason, the efficient methods or techniques are required to capable summarize and justify the essential information of the entire complex network structure. Alternatively, our research is relied on a different perspective for studying complex network. Instead of investigating the significant information from the entire network, only network's backbone is needed to pay more attention. In this way, the

G. Gan et al. (Eds.): ADMA 2018, LNAI 11323, pp. 413–427, 2018.
https://doi.org/10.1007/978-3-030-05090-0_35

minimization of complex and large-scale network to a smaller but effective called skeleton network is taken into account.

1.1   From Complex Network to Skeleton Network

Skeleton network is a product of a skeletonizing process after filtering the meaningless data from the complex network [2]. It is supposed to consider only a central-core network structure which aims to preserve the most important information of the original network. In this paper, the skeleton network is defined as "an essential network whose nodes and edges are extracted from complex network by its importance, unique and efficient for the whole network functions". To our knowledge, the construction of skeleton network consists of three consecutive phases: the determination of important nodes, the identification of associated interactive edges, and the reconstruction of skeleton network. Accordingly, how to quantify the importance of every nodes in complex network is a key step to skeletonize a network, and should be done at the beginning. Therefore, node importance identification is conducted in this literature in order to extract the skeleton nodes which are believed to carry the most useful information of complex network.

1.2   Node Importance Evaluation Inspired by Topology Potential

The determination of important nodes is recently relied on the most efficient node evaluation methods which can not only provide high accuracy results, but also require low computational demand. Based on topology characteristics, many scholars proposed the great significant methods to research the influence power of the important nodes such as PageRank [3], centrality measures, and topology potential. To our knowledge, Topology Potential (TP) is brought into a key technique to measure the node's importance over the recent years. TP is based on the concept that "the importance of nodes which adjacent to the important nodes" [4]. The efficiency of TP can be proved by lots of evidences that covers the analysis methods of social network, medical science and information search area [5–9]. According to its formula, the node mass m_j is an important parameter that directly affect to the value of a node's topology potential. However, most researches adopted $m_j = 1$, indicating all nodes has the same influence. Conversely, nodes in real-world networks always appear different quality. Each node exhibits different ability to connect with others. The selection of $m_j = 1$ for all nodes has been criticized to be unreasonable for measuring the value of nodes. Han et al. [10] shows that m_j can be used for describing intrinsic property of node. In [5], the entropy mapping method based on f function is employed to response for each node quality. For identifying communities, Wang et al. [11] introduced an improved m_j which is inspired by PageRank algorithm. They also argue that when suppose $m_j = 1$, the different interaction ability of nodes is omitted and the calculation of TP deviates from the actual value. In this way, the defined m_j is not only affect to the precision of community detection, but it could also directly affect to the accuracy of node importance evaluation. In our belief, TP is better to be conducted under the assumption of different mass of node reflects different interaction ability and distinctive importance. Applying TP may shade some light for our research to precisely extract the most important nodes with different mass. TP is also expected to preserve the most important information of complex network in a shape of skeleton.

The objective of this paper is to introduce the idea of m_j-Modified Topology Potential algorithm based on k-shell centrality (TP-KS) and to investigate the performance of TP-KS for identifying the most important nodes which can be viewed as the skeleton nodes. The important nodes encompass with the backbone information of complex network. They are believed to carry the specific characteristics that help us to understand network's behavior. Furthermore, network could remain its functions and survivability from a small portion of this important component. In order to evaluate the correctness of node's importance resulted from topology potential algorithms, the SIR model is implemented. We also use the Kendall's tau coefficient correlation to verify the relationship between the node's importance ranked by TP-KS and the ranking from the epidemic spreading model. The Monotonicity method is also employed to determine the ability of algorithm to distinguish the node's difference. The experimental analysis on six real-world networks and two artificial networks prove that TP-KS can effectively identify the importance of nodes with high accuracy.

The remainder of this paper is organized as follows. Section 2, we briefly review the concept of TP for node importance evaluation. Our propose TP-KS method is also introduced. In Sect. 3, we demonstrate the experiment setting which includes the details of datasets and the validation methods. Finally, experimental results and conclusions are discussed in Sects. 4 and 5, respectively.

2 Method

Topology potential was introduced based on the data field of cognitive physics [12]. According to the data field, any node in complex network is subject to an interaction with other nodes. Therefore, TP is defined as the prospective of any nodes in the network are influenced by its neighbors and its position [13].

In this work, the subject of complex network is social individuals which can be represented by nodes. While, edges illustrate the interactive relationships among every individual e.g. communication and knowledge transfer. Given an undirected complex network $G = (V, E)$ which consists of the set of nodes V and the set of edges E. $V = \{v_1, v_2, \ldots, v_n\}$ represents a unique node, while $E = \{e_1, e_2, \ldots, e_m\}$ is denoted as the interactive relation between two nodes u and v. Based on The Gaussian potential function [1], TP can be represented as the differential position of each node in the network. Each node's influence will quickly decay as distance increases. In term of mathematic definitions, TP of any node v_i is defined as follow.

$$\varphi(v_i) = \sum_{j \in N} \left(m_j * exp^{-\left(\frac{d_{ij}}{\sigma}\right)^2} \right) \tag{1}$$

where $\varphi(v_i)$ is the topology potential of node v_i, for $v_i \in V$. d_{ij} is denoted as the shortest distance between node v_i to node v_j [14]. Parameter σ is a factor used to control the influence region of each node. m_i is the mass of node.

The TP of each node can reflect node's importance in the topology by optimizing influence factor [13]. In Eq. (1), parameter σ control the range of influence of nodes v_i. Generally, the optimal choice of σ can be fixed and can be calculated by a minimum entropy. Since TP value of each node is different, the uncertainty of nodes in network topology space is the lowest accounting to the minimum entropy. Suppose the potential score of each node $v_1, v_2, \ldots, v_n$ be $\varphi(v_1), \varphi(v_2), \ldots, \varphi(v_n)$ respectively, the minimum potential entropy H of topology potential field can be defined as follow.

$$H = -\sum_{i=1}^{n} \frac{\emptyset(v_i)}{Z} \log \frac{\emptyset(v_i)}{Z} \tag{2}$$

where Z is the normalization factor, for any $\sigma \in (0, +\infty)$, H satisfy $0 \leq H \leq \log(n)$ and reach maximum value $\log(n)$ if and only if $\varphi(v_1) = \varphi(v_2) =, \ldots, = \varphi(v_n)$. For a given σ, the affecting area of every node is approximate to local area of $l = 3/\sigma\sqrt{2}$ hops. And when the distance is greater than $3/\sigma\sqrt{2}$, unit potential function will rapidly decay to 0 [9].

In data field, m_j (≥ 0) stands for the mass of node v_j ($i = 1, 2, \ldots, n$) which represents the strength of the data field from v_j and meets the normalization condition $\sum_{i \in n} m_i = 1$. Thus, all nodes in topology potential field are satisfy the normalization condition and has the same quality. It is in fact that the influence of a node in topology potential field increases with a node's position and its surroundings. However, the mass of a node itself which reflect its own ability should not be neglected. Enlighten by the idea of different nodes contains different ability, evaluating the importance of the nodes should be induced by different value of node mass parameter. In our belief, node mass is diverse and it can be derived from centrality indices. In this paper, k-shell decomposition (KS) proposed by Kitsak et al. [15] is selected as a representative among other centralities. Consequently, the mass of node i in topology potential field can be described as follow:

$$m_j = KS_v \tag{3}$$

KS is a maximal connected subgraph of G in which vertices have degree at least k. KS_v is the coreness ks of node v, indicates that node v belong to a k-shell but not to any $(k+1)$-shell. The KS classifies all nodes of G into k-shells by removing nodes iteratively. Starting from $k = 1$ nodes, and assign to the KS value $k = 1$ to the removed nodes until there are only nodes with degree $k > 1$. Then repeat the pruning for $k = 2$ nodes, until all nodes of G are assign to one of the k-shells. As a result, each node is associated with one KS index, and the network can be regarded as the unification of all k-shells. Then, the contribution of KS controlled by TP is further implemented based on Eq. (1). The more k-shell value is, the bigger node mass is. And the higher degree of node mass, the more apparent of node's importance and its ability.

3 Experimental Setting

3.1 Materials

In this study, we conduct well-known complex networks experiments to validate the ranking of node importance obtained by TP-KS. The experimental datasets are derived from the shared resources of social network analysis [16, 17]. The centrality values are calculated from a software plugin called CytoNCA [18] which is originally used in bioinformatics areas. The experiment was performed within JAVA programming in Windows environments. The Rstudio has been used to generate data visualization.

3.2 Data Sources

The experiments in this paper are performed in both artificial and real-world networks. We compare the performance of TP-KS with other methods on two artificial networks generated by the LFR model and BA model. LFR is a random network distribution model proposed by Lancichinetti-Fortunato-Radicchi. The LFR generator can produce the undirected and weighted networks with power laws degree distribution and over-lapping nodes. Moreover, the so-called Barabási-Albert (BA) network model is also employed. The syntactic random scale-free networks can be generated from BA model, and they are commonly embedded with the preferential attachment mechanism.

Beside the artificial networks, the different sizes of six real-world social networks has been conducted. The selected real-world networks are Zachary Karate club, Dolphin's social network, Les Miserables, US political books, Jazz Musician, and Email network. All real-world networks contain community structure and they are treated as undirected and unweighted network. The details of real-world network for our experiments are listed in Table 1.

Table 1. The topological features of six real-world networks. n and m are the number of nodes and edges. $\langle k \rangle$ and k_{max} denote the average degree and the maximum degree. c and d are the clustering coefficient and the average shortest path. *dens* represent the network density. β_{th} is the epidemic threshold of networks and β is the infection ratio used in the SIR epidemic model.

Network	n	m	$\langle k \rangle$	k_{max}	c	d	*dense*	β_{th}	β
Karate club	34	78	4.5882	17	0.2557	2.4082	0.3190	0.1287	0.15
Dolphin	62	159	5.1290	12	0.3088	3.3569	0.0841	0.1469	0.15
Les miserable	77	254	6.5974	36	0.4989	2.6411	0.0868	0.1515	0.17
Political books	105	441	8.4	25	0.3484	3.0787	0.0808	0.0837	0.10
Jazz musicians	198	2742	27.697	100	0.5203	2.2350	0.1406	0.0258	0.05
Email	1133	5451	9.6222	71	0.1662	3.6060	0.0085	0.0535	0.07

3.3 Evaluation Methodologies

To evaluate the performance of TP-KS, we rank all nodes according to its importance value in descendent order. For those with the same value, their relative order is random.

To quantify the correctness of the node importance ranking list obtained by TP-KS, we use the Susceptible-Infected-Recovered (SIR) model. As clarified in [19], the node importance raked list generated from one effective method should be consistent as much as possible with the one generated by the real spreading process. With this reason, we employ the SIR to estimate the importance of the nodes from the real spreading influence. We set the average number of SIR simulations = 1000 for all networks. Then, the number of recovered nodes at the end of spreading process is used to reflect the real influence of the node v.

Together with SIR, we adopt Kendall's tau correlation coefficient [20] as a method to measure the consistency of the node importance ranking lists. Kendall's tau can be computed by counting the number of pairs that agree in ranking and subtracting from the number of pairs that disagree in ranking. Let (x_i, y_i) and (x_j, y_j) be a set of joint observations from two ranking lists X and Y, respectively. Any pair of observation (x_i, y_i) and (x_j, y_j) are said to be concordant if the ranking for both elements agree with each other: that is, if $x_i > x_j$ and $y_i > y_j$ or $x_i < x_j$ and $y_i < y_j$. They said to be discordant if $x_i > x_j$ and $y_i < y_j$ or $x_i < x_j$ $y_i > y_j$. If $x_i = x_j$ or $y_i = y_j$, the pair is neither concordant nor discordant. The Kendall's tau τ coefficient can be defined as:

$$\tau = \frac{N_c - N_d}{\frac{1}{2}n(n-1)} \tag{4}$$

where N_c and N_d are the number of concordant and discordant pairs. The final value is normalized by the total number of (2^n) such pairs, resulting the range between -1 to 1. Where $\tau = 1$ means that the ranking of the two measures are identical, while a negative score signifies anti-correlation among rankings. The tau's value closer to 1 corresponds to the stronger ranking correlation.

Furthermore, we also employ the Monotonicity method to quantitatively measure the resolution of our method. As mentioned in [21], the good index in ranking the influence of nodes should takes a high resolution. Similarly, the good node importance evaluation method should also have a high resolution corresponding to the ability to characterize the node's difference. A Monotonicity M index of the ranking vector R is define by the following formula.

$$M(R) = \left[1 - \frac{\sum_{r \in R} N_r(N_r - 1)}{N(N-1)}\right]^2 \tag{5}$$

where N is the size of ranking vector R, N_r is the number of nodes with the same rank r. M index quantifies the fraction of nodes in the ranking list. If $M(R) = 1$, means that the node importance ranking method is perfectly monotonic due to each node is classified in a different index value. Otherwise, when $M(R) = 0$, means all node in R has identical rank.

4 Results and Discussion

4.1 Evaluate the Correlation Between SIR Spreading Ability and Different Methods

To evaluate the node importance ranking, we investigate the relationship between the real spreading influence of the nodes generated by SIR model and seven different node importance evaluation methods. The competitive methods are including the well-known PageRank (PR) [3], K-shell decomposition method controlled by gravity formula (G) [21], Degree and Importance of Line (DIL) [22], New Network Centrality (NC) [23] and Local Average Connectivity (LAC) [24], the classical topology potential (TP), and our proposed TP-KS. We examine the node importance value and the ranking obtained from six real-world networks listed in Table 1.

To employ SIR model, we first obtain the epidemic threshold β_{th} for each network. The β_{th} is given as $\beta_{th} \sim \langle k \rangle / \langle k^2 \rangle$, where $\langle k \rangle$ and $\langle k^2 \rangle$ denote the average degree and the second moment average degree, respectively. Afterwards, the infection probability β which influence the epidemic spreading process is set to be slightly larger than epidemic threshold. The value of β_{th} and β for different networks are illustrated in Table 1. The Kendall's tau coefficient correlation of the benchmarking methods is calculated by Eq. (4). As illustrates in Table 1. PR, G, LAC, DIL, NC, TP and TP-KS are all positively and highly correlate with the influence capability of the nodes evaluated by SIR model. Significantly, the node importance ranking calculated by TP-KS summarized in Table 2 has strongest correlation with the spreading influence by SIR in all networks. While PR, LAC, DIL, and NC are comparatively lower than other methods. It is remarkable that G and TP-KS provides closely results in tau's value, because both methods are based on KS. Additionally, the TP itself provide also a good result. Therefore, the performance of TP-KS controlled by the classical TP is enhanced with balancing between the local and global information of TP and KS, respectively.

Table 2. The correlation of PageRank, G, LAC, DIL, NC, TP and TP-KS. β_{th} is the spreading threshold of networks and β is the infection probability used in SIR spreading process. $\tau(\cdot)$ is the Kendall'tau coefficient correlation of the corresponding measures.

Network	β_{th}	β	$\tau(PR)$	$\tau(G)$	$\tau(LAC)$	$\tau(DIL)$	$\tau(NC)$	$\tau(TP)$	$\tau(TP_{ks})$
Karate club	0.1287	0.15	0.8021	0.8877	0.7148	0.8431	0.7433	0.9162	**0.9251**
Dolphins	0.1469	0.15	0.8377	0.8985	0.7615	0.9043	0.7705	0.9313	**0.9344**
Les miserables	0.1515	0.17	0.7973	0.9327	0.8360	0.8585	0.8855	0.9275	**0.9398**
Political books	0.0837	0.10	0.8223	0.9242	0.8233	0.7868	0.8249	0.8930	**0.9476**
Jazz musician	0.0258	0.05	0.8753	0.9494	0.8683	0.8543	0.8931	0.9350	**0.9618**
Email	0.0535	0.07	0.8427	0.9244	0.8119	0.8828	0.8127	0.9316	**0.9554**

To further evaluate the ability of our method in distinguishing the difference of the nodes, the Monotonicity is also performed. According to the results in Table 3, PR, G, TP, and TP-KS are denoted as the good methods which contains high monotonicity value. More importantly, the ranking from PR, TP and TP-KS are perfectly monotonic in Political book network. In other networks, almost all nodes are categorized into a different rank. This intuitively confirms that identifying mass of node based on KS do not lead the algorithm fail to uncover the difference of nodes in the core-like structured network like KS did.

Table 3. The ability of different methods to distinguish the node's difference by its importance value. $m(\cdot)$ is the Monotonicity index of the corresponding measures.

Network	$m(PR)$	$m(G)$	$m(LAC)$	$m(DIL)$	$m(NC)$	$m(TP)$	$m(TP_{ks})$
Karate club	0.95419	0.85258	0.77540	0.83946	0.71690	0.94377	0.95071
Dolphins	0.99788	0.92835	0.84959	0.96229	0.86231	0.99683	0.99789
Les miserables	0.95807	0.95006	0.82148	0.86733	0.82458	0.95673	0.95673
Political books	1	0.97488	0.95402	0.99963	0.97199	1	1
Jazz musician	0.99928	0.99887	0.99590	0.99805	0.99641	0.99887	0.99928
Email	0.99989	0.98934	0.85115	0.96286	0.85651	0.99989	0.99989

4.2 Evaluate the Top-k Important Nodes Ranked by Different Methods

In most situations of identifying node importance, people are more interested in the high influential nodes which subsist in a small portion of the entire network. This

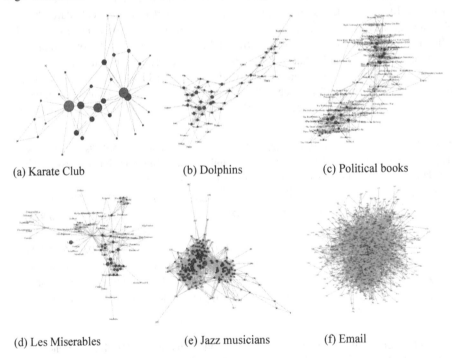

(a) Karate Club   (b) Dolphins   (c) Political books

(d) Les Miserables   (e) Jazz musicians   (f) Email

Fig. 1. The six real-world networks with node importance value identified by TP-KS. The important nodes in network are represented by the size of vertices corresponding to its importance value. The more important node is, the larger size the vertex is.

Table 4. The top-5 nodes importance ranked by the average number of infected nodes simulated by SIR are compared with the corresponding important nodes ranked by different methods.

Network	Average infected nodes	SIR	PR	G	LAC	DIL	NC	TP	TP_{ks}
Karate club	5.666	**34**	**34**	**1**	4	**34**	**1**	**34**	**1**
	5.426	**1**	**1**	**34**	8	**1**	**34**	**1**	**34**
	5.17	33	33	3	2	3	33	33	3
	4.887	3	3	33	14	33	2	3	33
	4.424	2	2	2	1	32	4	2	2
Dolphins	6.803	**38**	**15**	**15**	14	21	46	**15**	**15**
	6.465	**15**	**18**	46	10	**38**	14	**38**	**38**
	5.969	46	52	38	19	15	15	46	46
	5.743	41	58	34	46	37	58	34	21
	5.704	34	38	21	17	2	10	52	41
Les miserables	27.449	**11**	12	49	64	**11**	**11**	**11**	**11**
	26.117	**49**	1	**11**	66	56	**49**	**49**	**49**
	25.286	56	49	56	60	28	56	56	56
	25.103	28	56	59	62	49	59	28	59
	24.864	26	28	63	65	24	65	26	65
Political books	10.441	**13**	**13**	9	76	31	9	**13**	9
	10.312	**9**	**9**	13	85	4	85	**9**	13
	10.227	74	4	85	75	73	13	4	85
	9.814	85	85	4	74	13	73	85	4
	9.628	31	73	73	83	9	74	73	74
Jazz musicians	100	**7**	**67**	**7**	20	**67**	**7**	**67**	**7**
	99	**67**	**7**	20	**7**	93	**67**	**7**	**67**
	97	20	23	23	12	23	20	20	20
	96	23	20	18	18	90	23	23	23
	93	19	70	13	14	7	13	90	18
Email	129.797	**105**	**105**	**105**	886	333	**105**	**105**	**105**
	120.237	42	23	16	888	23	16	333	333
	116.078	16	333	42	887	105	196	23	42
	111.004	196	41	333	885	233	299	42	23
	110	333	42	23	788	41	434	41	76

section, we select top-5 nodes from each dataset to show the performance of different methods aiming at simplifying the representation of our results.

As shown in Table 4, the top-5 nodes from SIR is ranked according to the average number of infected nodes. The corresponding node importance ranking lists obtained by different methods are also sorted by their importance scores. For a specific discussion, we select Zachary Karate Club network to demonstrate how well does our algorithm perform. In Table 4, the top-2 nodes of all networks whose agree with the ranking from SIR are highlighted in bold. In Karate, value of two most important nodes: node 1 and 34 correlated with their characteristics illustrated in Fig. 1(a). Both of them undertakes a hub node responsible for exchanging information within communities. The information detailed in Table 4 confirms that TP-KS is in a good agreement with the ranking list from SIR. TP-KS express its accuracy results that are not deviate from other methods. These could give an evidence in the utility of TP-KS. It can be observed that the nodes which are categorized into the first and the second rank from each method are almost identical. While the remaining node ranks are slightly different. However, when the network become larger such as Email, the node's ranking is highly diverse. It is obvious that the size and the characteristics of different network has effects on identifying node importance. Therefore, the higher tau's value prized to TP-KS not only indicates a good ranking agreement with the real spreading model, but also to confirm the consistency of our method for dealing with various size of networks.

4.3 Evaluate the Performance of Different Methods to the Artificial Network

In this section, we evaluate the performance of different methods on two artificial networks: LFR and BA. The structural properties of LFR network can be characterized by specify many key parameters. We fix the size of LFR network $n = 100$, the average node degree $\langle k \rangle = 10$, the maximum node degree $max_{\langle k \rangle} = 50$, the power law exponent of degree distribution $\gamma = 2$. Furthermore, the mixing parameter μ that used to control the modularity of network is set to be 0.2. In order to validate the essential properties of complex network that affects to the performance of our algorithm, we change the power-law exponent and the mixing parameter to be $\gamma = 2.5$, and $\mu = 0.1$ as summarized in Table 5.

Table 5. The parameter setting for artificial network implementation; LFR and BA network. β_{th} is the spreading threshold of the generated networks and β is the infection probability used in the SIR spreading process. n is the number of nodes, m_t denotes the number of attachments in BA networks, $\langle k \rangle$ and $max_{\langle k \rangle}$ are the average node degree and maximum node degree, γ carry the power law exponent, and μ denotes the mixing parameter in LFR networks.

Network	β_{th}	β	n	m_t	$\langle k \rangle$	$max_{\langle k \rangle}$	γ	μ
LFR-1	0.0581	0.07	1000	-	10	50	2	0.2
LFR-2	0.0678	0.08	1000	-	10	50	2.5	0.2
LFR-3	0.0589	0.07	1000	-	10	50	2	0.1
BA-1	0.0744	0.10	1000	3	-	-	-	-
BA-2	0.0581	0.07	1000	4	-	-	-	-
BA-3	0.0474	0.07	1000	5	-	-	-	-

When performing the SIR simulations, we demonstrate the performance of different methods by considering only a particular infection probability ratio β. Simply say that the tau's value of the competing methods is evaluated under a single spreading probability. The results of tau which is corresponding to the choice of β are given in

Table 6. The correlation of node importance ranked by different node evaluation methods under the variation of LFR networks and BA networks. $\tau(\cdot)$ is the Kendall' tau coefficient correlation of the corresponding measures.

Network	$\tau(PR)$	$\tau(G)$	$\tau(LAC)$	$\tau(DIL)$	$\tau(NC)$	$\tau(TP)$	$\tau(TP_{ks})$
LFR-1	0.5694	0.8565	0.7997	0.5887	0.7664	0.7629	**0.9068**
LFR-2	0.5450	0.8462	0.7627	0.6110	0.7425	0.7713	**0.8906**
LFR-3	0.5835	0.8844	0.8227	0.6187	0.7962	0.7887	**0.9338**
BA-1	0.6684	0.7098	0.7278	0.7698	0.7321	**0.9164**	**0.9164**
BA-2	0.6919	0.7637	0.7476	0.7715	0.7545	**0.9140**	**0.9140**
BA-3	0.7420	0.8013	0.7462	0.7953	0.7570	**0.9064**	**0.9064**

Table 6. As a result, TP-KS perform much better than other methods in all LFR network. Follow with G that also provide a highly correlation. This implies that TP-KS and G can better capture the importance of the nodes in complex network that contains community structure.

Besides LFR network, we also employ BA network model to validate our algorithm. Starting with the connected network whose contains m_0 nodes, a new node is added to the network and is connected with other m_t nodes which proportional to the number of links of the existing nodes. We fix the number of nodes $n = 1000$, $m_0 = m$, and vary the number of attachments m_t equal to 3 and 5, respectively.

As demonstrates in Table 6, TP and TP-KS has highest correlation value in all of BA-1, BA-2 and BA-3. In contrast with PR that perform the worst comparing with other methods in both LFR and BA networks (<70% of ranking correlation). Note that KS centrality is limited to yield a significant of node importance ranking in BA. This because of the m_t of BA indicates every time step t, a new node will connect to m_0 existing nodes with identical value of k-shell. It is worth remarking that our proposed method defines the m_j parameter by KS. Therefore, mass of nodes cannot be differentiated. These depict that in BA network, TP-KS produces similar results as the classical TP, where the mass of all nodes is equal. To the end of this section, the results suggest that TP-KS not only be able to identify the node importance on the real-world networks, but also on the artificial syntactic networks.

4.4 Evaluate the Effect of Node Mass Parameter Identified by Different Centrality Measures

In order to confirm the power of the TP, we also examine the TP with m_j defined by various centrality measures as explained in Table 7. Different centralities indicate the interaction ability of any pair of nodes in different perspectives. Similar with node mass, centrality could reflect the importance value of the nodes accounting for the network structure.

The results in Table 8 illustrates that the integration of TP-DC, TP-LAC, TP-IC, and TP-NC provides comparatively good results (>90% of correlation in most cases). In contrast, both of TP-BC, and TP-SC contribute a smaller tau's correlation than others. TP-IC get the highest tau's value beyond others in Karate and Les Miserables.

Table 7. The eight-classical centrality measures which has been used to define m_j in topology potential algorithm.

Centrality measure	Information structure	Ref.
Betweenness Centrality (BC)	Global	[25]
Closeness Centrality (CC)	Global	[26]
Degree Centrality (DC)	Local	[27]
Eigenvector Centrality (EC)	Global	[28]
Information Centrality (IC)	Global	[29]
Local Average Connectivity (LAC)	Local	[24]
New Network Centrality (NC)	Local	[23]
Subgraph Centrality (SC)	Global	[30]

While TP compute the value of each node at $O(m * n)$. The complexity of IC costs up to $O(n^4)$, higher than other centrality indices. This caused TP-IC to perform the worst performance. It is obvious that the performance of centrality measures under the control of TP in Table 8, are higher than the performance of the distinctive centralities in Table 9. This implies that the method can balance the overlay of determination between the importance of a central node itself and the importance of neighbor's nodes by TP. However, the situation of EC and SC inverse with other centralities. The correlation of EC and SC perform better than TP-EC and TP-SC in all networks. EC and SC are based on the graph's adjacency matrix associated with the eigenvector. While EC focus on the relative high scoring of vertex in adjacency matrix, SC replaces the adjacency matrix with it trace on closed paths. Regardless the initial transformation of the adjacency matrix, SC can converge to EC when the discount parameter reached its maximal value. This reason may explain the closely results between these two methods. Nevertheless, the overall results summarized in Tables 8 and 9 manifest that when m_j parameter in TP is defined with centrality indices, TP algorithm can obtain the inherent importance value of the nodes. And the algorithm can improve its performance on identifying the most important nodes corresponding to the skeleton nodes.

Table 8. The relationship of the node importance ranked by different m_j-based centralities and the rank from SIR model. $\tau(\cdot)$ is the Kendall' tau coefficient of the corresponding measures.

Network	$\tau(TP_{BC})$	$\tau(TP_{CC})$	$\tau(TP_{DC})$	$\tau(TP_{EC})$	$\tau(TP_{IC})$	$\tau(TP_{LAC})$	$\tau(TP_{NC})$	$\tau(TP_{SC})$	$\tau(TP_{ks})$
Karate	0.7718	0.9180	0.8788	0.8948	**0.9465**	0.8824	0.9002	0.8610	0.9251
Dolphins	0.7916	0.9085	0.9291	0.8255	0.8905	0.9096	0.9075	0.8636	**0.9344**
Les Mis	0.7553	0.8900	0.9029	0.8944	**0.9497**	0.9299	0.9409	0.8595	0.9398
Jazz	0.8278	0.9530	**0.9713**	0.9250	0.9568	0.9602	0.9623	0.8892	0.9476
PolBooks	0.8051	0.8632	0.9415	0.7346	0.9562	0.9577	0.9440	0.8936	**0.9618**
Email	0.9141	0.9421	0.9451	0.9240	0.9462	0.9455	0.9460	0.8835	**0.9554**

Table 9. The relationship of the node importance ranked by different centrality indices and the rank from SIR model. $\tau(\cdot)$ is the Kendall' tau coefficient correlation of the corresponding measures.

Network	$\tau(BC)$	$\tau(CC)$	$\tau(DC)$	$\tau(EC)$	$\tau(IC)$	$\tau(LAC)$	$\tau(NC)$	$\tau(SC)$	$\tau(TP_{ks})$
Karate	0.8164	0.8699	0.8556	0.9234	0.8556	0.7148	0.7433	0.9162	**0.9251**
Dolphins	0.7731	0.8181	0.8805	0.8414	0.8847	0.7615	0.7705	0.9170	**0.9344**
Les Mis	0.6699	0.7761	0.8937	0.9111	0.8923	0.8360	0.8855	**0.9436**	0.9398
Jazz	0.7449	0.8617	0.9284	0.9287	0.9286	0.8683	0.8931	0.9287	**0.9476**
Polbooks	0.6766	0.6894	0.8703	0.7989	0.8689	0.8232	0.8249	0.9271	**0.9618**
Email	0.8227	0.9077	0.9075	0.9434	0.9066	0.8119	0.8127	0.9448	**0.9554**

5 Conclusion

In this work, the topology potential is applied to identify the important nodes. A set of the most important nodes is assumed to be responsible for the backbone information of the entire complex network structure. Under the assumption of different mass of node reflects different quality and reliability, we proposed a new method to identify important nodes with an improved m_j parameter of topology potential algorithm. With this assumption, the node importance value is derived from the topology potential function in which its mass is described by the k-shell decomposition. We employed our proposed method in both real-world and artificial networks. The Kendall's tau has been used to evaluate the node importance ranking relationship. From the experimental results, the node importance ranked by TP-KS is highly correlated with the ranking from SIR. Comparing with various existing methods, we can argue that TP-KS is significant in both theory and practice as it can improve the precision of node importance identification. The results are proved that the important nodes identified by TP-KS has stronger topology potential characteristics among neighbor nodes. Moreover, the minimal requirement to explain the substantial complex network information with a small number of important nodes that indicates skeleton nodes could be explored by our contribution.

Acknowledgement. This work was supported by the National Key Research and Development Program of China (No. 2016YFB0502600), The National Natural Science Fund of China (61472039), Beijing Institute of Technology International Cooperation Project (GZ2016085103), and Open Fund of Key Laboratory for National Geographic Census and Monitoring, National Administration of Surveying, Mapping and Geoinformation (2017NGCMZD03).

References

1. Li, D., Wang, S., Li, D.: Spatial Data Mining. Theory and Application. Springer, Heidelberg (2015). https://doi.org/10.1007/978-3-662-48538-5
2. Grady, D., Thiemann, C., Brockmann, D.: Robust classification of salient links in complex networks. Nat. Commun. 3(1), 864 (2012)

3. Kumari, T., Gupta, A., Dixit, A.: Comparative study of page rank and weighted page rank algorithm, vol. 2, no. 2, p. 9 (2007)
4. Zhang, D., Gao, L.: Virtual network mapping through locality-aware topological potential and influence node ranking. Chin. J. Electron. **23**(1), 61–64 (2014)
5. Wang, Y., Yang, J., Zhang, J., Zhang, J., Song, H., Li, Z.: A method of social network node preference evaluation based on the topology potential, pp. 223–230 (2015)
6. Sun, R., Luo, W.: Using topological potential method to evaluate node importance in public opinion. In: Presented at the 2017 International Conference on Electronic Industry and Automation, EIA 2017 (2017)
7. Han, Q., Wen, H., Ren, M., Wu, B., Li, S.: A topological potential weighted community-based recommendation trust model for P2P networks. Peer-Peer Netw. Appl. **8**(6), 1048–1058 (2015)
8. Lei, X., Zhang, Y., Cheng, S., Wu, F.-X., Pedrycz, W.: Topology potential based seed-growth method to identify protein complexes on dynamic PPI data. Inf. Sci. **425**, 140–153 (2018)
9. Ding, X., Wang, Z., Chen, S., Huang, Y.: Community-based collaborative filtering recommendation algorithm. Int. J. Hybrid Inf. Technol. **8**(2), 149–158 (2015)
10. Han, Q., et al.: A P2P recommended trust nodes selection algorithm based on topological potential. In: 2013 IEEE Conference on Communications and Network Security, CNS, pp. 395–396 (2013)
11. Wang, Z., Zhao, Y., Chen, Z., Niu, Q.: An improved topology-potential-based community detection algorithm for complex network. Sci. World J. **2014**, 1–7 (2014)
12. Wang, S., Gan, W., Li, D., Li, D.: Data field for hierarchical clustering. Int. J. Data Warehouse. Min. **7**(4), 43–63 (2011)
13. Han, Y., Li, D., Wang, T.: Identifying different community members in complex networks based on topology potential. Front. Comput. Sci. China **5**(1), 87–99 (2011)
14. Xiao, L., Wang, S., Li, J.: Discovering community membership in biological networks with node topology potential. In: 2012 IEEE International Conference on Granular Computing, GrC, pp. 541–546 (2012)
15. Kitsak, M., et al.: Identification of influential spreaders in complex networks. Nat. Phys. **6**(11), 888–893 (2010)
16. Network data. http://www-personal.umich.edu/~mejn/netdata/. Accessed 14 May 2018
17. Alex Arenas datasets. http://deim.urv.cat/~alexandre.arenas/data/welcome.htm. Accessed 14 May 2018
18. Tang, Y., Li, M., Wang, J., Pan, Y., Wu, F.-X.: CytoNCA: a cytoscape plugin for centrality analysis and evaluation of protein interaction networks. Biosystems **127**, 67–72 (2015)
19. Lawyer, G.: Understanding the influence of all nodes in a network. Scientific reports, vol. 5, no. 1, August 2015
20. Kendall, M.G.: THE treatment of ties in ranking problems. Biometrika **33**(3), 239–251 (1945)
21. Ma, L.-L., Ma, C., Zhang, H.-F., Wang, B.-H.: Identifying influential spreaders in complex networks based on gravity formula. Phys. Stat. Mech. Appl. **451**, 205–212 (2016)
22. Liu, J., Xiong, Q., Shi, W., Shi, X., Wang, K.: Evaluating the importance of nodes in complex networks. Phys. Stat. Mech. Appl. **452**, 209–219 (2016)
23. Wang, J., Li, M., Wang, H., Pan, Y.: Identification of essential proteins based on edge clustering coefficient. IEEE/ACM Trans. Comput. Biol. Bioinform. **9**(4), 1070–1080 (2012)
24. Li, M., Wang, J., Chen, X., Wang, H., Pan, Y.: A local average connectivity-based method for identifying essential proteins from the network level. Comput. Biol. Chem. **35**(3), 143–150 (2011)
25. Anthonisse, J.M.: The rush in a directed graph, January 1971

26. Sabidussi, G.: The centrality index of a graph. Psychometrika **31**(4), 581–603 (1966)
27. Tang, L., Liu, H.: Community detection and mining in social media. Synth. Lect. Data Min. Knowl. Discov. **2**(1), 1–137 (2010)
28. Bonacich, P.: Power and centrality: a family of measures. Am. J. Sociol. **92**(5), 1170–1182 (1987)
29. Estrada, E., Hatano, N.: Resistance distance, information centrality, node vulnerability and vibrations in complex networks. In: Estrada, E., Fox, M., Higham, D.J., Oppo, G.-L. (eds.) Network Science, pp. 13–29. Springer, London, London (2010). https://doi.org/10.1007/978-1-84996-396-1_2
30. Estrada, E., Rodríguez-Velázquez, J.A.: Subgraph centrality in complex networks. Phys. Rev. E **71**(5), 056103 (2005)

Nodes Deployment Optimization Algorithm Based on Energy Consumption of Underwater Wireless Sensor Networks

Min Cui[1], Fengtong Mei[1], Qiangyi Li[2(✉)], and Qiangnan Li[2]

[1] Zhengzhou University of Industrial Technology,
Henan Zhengzhou 451150, China
[2] Henan University of Science and Technology, Henan Luoyang 471023, China
{cjj198, cxl979}@yeah.net

Abstract. Underwater wireless sensor networks nodes deployment optimization problem is studied and underwater wireless sensor nodes deployment determines its capability and lifetime. If no underwater wireless sensor node is available in the monitoring area of underwater wireless sensor networks due to used up energy or any other reasons, the monitoring area where is not detected by any underwater wireless sensor node forms coverage holes. In order to improve the coverage of the underwater wireless sensor networks and prolong the lifetime of the underwater wireless sensor networks, based on the perception model, establish nodes detection model, combining with the data fusion. Because the underwater wireless sensor networks nodes coverage holes appear when the initial randomly deployment, a nodes deployment algorithm based on perception model of underwater wireless sensor networks is designed in this article. The simulation results show that this algorithm can effectively reduce the number of deployment underwater wireless sensor networks nodes, improve the efficiency of underwater wireless sensor networks coverage, reduce the underwater wireless sensor networks nodes energy consumption, prolong the lifetime of the underwater wireless sensor networks.

Keywords: Nodes deployment · Optimization algorithm
Underwater wireless sensor networks

1 Introduction

Because the wireless sensor network nodes coverage holes appear when the initial randomly deployment, a nodes deployment algorithm based on perception model of wireless sensor network is designed in this article [1–5]. In order to improve the coverage of the wireless sensor network and prolong the lifetime of the wireless sensor network, based on the perception model, establish nodes detection model, combining with the data fusion [6–10]. This algorithm can effectively reduce the number of deployment wireless sensor network nodes, improve the efficiency of wireless sensor network coverage, reduce the wireless sensor network nodes energy consumption, and prolong the lifetime of the wireless sensor network [11–15].

© Springer Nature Switzerland AG 2018
G. Gan et al. (Eds.): ADMA 2018, LNAI 11323, pp. 428–433, 2018.
https://doi.org/10.1007/978-3-030-05090-0_36

2 Assumption

To simplify the calculation, randomly deploy the quantity N_k of the k-th type mobile nodes in the monitoring region and mobile wireless sensor node s_j owns wireless sensor network ID number j.

The k-th type wireless sensor nodes in the network own the same sensing radius R_{sk}, the same communication radius R_{ck}, and $R_{ck} = 2R_{sk}$.

The wireless sensor nodes can obtain the location information of itself and its neighbor nodes.

The k-th type mobile node owns E_k energy and is sufficient to support the completion of the mobile node position migration process.

The k-th type mobile node sending 1 byte data consumes E_{sk} energy and receiving 1 byte data consumes E_{rk} energy.

The k-th type mobile node migration 1 m consumes E_{mk} energy.

3 Coverage Model

The monitored area owns A $\times$ B $\times$ C pixels which means that the size of each pixel is the $\Delta x \times \Delta y \times \Delta z$.

The perceived probability of the i-th pixel is perceived by the wireless sensor network is $P(p_i)$, when $P(p_i) \geq P_{th}$ (P_{th} is the minimum allowable perceived probability for the wireless sensor network), the pixels can be regarded as perceived by the wireless sensor network.

The i-th pixel is whether perceived by the wireless sensor node perceived to be used $P_{cov}(P_i)$ to measure, i.e.

$$P_{cov}(P_i) = \begin{cases} 0 & if \quad P(p_i) < P_{th} \\ 1 & if \quad P(p_i) \geq P_{th} \end{cases} \tag{1}$$

The coverage rate is the perceived area and the sum of monitoring area ratio is defined in this article, i.e.

$$R_{area} = \frac{P_{area}}{S_{area}} = \frac{\Delta x \times \Delta y \times \Delta z \times \sum_{x=1}^{A} \sum_{y=1}^{B} \sum_{z=1}^{C} P_{cov}(p_i)}{\Delta x \times \Delta y \times \Delta z \times A \times B \times C} \tag{2}$$

Among them, P_{area} is the perceived area while S_{area} is the sum of monitoring area.

4 Perception Model

The event r_{ij} is defined that the i-th pixel p_i which is perceived by the ID number j wireless sensor nodes, the probability of occurrence of the event is $P(r_{ij})$ which is the perceived probability $P(p_i, s_j)$ that the pixel p_i is perceived by wireless sensor node s_j, i.e.

$$P(p_i, s_j) = \begin{cases} 1 & if & d(p_i, s_j) \leq R_{sk} - R_{ek} \\ \ln\{1 - \frac{e-1}{2R_{ek}}[d(p_i, s_j) - R_{sk} - R_{ek}]\} & if & R_{sk} - R_{ek} < d(p_i, s_j) < R_{sk} + R_{ek} \\ 0 & if & d(p_i, s_j) \geq R_{sk} + R_{ek} \end{cases}$$

$$(3)$$

Among them, the $d(p_i, s_j)$ is the distance between the i-th pixel p_i and the j-th wireless sensor node s_j, the sensing radius of the k-th type wireless sensor node is R_{sk}, the perceived error range of the k-th type wireless sensor node is R_{ek}.

This article used a number of wireless sensor nodes cooperative sensing monitoring method and the pixel p_i is perceived by all wireless sensor nodes collaborate perceived probability is

$$P(p_i) = 1 - \prod_{j=1}^{N} [1 - P(p_i, s_j)]$$

$$(4)$$

5   This Article Algorithm

The position of wireless sensor network node S_i is (x_i, y_i, z_i), the perception model is in the following:

$$P(S_i, T) = \begin{cases} 0 & d(S_i, T) \geq R + R_e, \\ \frac{E_{rem}}{E_{ini}} e^{-\gamma\alpha^\beta} & R - R_e < d(S_i, T) < R + R_e \\ 1 & d(S_i, T) < R - R_e \end{cases}$$

$$(10)$$

Among them, $P(S_i, T)$ is the perception probability of the wireless sensor network node S_i to target point T, $d(S_i, T)$ is the distance between sensor node S_i and the target point T, R_e is a uncertainty perception measure of wireless sensor node S_i, E_{ini} is the initial energy of the wireless sensor node S_i, E_{rem} is the remaining energy of the wireless sensor node S_i, α, β, γ are the perception of the wireless sensor node within the scope of monitoring quality coefficient.

6   Simulation Result

MATLAB software is used as simulation in this article. Assume that require p-reliability coverage in the monitoring area, among them p = 0.9, and do not consider the effect of target distribution and other environmental factors. According to the characteristics of the passive sonar and underwater sensor networks node and the related definitions, respectively the simulation random deployment algorithm and based on the "virtual force" deployment algorithm under monitoring area coverage performance and deployment algorithm based on perception model after monitoring area coverage performance and target node test results.

The simulation results are shown in Figs. 1, 2 and 3.

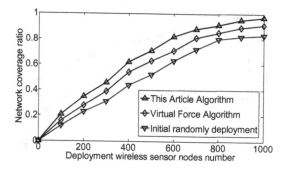

Fig. 1. Relations of network coverage and nodes number

Figure 1 is under the initial randomly deployment, virtual force algorithm and this article algorithm, in the same test area with an increase in the number of nodes. In contrast to initial randomly deployment and virtual force algorithm, this article algorithm has more effective coverage, a single node perception is more efficient, this is a direct result of the node deployment, compared virtual force algorithm and initial randomly deployment reduced the scope of sensor node overlapping sense perception, and because this article algorithm compared virtual force algorithm adopted data fusion algorithm, and reduced the perceived blind area, therefore, this article algorithm under the effective coverage of the sensor network greater than virtual force algorithm, virtual force algorithm is higher than the initial randomly deployment.

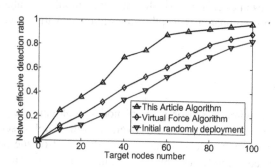

Fig. 2. Relations of network effective detection ratio and target nodes number

Figure 2 is under the initial randomly deployment, virtual force algorithm and this article algorithm, in the same detection area to deploy the same number of sensor node, respectively, with different number of effective detection rate of the target node contrast figure. By the graph, this article algorithm, the wireless sensor network for effective detection of the target node rate than initial randomly deployment and virtual force algorithm, this is because the this article algorithm deployment under the data fusion algorithm is adopted to perception results effective fusion of different sensors, improve the detection probability of the target node, increase the effective coverage.

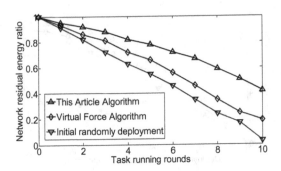

Fig. 3. Relations of network residual energy and task running rounds

Figure 3 is under the initial randomly deployment, virtual force algorithm and this article algorithm, at the same detection area under the same number, under the same coverage performance, the perception node residual energy contrast figure of sensor networks. Because the virtual force algorithm and this article algorithm in the initial stage of perceptual mobile node, and this article algorithm under the movement number is greater than the virtual force algorithm, so the energy consumption is larger, and the energy consumption is greater than the virtual force algorithm this article algorithm is presented. However due to both used the redundancy node dormancy mechanism, after complete the deployment, virtual force algorithm and this article algorithm energy per unit time is less than initial randomly deployment, due to the redundancy this article algorithm under network is greater than the virtual force algorithm, therefore, this article algorithm consumes energy is smaller than the virtual force algorithm. Therefore this article algorithm compared with other algorithm own the longer lifetime of the network.

7   Conclusion

This article aims at wireless sensor network nodes deployment algorithm based on perception model, in order to improve the coverage of the wireless sensor network and prolong the lifetime of the wireless sensor network, based on the perception model, establish nodes detection model, combining with the data fusion. This article algorithm can effectively reduce the number of deployment wireless sensor network nodes, improve the efficiency of wireless sensor network coverage, reduce the wireless sensor network nodes energy consumption, and prolong the lifetime of the wireless sensor network.

References

1. Song, X.L., Gong, Y.Z., Jin, D.H., Li, Q.Y., Jing, H.C.: Coverage hole recovery algorithm based on molecule model in heterogeneous WSNs. Int. J. Comput. Commun. Control **12**(4), 562–576 (2017)
2. Song, X.L., Gong, Y.Z., Jin, D.H., Li, Q.Y., Zheng, R.J., Zhang, M.C.: Nodes deployment based on directed perception model of wireless sensor networks. J. Beijing Univ. Posts Telecommun. **40**, 39–42 (2017)
3. Zhao, M.Z., Liu, N.Z., Li, Q.Y.: Blurred video detection algorithm based on support vector machine of Schistosoma Japonicum Miracidium. In: International Conference on Advanced Mechatronic Systems, pp. 322–327 (2016)
4. Jing, H.C.: Node deployment algorithm based on perception model of wireless sensor network. Int. J. Autom. Technol. **9**(3), 210–215 (2015)
5. Jing, H.C.: Routing optimization algorithm based on nodes density and energy consumption of wireless sensor network. J. Comput. Inf. Syst. **11**(14), 5047–5054 (2015)
6. Jing, H.C.: The study on the impact of data storage from accounting information processing procedure. Int. J. Database Theory Appl. **8**(3), 323–332 (2015)
7. Jing, H.C.: Improved ultrasonic CT imaging algorithm of concrete structures based on simulated annealing. Sens. Transducers **162**(1), 238–243 (2014)
8. Zhang, J.W., Li, S.W., Li, Q.Y., Liu, Y.C., Wu, N.N.: Coverage hole recovery algorithm based on perceived probability in heterogeneous wireless sensor network. J. Comput. Inf. Syst. **10**(7), 2983–2990 (2014)
9. Jing, H.C.: Coverage holes recovery algorithm based on nodes balance distance of underwater wireless sensor network. Int. J. Smart Sens. Intell. Syst. **7**(4), 1890–1907 (2014)
10. Wu, N.N., et al.: Mobile nodes deployment scheme design based on perceived probability model in heterogeneous wireless sensor network. J. Robot. Mechatron. **26**(5), 616–621 (2014)
11. Li, Q.Y., Ma, D.Q., Zhang, J.W.: Nodes deployment algorithm based on perceived probability of wireless sensor network. Comput. Meas. Control **22**(2), 643–645 (2014)
12. Jing, H.C.: Improving SAFT imaging technology for ultrasonic detection of concrete structures. J. Appl. Sci. **13**(21), 4363–4370 (2013)
13. Li, S.W., Ma, D.Q., Li, Q.Y., Zhang, J.W., Zhang, X.: Nodes deployment algorithm based on perceived probability of heterogeneous wireless sensor network. In: International Conference on Advanced Mechatronic Systems, pp. 374–378 (2013)
14. Zhang, H.T., Bai, G., Liu, C.P.: Improved simulated annealing algorithm for broadcast routing of wireless sensor network. J. Comput. Inf. Syst. **9**(6), 2303–2310 (2013)
15. Li, Q.Y., Ma, D.Q., Zhang, J.W., Fu, F.Z.: Nodes deployment algorithm of wireless sensor network based on evidence theory. Comput. Meas. Control **21**(6), 1715–1717 (2013)

A New Graph-Partitioning Algorithm for Large-Scale Knowledge Graph

Jiang Zhong[(⊠)], Chen Wang, Qi Li, and Qing Li

College of Computer Science, Chongqing University, Chongqing 400030, China
zhongjiang@cqu.edu.cn

Abstract. Large-scale knowledge graph is finding widely practical applications in many fields such as information retrieval, question answering, health care, and knowledge management and so on. To carry out computations on such large-scale knowledge graphs with millions of entities and facts, partitioning of the graphs is necessary. However, the existing partitioning algorithms are difficult to meet the requirements on both partition efficiency and partition quality at the same time. In this paper, we utilize the community-based characteristic that real-world graphs are mostly power-law distribution, and propose a new graph-partitioning algorithm (called MCS) based on message cluster and streaming partitioning. Compared with the traditional algorithms, MCS is closer to or even surpasses Metis package in the partition quality. In the partition efficiency, we use the PageRank algorithm in the spark cluster system to compute the Twitter graph data, and the total time of MCS is lower than that of Hash partitioning. With an increasing number of iterations, the effect is more obvious, which proves the effectiveness of MCS.

Keywords: Large-scale knowledge graph · Graph partitioning
Streaming partitioning · Community detection · Parallel computing

1 Introduction

In 2012, Google proposed knowledge graph to enhance its search engine's search results [1, 2]. Knowledge graphs usually contain two layers: schema and instances. The schema layer contains concepts, property and relationships between concepts. While the instances layer contains entities and relationships between different entities. Schema is usually expressed as an ontology. While instances can be represented as a set of RDF [3] (Resource Description Framework, a standard model for data interchange on the Web) triples, which constitute many large graphs.

Large-scale knowledge graphs contain massive entities and abundant relations among the entities, and they can be widely applied to many practical applications such

This work gets supports from the National Key Research and Development Program of China under Grant 2017YFB1402401, Technology Innovation and Application Demonstration project of Chongqing under Grant cstc2018jszx-cyzdX0086, Fundamental Research Funds for the Central University under Grant 2018CDYJSY0055 and CERNET Next Generation Internet Technology Innovation Project under Grant NGII20150706.

© Springer Nature Switzerland AG 2018
G. Gan et al. (Eds.): ADMA 2018, LNAI 11323, pp. 434–444, 2018.
https://doi.org/10.1007/978-3-030-05090-0_37

as semantic search, question answering and recommender systems, information retrieval, health care, knowledge management and so on [4].

By the end of 2012, the knowledge graph of Google has contained over 570 million entities and more than 18 billion facts about relationships between different entities [5]. Other public knowledge graphs, such as DBpedia [6] and YAGO [7] contain millions of entities and hundreds of millions of facts. All the above famous knowledge graphs would continuously grow, as more facts would be automatically discovered from the underlying sources or manually created by human.

The fast evolving on large-scale knowledge graphs poses some challenges to the effective usage of them. One challenge is how to carry out effectively computations on large-scale knowledge graphs, such as knowledge fusion, knowledge acquisition and knowledge reasoning.

To carry out computations on such large-scale knowledge graphs with millions of entities and facts, the first step is partition the graph. In order to partition such large-scale knowledge graphs, distributed iterative processing systems (e.g., Spark [8], Pregel [9], and GiraphLab [10]) have been developed. These systems mainly adopt the Hash function to send the vertices to each processing unit, which refers to the computing units in cluster system. Although this method has low time complexity, the communication traffic among processing units will be large in the iterative process. If using an algorithm with better partition quality (as Metis [11]) instead of the Hash method, for the high time complexity of Metis, the total cost of time is much larger than suing the Hash method. Therefore, the more efficient partitioning algorithm becomes the existing urgent problem in the distributed graph computing system.

The graph partitioning belongs to NP complete problem [12]. The current research work mainly covers into two categories: centralized partitioning and streaming partitioning. The centralized partitioning algorithm has been studied for a long time, but with its high time complexity, it can handle graphs with a few vertices and edges. The biggest advantage of streaming partitioning is that the streaming method only processes one vertex at a time. The information used is neighbors, so the efficiency of centralized partitioning algorithm is lower than that of streaming partitioning.

However, the existing partitioning methods have ignored the structure of the graph itself. The paper utilizes the community-based feature of most graphs to propose a message cluster and streaming partitioning (MCS) algorithm, which has great improvement in partition quality and efficiency compared with the traditional graph partitioning algorithms.

The main contributions of the paper are as follows:

(1) We introduce a message cluster and streaming partitioning (MCS) algorithm by using characteristics of the community in most graphs and the low complexity of streaming algorithm. The partition efficiency is obviously improved.
(2) We apply the idea of label propagation to preprocessing of graph partitioning with the design of restrictive propagation process, and quantitatively analyze its stability.
(3) We demonstrate the experimental results on different types of real and synthetic graphs by comparison with traditional algorithms in different aspects, to prove the effectiveness of the MCS algorithm.

2 Streaming Partitioning

In order to increase the efficiency of streaming partitioning, we try to use distributed streaming algorithm to partition the graph [13]. This approach increases the complexity of hardware and program. Restreaming partitioning algorithm is provided forward in [14] and its partition quality has obviously increase after several iterations.

The graph data $G = (V, E)$, V and E denote sets of vertices and edges respectively. $|V| = n$. $|E| = m$. k is the number of subsets. $N(v)$ refers to the set of vertices that v neighbors. A k-partitioning of G can be defined as a mapping (partition function) $\pi : V \rightarrow \{1, 2, \ldots, k\}$ that distributes the vertices of V among k disjoint subsets $(S_1 \cup S_2 \cup \ldots \cup S_k = V)$ of roughly equal size and make the number of edges cut (edges whose endpoints belong to different subsets) as little as possible. Equation 1 demonstrates the load coefficient (ρ) and the fraction of edges cut (λ), n/k is the average load.

$$\lambda = \frac{\#\text{the number of edges cut}}{\#\text{total edges}} \tag{1}$$

$$\rho = \frac{\#\text{maximum load}}{\#(n/k)} \tag{2}$$

S_i^t represents the set of vertices in subset $S_i (i \in [1, k])$ at time t. Streaming algorithm is to traverse the graph and send the vertex v to the subset S_i one by one according to streaming function. Different algorithms has different heuristic rules. For example, the heuristic rule of the Non-Neighbors algorithm is minimum $|S_i \backslash N(v)|$. The heuristic rule of Deterministic Greedy is maximum $|N(v) \cap S_i|$. The heuristic rule of Exponentially Weighted Deterministic Greedy is maximum $|N(v) \cap S_i|(1 - \exp(|S_i| - n/k))$.

The following details describe the common streaming algorithm. The specific process of these algorithms in [6] are as following:

(1) Hash (H): $(|V| \mod k) + 1$
(2) Balance (B): $(|V| \mod k) + 1$
(3) Exponentially Weighted Deterministic Greedy (EDG): $max_{i \in [1,k]}\{|S_i^t \cap N(v)| \times (1 - exp(|S_i| - c))\}$
(4) Exponentially Weighted Triangles (ET): $max\{|S_i^t| \times (1 - exp(|V_i| - c))\}$
(5) Linear Weighted Triangles (LT): $max\{|S_i^t| \times (1 - |V_i|/c)\}$
(6) Triangles (T): $max\{|S_i^t|\}$
(7) Chunk (C): $\lceil |V|/c \rceil$

Stanton and Kliot analyzed [15] the performance of a series of heuristic streaming algorithms. The best is Linear Weighted Deterministic Greedy (LDG). Equation 2 demonstrates its rule of sending the vertex v to the subset S_{ind} one by one.

$$S_{ind} = \underset{i \in \{1,\ldots,k\}}{\arg \max}\{|N(v) \cap S_i^t|w(t, i)\} \tag{3}$$

$$w(t, i) = 1 - |S_i^t|/(n/k) \tag{4}$$

here $|N(v) \cap S_i^t|$ represents the number of v neighbors in S_i at time t. In order to reach load balancing, it adds penalty function $w(t, i)$ to punish the subset which has too many vertices. In the initial phase of the algorithm, many vertices' value calculated by function is zero, because their neighbors have not been calculated.

3   Method Description

3.1   MCS Algorithm Description

We introduce the MCS algorithm in three steps. We cluster the input graph and sort the clustering results, and then select the vertices in the order of community to send them to the corresponding processing unit. The following steps describe the specific details.

Step one: Graph Clustering. The Label Propagation algorithm (LPA) [15] is used for community detection. Here, we use it for two reasons. First, LPA time complexity is very low and is close to the linear complexity t × |E|, where t denotes the number of iterations. Second, LPA can discover dense structures in complex networks quickly. We carefully analyzed the propagation process of LPA and found that most vertex labels do not change after the fourth iteration. Figure 1 records the percentage change in vertex labels for each iteration.

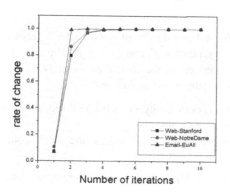

Fig. 1. Convergence of real-world graphs

According to this feature, we set iteration time $t \le 4$. Due to the randomness of LPA, it is very easy to create a large community, which is not allowed. Therefore, mechanisms must be adopted to limit the size of the community. We set the size of the single community $|F_i| \le |V|/a \times k (a \in N^+)$. Two properties are added to each vertex. The first is the vertex label (*flag Message*) and the second is the vertex status (*state Message*). In the initial step of MCS algorithm. The content of *flag Message* is the vertex number itself (label = Id) and the content of *state Message* is false (lock = false). A vertex in the state that lock is false can send its *flag Message* to its neighbors and receive *flag Message* from its neighbors. When a vertex in the state that lock is true, the vertex can neither send nor receive *flag Message*. Each vertex takes the maximum number of the same label as its own label according to its *flag Message* received. When the size of one community is $|F_i| = |V|/2 \times k$, change the *state Message* of its vertices to true.

The pseudo code is shown below.

Algorithm 1. Graph Clustering Algorithm
Input: Initial Graph $G = (V, E)$
Output: Clustering result $\{F_1, F_2, ..., F_n\}$
Initialize the parameters: the number of iterations *Iter*, the number of processing units k, initiate the *flag Message* and *state Message* of all vertices.
Begin
    1. repeat
    2. *Iter:* = *Iter* + 1
    3. for each vertex in Graph
    4.{ Each vertex takes the maximum number of the same label as its own label according to the *flagMessage* received }
      5.   if $|F_i| = |V|/2 \times k$
      6.   each vertex in F_i
      7.   the vertex's lock changes from false to true
      8.   else if
      9. end for
     10. until stop condition not met

Step 2: Determining the community sequence. edge$_{i,j}$ denotes the number of edges between the different communities F_i and F_j ($i \neq j$). The average capacity of each unit is $c = \lceil |V|/k \rceil$. c_{left} is defined as the remaining capacity of the processing unit. The order of communities is determined as follows:

(1) Randomly select one community Fa as the first sequence item, and follow Step 3 to load it into the processing unit.
(2) If $c_{left} \geq max_p\{edge_{p,F_a}\}(p \neq F_a)$, select the next community $|F_b| = max_p \{edge_{p,F_a}\}$.
(3) If $c_{left} < min_p\{edge_{p,F_a}\}(p \neq F_a)$, select the next community $|F_b| = min_p \{edge_{p,F_a}\}$.

Step 3: Streaming partitioning. For vertices in the same community, loaded the same unit in principle is optimal. Because of the load balancing requirement, vertices in the same community may not be in same processing unit. We use LDG algorithm to further optimize the fraction of edges cut.

Figure 2 depicts the MCS partitioning process (a) and the LDG partitioning process (b), $k = 2$. Our method detect the dense structure by graph clustering first and then using streaming algorithm to partition the specified community sequence. MCS algorithm is easier to partition the dense structure into subsets compared with LDG. The superiority of the proposed method is obvious from Fig. 2.

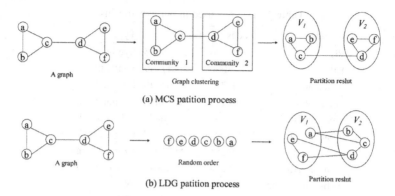

(a) MCS patition process

(b) LDG patition process

Fig. 2. The partitioning process of MCS algorithm and LDG algorithm.

3.2 Time Complexity Analysis

We analyze the time complexity for the MCS algorithm. In the first step, the complexity for traversing the whole graph is approximately $O(t \times |E|)$, where t denotes the number of iterations ($t \leq 4$). In the second step, the determination of the community sequence requires time complexity at less than $O(\log(|V|))$. In the third step, the streaming algorithm requires $O(\log(|E|))$, and the time complexity of the whole MCS algorithm requires $O(t \times |E| + \log(|V| \times |E|))$.

4 Experimental Results and Analysis

4.1 Experimental Setup

The information of the real-world graph and the synthetic graph used in the experiment is shown in Table 1. All the experiments were carried out in a stand-alone Intel Xeon processor with 2.67 GHz, 32 G memory.

4.2 Algorithm Comparison

It has been generally observed that presenting the graph data in either breadth-first search (BFS), depth-first search (DFS), or Random search does not greatly alter performance [15]. Of these orders, a random ordering is the simplest to guarantee in large-scale streaming data scenarios, and so we restrict our analysis to only consider random vertex orders for simplicity.

Figure 3 shows the partitioning results of Amazon0312 ($k = 32$). For the fraction of edges cut, we can see that MCS is better than the traditional approaches and is closer to the result of Metis. The MCS equilibrium coefficient is 1.00.

Table 1. Datasets used in our experiments.

Graph	Nodes	Edges	Type
PL	1000	9895	Synthetic
ER	5000	1247739	Synthetic
amazon0312	400727	3200440	Co-purchasing
amazon0302	262111	1234877	Co-purchasing
amazon0505	410236	3356824	Co-purchasing
amazon0601	403394	3387388	Co-purchasing
Wiki-Talk	2394385	5021410	Communication
email-EuAll	265214	420045	Communication
web-NotreDame	325729	1117563	Web
Web-Stanford	281903	1992636	Web
soc-LiveJournal1	4847571	68993773	Social
Twitter-2010	41652230	1468365182	Social

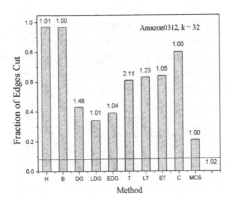

Fig. 3. The partitioning results of Amazon0312, k = 32. (Color figure online)

The red line represents the result of Metis (0.08%). The numerical values on the bar chart represent the equilibrium coefficients of the corresponding algorithms. The equilibrium coefficient 2.11 of triangle algorithm (T) is the maximum. The equilibrium coefficient of the proposed algorithm is 1.00, and its fraction of edges cut is 21%, which is the closest to that of Metis.

We use the graphs with different characteristics to analyze the applicability of the MCS algorithm. The power-law graph (PL), dense graph and non-power-law graph (ER) generated by NetworkX software package. The upper right corners of Fig. 4 shows the frequency of degree. Figure 4 shows the partitioning results of the various heuristic algorithms on the ER (left) and PL (right) graph, respectively. The vertical axis shows the fraction of edges cut. The red line shows the partitioning results of Metis. A closer value to Metis means the partitioning is of better quality. The partitioning result of ER is shown in Fig. 4(left), and the quality of the Linear Weight Deterministic Greedy algorithm is the best, which reached 83%. The hash result of 92%

is the worst, and our proposed algorithm in this paper is 85%, with Metis at 81%. Figure 4(right) shows the PL graph partitioning result. The fraction of edges cut by our method is 49%, with Metis at 50%. Because the characteristic 'world let' of the power-law network graphs are very suitable for the initial clustering, our method is very suitable for the partitioning of the power-law graphs.

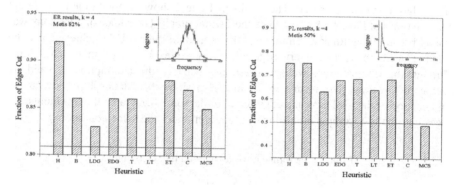

Fig. 4. ER (left) and PL (right) results, k = 4.

The first step of the MCS algorithm is the label selection problem for the vertex. If the same number of neighbors labels more than one category, which includes randomness. Therefore, we need to verify whether the randomness has an influence on the partitioning results. In Fig. 5, we show the stability results of the MCS algorithm on three real-world graphs (amazon0505, amazon0312, and amazon0601). The vertical axis shows the Difference = $\lambda_t - \lambda_{t+1}$ (the difference between the fraction of edges cut from the previous experiment and the fraction of edges cut from the next experiment), and the abscissa indicates the number of running times. It is obvious that the fluctuation rate is small and it can even be negligible for large-scale graphs from Fig. 5.

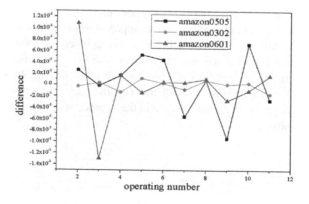

Fig. 5. Stability testing of our proposed approach.

Finally, the efficiency we need to verify of our algorithm mainly refers to the time of graph partitioning. However, because the time of partitioning is obviously higher than that of Hash and the purpose of the algorithm is to reduce the time of graph computation. The efficiency here is mainly embodied in the total time of the three phases of the graph loading, the graph partitioning and the graph computation. If the total time is lower than the most widely used Hash partitioning, then the effectiveness of the algorithm can be proved. The results in Table 2 show the total time for different iterative times of PageRank to compute the Twitter graph. The choice of PageRank was mainly because PageRank is familiar and the procedure can easily be achieved, and the iteration needs to traverse every edge and every vertex in the graph for each time. Through the optimization, we can determine whether the assumption is valid by computational time. The results in Table 2 also show the relationship between the computation of the total time and edges cut. For the same size graph, the less number of edges cut, the less computation time of the graph under the same condition.

Table 2. Performance of the partitioning algorithms.

Data set	Machines(k)	Iterations	Runtime (min)	
			Hash	MCS
Twitter-2010	20	6	39.2	**37.8**
		9	65.42	**61.7**
		12	85.8	**76.69**
		15	98.4	**83.21**
	50	6	17.72	21.7
		9	26.16	**26.63**
		12	34.09	**32.09**
		15	42.3	**39.6**

From Table 2, it is obvious that when the number of iterations is small, the total time is more than the hash method due to the high initial partition time complexity of MCS method. However, as the number of iterations increases, the advantage of the less number of edges cut (traffic) become more apparent relative to hash method. For example, when the number of iterations is 15 on 20 machines, the total time is reduced by nearly 15%. The experimental results formulate the verity the effectiveness of the MCS algorithm. The total time for different iterative times of PageRank computation on the Twitter-2010 graph, Hash and MCS. The difference between the two experiments gradually increases. With the increase in the number of iterations, the advantages of MCS is more obvious.

5 Conclusion

In this paper, for the computation of large-scale knowledge graph, we propose a new graph-partitioning algorithm (MCS) based on message cluster and streaming partitioning. The process of MCS algorithm is to cluster the original input graphs quickly and then using the streaming algorithm to send the vertices to the specified processing unit. MCS is a great improvement on the traditional partitioning algorithms in terms of the fraction of edges cut as well as the low time complexity. Compared with the traditional algorithms, in partition quality MCS is closer to or even surpasses Metis package. In partition efficiency, we use the PageRank algorithm in the spark cluster system to compute the Twitter graph data, and the total time of MCS is less than that of Hash partitioning. With an increasing number of iterations, the advantage is more obvious.

Research and the widespread application of the distributed system have posed a severe challenge to the effect and efficiency of the graph-partitioning algorithm. In future research, we will continue to study the graph partitioning, hoping to improve the fraction of edges cut and the time complexity.

References

1. Wei, Y., Luo, J., Xie, H.: KGRL: an OWL2 RL reasoning system for large scale knowledge graph. In: 12th International Conference on Semantics, Knowledge and Grids, pp. 83–89. IEEE, Piscataway (2017)
2. Chen, J., Chen, Y., Du, X., et al.: SEED: a system for entity exploration and debugging in large-scale knowledge graphs. In: 32nd IEEE, International Conference on Data Engineering, pp. 1350–1353. IEEE, Piscataway (2016)
3. Passant, A.: dbrec—Music Recommendations Using DBpedia. The Semantic Web. Springer, Heidelberg (2010). https://doi.org/10.1007/978-3-642-17749-1_14
4. Tan, Z., Zhao, X., Fang, Y., et al.: GTrans: generic knowledge graph embedding via multi-state entities and dynamic relation spaces. IEEE Access 6(99), 8232–8244 (2018)
5. Dong, X., Gabrilovich, E., Heitz, G., et al.: Knowledge vault: a web-scale approach to probabilistic knowledge fusion. In: Proceedings of the 20th ACM SIGKDD International Conference on Knowledge Discovery and Data Mining, pp. 601–610. ACM, New York (2014)
6. Auer, S., Bizer, C., Kobilarov, G., et al.: Dbpedia: A Nucleus for a Web of Open Data. The Semantic Web. Springer, Heidelberg (2007). https://doi.org/10.1007/978-3-540-76298-0_52
7. Hoffart, J., Suchanek, F.M., Berberich, K., et al.: YAGO2: A spatially and temporally enhanced knowledge base from Wikipedia. Artif. Intell. 1(194), 28–61 (2013)
8. Zaharia, M., Chowdhury, M., Franklin, M.J., et al.: Spark: cluster computing with working sets. HotCloud 10(95), 10 (2010)
9. Malewicz, G., Austern, M.H., Bik, A.J., et al.: Pregel: a system for large-scale graph processing. In: Proceedings of the 2010 ACM SIGMOD International Conference on Management of data, pp. 135–146. ACM, New York (2010)
10. Low, Y., Gonzalez, J.E., Kyrola, A., et al.: Graphlab: a new framework for parallel machine learning. arXiv preprint arXiv,1408-2041 (2014)
11. Karypis, G., Kumar, V.: A fast and high quality multilevel scheme for partitioning irregular graphs. SIAM J. Sci. Comput. 1(20), 359–392 (1998)

12. Dutt, S.: New faster kernighan-lin-type graph-partitioning algorithms. In: Proceedings of 1993 International Conference on Computer Aided Design (ICCAD), pp. 370–377. IEEE, Piscataway (1993)
13. Battaglino, C., Pienta, P., Vuduc, R.: GraSP: distributed streaming graph partitioning. In: Proceeding of first High performance Graph Mining Workshop, Sydney (2015)
14. Nishimura, J., Ugander, J.: Restreaming graph partitioning: simple versatile algorithms for advanced balancing. In: Proceedings of 19th ACM SIGKDD International Conference on Knowledge Discovery and Data Mining, pp. 1106–1114. ACM, New York (2013)
15. Stanton, I., Kliot, G.: Streaming graph partitioning for large distributed graphs. In: Proceedings of 18th ACM SIGKDD International Conference on Knowledge Discovery and Data Mining, pp. 1222–1230. ACM, New York (2012)
16. Chen, L., Li, X., Sheng, Q.Z., et al.: Mining health examination records—a graph-based approach. IEEE Trans. Knowl. Data Eng. 9(28), 2423–2437 (2016)

SQL Injection Behavior Mining Based Deep Learning

Peng Tang, Weidong Qiu$^{(\boxtimes)}$, Zheng Huang, Huijuan Lian,
and Guozhen Liu

School of Cyber Security, Shanghai Jiao Tong University, Shanghai, China
{tangpeng, qiuwd}@sjtu.edu.cn

Abstract. SQL injection is a common network attack. At present, filtering methods are mainly used to prevent SQL injection, yet risks of incomplete filtering still remains. By deep learning, we detect whether the user behaviors contain SQL injection attacks. The scheme proposed in this article extracts the characteristics of the HTTP traffic in the training sets and uses the deep neural network LSTM and the MLP training data sets, the final predictive capacity of the testing sets is over 99%. The deep neural network uses ReLU as the activation function of the hidden layer, continuously updates the weight parameters through gradient descent algorithm, and finally completes the training within 50 epoch iterations.

Keywords: SQL injection · Deep learning · MLP · LSTM

1 Introduction

Network security situation has become increasingly complex and novel attack means presenting fast, wide and automatic emerge in endlessly. Threats in telecommunication massive data grow increasingly gravely and the threat intelligence detection platforms should also develop towards fine-grained, integrity, and accuracy.

Reference [1] introduces a new web vulnerability scanner that uses penetration testing and combinative evasion techniques to evade firewalls and filters and to discover injection. Reference [2] proposes an effective, light, and fully automated tool, SQL injection Prevention by Input Labeling (SQLPIL), leverages prepared statements to prevent SQLIAs at runtime. And Ref. [3] analyzes the advantages and disadvantages of the existing techniques against SQL Injection and proposes a novel and effective solution to avoid attacks on login phase. Reference [4] proposes a detection model for detecting and recognizing the web vulnerability which is able to decrease the possibility of the SQL Injection attack that can be launch onto the web application. Reference [5] analyzes the performance and detection capabilities of latest black-box web application security scanners against stored SQLI and stored XSS and develops the custom test-bed to challenge the scanners' capabilities to detect stored SQLI and stored XSS.

Based on these studies, this article accesses more telecommunication pipeline data, utilizes data analysis and deep learning algorithms to design models, mining richer and

G. Gan et al. (Eds.): ADMA 2018, LNAI 11323, pp. 445–454, 2018.
https://doi.org/10.1007/978-3-030-05090-0_38

more comprehensive threat intelligence, thus constructing a robust threat detection platform to enhance the threat perception and information security.

2 The Architecture

2.1 System Framework

The SQL injection attack is to construct malicious SQL statements and then submits them to the server in form on the Web page. After executing the malicious SQL statement, the server will return the private data that the attacker wants or achieves other malicious purposes. One of the submitted forms is to use the method of HTTP's GET, where the form data is added to the original URL as a parameter.

SQL injection attacks do great harm to websites. On the one hand, the attacker can simply inject statements into the user input box, making the logical judgment constant, thus bypassing server access control to unauthorized users, not only causing leakage of the user's private data, but may even bring huge economic loss to users; On the other hand, the attacker can estimate the injection of incorrect SQL statements, resulting in syntactic or logical errors in the database, causing the database to crash and unable to provide data to the website normally.

In this article, behavior characteristics of SQL injection were extracted, and a SQL injection detection model based on deep learning was designed. The overall framework of the model is shown in Fig. 1:

The data acquisition layer collects the information through the data collection interface, and performs pre-processing such as filtering, desensitization and standardization on the collected data. The data collection sources include DPI equipment, and the data collection objects include DPI data and other information.

The data storage layer uses the distributed file system HDFS to realize the long-term storage of massive security basic information such as the original DPI data collected. At the same time, it is stored separately in the server according to different detection requirements. Based on the relational database, it realizes the storage and management of the collected data such as attack statistics.

The data calculation and analysis layer is based on the computing platform such as Mapreduce. Through the security event mining capabilities such as special extraction and deep learning analysis of distributed storage data, the SQL injection detection function is realized, and the abnormal and illegal network behaviors of the large network and the customer network are discovered in time.

The process of the SQL injection detection model based on deep learning is shown in Fig. 2.

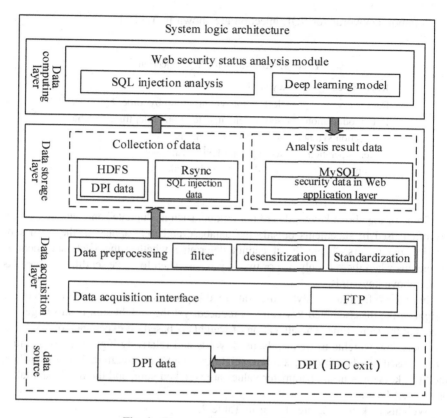

Fig. 1. The overall framework of system.

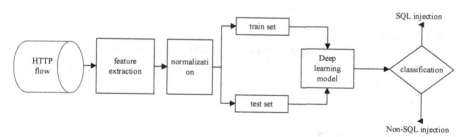

Fig. 2. The process of the SQL injection detection model.

2.2 Feature Engineering

(1) Keyword weights sum

SQL statement keyword is the key to SQL injection statements. Injection statements without keywords can hardly affect the normal operation of the database even if they are passed to the back end. Therefore, this article primarily analyzes whether the URL contains SQL statement keywords.

Common keywords for SQL injection are shown in Table 1.

Table 1. Keywords for SQL injection.

Keyword category	Keyword
Data query	select, union, count, group by, order by
Data modification	insert, delete, update, drop table, truncate table
Connection symbol	and, or, where, from, into
System operation	exec, xp_cmdshell, master, net
File operation	load_file, outfile, dumpfile

Some of the above keywords are not unique, such as "update" and "count", which are often used in statements outside the database and may appear in the form submission parameters. Typically, it cannot determine whether the URL statement is suspicious when there is only one or two keywords. Therefore, we set different weights to the above keywords.

In order to facilitate analysis, the combined keywords are disassembled, so that the weight of the disassembled keywords is reduced, yet the weight sum is still large if it occurs at the same time. The allocation of weights is based on experience of assigning relatively low weights to frequently used words and relatively high weights to infrequently used words, with a maximum weight of 5 and a minimum of 1. In this article, the SQL keywords in the parameter values are weighed sum, and the results are taken as the first feature.

Weights of keywords are shown in Table 2.

Table 2. Weights of keywords.

Weight	Keyword
5	union, truncate, xp_cmdshell, load_file, outfile, dumpfile, exec
3	select, update, insert, delete, count, where, group, order, drop, table, master, net
1	and, or, by, from, into

(2) **Percentage of spaces**

In most cases, the URL parameter is a number or a simple string with zero or fewer spaces. Although there are also a few cases where long information is passed using URL, the inclusion of SQL statements in the URL makes the number of spaces significantly more than the normal URL.

Under normal circumstances, if the URL parameter contains non-alphanumeric characters, these special characters will be converted into % prefix characters, and the data will be transmitted to the server for decoding. For spaces, it will be converted to "%20", and then we analysis the ratio of the number of "%20" strings and spaces to the length of the URL parameter value.

(3) Percentage of special characters

Normal URLs rarely contain special characters, and SQL attackers often construct statements containing malicious SQL in order to confuse the combinations of the SQL statement by the server. It contains some of the commonly used operations and conditional symbols for SQL statements, such as equal signs, parentheses, and single quotes.

On the other hand, the attacker constructs SQL statements with annotations, which change the structure of the original SQL when the parameters are passed to the server, making the query conditions smaller and thus gaining greater authority. Commonly used comments in SQL statements include "–", "#", "/**/", etc. These symbols will increase the possibility of URL injection for SQL to some extent. Therefore, we analyze the percentage of special characters in the data as a percentage of the length of the URL parameter values.

Special characters are shown in Table 3

Table 3. Special characters.

Characters	Instructions
+, -, *,/	For operation
=, ! = , ^ = , <>, >=, <=	For assignment and conditional judgment
-, #,/*, */	For notes
', ", @, \, ()	Other

2.3   Model Selection

In this article, deep neural network is adopted to train the extracted URL characteristic values. Deep neural networks include multiple Sensor Models, LSTM networks, and CNN networks. This article uses LSTM network and MLP network for training and experimental results are compared.

(1) The network structure of the model

The LSTM network structure is shown in Fig. 3:

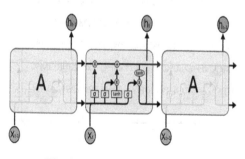

Fig. 3. LSTM network structure.

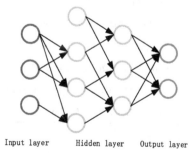

Input layer   Hidden layer   Output layer

Fig. 4. MLP network structure.

The LSTM model includes an input layer, a hidden layer, and an output layer. According to the structure of the LSTM network, the formula for each LSTM unit is shown in below. Ft represents the forgetting threshold, It represents the input threshold, Ct represents the cell state at the previous moment (where the loop occurs), Ht represents the output of the current unit, and Ht-1 represents the output of the unit at the previous moment.

$$f_t = \sigma\left(W_f \bullet [h_{t-1}, x_t] + b_f\right) \tag{1}$$

$$i_t = \sigma(W_i \bullet [h_{t-1}, x_t] + b_i) \tag{2}$$

$$\tilde{C}_t = \tanh(W_C \bullet [h_{t-1}, x_t] b_C) \tag{3}$$

$$C_t = f_t * C_{t-1} + i_t * \tilde{C}_t \tag{4}$$

$$O_t = \sigma(W_o[h_{t-1}, x_t] + b_o) \tag{5}$$

$$h_t = O_t * \tanh(C_t) \tag{6}$$

The MLP network structure is shown in Fig. 4.

3 Experiment and Result Evaluation

3.1 Experimental Data

The malicious samples used for training test modeling come from SQL injection samples provided by information security enthusiasts on the open source websites; Normal data comes from the HTTP access information of common Internet users captured by the operator's critical network node. The data of training sets and test sets are distributed in a ratio of 3:7; The normal data and malicious data in the training sets and test sets are 1:1.

The malicious data samples are payloads of the parameters after URL parsing, which are as follows:

```
323%27%20AND%20%28%20SELECT%202937%20FROM%28%20SELECT%20COUNT%28%
2A%29%2C%20CONCAT%280x3a6d70663a%2C%28%20SELECT%20MID%28%28%20IFNUL
L%28%20CAST%28%20database%28%29%20AS%20CHAR%20%29%2C0x20%29%29%2C1%
2C50%29%29%2C0x3a736e623a%2CFLOOR%280%29%2A2%29%29x%20FROM%
20INFORMATION_SCHEMA.CHARACTER_SETS%20GROUP%20BY%20x%29a%29%29%20AN
D%20%27rmHN%27=%27rmHN
```

After the URL is decoded, you can see the obvious SQL injection statement.

323' AND (SELECT 2937 FROM(SELECT COUNT(*), CONCAT(0x3a6d70663a,(SELECT

MID((IFNULL(CAST(database() AS

CHAR),0x20)),1,50)),0x3a736e623a,FLOOR(RAND(0)*2))x FROM

INFORMATION_SCHEMA.CHARACTER_SETS GROUP BY x)a) AND 'rmHN'='rmHN

The full URL of the normal data sample is shown below:

http://nl.rcd.iqiyi.com/apis/mbd/upload.action?agent_type=20&version=8.11.0&ua=iPhone8,2&net

work=1&os=11.0.3&com=1&wsc_istr=3CE76AD3-3291-4625-9DC6-5163437FE4F6&wsc_lgt=11

8.83249&wsc_ltt=31.91966&wsc_tt=02&wsc_ost=13&wsc_osl=zh-Hans-CN&wsc_st=iQiYiPhone

The analysis of URL text is mainly to analyze the URL parameter strings. According to URL naming rules, a complete URL includes protocol, domain name, port, directory, file, parameter. Where the parameter part is represented in the form of a key-value pair. The key and value are separated by "=" and different key pairs are separated by "&". Take the following URL as an example:

http://hdns.ksyun.com/d?dn=jsmov2.a.yximgs.com&ttl=1

The parameter part is "dn=jsmov2.a.yximgs.com&ttl=1", indicating that the URL contains two parameters "dn" and "ttl", and the parameter values are "jsmov2.a.yx-imgs.com" and "1" respectively. The analysis of URL text in this article is the analysis of each parameter value.

3.2   The Training Method of the Model

The process of training the model is shown in Fig. 5.

The flow of SQL injection attack detection is shown in Fig. 6.

3.3   Training Results

The MLP model as well as the LSTM model were trained with the training sets, and the model effect was detected with the test sets. The accuracy rate and recall rate of the model are shown below (Table 4):

The experimental results show that the eigenvalues selected in this paper combined with MLP and LSTM networks have higher detection accuracy for SQL injection. As can be seen from the chart, the MLP network is better than the LSTM network (Figs. 7 and 8).

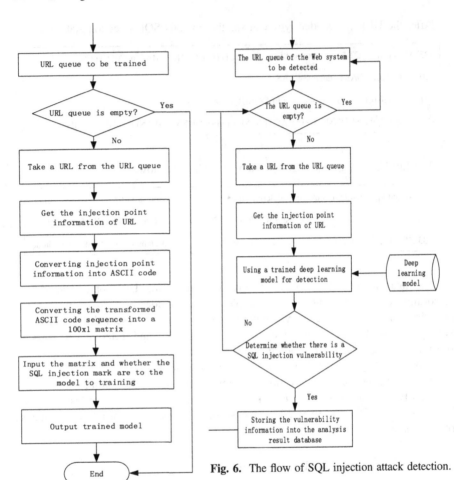

Fig. 5. The process of training the model.

Fig. 6. The flow of SQL injection attack detection.

Table 4. Training Results.

Model	Acc	Precision	Recall	FAR
MLP	99.55%	97.44%	97.21%	0.24%
LSTM	95.00%	99.25%	90.69%	0.69%

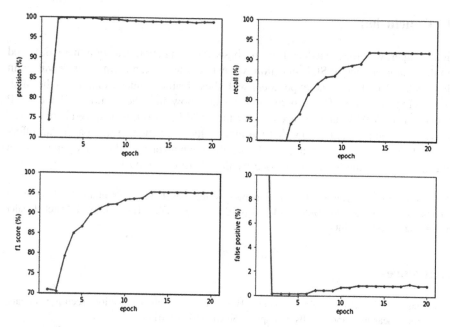

Fig. 7. MLP training results.

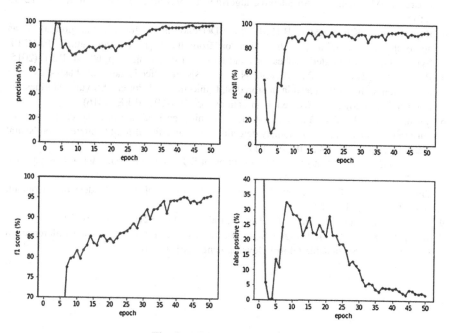

Fig. 8. LSTM training results.

4 Conclusion

In this paper, we propose a deep learning-based SQL injection detection method. Based on the characteristics of SQL injection, we extract the corresponding feature values in the user's HTTP traffic and experiment with these feature values as inputs to the MLP and LSTM networks. The experimental results show that the accuracy of the MLP network is over 99%, and the accuracy of the LSTM network is achieved above 95%. The experimental results show that the eigenvalues extracted in this paper can effectively detect SQL injection attacks. Using the method proposed in this paper, it can effectively avoid the loopholes caused by the filtering method.

Acknowledgments. This work was supported by the Development Program of China under Grants Complexity 2017YFB0802704 and program of Shanghai Technology Research Leader under grant 16XD1424400.

References

1. Huang, H.C., Zhang, Z.K., Cheng, H.W., et al.: web application security: threats, countermeasures, and pitfalls. Computer **50**(6), 81–85 (2017)
2. Masri, W., Sleiman, S.: SQLPIL: SQL injection prevention by input labeling. Secur. Commun. Netw. **8**(15), 2545–2560 (2015)
3. Bhardwaj, M., John, A.: An adaptive algorithm to prevent SQL Injection **4**(3–1), 12–15 (2015)
4. Buja, G., Jalil, K.B.A., Ali, F.B.H.M., et al.: Detection model for SQL injection attack: an approach for preventing a web application from the SQL injection attack. In: IEEE Symposium on Computer Applications and Industrial Electronics, pp. 60–64. IEEE (2015)
5. Parvez, M., Zavarsky, P., Khoury, N.: Analysis of effectiveness of black-box web application scanners in detection of stored SQL injection and stored XSS vulnerabilities. In: Internet Technology and Secured Transactions, pp. 186–191. IEEE (2016)
6. Yuan, G., Li, B., Yao, Y., et al.: A deep learning enabled subspace spectral ensemble clustering approach for web anomaly detection. In: International Joint Conference on Neural Networks, pp. 3896–3903. IEEE (2017)
7. Kumar, M., Indu, L.: Detection and prevention of SQL injection attack. Int. J. Comput. Sci. Inf. Technol. **5**, 374–377 (2014)
8. Shi, C.C., Zhang, T., Yu, Y., et al.: A new approach for SQL-injection detection. Comput. Sci. **127**, 245–254 (2012)
9. Lecun, Y., Bengio, Y., Hinton, G.: Deep learning. Nature **521**(7553), 436 (2015)
10. Kaur, N., Kaur, P.: Modeling a SQL injection attack. In: International Conference on Computing for Sustainable Global Development. IEEE (2016)

Evaluation Methods of Hierarchical Models

Abdulqader M. Almars[1,2]([⊠]), Ibrahim A. Ibrahim[1,3]([⊠]), Xin Zhao[1]([⊠]), and Sanad Al-Maskari[1]([⊠])

[1] The University of Queensland, Brisbane, QLD, Australia
{a.almars,i.ibrahim,x.zhao,s.almaskari}@uq.edu.au
[2] Taibah University, Madinah, Saudi Arabia
[3] Minia University, Minia, Egypt
i.ibrahim@minia.edu.eg

Abstract. In this paper, we consider the problem of evaluating the quality of hierarchical models. This task arises due to the current researchers use subjective evaluation, such as a survey to test the goodness of a hierarchy discovered by their models. We propose three methods to evaluate the quality of hierarchy extracted from unstructured text. These methods are used to reflects three important characteristics of an optimal tree: (1) *Coverage* which reflects a topic on a high level, close to the root node, should cover a wider range of sub-concepts than those on a lower level; (2) *Parent-child relentless* which means the parent topic in the tree should be semantically related to its children rather than to its non-children; (3) *Topic coherence* that identifies all words within a topic should be semantically related to the other words. Moreover, we introduce a new metric called, *Interest-based coherent* to evaluate the hierarchical tree extracted from structured data like relational data. We compare different state-of-art methods and perform extensive experiments on three real datasets. The results confirm that the proposed methods can properly evaluate the quality of the hierarchy discovered by several models.

Keywords: Hierarchical models · Ontology learning
Evaluation methods · Structured data · Unstructured data

1 Introduction

Hierarchical modeling is a classic text mining task which is used to discover and organize the hidden topics that appear in a document collection. For example, in laptop review, users discuss the main topics of a laptop, such as the overall design, battery life, and screen quality. Extracting a hierarchy from these data can help people grasp essential information and understand key contents at different granularity. The discovery of hierarchy also facilitates many applications like user recommendation, summarization, and sentiment analysis [3,7,9] and so on.

© Springer Nature Switzerland AG 2018
G. Gan et al. (Eds.): ADMA 2018, LNAI 11323, pp. 455–464, 2018.
https://doi.org/10.1007/978-3-030-05090-0_39

There is a significant body of research has been done to develop a sophisticated hierarchical topic models [3,4,12]. Blei et al. [4] proposed a hierarchical model called the nested Chinese restaurant process (nCRP) to learn latent structures from data. The topic hierarchy is evaluated in term of held-out perplexity. Held-out perplexity is a common way to measure topic models, but this method is not appropriate because the semantic quality of a topic hierarchy is not considered. Wang et al. [16] proposed a novel phrase mining approach to recursively construct topical from a content-representative document. However, the existing hierarchical models use subjective methods (e.g, surveys), to measure the goodness of hierarchy they discover. Consequently, participants may not feel encouraged to provide accurate, honest answers.

To the best of our knowledge, we the first specifically address the issue of evaluating hierarchical models. There is a need for a universal method that measures the capability of a hierarchical model in discovering an optimal tree. To address the problem in current research, in this paper we propose three methods to evaluate the goodness of the tree extracted from textual data, topic coherence, coverage, and parent-child relatedness. This method aims at measuring the three important properties of an ideal tree. First, all the top words represent a topic in the tree should be coherent. Second, a topic on the tree should be organized from general, close to the root node, to specific near the leaf node. Third, a topic in a tree should be organized as parent and children topics, where the parent topic is semantically related to its children rather than to its non-children. In addition, we propose a new metric, namely, interest-based coherent to evaluate the quality of hierarchical tree discovered from relational data. We conducted experiments on three datasets and compare the quality of a hierarchy discovered by several state-of-art methods. Based on the proposed methods, the results show some existing methods succeed to extract a high-quality tree from the text. In summary, the main contribution of this paper is as follow:

The rest of the paper is organized as follows: in Sect. 2, we review the related work of hierarchical models. Section 3 introduces our proposed methods to evaluate the quality of the tree. Experimental results are presented in Sect. 4. We finally present conclusions and future work in Sect. 5.

2   Related Works

Several approaches have been proposed to address the problems of hierarchical extraction [2,4,12]. For example, the hierarchical Pachinko allocation model (hPAM) [13] was developed to discover the hidden topics and the correlations between them. However, the quality of hierarchical structure produced by the model is not objectivity measured. Blei et al. [4] proposed a generative probabilistic model known as the nested Chinese restaurant process (nCRP) to learn latent structures from data. Blei et al. suggested tow metrics to evaluate the topic hierarchy, node specialization, and hierarchy affinity.

Kim et al. [12] applied the same metrics proposed for structure evaluation. However, these metrics are only proposed to evaluate complicated topic models. In [18], Xu et al. propose a novel knowledge-based hierarchical topic model

(KHTM), which consider prior knowledge to construct a topic hierarchy. However, the main limitation of these existing models [6,12,18] is that they only evaluate the quality of topics, other characteristics of an optimal tree are not measured like the relatedness of parent topic and child topic and the coverage of the tree.

Zhao and Li [20] developed to extract *hot features* from short texts, and organize them hierarchically in a tree. However, the evolution of the tree's quality and the usefulness of its output were not studied. Moreover, Wang et al. [16] proposed a novel phrase mining approach to recursively construct topical hierarchy from the text. In our previous work [3], we developed an LDA-based method called Structured Sentiment Analysis (SSA) approach which recursively learns the hierarchical tree of the topics from the short text. However, the main limitation of the research mentioned above is that they apply subjective methods to evaluate the quality of the hierarchy.

3   The Proposed Methods

In literature, held-out perplexity is a widely used metric in topic modeling. However, this method is not appropriate here because neither the quality of topics nor the semantic relatedness between topics is considered. Following the same idea as that of prior research on hierarchical topic modeling [11], we introduce three measures to quantitatively evaluate the quality of tree discovered from the text, topic coherence, coverage, and parent-child relatedness. In addition, we propose another method to evaluate the tree extracted from structured data (e.g, relational database). In the following sections, we describe these methods in details.

3.1   Evaluation Methods in Text

In this section, we discuss three proposed methods to evaluate the goodness of a hierarchical tree extracted from unstructured text, topic coherence, coverage, and parent-child relatedness.

Topic Coherence. In the topic hierarchy, each topic is represented by a list of top words. The topic coherence is based on the idea that all words in this topic should be consistent with the semantic meaning of other words. For example, the topic *picture, pic, image* is coherent, because all three words refer to the same thing (i.e., *pic* is an abbreviation of *picture*, and *image* is an alternative word for *picture*). In order to evaluate the coherence of topics, we utilize an automated metric, namely coherence topic, proposed by Mimno et al. [15]. Topic Coherence has been used popularly by other researchers [17,19]. Suppose a topic z is characterized using a list $z = \{w_1^z, w_2^z, ..., w_n^z\}$ of n words . The coherence score of t is given by:

$$Ch(t) = \sum_{i=2}^{n} \sum_{j=1}^{i-1} log \frac{D(w_i^z, w_j^z) + 1}{D(w_j^z)}. \tag{1}$$

where $D(w_i, w_j)$ is the number of documents containing two topical words w_i and w_j in corpus. $D(w_j)$ is the number of documents containing a word, w_j. As a way of smoothing, we added by 1 to the numerator, $D(w_i, w_j)$, to avoid the value to be 0. A higher coherent score implies the all words within the topic are semantically related and refer to the same thing.

Coverage. Topics in a hierarchy should be organized from general to specific. The Topics near the root node must cover many documents, while those close to the leaf-nodes should have lower coverage. For example, the parent topic "battery" in a hierarchy has better coverage than its children, *life, usage* and *capacity*. Given the n top words of a topic, $z = \{w_1^z, w_2^z, ..., w_n^z\}$, we first replace the top words in the whole document with the first word. For example, if the top words of a topic are *picture, image, pic and photo*, we replace every document contain any of these words with the first word. We assume that all words under the same topic talk about the same thing. Then, the coverage score is calculated as follows:

$$Cov(L) = \frac{1}{k} \sum_z PMI(t_z). \tag{2}$$

$$PMI(z) = \frac{1}{n} \sum_j log \frac{p(w_1^z, w_j)}{p(w_1^z)p(w_j)}. \tag{3}$$

where $p(w_1^z, w_j)$ is the frequency of first words w_1^z with word w_j in the whole documents. $p(w_1^z)$ and $p(w_j)$ is the number of times word w_1^z and w_j respectively. k is the number of topics in level L. Pointwise mutual information (PMI) [8] is employed to calculate the similarity of pairs in the whole document. The PMI score for each topic t_z in the tree is computed. Then, the average coherence of all topics at the same level is used to reflect the coverage of the model. As the level increase, the coverage score increase, which means the model assumption is correctly reflected.

Parent-Child Relatedness. The goal of the third evaluation is to assess parent-child relatedness. All root topics of a hierarchy should not only be semantically related to their direct children but also to all offspring. For example, the root node *iPhone* should be related to its sub-topics *(camera, headphone, etc.)* and its sub-sub-topics *(picture, adapter, etc.)* Given a topic z, we measure the topic's relatedness score to its children and compare it to its non-children using Eqs. (4) and (5):

$$Child(z) = \frac{1}{k} \sum_k \frac{D(z, chn_k^z)}{D(z), D(chn_k^z)} \tag{4}$$

$$Non - Child(z) = \frac{1}{k} \sum_t \sum_k \frac{D(z, chn_k^t)}{D(z), D(chn_k^t)}, t \neq z \tag{5}$$

where $D(z, chn_k^z)$ is the number of times parent topic z appears with its child topic, chn_k^z and $D(z, chn_k^t)$ is the number of times parent topic z appears with its non-child topic chn_k^t. Similarly to the coverage calculation, all parent, child topics and non-child topics are represented by the first word. The overall parent-child relatedness is measured by taking the average score of relatedness to children and non-children topics for all parent topics at the same level.

3.2 Evaluation Methods in Structured Data

In this section, we discuss a method to evaluate the goodness of a hierarchical tree extracted from structured text like relational data.

Interest-Based Coherent. We introduce a special case for evaluating the hierarchical tree generated based on a set of interest-based visualizations produced by our previous work [1,10]. Our goal is to measure interest-based coherent score of each node on the tree. This can be achieve by measuring the interestingness (importance) of each attributes in the dataset. The interestingness of each node and can be computed using the utility score plus the frequency score. Ibrahim el at. [1] propose a metric to calculate the utility score. The utility score defines the interesting score of a specific visualization. The frequency indicates the importance of a dimensions, measures attributed or aggregate functions in the whole dataset. To draw the coherent score of each node, we use the equations below 6 and 7:

$$Leaf = (u) \times \alpha + (1 - \alpha) \times (f) \tag{6}$$

$$Non - leaf = \frac{1}{N} \sum_{j=1} coherent_j, j \in children. \tag{7}$$

where u represents the utility score, f the frequency score and alpha α is a hyperparameter to control either utility score or frequency should have a higher impact on the representation of the tree.

4 Experiments

In this section, we first introduce the dataset and the methods used for evaluation and then demonstrate the experimental results.

4.1 Datasets

In the following section, we give brief descriptions real-world dataset used for the experiment.

- **Laptop.** A collection of more than 10,014 distinct reviews crawled from Amazon [12]. To reduce the noise, we pre-processed the data to ignore common words that carry less important meaning like stop words, non-English characters and URLs are removed from the texts.

- **DBLP**. A collection of 33,313 titles was retrieved from a set of recently published papers in computer science. This dataset was prepared and has been previously used in [5,16]. We performed standard data preprocessing, including stop word removal. We also removed words with fewer than three characters.
- **MIMICIII**. MIMIC III is freely available, which has 397 million records from different 26 tables such as admissions, chart events, input events, output events,.. etc. This dataset has been previously used in our previous work [1]. In this experiment, we use two tables, Admissions and Input events. Admissions table contains 58,976 distinct hospital admissions records with 19 dimensions. Input events table contains 17.5 million tuples with 22 dimensions using Philips CareVue system.

4.2   Experiment on Hierarchical Topic Models

In this experiment, We mainly compare the quality of a topic hierarchy extracted from four well-known approaches.

- **rCRP** [11]. A non-parametric hierarchical topic model that infers the hierarchical structure of topics from discrete data. To generate a tree, we tune its hyperparameters to generate an approximately identical number of topics as other methods.
- **hPAM** [14]. A parametric hierarchical model that takes a document as input and generates a three levels of a topic hierarchy.
- **SSA** [3]. This is a recursive state-of-the-art LDA-based hierarchical model that discovers a tree with a specified depth and width. In the experiment, We tune the hyperparameters to generate the same shape.
- **HASM** [12]. A hierarchical aspect sentiment model that discovers a hierarchy of topics with corresponding sentiment polarity. In the experiment, we tune its hyperparameters to generate an approximately identical number of topics as those of other methods.

4.3   Evaluation Measures

We use three measures introduced in this paper to quantitatively evaluate the quality of the tree, topic coherence, coverage, and parent-child relatedness of the tree. We then use these metrics to compare the characteristics of a topic hierarchy constructed by state-of-art models.

Topic Coherence. In our experiments, we set the number of words in each topic to five. Since some baseline models can produce topics with less than five words, we only evaluate topics that contain five words. To evaluate the overall quality of a topic set, we calculate the average topic coherence score for each method. Here, we only show the score related to three levels of a topic hierarchy. The results are illustrated in Table 1. A higher coherence score indicates a better quality

Table 1. Average coherence score

	Laptop			DBLP		
	Level 1	Level 2	Level 3	Level 1	Level 2	Level 3
rCRP	−3.30	−2.16	−2.54	−3.18	−3.23	−3.14
hPAM	−1.60	−2.56	-	−2.99	−3.05	-
HASM	−2.30	−2.25	−3.51	−3.17	-	-
SSA	-	-	-	-	-	-

topic discovered by the model. On the DPLP dataset, the HASM created a flat hierarchy. In the SSA mode, we did not evaluate the topic coherence because the topic in the tree was represented by a single word.

Table 2. Average coverage score

	Laptop			DBLP		
	Level 1	Level 2	Level 3	Level 1	Level 2	Level 3
rCRP	−0.36	−0.40	−0.65	−0.32	−0.50	−0.43
hPAM	−0.33	−0.41	-	−0.64	−0.51	-
HASM	−0.28	−0.45	−0.65	-	-	-
SSA	−0.33	−0.46	−0.49	−0.72	−0.83	−0.87

Coverage. To investigate the coverage of topics discovered by all the test methods, suppose a topic t is characterized using a list of top words $t = \{w_1, w_2, w_3, ...w_N\}$ where N is the top words for the topic t. Before we measure the coverage score, we replace all words in the corpus with the 1st word. We assume all words within the same topic refer to the same thing. A higher coverage score indicates a better quality topic. The results are illustrated in Table 2. For all datasets, SSA clearly shows a decrease in the coverage score when the depth of the tree increases, which means the topics near the root nodes are general topics, while those near the leaf-nodes are specific topics. Unlike our model, the patterns in rCRP and the hPAM are different. For example, in rCRP, the coverage of topics at the third level is always higher than that of the topics at the second level, which means the topics extracted by the model are not organized from general to specific. In DBLP, due the sparsity and shortness of the text, HASM fails to discover a hierarchical tree.

Parent-Child Relatedness. We assume that parent topic t should be more similar to its descend children than to the children that from other topics. In this experiment, we only measure the relatedness score for a parent topic at

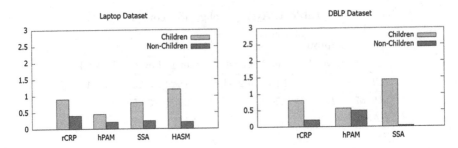

Fig. 1. Parent-Child relatedness. The average score between parent topics and their children topics compared with non-children topics at the second level. A higher score means that the parent topic are more similar.

the second level, with the children topics at the third level. We replace all top words that represent the topic with the first word. The overall relatedness score is measured by taking the average score to children and non-children topics for all parent topics at the second level. Figure 1 illustrates the parent-child relatedness of four models. The higher scores for children indicate that a parent is more similar to its children, compared to non-children nodes at the same level. All models show significant differences between children and non-children on Laptop dataset. The relatedness of the HASM is not considered for the DBLP dataset because it generated a flat tree.

4.4   Experiment on Relational Data

In this section, we use MIMICIII relational data to create a hierarchical tree of interesting visualization. To evaluate the quality of the tree, we use the Interest-Based Coherent method.

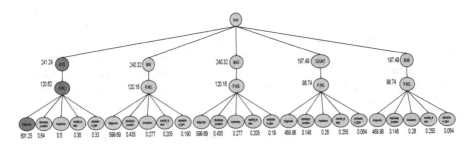

Fig. 2. Hierarchical representation for admission

4.5   Interest-Based Coherent

The creation of the tree is explained in our previous work [1]. Figures 2 and 3 show the hierarchical representation of two tables, admission and input events. Each node on the tree represents the importance (effect) of attributes in the

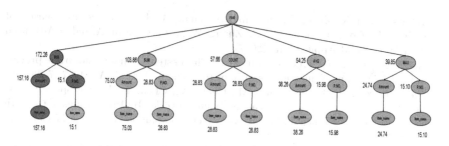

Fig. 3. Hierarchical representation for inputevents

database. The quality of the tree is measured according to the interestingness of attributes in the database. The first level of the tree shows aggregate functions, the second level represents measure attributes and leaf nodes show dimensions attributes. The most important or interesting attributes appear in the left and least one in the rightmost position. The coherence score of each node is calculated using the Eq. (6). The higher coherence score indicates more interesting attributes.

5 Conclusion

The evaluation methods currently used in the hierarchical model are generally inaccurate and cannot be applicable to all models. In this paper, we propose three methods to evaluate the structure of the topics discovered from the text, topic coherence, coverage, and parent-child relatedness. Moreover, we introduce new methods called interest-based coherent to reflects the quality of three discovered from relational data. We conduct experiments on three datasets. The result provides empirical evidence that our proposed methods can be used to evaluate any hierarchical models, and can fully reflect the main characteristics of an optimal tree. In future work, we would like to conduct more experiments on different and compare the effectiveness of several models in hierarchy discovery.

Acknowledgement. The author would like to thank the University of Thibah for the scholarship. This research is partially supported by Nature Science Foundation of China (Grant No. 61672284) and Australian Research Council (ARC) Discovery Project DP160104075. We would like to thank Lemma solutions www.Lemma.com.au for their helps and valuable advice.

References

1. Ibrahim, I.A., Almars, A., Pokharel, S., Zhao, X., Li, X.: Interesting recommendations based on hierarchical visualizations of medical data (2018)
2. Almars, A., Li, X., Ibrahim, I.A., Zhao, X.: Learning concept hierarchy from short texts using context coherence. In: Hacid, H., Cellary, W., Wang, H., Paik, H.-Y., Zhou, R. (eds.) WISE 2018. LNCS, vol. 11233, pp. 319–329. Springer, Cham (2018). https://doi.org/10.1007/978-3-030-02922-7_22

3. Almars, A., Li, X., Zhao, X., Ibrahim, I.A., Yuan, W., Li, B.: Structured sentiment analysis. In: Cong, G., Peng, W.C., Zhang, W.E., Li, C., Sun, A. (eds.) Advanced Data Mining and Applications (2017)

4. Blei, D.M., Griffiths, T.L., Jordan, M.I.: The nested Chinese restaurant process and Bayesian nonparametric inference of topic hierarchies. J. ACM **57**, 7 (2010)

5. Chen, H., Yin, H., Li, X., Wang, M., Chen, W., Chen, T.: People opinion topic model: opinion based user clustering in social networks. In: Proceedings of the 26th International Conference on World Wide Web Companion, WWW 2017 Companion (2017)

6. Chen, P., Zhang, N.L., Liu, T., Poon, L.K.M., Chen, Z.: Latent tree models for hierarchical topic detection. CoRR abs/1605.06650 (2016)

7. Chen, W., et al.: EEG-based motion intention recognition via multi-task RNNs, pp. 279–287. SIAM (2018)

8. Church, K.W., Hanks, P.: Word association norms, mutual information, and lexicography. Comput. Linguist. **16**, 22–29 (1990)

9. Gerani, S., Carenini, G., Ng, R.T.: Modeling content and structure for abstractive review summarization. Comput. Speech Lang. (2016)

10. Ibrahim, I.A., Albarrak, A.M., Li, X.: Constrained recommendations for query visualizations. Knowl. Inf. Syst. **51**, 499–529 (2017)

11. Kim, J.H., Kim, D., Kim, S., Oh, A.: Modeling topic hierarchies with the recursive Chinese restaurant process. In: Proceedings of the 21st ACM International Conference on Information and Knowledge Management, pp. 783–792. ACM (2012)

12. Kim, S., Zhang, J., Chen, Z., Oh, A.H., Liu, S.: A hierarchical aspect-sentiment model for online reviews. In: AAAI (2013)

13. Li, W., McCallum, A.: Pachinko allocation: dag-structured mixture models of topic correlations. In: Proceedings of the 23rd International Conference on Machine Learning, ICML 2006, pp. 577–584. ACM, New York (2006)

14. Mimno, D., Li, W., McCallum, A.: Mixtures of hierarchical topics with pachinko allocation. In: Proceedings of the 24th International Conference on Machine Learning, ICML 2007, pp. 633–640 (2007)

15. Mimno, D., Wallach, H.M., Talley, E., Leenders, M., McCallum, A.: Optimizing semantic coherence in topic models. In: Proceedings of the Conference on Empirical Methods in Natural Language Processing, EMNLP 2011, pp. 262–272. Association for Computational Linguistics, Stroudsburg (2011)

16. Wang, C., et al.: A phrase mining framework for recursive construction of a topical hierarchy. In: Proceedings of the 19th ACM SIGKDD International Conference on Knowledge Discovery and Data Mining, KDD 2013, pp. 437–445 (2013)

17. Wang, S., Chen, Z., Liu, B.: Mining aspect-specific opinion using a holistic lifelong topic model. In: Proceedings of the 25th International Conference on World Wide Web, WWW 2016, pp. 167–176 (2016)

18. Xu, Y., Yin, J., Huang, J., Yin, Y.: Hierarchical topic modeling with automatic knowledge mining. Expert Syst. Appl. **103**, 106–117 (2018)

19. Yan, X., Guo, J., Lan, Y., Cheng, X.: A biterm topic model for short texts. In: Proceedings of the 22nd International Conference on World Wide Web, WWW 2013 (2013)

20. Zhao, P., Li, X., Wang, K.: Feature extraction from micro-blogs for comparison of products and services. In: Lin, X., Manolopoulos, Y., Srivastava, D., Huang, G. (eds.) WISE 2013. LNCS, vol. 8180, pp. 82–91. Springer, Heidelberg (2013). https://doi.org/10.1007/978-3-642-41230-1_7

Local Community Detection Using Greedy Algorithm with Probability

Xiaoxiang Zhu and Zhengyou Xia[✉]

College of Computer Science and Technology,
Nanjing University of Aeronautics and Astronautics, Nanjing 210016, China
zhengyou_xia@nuaa.edu.cn

Abstract. With the arrival of the era of big data, the scale of network has grown at an incredible rate, which has brought challenges to community discovery. Local community discovery is a kind of community discovery that does not need to know global information about network. A quantity of local community discovery algorithms have been put forward by researchers. Traditional local community discovery generally needs to define local community modularity Q, and greedily add nodes to the community when $\Delta Q > 0$, which is easy to fall into the local optimal solution. Inspired by the ideal of simulated annealing, greedy algorithm with probability LCDGAP is proposed to detect local community in this paper, which can be applied to all algorithms that perform greedy addition. We permit that the node can be aggregated into the community with a certain probability when $\Delta Q < 0$. At the same time, we guarantee that this probability will be getting smaller with the increase of program running time, ensuring the program's convergence and stability. Experimental result proves that LCDGAP performs effectively not only in real-world dataset but also computer-generated dataset.

Keywords: Greedy algorithm · Local community detection · Probability

1 Introduction

Community discovery is an indispensable part of big data mining and analysis [1–3], which can be applied in many fields, such as social network, World Wide Web (WWW), paper cited network and biological network. Community structure is a key tool for studying complex system, but there is no standard definition for community currently. The widely accepted definition points to that the connection between nodes within same community is denser than nodes in different communities.

There exist many traditional algorithms for global community discovery, such as modularity optimization [4, 5], hierarchical clustering [6] and label propagation [7, 8].

However, traditional approaches were designed to detect community structure with the knowledge of entire network, which means that it can not work when the network is too large to deal with or too dynamic to acquire. Local community detection algorithm has been proposed to solve the problem, exploring one vertex at a time with local information. Local modularity measure R [9] was posed to discover local community by Clauset. Similarly, Luo et al. put forward modularity measure M [10] to discover

© Springer Nature Switzerland AG 2018
G. Gan et al. (Eds.): ADMA 2018, LNAI 11323, pp. 465–473, 2018.
https://doi.org/10.1007/978-3-030-05090-0_40

local community. M takes the ratio of inner edges and external edges into consideration. The method is a two-phrase algorithm, which includes greedy addition phrase and
deletion phrase. Vertices will be greedily added or removed from C only if it can cause
an increase in modularity M, until ΔM has not increased any more. In order to promote
the accuracy and reduce outliers, L [11] was presented by Chen et al. LFM [12] was
given by Lancichinetti et al., which uses measure Fitness to describe the relationship
between node and community. ILCDSP [13] was proposed by S Xia et al., which tries
to break greedy addition based on the measure M.

The remainder of paper is organized as follows. Our algorithm LCDGAP will be
discussed in detail in Sect. 2. Open standard dataset and synthetic benchmark network
dataset are taken to validate our method in Sect. 3. Finally, we summarizes our paper in
Sect. 4.

2 Our Algorithm

Classic community discovery algorithm needs to go through at least two steps. First,
the local community metric Q is defined, then node will be added into community
greedily according to the measure Q, and the above process is repeated continuously
until the convergence condition is reached. Many algorithms we introduced earlier
follow this framework. Most of improvements on local community detection focus on
how to choose better starting point or define better metric. Our paper starts from a
unique perspective and executes greedy addition with probability, which is helpful to
escape local optimal solution.

2.1 Theory of Simulated Annealing

We notice that greedy addition operator adopted by many local community discovery
algorithms is belong to hill-climbing algorithm which moves in the direction of local
optimum. The simulated annealing algorithm is also a greedy algorithm based on the
theory of solid annealing, random factor was introduced into search space, which
means that it accepts a solution that is worse than the current period with a certain
probability, so it has a chance to jump out of this local optimal solution.

The simple description of the simulated annealing is as follows:

(1) Initialization: Initial temperature T, lower temperature limit Tmin, initial state x,
    number of iterations L for every T.
(2) Generate new solution Xnew = X + Δx.
(3) Calculate the increment $\Delta f = f(Xnew) - f(X)$, f(x) is the optimization goal.
    Determine whether to accept the new solution according to the metropolis crite-
    rion p.

$$p = \begin{cases} 1, \Delta f > 0 \\ e^{\frac{-\Delta f}{T}}, \Delta f < 0 \end{cases} \tag{1}$$

(4) Reduce the temperature T, then terminate the program execution when it drops to Tmin or loop execution step 2 and 3 for L times.

2.2 Greedily Addition with Probability

The basic idea of LCDGAP is to carry on greedy addition with probability, which is inspired by the idea of simulated annealing. In our paper, we use the measure M as local modularity. We add node into the community like the previous algorithms when $\Delta m > 0$. Meanwhile, we need to accept the node to join the community with a certain probability when $\Delta m < 0$. Here we rewrite the metropolis criterion to accept a solution with a certain probability p.

$$p = \begin{cases} 1, \Delta m > 0 \\ e^{\frac{\Delta m}{T}}, \Delta m < 0 \end{cases} \tag{2}$$

The algorithm is relatively sensitive to parameters, which means that the initial value of temperature T and the rate of temperature drop α are one of the important factors affecting the global search performance. In order to cut down the number of parameters and accelerate convergence on local community detection, we present a simple and effective probabilistic model to balance performance and execution time.

$$p = \begin{cases} 1, \Delta m > 0 \\ k^{-t}, \Delta m < 0 \end{cases} \tag{3}$$

Where k represents a constant greater than 1, which is used to control the rate of convergence, the symbol t represents the number of times the solution is accepted when $\Delta m < 0$, the probability of being able to be accepted is getting smaller and tends to zero gradually with the increase in the number of acceptance, which is proved to be easy and feasible by experiments.

2.3 Steps of LCDGAP

Our algorithm LCDGAP is composed mainly of two stages: greedy addition with probability stage and deletion stage. In the greedy addition with probability stage, we calculate Δm when nodes in N are added into C and determine whether to add them into the community according to Eq. (3). In the deletion stage, all the nodes in C will be judged to decide whether to remove them from C according to Δm. These two phrases are repeated until no vertex can be added into community C. The pseudo code of proposed algorithm is shown as belows.

1.**Our Algorithm LCDGAP**
2. **Input**: G=(V,E) and a given source node v_source
3. **Output**: a local community C contains source node v_source
4. Add v_source to local community C, add the neighbors of v_source to N
5. Add v_source to list Q, iter=1
6. **While** Q is not empty **do**:
7.   compute local modularity m
8.   //greedy addition with probability stage phrase
9.   for node_i in N **do**:
10.   compute Δm
11.   if Δm>0 or (Δm<0 and p>random.random()) **do**
12.   add node_i to C and Q , remove node_i from N, compute current m
13.   //deletion phrase
14   for vi in C **do**
15.   compute Δm
16.   if Δm>0 and removing of vi does not disconnect C **do**
17.   remove vi from C and remove vi from Q if vi in Q.
18.   update C,N and m
19. **if** given node v_source is not in C then return no community
20. return C

LCDGAP can detect local community from a given source vertex in $O(k^2d)$ time, where K is the number of nodes to be added into community C and d denotes the average degree of the nodes in community C.

3   Experimental Result and Analysis

In order to verify the feasibility of our proposed algorithm, we compare LCDGAP with LWP, LFM and ILCDSP by real-world and synthetic dataset. The evaluation methods we take refers to precision, recall and F-score. LCDGAP is implemented in python (Python 3.4) language, and the program runs as window 7 operating system, 2.53 GHZ, 3 GB memory. In the experiment, we fix the parameter k to 2.

3.1   Real Network Experiment

We choose Karate, Dolphins, Football and Polbooks as public standard dataset. The specific information of real-world network is listed in Table 1. Due to the small scale of the real dataset, we discovery local community for each node in every dataset.

Table 1. Dataset for real network

Name	Nodes	Edges	Average degree
Karate	34	78	4.59
Dolphins	62	159	5.13
Polbooks	105	441	8.40
Football	115	613	10.66

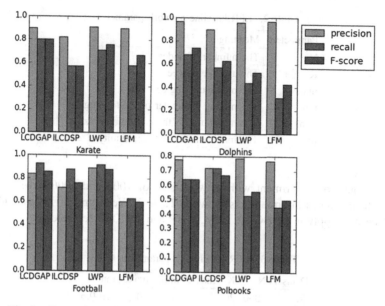

Fig. 1. Comparison of three evaluations for different methods on four dataset

The experimental result is shown in Fig. 1 for four real-world network, we can see intuitively that our algorithm LCDGAP improves recall value greatly in Karate, Dolphins and Football compared with other algorithms. For Polbooks, the recall of our method is just weaker than ILCDSP, the precision of our method LCDGAP is better than ILCDSP and very close to LWP. In terms of F-score, our method is ahead of other algorithms generally. It can be seen clearly from Fig. 1 that most of the maximum values are occupied by our algorithm LCDGAP. We can deduce that our approach works best in these four algorithms for four real-world dataset generally.

Our method and ILCDSP are all based on LWP, but LCDGAP adopts different greedily addition with probability to promote stability and performance, which may explains why our algorithms LCDGAP behaves better than ILCDSP.

3.2 Synthetic Network Experiment

The LFR benchmark is used for network community division performance test, which can generate the specified distribution network of the real situation. The meaning of the parameters is shown in the following Table 2.

Table 2. LFR baseline network parameters

Parameters	Meaning
N	The number of nodes
K	The average degree
maxK	The maximum degree
mu	The mixing parameter of the topology
minc	The minimum for the community size
maxc	The maximum for the community size

Two groups of experiment with network size of 1000 and 5000 have been set up. Specific parameter information is shown in Table 3. We sample 50 nodes randomly from dataset to apply in LCDGAP, LFM, LWP and ILCDSP separately.

Table 3. LFR benchmark parameters setup

Dataset	N	K	maxK	minc	maxc	mu
D1	1000	20	50	10	50	0.1–0.6
D2	5000	20	50	10	50	0.1–0.6

In Fig. 2, we can see that LCDGAP is far greater than other algorithms in terms of recall even if the community has become blurred. When mu $\leqslant 0.2$, the value of precision and F-score of our method is lower than other methods. As the mu value gets larger, which means that the difficulty of community discovery is constantly growing, the precision of our approach is close to other methods, and the F-score of our method is ahead of other algorithms, which is due to the fact that corresponding recall is larger than others.

In Fig. 3, when mu $\leqslant 0.2$, we find that the precision and F-score of LCDGAP are the lowest. However, LCDGAP is superior to other algorithms with the increase of mu value, which is similar with the previous network data D1. Meanwhile, LCDGAP always maintains absolute superiority on recall, no matter how the value of mu changes.

In view of dataset D1 and D2, the advantage of our algorithm can not be seen when mu is small, as the value of mu increases, LCDGAP is in the highest flight compared with other algorithms. In summary, LCDGAP is ahead of other methods in recall absolutely and does not harm hugely precision, which explains why the F-score of LCDGAP is in the first place. Experimental results show that our ideal of greedily addition with probability is very effective.

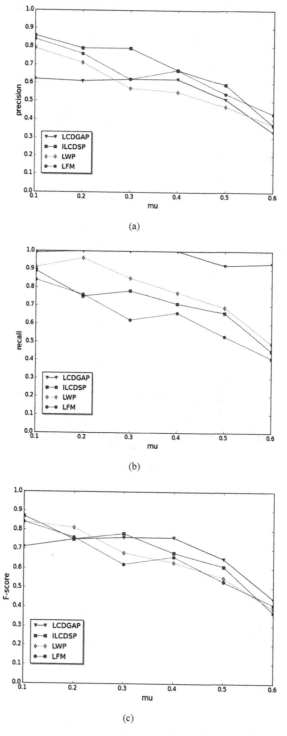

(a)

(b)

(c)

Fig. 2. Comparison of three evaluating index for all algorithms in D1

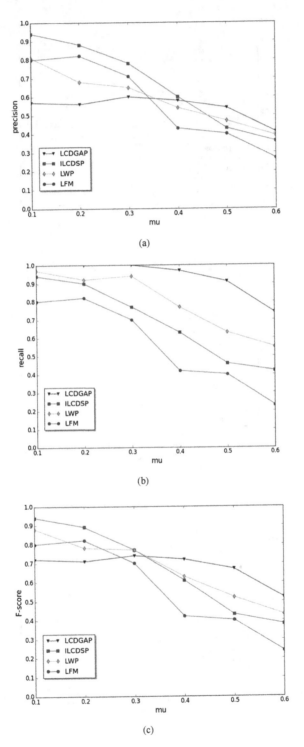

Fig. 3. Comparison of three evaluating index for all algorithms in D2

4   Conclusion

Inspired by the ideal of simulated annealing, LCDGAP is proposed. Which carries on greedy addition with probability. At the same time, in order to keep balance between performance and time efficiency, we present a simple and effective probability model instead of the model inspired by simulated annealing.

LCDGAP introduces greedily addition with controlled probability, trying to avoid falling into local optimality. Experimental result shows that LCDGAP behaves better than other algorithms both in real-world and synthetic network.

Though our method has higher recall than other algorithms, precision value is slightly lower than some algorithms in some cases. We are anxious to transplant greedy addition with probability to other similar algorithms in the future.

References

1. Chen, W., Wang, S., Zhang, X., et al.: EEG-based motion intention recognition via multi-task RNNs. In: pp. 279–287. Society for Industrial and Applied Mathematics (2018)
2. Yue, L., Chen, W., Li, X., et al.: A survey of sentiment analysis in social media. Knowl. Inf. Syst. 1–47 (2018)
3. Xia, Z., Bu, Z.: Community detection based on a semantic network. Knowl.-Based Syst. **26**, 30–39 (2012)
4. Newman, M.E.J., Girvan, M.: Finding and evaluating community structure in networks. Phys. Rev. E **69**(2), 026113 (2004)
5. Bu, Z., Zhang, C., Xia, Z.: A fast parallel modularity optimization algorithm for community detection in online social network. Knowl.-Based Syst. **176**(3) (2013)
6. Blondel, V.D., Guillaume, J.-L., Lambiotte, R., Lefebvre, E.: Fast unfolding of communities in large networks. J. Stat. Mech. **2008**(10) (2008)
7. Raghavan, U.N., Albert, R., Kumara, S.: Near linear time algorithm to detect community structures in large-scale networks. Phys. Rev. E **76**(3), 036106 (2006)
8. Zhang, X.-K., Fei, S., Song, C., et al.: Label propagation algorithm based on local cycles for community detection. Int. J. Mod. Phys. B **29**(05), 1550029 (2015)
9. Clauset, A.: Finding local community structure in networks. Phys. Rev. E **72**(2), 026132 (2005)
10. Luo, F., Wang, J.Z., Promislow, E.: Exploring local community structures in large networks. Web Intell. Agent Syst. **6**(4), 387–400 (2008)
11. Chen, J., Zaïane, O., Goebel, R.: Local community identification in social networks. In: Advances in Social Network Analysis and Mining, pp. 237–242 (2009)
12. Lancichinetti, A., Fortunato, S., Kertész, J.: Detecting the overlapping and hierarchical community structure of complex networks. New J. Phys. **11**(3), 19–44 (2009)
13. Xia, S., Zhou, R., Zhou, Y., Zhu, M.: An improved local community detection algorithm using selection probability. Math. Prob. Eng. **2014**(2), 1–10 (2014)

A Player Behavior Model for Predicting Win-Loss Outcome in MOBA Games

Xuan Lan[1], Lei Duan[1(✉)], Wen Chen[1], Ruiqi Qin[1], Timo Nummenmaa[2],
and Jyrki Nummenmaa[2]

[1] School of Computer Science, Sichuan University, Chengdu, China
xuan_lan@163.com, richforgood@163.com, {leiduan,wenchen}@scu.edu.cn
[2] Faculty of Natural Sciences, University of Tampere, Tampere, Finland
{timo.nummenmaa,jyrki.nummenmaa}@uta.fi

Abstract. Multiplayer Online Battle Arena (MOBA) game is currently one of the most popular genres of online games. In a MOBA game, players in a team compete against an opposing team. Typically, each MOBA game is a larger battle composed of a series of combat events. During a combat, the behavior of each player varies and the outcome of a game is determined both by the variation of each player's behavior and by the interactions within each instance of combat. However, both the variation and interaction are highly dynamic and difficult to master, making it hard to predict the outcome of a game. In this paper, we present a player behavior model (called pb-model). The model allows us to predict the result of a game once we have collected enough data on the behaviour of the players. We first use convolution to extract the features of player behavior variation in each combat and model them as sequences by time. Then we use a recurrent neural network to process the interaction among these sequences. Finally, we combine these two structures in a network to predict the result of a game. Experiments performed on typical MOBA game dataset verify that our pb-model is effective and achieves as high as 87.85% prediction accuracy.

Keywords: Game outcome prediction · Data mining · Deep learning

1 Introduction

Online gaming is increasingly popular as entertainment for more and more people. It is reported that, in 2015, young people spent 5 h on online gaming per day on average[1]. Among various kinds of online games, such as first-person shooter game, real-time strategy game and massively multiplayer online game, the *multiplayer online battle arena* (MOBA) game attracting over millions of concurrent

---

This work was supported in part by the National Natural Science Foundation of China under grant No. 1572332, the Fundamental Research Funds for the Central Universities under grant No. 2016SCU04A22, and the China Postdoctoral Science Foundation under grant No. 2016T90850.
[1] http://caas.raptr.com/category/most-played/.

G. Gan et al. (Eds.): ADMA 2018, LNAI 11323, pp. 474–488, 2018.
https://doi.org/10.1007/978-3-030-05090-0_41

users is a fusion of role-playing game, action game and real-time strategy game. Table 1 lists some representative MOBA games including *League of Legends*, which is the most popular one in the US and Europe during 2017[2].

Table 1. Some popular MOBA games

Game	Developer	Popular rank (see footnote 2)
League of Legends (LoL)	Riot games	1
Defense of the Ancients (DotA) 2	Valve corporation	10
Heroes of the Storm (HotS)	Blizzard entertainment	15

Typically, in a MOBA game, players are divided into two teams. Each player controls a single character in a team who competes versus another team of players. There are various abilities and advantages for each player character to contribute to the overall strategy of the team. The team destroying the opposing team's main structure is the winner. Moreover, there are many resources dispersed in the game map. Therefore, the players also have to fight with the adversaries to acquire these resources to enforce their fighting ability.

It should be noted that most of the games offered by the *World Cyber Games* (WCG) are MOBA. It is reported that a typical online MOBA game should has the following properties: (1) multiplayer: players from all over the world are allowed to challenge each other in the games. (2) competition: the purpose of the players (or teams) is to compete against others to win the games.

Recently, more and more game developers have recognized the popularity of MOBA games in the market. Thus, with its current status in the market, analysis of player behavior is an indispensable design step in the development of a succesfull MOBA game. Artificial intelligence and data mining techniques are highly useful to build a model of players behavior. For example, Erickson *et al.* [1] evaluated players' expertise level in game *DotA2* (a typical MOBA game).

Correspondingly, modeling player behavior to discover the unknown patterns and knowledge of games based on players' interactions or even possible interactions is attracting attention from not only industry but also academic fields. The analysis results can be utilized to predict the outcomes of online matches or to improve the game itself, e.g. by enhancing the intelligence of game AI, to improve defence against Bots, or to improve game balance. Gameplay simulations based on modeling player strategies based on player character interactions have also been used in game rule analysis [2].

By modeling player behaviors, the win-loss outcome of a MOBA game, i.e., the winning team, can be predicted. For example, Erickson *et al.* [3] used player behaviors to predict the outcome of the *Starcraft* game, and Rioult *et al.* [4] also used player behaviors to predict the outcome of game *DotA2*. However, to the

---

[2] https://newzoo.com/insights/rankings/top-20-core-pc-games/.

best of our knowledge, there is no previous work developing a general prediction model for MOBA games. In other words, a prediction model that can be used for all kinds of MOBA games without apriori knowledge.

Typically, a MOBA game is composed of a series of combat instances. This means, players interact with each other and the game environment in a combat situation. A player may be victorious or be defeated in combat. Victory or defeat in a combat situation will affect the behavior of the player (e.g., experience, equipments, coins). The variation of player behaviors is highly dynamic and quite different during combats. Therefore, it is necessary to analyze the relationships between player behaviors and combat outcomes.

In this paper, we propose a player behavior model representing the behavior variations, to predict the win-loss outcome of MOBA games. The main contributions include: (1) Developing a convolutional neural network, called pb-CNN, to extract the variation of player behaviors in each combat during a MOBA game; (2) Building a recurrent neural network, called pb-RNN, to model the sequences of player behavior variations in a MOBA game; (3) Constructing a player behavior model, called pb-model, to predict the win-loss outcome of a MOBA game based on both pb-CNN and pb-RNN; (4) Testing the effectiveness of pb-model based on game data set of *DotA2*.

The rest of this paper is organized as follows. Section 2 reviews the related works. Section 3 gives a detailed description of our behavior model, and Sect. 4 verifies our model. We conclude our work in Sect. 5.

2   Related Work

Recently, it has been recognized that the analysis of player behavior is important for the design of MOBA games. Some artificial intelligence and data mining techniques are applied to model player behavior. In general, there are two kinds of works on player behavior analysis.

The first category uses player behavior to analyze their influence on the game. Pobiedina *et al.* [5] ranked the players behaviors to find important factors influencing the outcome of game *DotA2*. Tinnawat [6] analyzed the behavior relationship between the role in the *DotA* and the leadership of the same person in real life. He found that people who have stronger leadership skills in real life tend to act the role of "carry" in games. The "carry" is a role that has large power and leads other members to fight in the game. Drachen *et al.* [7] used player behavior to analyze the spatiotemporal relationship among skills and the actions of the team. Castaneda *et al.* [1] studied different behaviors between the novices and the experts in *DotA2* and spread this study to other domains. Li *et al.* [8] used player behaviors to visualize snowballing and comeback of MOBA games. They developed a visual analytic system which can find key events and game parameters resulting in snowballing or comeback occurrences in MOBA games. Suznjevic *et al.* [9] used player behaviors to evaluate player skills in the game, calculating the player rating based solely on the match outcome.

The second category predicts the outcome of the game based on player behaviors. Erickson *et al.* [3] extracted behaviors from game logs, and constructed a model based on the current game state to predict the outcome of the *Starcraft*. Rioult *et al.* [4] used the topological player behaviors to predict the outcome of game *DotA2*. These behaviors consist of area of polygon described by the players, inertia, diameter, and distance to the base. Dereszynski *et al.* [10] used the hidden markov model, Weber *et al.* [11] used the decision tree and Synnaeve *et al.* [12] used the Bayesian model to predict strategies in Real-Time Strategy Games. Wang *et al.* [13] aimed at lineups using Naïve Bayes classifier to predict the outcome of *DotA2*. Cleghern *et al.* [14] used a value-split model for health of players to predict future states in time series attribute data. Yang *et al.* [15] proposed that combats in the game determined the ultimate outcome, by constructing combat sequences to identify the patterns in graph, and using these patterns to predict the outcome.

Deep learning is a powerful and effective tool with increasingly growing application in many domains. So far, some work has been done in analyzing game player behaviors using deep learning methods. Park *et al.* [16] proposed a deep learning based player evaluation model by combining both quantitative game statistics and the qualitative analysis provided by news articles. This model was based on the player behaviors during certain periods and the news articles in the same period. Leibfried *et al.* [17] used deep learning to predict rewards in Atari Games. Their work demonstrated that it was possible to learn system dynamics and the reward structure jointly in Atari Games. Oh *et al.* [18] also used deep learning to perform action-conditional video prediction in Atari Games. Their work was the first to make and evaluate long-term predictions on high-dimensional videos by controlling inputs.

3   Player Behavior Model for Outcome Prediction

In this section, we introduce our pb-model for MOBA game outcome prediction. The proposed pb-model consists of three parts: (1) player behavior extraction, (2) behavior sequence modeling, and (3) game outcome prediction.

3.1   Player Behavior Extraction

During each combat in a MOBA game, many factors with respect to player behaviors, such as abilities, equipments, levels, and experiences, are changing during the game. In a combat of a MOBA game, the variation of player behaviors are closely related to the win-loss outcome of the combat. For example, if a team wins a combat, the player behaviors of this team will vary more quickly than the losing one. For example, in game *DotA2*, the winning team will have more golds, more abilities, and more equipment. Clearly, the player behavior variation is the most important factor to the win-loss outcome of a combat.

In a MOBA game, we denote B the set of player behaviors, B_i the i-th player behavior variations between every two consecutive combats, p the feature of behavior variations that we want to extract, and $S(\cdot)$ the relevant function among all behaviors in a behavior set. The feature model is formulated as follows.

$$p = S(B_1, B_2, ..., B_n) \tag{1}$$

A proper relevance function is the key of our feature model. There are two problems should be addressed. The first is that disposing these behaviors jointly may result in exploding parameters. A straightforward solution to this problem is fusing some behaviors together as the best feature. However, the impacts of behaviors not chosen are ignored completely. The second is that extracting dynamic behavior variation is difficult due to their varying over time.

In Deep Learning, the CNN model is adept at extracting features, decreasing the amount and dimension of parameters and saving the runtime of processes. Based on CNN, we develop pb-CNN, which takes all player behaviors in a combat as input, and embeds them into input matrices to get the feature of behavior variations. We apply following convolution to pb-CNN.

$$R_i = g(WB_i + b) \tag{2}$$

where g is a non-linear active function, b is a bias vector, and W is the convolution kernel. In pb-CNN, we convolute each behavior for 100 times. We denote by R_i the feature map of the i-th player behavior. By convoluting all the behaviors in the behavior set, we can get a series of feature maps.

In pooling, pb-CNN uses max-pooling to get the max value from all feature maps, which is the most important feature. Then pb-CNN utilizes various sizes of pooling kernels aiming at different sizes of feature maps in order to get the same size of vector results. Each behavior can get its own feature vector O_i by pooling, and all of them combined is the feature of behavior variations. The relevant function is formulated as follows.

$$p = S(B_1, B_2, ..., B_n) = \sum_{n=1}^{i} O_i \tag{3}$$

In the process of combination, we combine these vectors in order to get the feature of behavior variations. Figure 1 shows the process of combination.

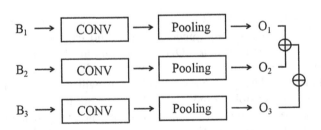

Fig. 1. Structure of pb-CNN

3.2   Behavior Sequence Modeling

As described above, in a MOBA game, a team has to compete for the resources and protect its buildings in order to win a game. There are several combats in a game, each result of the current combat will influence the outcome of next combat, and eventually the outcome of the whole game. Clearly, the game outcome is determined by the results of a series of combats that contained in the game.

We can combine all the combats into a sequence. Let $C = \{C_1, C_2, ..., C_i\}$ be the combat sequence in a game, where C_i is the outcome of the i-th combat. Considering the feature of a previous combat will influence the outcome of later combats, the interactions between different combat features should also be considered, besides for the features considered in pb-CNN. Let p_i be the feature of the variation from the i-th combat to the $(i + 1)$-th combat. We combine these features into a sequence in the order of combats. Let $P = \{p_1, p_2, ..., p_i\}$ be the feature sequence of combat variations, H the relevant function between all features in the combat sequence, and W the prediction vector. The feature sequence model is formulated as follows.

$$W = H(p_1, p_2, ..., p_i) \tag{4}$$

Different from traditional networks, RNN utilizes a recurrent structure to dispose interactions between sequences. LSTM is one kind of RNN, inheriting the peculiarity of RNN in disposing sequences, but solves the vanishing and exploding gradients by "gate". We input the feature sequence P to the LSTM, and formulate pb-RNN as follows.

$$f_t = \sigma(W_f p_t + U_f h_{t-1} + b_f) \tag{5}$$

$$i_t = \sigma(W_i p_t + U_i h_{t-1} + b_i) \tag{6}$$

$$\tilde{C} = tanh(W_c p_t + U_c h_{t-1} + b_c) \tag{7}$$

$$C_t = f_t \cdot C_{t-1} + i_t \cdot \tilde{C} \tag{8}$$

$$O_t = \sigma(W_o p_t + U_o h_{t-1} + b_o) \tag{9}$$

$$h_t = O_t \cdot tanh(C_t) \tag{10}$$

where W, U, b are the parameter matrices, $\sigma(\cdot)$ is the sigmoid function, p_t is the t-th feature, h_t is the network state of t-th feature and C_t is the cell of the t-th feature. Equation 5 is the forget gate, Eq. 6 is the input gate and Eq. 9 is the output gate, these three gates are used to process data. Equations 7 and 8 are the cells, which remove, save and update data from three gates.

In practice, we discover that in the process of most combats, there is a situation called "kill", where the player who is "killed" will lose the golds and cause

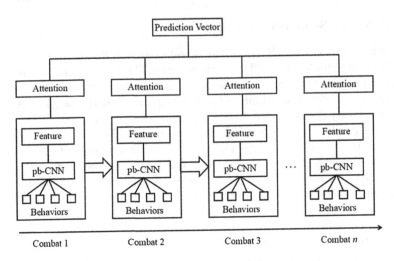

Fig. 2. Framework of pb-RNN

slight effects on the teams, and also influence the outcome of the game. There-fore, we try to add this behavior to our model. Different from the other behaviors, "kill" only happens during a combat and is not time-varying, so it cannot be added to pb-CNN to extract features. We add it to pb-RNN as attention, and the attention is formulated as follows.

$$a_t = \frac{\sum d_t}{D_t} \tag{11}$$

where d_t is the sum of players "killed" in a team at the t-th combat and D_t is the sum of players who are "killed" at the t-th combat. If no "kill" happens in a combat, the value of the attention is 0.5.

For the output of pb-RNN, we apply attention to them to get the state h_t, and finally we can get vector h for prediction by the weighted average of h_t. Vector h contains the relationship of each behavior and interactions among all features in the sequence. Then, we have:

$$W = H(p_1, p_2, ..., p_t) = h = \sum \overline{a_t h_t} \tag{12}$$

By pb-RNN, whenever the game is going on, we can use behaviors and "kill" of each combat to predict the game outcome. We can also update pb-RNN at any time as the match goes on to improve the prediction accuracy. Figure 2 shows the framework of pb-RNN.

3.3   Game Outcome Prediction

We extract features in pb-CNN, and put the feature sequence in pb-RNN. Each of the two teams can get the prediction vectors h_1 and h_2, respectively. Then,

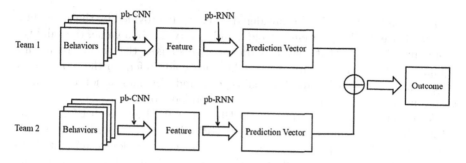

Fig. 3. Framework of the behavior model predicting MOBA game outcome

we combine these vectors and use softmax function to forecast the outcomes. Figure 3 shows the prediction process.

We construct the game outcome prediction model, called pb-model, by combining the output vectors of pb-RNN, h_1 and h_2, as the input of the full connect layer and softmax layer. In full connect layer, pb-model combines the features and maps them to the label. In softmax layer, pb-model gets the classification probability and chooses the max probability as the outcome of the game.

$$r = W(h_1, h_2) + b \tag{13}$$

$$p = softmax(r) \tag{14}$$

where W is a parameter matrix, and b is the bias vector. r is the result of the full connect layer. The objective of pb-model is to minimize the difference between the outcome p we calculate and the real outcome y. In pb-model, we use cross entropy function as our loss function:

$$J(\theta) = -\frac{1}{m} \sum_{i=0} (y^i \log p^i + (1 - y^i) \log(1 - p^i)) \tag{15}$$

4 Experiment

4.1 Setup

Considering that the game *DotA2* is the most popular MOBA game, we collected *DotA2* game data from Kaggle[3] to evaluate the effectiveness of our proposed pb-model. We chose data that contains temporal information, organized it by time and classified it into each game. In total, 20,000 game instances for model training were collected from 998 *DotA2* games. For each game, we chose the first 20 combats as training data, since the win-loss outcomes of most *DotA2* games are fixed in the first 20 combats.

---

[3] https://www.kaggle.com/.

In our empirical study, we considered 4 behaviors provided by the game data set, which are "ability", "item", "gold", and "subtype". In a game, the "ability" represents the skills that player character used or got, the "item" represents equipments that player character bought, the "gold" refers to the money that player character awarded and the "subtype" stands for information that broadcasted in the public channel, e.g., barracks were destroyed.

All experiments were performed on Ubuntu 16.04, Intel Xeon E5-2683 2.00 GHz CPU, 64G memory, and GTX1080 GPU. We used tensorflow to encode our program. In our pb-model, the parameters were randomly initialized. In pb-CNN, we chose RELU function as the active function, initialized embedding nominal randomly and set dimension to 200. The size of convolutional kernels was set as 3, 4, and 5, respectively while the sum of kernels was set as 100. In the training step, we set the min-batch as 64 and the learning rate as 0.001. We used Adam as the optimizer and SGD to optimize the loss function. We set the epoch of training to 500 and used early-stop for training.

4.2   Test on Prediction Accuracy

We evaluated the prediction accuracy of pb-model by 10 fold-cross-validation using *DotA2* game data set. In addition, to verify pb-model is insensitive to the game process, we define 3 kinds of game process as follows.

- "First 3": Refers to the first 3 combats in the beginning of the game, by which we test whether or not the outcomes of the game can be determined.
- "First 10": Refers to the first 10 combats of the game which comes to the middle of the game. Two teams have already had several combats so far, so the variation of behaviors has become obvious to some degree. At this time, the result of prediction is of great actual value.
- "First 20": Refers to the first 20 combats of the game, i.e., the end of the game, where the variation of behaviors is completely obvious. We get the prediction result of pb-model, and check if it is correct.

Table 2. Prediction accuracy w.r.t. the number of behaviors and the number of combats in *DotA2* game data

Player behaviors	Number of combats						
	1	3	10	14	16	18	20
Ability	0.5234	0.5468	0.5625	0.58	0.611	0.6032	0.6054
Ability, item	0.5273	0.6875	0.7265	0.7281	0.7226	0.7281	0.7375
Ability, item, gold	0.5078	0.5468	0.7312	0.71875	0.75	0.75	0.7549
Ability, item, gold, subtype	0.5062	0.6436	0.8056	0.8125	0.8556	0.875	0.8785

Table 2 lists the prediction accuracy of pb-model with respect to the number of behaviors and the number of combats. We can see that the accuracy got by the first 10 combats are very close to the accuracy got by all combats. In addition, it is clear to see that the accuracy of prediction has increased when considering more behaviors.

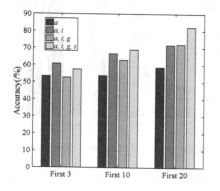

(a) Average prediction accuracy

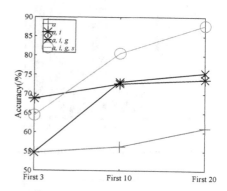

(b) Maximum prediction accuracy

Fig. 4. Performance of pb-model (*a*: "ability", *i*: "item", *g*: "gold", *s*: "subtype")

Figure 4 illustrates the average and maximum prediction accuracies of pb-model at different game process stages. From Fig. 4, we can see that:

(i) The win-loss outcome of the first 3 combats will not influence the outcome of a game. It is determined by the interaction of all combats rather than by any sole combat.

(ii) We used 4 behaviors in "First 20" and got an accuracy of prediction of 87.85%, which shows the high accuracy of our pb-model. Besides, the pb-model can be universally applied to all MOBA games, because the behaviors and combats produced in all MOBA games are easy to get.

(iii) Compared with the "First 20", the result of the "First 10" is of greater value despite its accuracy is lower. This is because the outcome is expected to be predicted as early as possible. The result of the "First 20" has higher accuracy but comes too late, since the final outcome can be directly observed in the end of a game without any need for prediction. The result of the "First 10" is what expected with an accuracy of 80.56%. Its D-value is acceptable, and it demonstrates that pb-model is of value in application and prediction.

Figure 5 shows the training of pb-model with respect to the behaviors. We can see that the runtime increases greatly when involving behavior "gold", since behavior "gold" is numeric, while the other three behaviors are nominal.

4.3 Test on Behaviors

In Sect. 3.1, we found that all behaviors have a different contribution to the game outcome. In this experiment, we test the contributions of behaviors in the result of "First 20".

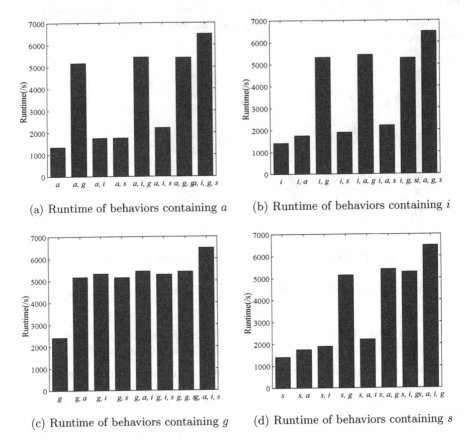

(a) Runtime of behaviors containing a (b) Runtime of behaviors containing i

(c) Runtime of behaviors containing g (d) Runtime of behaviors containing s

Fig. 5. Training time of pb-model w.r.t. behaviors (a: "ability", i: "item", g: "gold", s: "subtype")

Table 3. Prediction accuracy with respect to behavior

Player behavior	Accuracy (%)	Player behavior	Accuracy (%)
Ability	60.55	Item, subtype	83.75
Item	74.56	Gold, subtype	79.68
Gold	68.75	Ability, gold, item	75.5
Subtype	78.13	Ability, item, subtype	84.06
Ability, item	73.75	Ability, gold, subtype	81.71
Ability, gold	62.5	Item, gold, subtype	84.38
Ability, subtype	81.25	Ability, item, gold, subtype	87.85
Item, gold	78.79		

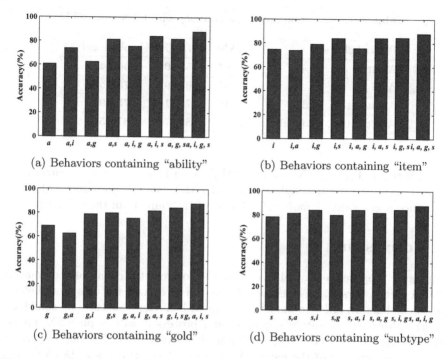

Fig. 6. Prediction accuracies w.r.t. behaviors (a: "ability", i: "item", g: "gold", s: "subtype")

We used all orders of behaviors as the input and gave the prediction accuracy in Table 3. As shown in the table, we can find that different behaviors have different contribution to the result, due to the same number of different behaviors as input but get the different result. Figure 6 illustrates the prediction accuracy contribution by different behaviors. We can see that the prediction accuracy increased slightly when considering "ability". That is, "ability" has the least accuracy contribution compared with other behavior. From Fig. 6, we can see that the rank of behavior contribution: "subtype", "item", "gold", "ability".

4.4   Comparison with CNN and LSTM

In our pb-model, we used pb-CNN to extract behaviors features and pb-RNN to process feature sequence. In this experiment, we will verify the effectiveness of our pb-model and compare it with the ones only using CNN or LSTM. When only using CNN, we only extracted features and combined these features to predict outcome directly. When only using LSTM, we only processed sequences and combined all behaviors without extracting the features from them. We compared our pb-model with CNN and LSTM in different number of behaviors and listed the prediction accuracy in Table 4.

Table 4. Comparation prediction accuracy with CNN and LSTM

Player behavior	CNN(%)	LSTM(%)	pb-model(%)
Ability	46.87	53.90	**60.54**
Ability, item	48.44	61.09	**73.75**
Ability, item, gold	52.18	70.31	**75.49**
Ability, item, gold, subtype	56.25	80.47	**87.85**

As listed in Table 4, CNN and LSTM disregard the influence on each other, so their accuracies are not high. Compared with them, our pb-model is the best one, since it can capture more information of the game than the other two.

5   Conclusions

This paper proposed a player behavior model to predict the win-loss outcome of MOBA games. Firstly, we constructed pb-CNN to extract features of all behaviors in combats. In order to model this combat-behavior structure, we combined these features into a feature sequence, and process it with a pb-RNN model. pb-RNN takes prediction vector as output, and uses softmax function to predict the win-loss outcome of the game. We performed experiments in DotA2 game data set and verified the proposed pb-model is effective. Our player behavior model can get acceptable prediction accuracy using the middle stage of game process. Thus, the pb-model is of great value in application.

The proposed behaviour model can be applied to at least three different purposes, with further research improving its usefulness. The first purpose is predicting the outcome of a match during a live eSports event. Here, the prediction accuracy grows throughout the game, because more combat instances are available for analysis as the game progresses. However, this would require the analysis to be faster than what is currently possible. The second purpose is making changes to the game design based on the model. An example of this need would be the finding that some player choices can provide an unwanted edge in the game, resulting in an unbalanced gameplay experience. The third purpose is to use the data to achieve more realistic gameplay simulations. Modelling player strategies based on the probability of players to engage in certain game features can be used to evaluate and improve a game rules and its design [2]. This model provides data that can be used in such simulations.

There are several interesting issues that deserve further research. First, having both numeric and nominal values together in the current setup in pd-CNN increases the time needed for the computations. We should consider a way to rearrange this differently to improve the efficiency. Second, in pd-CNN, the nominal behaviors were randomly initialized in embedding. We plan to adapt another embedding methods to get the relationship among such behaviors before training, to get a better starting point in the gradient descent, which can further contribute to making a prediction with higher accuracy.

References

1. Castaneda, L.M., Sidhu, M.K., Azose, J.J., Swanson, T.: Game play differences by expertise level in dota 2, a complex multiplayer video game. IJGCMS **8**(4), 1–24 (2016)
2. Nummenmaa, T.: Executable formal specifications in game development: design, validation and evolution. University of Tampere, Tampere University Press (2013)
3. Erickson, G.K.S., Buro, M.: Global state evaluation in StarCraft. In: Digital Entertainment (2014)
4. Rioult, F., Mtivier, J.P., Helleu, B., Scelles, N., Durand, C.: Mining tracks of competitive video games. AASRI Procedia **8**, 82–87 (2014)
5. Pobiedina, N., Neidhardt, J., del Carmen Calatrava Moreno, M., Werthner, H.: Ranking factors of team success. In: Proceedings of 22nd International World Wide Web Conference, pp. 1185–1194 (2013)
6. Nuangjumnong, T.: The influences of online gaming on leadership development. Trans. Comput. Sci. **26**, 142–160 (2016)
7. Drachen, A., et al.: Skill-based differences in spatio-temporal team behaviour in defence of the ancients 2 (dota 2). In: Proceedings of 2014 IEEE Games Media Entertainment, pp. 1–8 (2014)
8. Li, Q., et al.: A visual analytics approach for understanding reasons behind snowballing and comeback in MOBA games. IEEE Trans. Vis. Comput. Graph. **23**(1), 211–220 (2017)
9. Suznjevic, M., Matijasevic, M., Konfic, J.: Application context based algorithm for player skill evaluation in MOBA games. In: Proceedings of 2015 International Workshop on Network and Systems Support for Games, pp. 1–6 (2015)
10. Dereszynski, E.W., Hostetler, J., Fern, A., Dietterich, T.G., Hoang, T., Udarbe, M.: Learning probabilistic behavior models in real-time strategy games. In: Digital Entertainment (2011)
11. Weber, B.G., Mateas, M.: A data mining approach to strategy prediction. In: Proceedings of the 2009 IEEE Symposium on Computational Intelligence and Games, pp. 140–147 (2009)
12. Synnaeve, G., Bessière, P.: A Bayesian model for opening prediction in RTS games with application to StarCraft. In: Proceedings of 2011 IEEE Conference on Computational Intelligence and Games, pp. 281–288 (2011)
13. Wang, K., Shang, W.: Outcome prediction of DOTA2 based on Navïe Bayes classifier. In: Proceedings of 16th IEEE/ACIS International Conference on Computer and Information Science, pp. 591–593 (2017)
14. Cleghern, Z., Lahiri, S., Özaltin, O.Y., Roberts, D.L.: Predicting future states in DOTA 2 using value-split models of time series attribute data. In: Proceedings of the International Conference on the Foundations of Digital Games, pp. 5:1–5:10 (2017)
15. Yang, P., Harrison, B.E., Roberts, D.L.: Identifying patterns in combat that are predictive of success in MOBA games. In: Proceedings of the 9th International Conference on the Foundations of Digital Games (2014)
16. Park, Y.J., Kim, H.S., Kim, D., Lee, H., Kim, S.B., Kang, P.: A deep learning-based sports player evaluation model based on game statistics and news articles. Knowl.-Based Syst. **138**, 15–26 (2017)

17. Leibfried, F., Kushman, N., Hofmann, K.: A deep learning approach for joint video frame and reward prediction in Atari games. CoRR abs/1611.07078 (2016)
18. Oh, J., Guo, X., Lee, H., Lewis, R.L., Singh, S.P.: Action-conditional video prediction using deep networks in Atari games. In: Proceedings of Annual Conference on Neural Information Processing Systems 2015, pp. 2863–2871 (2015)

An Improved Optimization of Link-Based Label Propagation Algorithm

Xiaoxiang Zhu and Zhengyou Xia[✉]

College of Computer Science and Technology,
Nanjing University of Aeronautics and Astronautics, Nanjing 210016, China
zhengyou_xia@nuaa.edu.cn

Abstract. Community Detection has been an important tool for network and overlapping Community exists in real work ubiquitously. In this paper, an improved optimization of link-based label propagation algorithm rather than node called LinkLPAm is proposed to detect overlapping community. We briefly introduce our main work. Firstly, the initialization on edge labels by rough core is presented to speed up the process of detecting overlapping community, which is a big timesaver for the link network that magnified a lot of times compared with original node network. Secondly, an optimization algorithm of label propagation on link is given to update label on edge. Thirdly, in order to restrict the number of communities and the number of nodes in the community, the metric community similarity between communities is defined and greedy mergence algorithm is taken to merge communities according to community similarity. Finally, experimental result shows that our method LinkLPAm is serviceable for find reasonable overlapping community.

Keywords: Overlapping community · Optimization · Link · Label propagation

1 Introduction

Community discovery is an indispensable part of big data mining [1–3]. Previous researches on non-overlapping global community fail to reflect multi and complex relationship between the individuals. Recently, lots of overlapping community detection approaches has been proposed. Generally, they can fall into two basic categories: node-based algorithms and link-based algorithms.

Node-based methods account for major proportion in the methods of discovering overlapping community. In 2009, Lancichinetti, Fortunato and kertesz presented Local Fitness Maximization LFM [4]. As we know, label propagation algorithm is an efficient approach for community discovery. COPRA [5] was put forward by Gregory in 2010. A concept of community coefficient is given to measure the weights of communities a node has. SLPA [6] was raised by Raghavan. FPMQA [7] is a fast parallel modularity optimization algorithm.

Link-based algorithms have their distinctive superior to node-based methods in multi-scale network. Shi et al. put forward an algorithm GaoCD [8] under the framework of genetic algorithm. Sun proposed LinkLPA [9] based on the skeleton of label propagation with preference on link to discover overlapping community.

© Springer Nature Switzerland AG 2018
G. Gan et al. (Eds.): ADMA 2018, LNAI 11323, pp. 489–498, 2018.
https://doi.org/10.1007/978-3-030-05090-0_42

The remainder of paper is organized as follows. The preliminaries about basic notation and framework of label propagation on link are mentioned in Sect. 2. In Sect. 3, LinkLPAm will be discussed in detail, followed by experiments in Sect. 4. Finally, we conclude the paper in Sect. 5.

2 Preliminaries

2.1 Notation

Given a network $G = \{V, E\}$, where V means set of vertices, E represents all edges.

Definition 1 (Incident Edges of Node). For every node $v \in V$, the incident edges of node is as:

$$I(v) = \{(u, v)|(u, v) \in E\}$$

Definition 2 (Neighbor of Edge). For any edge $e \in E$, the edge e can be described as symbol of $e = (u, v)$, where u and v are the vertices attaching to the edge e. Neighbor of edge is calculated by:

$$N(e) = I(v) \cup I(u) - e$$

Definition 3 (Degree of Edge). The degree of edge is depicted by:

$$D(e) = Len(N(e))$$

Where N(e) represents the set of the neighbors of edge, the symbol of Len means the number of the set.

2.2 Framework of Link-Based Label Propagation

The framework of link-based label propagation is described as follows:

(1) Allocating every edge with unique label, which matches the id of edge.
(2) Sorting the edge list by random.
(3) For every edge e in edges, the edge e chooses the label with special strategy, such as the most frequent label from the neighbor edge.
(4) If labels for all edges are not change at all, then go to step (5), otherwise, the process from step 2 to step 4 is repeated.
(5) Transforming the link communities into node communities.

3 LinkLPAm

3.1 The Initiation on Edge Label

The graph $G = (V, E)$ is consist of n nodes and m links, usually satisfying the condition $m \gg n$. When we map the network into link graph G', we find that the size of link graph G' is enlarged too many times compared with original network.

So, it is very necessary for large-scale network to initialize edge label by taking the ideal of rough core. The pseudo code is shown as below.

Algorithm initEdgeLabel
___
Input: G=(V,E);
Output: rough cores rc;
  1: rc={} #init rough cores
  2: sort the edges E according to their edge degree in descending order;
  3: **for** e in E **do**:
  4: core={} #save rough cores;
  5: **if** e.free and degree of edge e>=4 **do**:
  6:    find max_nei_e which has the largest edge degree in N(e) and max_nei_e.free;
  7:    **if** max_nei_e is exists **do**:
  8:      add max_nei_e and e into the core;
  9:    **end if**
10:    comNei=intersection(N(max_nei_e), N(e));
11:    **while** comNei is not Empty **do**
12:      remove the edges which are not free in comNei;
13:      get the edge max_edge in comNei according to edge degree;
14:      add the edge max_edge into the core;
15:      comNei= intersection(N(max_edge), comNei);
16:    **end while**
17:    add core into rc;
18: **end if**
19: **end for**
20: return rc

3.2 An Improved Optimization of LPA on Link

The simplest strategy of updating label for label propagation on link is to choose the most frequent label from neighbor edges for given edge. However, the strategy exists randomness when the most frequent label is not the only one. Inspired by LPAm [10], an improved optimization of LPA on Link is proposed.

LPA optimization algorithm on link is shown as:

$$Lu = \arg\max_{l} \sum_{v=1}^{n} Buv\delta(l, lv) \tag{1}$$

Where B is a matrix representing the adjacency between edges, δ is Kronecker delta. If edge u is adjacent to the edge v, then the value of Buv is one, otherwise zero.

Next, we adopt an objective function H that is maximized by optimization rule.

$$H = \frac{1}{2} \sum_{v=1}^{n} \sum_{u=1}^{n} Buv\delta(lu, lv) \tag{2}$$

In order to separate edge x from all edges, we can rewrite Eq. (2), yielding

$$H = \frac{1}{2}(\sum_{v \neq x}\sum_{u \neq x} Buv\delta(lu, lv) + \sum_{u=1}^{n} Buv\delta(lu, lx) + \sum_{v=1}^{n} Buv\delta(lx, lv) - Bxx) \quad (3)$$

We still simplify Eq. (3) by taking advantage of the symmetry of the adjacency matrix B, giving

$$H = \frac{1}{2}(\sum_{v \neq x}\sum_{u \neq x} Buv\delta(lu, lv) - Bxx) + \sum_{u=1}^{n} Buv\delta(lu, lx) \quad (4)$$

Because the value of left term of Eq. (4) do not change with edge x, the right term is just the shape of maximizing the LPA optimization rule written in Eq. (1).

We know that Eq. (4) may generate unreasonable partitions, so we give the heuristic objective function H' by adding penalty term G.

$$H' = H - \varepsilon * G \quad (5)$$

$$K_L = \sum_{i=1}^{n} k_i\delta(L_i, L) \quad (6)$$

$$G = \frac{1}{2}\sum_{l=1}^{n} K_L^2 \quad (7)$$

Where ki represents the degree of edge, K_L denotes the total degree of edge with the same label L. As with G is minimal when all edges have unique labels and maximal when all edges have the same label, trying to avoid local optimization.

We get final objective function H' by incorporating G into H'.

$$H' = \frac{1}{2}\sum_{u=1}^{n}\sum_{v=1}^{n} (Buv - \varepsilon KuKv)\delta(lu, lv) \quad (8)$$

The final rule for updating edge label by combining Eqs. (8) and (4) is shown:

$$Lu = \arg\max_{l} \sum_{v \neq u} (Buv - \varepsilon KvKu)\delta(L, Lv) \quad (9)$$

Where ε is a constant $\varepsilon = \frac{1}{2M}$, 2M represents the sum of degree of all edges.

Next, we explain the execution procedure of label propagation with preference on link according to the formula (9).

We randomly choose the update link sequence from Fig. 1, which is bdaecghf. Initial label of every edge is its own edge number. For edge b, we choose the label by the rule of Eq. (9), then find that the values of label a, d and e are 9/11, 9/11 and 5/11. So we choose label a or d as the label of edge b. we continue to execute the above step according to the remainder sequences daec. The links {a, b, c, d, e} are classified into the same community. When we update the edge g, we find the edge g is also belong to

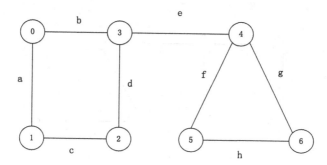

Fig. 1. Simple network

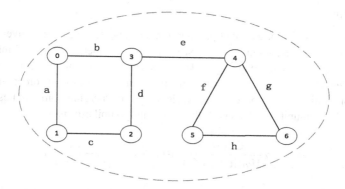

Fig. 2. Monster community

the community of {a, b, c, d, e}. Finally, all links belong to one community, which causes a terrible monster community. The final result is shown in Fig. 2.

From the above analysis, we can clearly see that the method is too time-consuming, because updating a link label must calculate the information of all edges. Furthermore, above formula is easy to generate community monster for small-scale network.

Due to the characteristic of label propagation, we suppose that a label of edge will be decided by the neighbor edges. So, we only take the edges adjacent to the target edge into account instead of all edges. The refined formula is shown as follow:

$$Lu = \arg\max_{L} \sum_{v \in N(u)} (Buv - \lambda KvKu)\delta(L, Lv) \tag{10}$$

Now, we take the refined formula to update link label, we will find that two link communities {a, b, c, e, d} and {f, g, h} can be detected correctly. Link communities will be transformed into node communities, so we get the right overlapping communities, which are {0, 1, 2, 3, 4} and {4, 5, 6} separately in Fig. 3.

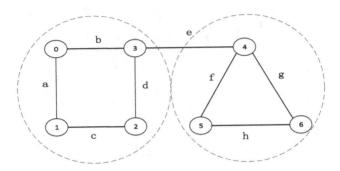

Fig. 3. Right communities by our algorithm

3.3   Postprocess

After the phase of link-based label propagation terminates, we should convert the link communities to node communities. Firstly, we use the metric ANE in LinkLPA to restrict the number of overlapping nodes. It is very significant for us to merge the communities so as to reduce the petty and fragmented modules. In order to process those communities, we propose a greedy algorithm for merging communities.

Given 2 communities, c1 and c2, the community similarity is:

$$CS = \alpha \frac{|C1 \cap C2|}{Min(|C1|, |C2|)} + (1 - \alpha) * \frac{2|E|}{|V|(|V| - 1)} \tag{11}$$

|V| is the sum of the number of nodes for community c1 and community c2, |E| means the number of edges between community c1 and community c2, α denotes variable weight parameter.

The process of greedily merging communities is described as below:

(1) Calculating the community similarity between the pairs of communities.
(2) If maximum community similarity is greater than the given threshold of community similarity, then go to the step 3, otherwise end the process.
(3) Merging the two communities with maximum community similarity.
(4) Recalculating the community similarity between the remainder of communities each other.

4   Experimental Result and Analysis

We compare LinkLPAm with some classic algorithms. In the experiment, we fix the parameters α and SimilarityRatio for our algorithm to 1 and 0.3.

4.1 Dataset and Measure

We choose Karate, Dolphins, Football and Polbooks as public standard dataset. The information of dataset is listed in Table 1. To evaluating the performance of overlapping community, Nicosia et al. [11] proposed a variant modularity Qov.

Table 1. Data set for real network

Name	Nodes	Edges	Average degree
Karate	34	78	4.59
Dolphins	62	159	5.13
Polbooks	105	441	8.40
Football	115	613	10.66

4.2 Experimental Analysis

The modularity Qov for Karate is shown as Fig. 4. We can clearly see that LinkLPAm gets the best overlap modularity of 0.7080 in Karate network, which is superior to other five algorithms. LinkLPA makes worse effect on overlapping community detection, which is only better than GaoCD.

The second real network is Dolphins. From Fig. 5, we know that LinkLPAm performs better than other algorithms. COPRA, SLPA get the similar overlap modularity with almost 0.72. However, LinkLPA and LFM also behave worse than LinkLPAm we proposed, which are 0.6796 and 0.6402 separately. GaoCD is also on the bottom of those algorithms.

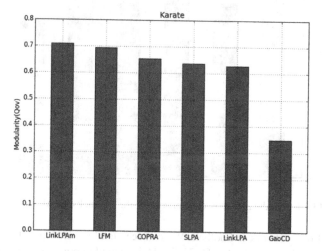

Fig. 4. Modularity Qov comparison for the Karate social network

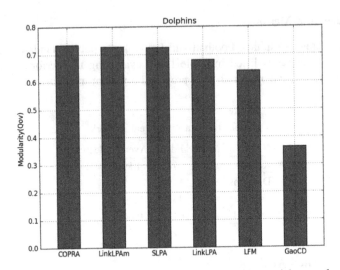

Fig. 5. Modularity Qov comparison for the Dolphins social network

The third real network is Polbooks, of which the result corresponds to Fig. 6. All algorithms also get high overlap modularity except GaoCD, proving that the community structure of Polbook is very easy to detect compared with other real networks.

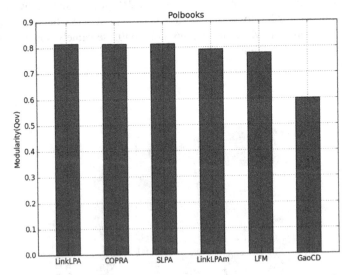

Fig. 6. Modularity Qov comparison for the Polbooks social network

The final real network is Football. We can see clearly from Fig. 7 that LinkLPAm gets the best overlap modularity of 0.71, SLPA is second to our algorithm obtaining the value of 0.70. COPRA gets the modularity value of 0.68. However, the remainder of three algorithms LinkLPA, LFM and GaoCD perform very badly.

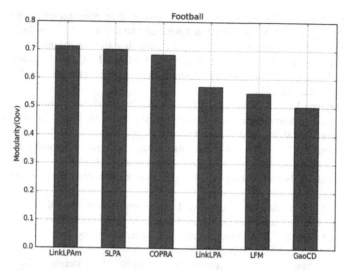

Fig. 7. Modularity Qov comparison for the Football social network

From the perspectives of the overlap modularity, LinkLPAm performs the best on the four networks on the whole compared with other algorithms, which proves that our improvement is very effective.

5 Conclusion

In the paper, an improved optimization algorithm based on label propagation with link is proposed to detect overlapping community detection.

LinkLPAm generates initial edge label by rough core to accelerate the process. But beyond that, an improved optimization of link-based label propagation with local information to reduce running time and achieve better performance relatively. Finally, a greedy mergence community operator based on the defined measure community similarity is proposed to merge close and fragmented communities, which can avoid many unreasonable combination and make community structure more reasonable.

Experimental result shows that LinkLPAm performs best in dataset. However, LinkLPAm costs more time than other algorithms using node-based network when the size of network is large. A parallel version of LinkLPAm can be developed to discover overlapping partitions in acceptable time in the future.

References

1. Lin, Y., Chen, W., Li, X., et al.: A survey of sentiment analysis in social media. Knowl. Inf. Syst. 1–47 (2018)
2. Xia, Z., Bu, Z.: Community detection based on a semantic network. Knowl.-Based Syst. **26**, 30–39 (2012)

3. Chen, W., Wang, S., Zhang, X., et al.: EEG-based motion intention recognition via multi-task RNNs. In: Society for Industrial and Applied Mathematics, pp. 279–287 (2018)
4. Lancichinetti, A., Fortunato, S., Kertész, J.: Detecting the overlapping and hierarchical community structure of complex networks. New J. Phys. **11**(3), 19–44 (2008)
5. Gregory, S.: Finding overlapping communities in networks by label propagation. New J. Phys. **12**(10), 2011–2024 (2009)
6. Xie, J., Szymanski, B.K., Liu, X.: SLPA: uncovering overlapping communities in social networks via a speaker-listener interaction dynamic process. In: International Conference on Data Mining Workshops IEEE Computer Society, pp. 344–349. IEEE (2011)
7. Bu, Z., Zhang, C., Xia, Z.: A fast parallel modularity optimization algorithm for community detection in online social network. Knowl.-Based Syst. **76**(3) (2013)
8. Shi, C., et al.: A link clustering based overlapping community detection algorithm. Data Knowl. Eng. **87**(9), 394–404 (2013)
9. Heli, S., et al.: LinkLPA: a link-based label propagation algorithm for overlapping community detection in networks. Comput. Intell. **33**, 308–331 (2016)
10. Barber, M.J., Clark, J.W.: Detecting network communities by propagating labels under constraints. Phys. E **80**(2), 026129 (2009)
11. Nicosia, V., Mangioni, G., Carchiolo, V., Malgeri, M.: Extending the definition of modularity to directed graphs with overlapping communities. J. Stat. Mech. **2009**(3), 3166–3168 (2008)

Research on Commodity Recommendation Algorithm Based on RFN

Kai Wang[1], Bohan Li[1,2,3(✉)], Shuo Wan[1], Anman Zhang[1],
and Donghai Guan[1,2]

[1] College of Computer Science and Technology,
Nanjing University of Aeronautics and Astronautics, Nanjing 211106, China
hxyrqx123@sina.com, {bhli,shuowan,dhguan}@nuaa.edu.cn
[2] Collaborative Innovation Center of Novel Software Technology
and Industrialization, Nanjing 210016, China
[3] Jiangsu Easymap Geographic Information Technology Corp., Ltd,
Yangzhou 225000, China

Abstract. The recommendation system is one of the most widely used applications in E-commerce. By studying the user's preferences, we can recommend underlying contents for the user from the mass merchandise information. However, most recommendation systems pay much attention on popular products, just ignore those products that are currently not popular but potential for excavation. Our recommendation system based on RFN (Reverse Furthest Neighbor) queries follows the idea of mining popular products in the niche market. We improve the traditional collaborative filtering recommendation algorithm and adopt a collaborative filtering algorithm based on expert users. The modified algorithm can recommend products with potential value based on the power law, which make the distribution of minority mined more adequately by the users. The experimental results show that the recommendation system has high recommendation quality and practical value.

Keywords: Recommendation system · Collaborative filtering
Reverse furthest neighbor · Power law

1 Introduction

The problem of information overload has become increasingly serious in the era of big data [1, 2]. To analyze user preferences and recommend targeted products to users, many recommendation algorithms and related systems have emerged. From users' perspective, there are context aware recommendation systems [3], mobile recommendation systems [4], etc. Most of these recommendation systems are based on user preferences for commodity recommendation. However, from a merchant's point of

This work is supported by the National Natural Science Foundation of China (61672284, 41301407), the Funding of Security Ability Construction of Civil Aviation Administration of China (AS-SA2015/21), the Innovation Funding of Nanjing University of Aeronautics and Astronautics (NJ20160028, NT2018028, NS2018057).

G. Gan et al. (Eds.): ADMA 2018, LNAI 11323, pp. 499–513, 2018.
https://doi.org/10.1007/978-3-030-05090-0_43

view, what they hope more is to promote products with low sales volume and no popularization to users and realize diversified commodity strategies. The "long tail" theory explains this phenomenon, if the channels for storage and circulation of products are large enough, the market share of products with weak demand or poor sales can rival or even be larger than those of the few hot-selling products. That is, many small markets converge to produce the market energy that matches the mainstream. That is, the sales of the enterprise are not the head of the "best-selling" on the traditional demand curve, but the long tail that represents the unpopular good and is often forgotten by people. Therefore, the commodities located at the tail have greater excavation value, since the commodities located at the front of the tail have greater potential and higher popularity, they have further promoted value [5]. Take movies as an example, the more popular movies are accompanied by features such as action, comedy, and science fiction, but documentaries do not have these attributes, so most people think that documentaries are boring. Further mining of popular products in niche can discover the original value of products. For this purpose, we focus on exploring popular products in niche is recommended to the user to improve the diversification of sales of products located at the tail and combines with the power law to improve the recommendation algorithm.

Power law distribution is widespread in many fields and spreads across various disciplines such as physics, finance and computer science. It is Zipf and Pareto which make important contributions to the power law distribution research. Zipf found word according to frequency in descending order, the frequency of each word is inversely related to the constant power of its noun. Pareto studied the statistical distribution of personal income and found that the income of a few people is much higher than that of most people, so the famous 80/20 rule is proposed [6].

The existing recommendation system uses Top-K query to recommend products to users [7–9]. In the field of spatial databases, there have been many research workers who have conducted research on KNN, RNN, RKNN and other related technologies. As a novel query technology, RFN has relatively few people to study it, and RFN queries can effectively solve the problem of the weakest influence set [10–12]. Due to the complexity and diversity of product information, these recommendation systems need to do a lot of information processing to filter out products that meet user preferences. As we know, no research has put forward the application of combining RFN with recommendation system so far. Therefore, we initiate proposes combining RFN with recommendation instead of the traditional Top-k query recommendation.

Based on the above considerations, the main work of this paper is as follows:

- Firstly, the traditional preference prediction model is summarized, and the expert preference prediction mechanism and expert user extraction algorithm based on the new standard are proposed.
- The proposed algorithm is mainly aimed at effectively alleviating the "big-tailed" phenomenon. To find a threshold point in the power law distribution, the head is ignored, and the commodities in the tail are mainly studied. The scope of the original huge commodity information is reduced. That is, the scope of the original data set needed for the query work is reduced, which indirectly reduces the time and space consumption of the search query algorithm.

- RFN's spatial database query idea and recommendation system are combined to recommend K niche products to users, and a recommendation algorithm based on RFN is proposed.

The rest of this paper is organized as follows. Section 2 presents an overview of the related work. We formulate a series of related issues in Sect. 3. The extraction of expert users and the continuity of commodity label attributes are described in detail in Sect. 4 and how to judge the power law, solve the relevant parameters and use the power law to optimize the query space, to build a recommendation system based on RFN in Sect. 4. An extensive experimental study is provided in Sect. 5. Section 6 concludes the paper.

2 Relation Work

2.1 Preference Prediction

Collaborative filtering proposed by Goldbeg et al. [13] is one of the most successful technologies in the recommendation system. Most recommendation systems usually adopt this collaborative filtering method for user preference prediction. Sarwar et al. [14] put forward collaborative filtering algorithm based on goods. Peng et al. [15] improved the traditional collaborative filtering algorithm. Collaborative filtering is mainly based on similarity computation, and the objects with high similarity are used to represent the preference attributes of target objects. Collaborative filtering concludes user-based collaborative filtering [16] and item-based collaborative filtering [17]. The basic idea of the user-based collaborative filtering algorithm is that when a user A needs a personalized recommendation, he can first find a user group G with similar interests. The item-based collaborative filtering algorithm is to find items similar to the target user's favorite items, and then recommend the similar items to the target user. Vozalis et al. [18] compare these collaborative filtering strategies through experiments. Both UBCF and IBCF involve similarity computation. The most common method for calculating similarity is based on the calculation of the correlation coefficient between vectors. Pearson correlation coefficient is a very suitable method for similarity evaluation. Specifically, the Pearson correlation coefficient uses the linear correlation between two vectors to define the similarity between vectors. In the user-based method, the definition of similarity between users is as follows:

$$P_{u,i} = \frac{\sum_{v \in S} sim(u, v) r_{v,i}}{\sum_{v \in S} |sim(u, v)|} \tag{1}$$

The similarity computation method based on collaborative filtering of items is similar to (1).

2.2   Continuous Product Tag Attributes

The tag-based recommendation system has been widely used in various websites, especially social networking, music, and video etc. For example, the combination of labels and social relations is applied to micro-blog's recommendation [19], and the content characteristics of items are accurately described by tags. [20] is a recommendation research based on personalized tags, and proposes a recommendation algorithm that fuses tag popularity, time weights, and matrix decomposition. The continuous attribute tag representation has the following advantages:

(1)   It represents the degree to which an item is accompanied by a certain attribute and is no longer a single choice. When the user wants a product with certain attributes, the product with the attribute label 1 is selected, if users feel this attribute is not obvious after personal experience, the recommendation effect may not be ideal. With the use of a continuous label, it is intuitive to understand the extent that the product carries certain attributes so that the user can make better choices.

(2)   When the attribute labels are continuous, the products can be abstracted as points in the space. For example, if a product has n attributes, the product can be abstracted as a point in an n-dimensional space. Each attribute value is a measure on each axis. In this way, the search query algorithm in the space can be applied to the recommendation system.

2.3   Power Law Solution with Spatial Attributes

Many researchers discovered the phenomena that obey the power law distribution in natural science. Faloutsos et al. [21] studied the topology of the Internet, and found that the degrees of the routing nodes between the networks obey power law distribution. Clementi et al. [6] found that the income of Italy residents showed a power law distribution. Newman et al. [22] compared and summarized Zipf, Pareto and power-law. Clauset et al. [23] used the empirical model to estimate the relevant parameters of the power law. In this paper, we mainly use the power law to filter the original data set. The filtered data sets are spatially processed according to their attributes, and then the spatial query strategy is used to screen and obtain the recommended results that meet user preferences. In recent years, spatial distance query strategy has been studied extensively, and these queries can be summarized as follows: nearest neighbor and k nearest neighbors, reverse nearest neighbor and reverse k nearest neighbors, reverse furthest neighbor. However, most researchers pay more attention on the nearest neighbor and k nearest neighbors, reverse nearest neighbor and reverse k nearest neighbors. In contrast, few researchers are conducting research on RFN. The well-known Yao et al. [10] first proposed methods for determining RFN queries such as the PFC algorithm and the CHFC algorithm. Liu et al. [11] proposed a PIV algorithm based on the security domain, which avoids redundant queries in some cases and improves the efficiency of algorithm execution. This paper combines RFN query and recommendation system to build a recommendation system different from Top-K.

3 Problem Formulation

Traditional user-based collaborative filtering algorithm uses the user's rating of the project to calculate user similarity, if the number of user sets is huge, it will be very complicated for new users to compute similarity with each old user. If a group of representative users can be selected when new users join, first calculate the similarity with these representative users and then predict the preferences, which can reduce the computation time. This article uses expert users to represent these special users.

Definition 1 (Expert user). In the recommendation system [24], when a user satisfies the following three conditions, such a user is called an expert user:

(1) Users in their areas of expertise are more active than most users and will give further evaluation in this field.
(2) Users are more professional than most users, and their evaluation is generally more accurate.
(3) Users in the social field have a higher reputation value than most users, that is users who have more trust in it.

Definition 2 (Attribute expert user). For all the items that have been evaluated for the user U, which contains the number of attribute tags T is not less than S, then the user U is an attribute expert user of the T attribute. S is a variable parameter that is used to represent the minimum reliability of the attribute expert user and it can be set according to different application scenarios.

The similarity between the user u and the expert e is calculated as follows:

$$sim(u, e) = \frac{\sum\limits_{s \in S_{ue}} (R_{u,s} - \overline{R_u})(R_{e,s} - \overline{R_u})}{\sqrt{\sum\limits_{s \in S_{ue}} (R_{u,s} - \overline{R_u})^2 * \sum\limits_{s \in S_{ue}} (R_{e,s} - \overline{R_u})^2}} * \frac{2|I_{ue}|}{|I_u| + |I_e|} \tag{2}$$

Where $|I_u|$ and $|I_e|$ respectively represent the collection of comment items for user u and expert e. User u's prediction score for item i:

$$R_{u,i} = \overline{R_u} + \frac{\sum\limits_{e \in S(u)} [sim(u, e) * (R_{e,i} - \overline{R_e})]}{\sum\limits_{u \in S(u)} |sim(u, e)|} \tag{3}$$

Where $\overline{R_u}$ and $\overline{R_e}$ respectively represent the average score of a user u and expert e, and $R_{e,i}$ indicates the score of the expert e on the item i.

Before calculating the similarity between users and experts, we need to know which users are expert users. Several criteria are as followed:

(1) Activity: One of the differences between ordinary users and expert users is that expert users are very knowledge-based in some respects, therefore expert users will give more and more accurate evaluation. The more users evaluate products, the more active users will be. The activity of the user can be calculated by the following formula:

$$ACT(u) = \frac{N(u)}{N(U)} \tag{4}$$

Where $N(u)$ represents the number of evaluations of user u and $N(U)$ represents the number of evaluations of all users.

(2) Another difference between ordinary users and expert users is that the accuracy of the evaluation of the product by the expert user is usually higher than that of the ordinary user, that is the expert rating is almost the same as the average rating of the item. To measure the accuracy of user ratings, we first need to know the average score of item I. If item I is scored by N users, the average score of items I can be obtained from the following:

$$\overline{R_i} = \frac{1}{N} * \sum_{u=1}^{N} R_{u,i} \tag{5}$$

The user's rating deviation for each item is $|R_{u,i} - \overline{R_i}|$.

To limit the score deviation to 1 for better analysis, we do the following:

$$D(R_{u,i}) = 1 - \frac{|R_{u,i} - \overline{R_i}|}{\sum_{i=1}^{N} |R_{u,i} - \overline{R_i}|} \tag{6}$$

Finally, the average of the user's rating deviation is used to indicate the accuracy of the user rating:

$$C(u) = \frac{1}{M} \sum_{i=1}^{M} D(R_{u,i}) \tag{7}$$

Where M is the number of evaluations given by the user.

We use the following formula to evaluate expert users with the above two criteria:

$$EVAL(u) = \alpha * ACT(u) + \beta * C(u) \tag{8}$$

Where $\alpha + \beta = 1$, α and β respectively represent the weight of each part.

Lemma 1: For a user set U, the expert set is $E = \{u | \forall u \in U, EVAL(u) > t\}$, where t represents the threshold, when EVAL (U) > t, the user u can be identified as an expert user, so we can control the number of expert users by controlling t.

Suppose that D represents a set of data in n-dimensional space. The reverse furthest neighbor technology is concerned with Euclidean distance, namely $\|p - q\|$ (p, q is any two points in space).

Definition 3. The furthest neighbor of any point q about the set point $P : FN(q, P) = p$ s.t. $p \in P$, for $\forall p' \in P$ and $p' \neq p, \|p - q\| \geq \|p' - q\|$.

Definition 4. $q \in Q$, the RFN is a collection point where the point in the data set P is relative to the query set Q with q as the most furthest neighbor, i.e. $RFN(q, Q, P) = \{p | p \in P, FN(p, Q) = q\}$ (Table 1).

Table 1. Symbol definition

Symbol	Description
U, I	Representing the collection of users and the collection of goods
Sim(u, e)	Similarity between user u and e
ACT(u)	Activity of user u
C(u)	Rating accuracy of user u
Ru, i	User u's rating for commodity i
$\overline{R}_i$	Average score of commodities i
DAVi	Discrete vectors for commodity i
CAVi	Continuity vector of commodity i
P(x)	Power law probability distribution function
Total(u, i)	User u commented on the number of commodities with i attribute

4   Recommendation Method Based on RFN

The previous section defined expert users. In this section, expert users are extracted, and the product attributes are continuously spatially mapped.

4.1   Expert User Extraction Algorithm

For new users, the system does not know their preferences. It can only find new users' nearest neighbors through collaborative filtering and predicts new users' preferences through the preferences of their nearest neighbors.

We calculate the similarity between new users and expert users to find K nearest neighbors. If there is no expert user or the number of expert users is insufficient, the missing neighbors are looked for among ordinary users, and then the predicted score of the new user is acquired by the predictive scoring model.

Algorithm 1: Expert user extraction algorithm *EEU*

Input: User -Item Rating Matrix R, User U

Output: Expert User Collection EU

1. $EU \leftarrow \emptyset$ //Initialize the set EU to null
2. **for** $i \leftarrow 0$ **to** *U.number* **do**
3. $ACT \leftarrow act(u)$ //Computing the activity of user u
4. $accuracy\_rate \leftarrow E(u)$ //Calculating the accuracy of user u
5.   **if** $Eval(u) > t$ //If the user's final score is greater than the threshold
6. $EU.add(u)$ //Add the user u to the expert set EU
7.   **end if**
8. **end for**
9. **return** EU //return expert user set

4.2   Label Attribute Continuity Algorithm

Attribute Expert User's Acquisition

In the recommendation system, how to extract attribute expert users when we have a user-item rating matrix. First of all, we need to get the user's type matrix UTM to define the user in detail through UTM. Table 2 captures part of UserA's score in the scoring matrix. Thus, UserA has watched the five films in the table and gave relevant evaluations. Combined with discretization of movie label properties, we can get the user type matrix UTM, as shown in Table 3.

Table 2. UserA's movie rating matrix

	Star	Heat	Titanic	Spawn	Lion
UserA	5	2	4	3	3

Table 3. UserA's type matrix UTM

	Action	Adventure	Children's	Comedy
Star	1	1	0	0
Heat	1	0	0	0
Titanic	1	0	0	0
Spawn	1	1	0	0
Lion	0	0	1	0
Total	4	2	1	0

Lemma 2: For a user set U, attribute expert set $AE(i) = \{u|\forall u \in U, Total(u, i) > t\}$, where t represents the threshold, when the user u commented that the number of items with attribute i exceeds t, the user can be identified as an expert user of attribute i. So, we can control the number of expert users by controlling t.

Discrete Attribute Label Continuity Algorithm

When a certain attribute label of an item is evaluated, experts in this field often have more reference value to the item. We have got the expert user who has the tag attribute by processing the user rating matrix in the previous section. Next, the attribute label of the item will be continuous by the DASC algorithm.

Firstly, we obtain the item i's discrete attribute vector DAVi which is required to be continuous in attribute space. $DAV_i = (da_{i,1}, da_{i,2}, ..., da_{i,n})$, $da_{i,m} = 0|1$, $m <= n$, in addition, we need to get a user set user_set$_i$ that has rated i, then for each vector attribute a with a value of 1, we get the expert set_set$_{i,a}$ that contains all expert user about this attribute among user_set$_i$. For each user in expert_set$_{i,a}$, the user's evaluation about i is obtained through the following formula:

$$a_{u,i,a} = \frac{max\_rating * R_{u,i} + min\_rating * R_{u,i}}{4 * \overline{R_{u,a}}} \tag{9}$$

Where ar$_{u,i,a}$ is the user u's expert rating for the a attribute of item i, max_rating is the maximum rating value of the recommendation system, $R_{u,i}$ is the true rating of user u for item i, $\overline{R_{u,a}}$ is the average rating of user for all items with the tag attribute a.

Because the scoring scales of different users are different, some users are used to give high scores, while others are used to give low scores. In order to prevent this situation from affecting the decision results, (9) is applied to the unified processing of each user's rating.

$$ca_{i,a} = \sum_{u \in expert\_set_{i,a}} \frac{R_{u,i}}{max\_rating\_a * \overline{R_{u,a}}|expert\_set_{i,a}|} \tag{10}$$

Where ca$_{i,a}$ is the continuous value of the item i with respect to the attribute tag a, max_rating_a is the maximum score for attribute a. Based on the above formula, the discrete attribute vector DAV_i of the item i can be converted into a continuous attribute vector CAV_i. $CAV_i = (ca_{i,1}, ca_{i,2}, ..., ca_{i,n})$, $0 <= ca_{i,m} <= 1$, $m <= n$. The calculation algorithm is shown in [24].

4.3   Recommendation Combined with Power Law

Because the original product information is complex and huge, increasing the execution time of the query algorithm, we want to eliminate the popular products in the original product information, leaving niche products, and then selecting the popular products in the niche to recommend to the users.

The probability density function of the power law distribution can be written as: $y = cx^{-a}$. The characteristics of this distribution are: the scale is very small for most of time, only a small amount of time is very large. The tail part of the cumulative function of the power law distribution decays slowly with its probability density value. It's the basis for judging whether two random variables are consistent with the power law distribution. The formula proves that the probability density function of the power law distribution:

$$p(x) = \frac{\alpha - 1}{x_{min}} \left(\frac{x}{x_{min}} \right)^{-\alpha} \tag{11}$$

$$p(x|\alpha) = \prod_{i=1}^{n} \frac{\alpha - 1}{x_{min}} \left(\frac{x_i}{x_{min}} \right)^{-\alpha} \tag{12}$$

We use the maximum likelihood method to solve the power law index. When the maximum likelihood (MIE) function is used, the maximum point of original function and logarithmic function is the same.

$$L = \ln p(x|\alpha) = \ln \prod_{i=1}^{n} \frac{\alpha - 1}{x_{min}} \left(\frac{x_i}{x_{min}} \right)^{-\alpha}$$
$$= \sum_{i=1}^{n} \left[\ln(\alpha - 1) - \ln x_{min} - \alpha \ln \frac{x_i}{x_{min}} \right] \tag{13}$$
$$= n \ln(\alpha - 1) - n \ln x_{min} - \alpha \sum_{i=1}^{n} \ln \frac{x_i}{x_{min}}$$

Find the partial derivative of α to get $\partial L/\partial \alpha = 0$.

$$\hat{\alpha} = 1 + n \left[\sum_{i=1}^{n} \ln \frac{x_i}{x_{min}} \right]^{-1} \tag{14}$$

Finally, a model of approximate solution of α is obtained. Where X_{min} is the minimum value of the power law distribution variable. There have been related studies that found that the time interval between consecutive reviews of customers obeys the power law distribution. To reduce the original data space, we investigate a threshold as the boundary, which narrows the scope of the commodity. Since the longer the comment interval, the lower the popularity of the goods is. To recommend popular products in the niche, we should first select products with longer comments intervals, and then recommend products with shorter intervals for these products.

Because curvature represents the degree of variation of the curve, the larger the curvature, the greater degree of bending of the curve. Therefore, the point with the largest curvature can be used to represent the threshold.

$$\rho = \frac{1}{k} = \frac{[1+f'^2(t)]^{\frac{3}{2}}}{|f''(t)|} \tag{15}$$

The threshold point can be obtained by this curvature formula. Because the curve above the threshold is steeper and the time interval is shorter, it can fully show that this commodity is quite popular. Products with an interval below the threshold can indicate the sales are good. Therefore, we only study commodities whose time interval is greater than this threshold.

4.4 Spatialization Based on User Preferences

Since the product attributes have been labelled continuously and the processed product information vector has been mapped into space in the previous sections, to use the spatial search algorithm to recommend products to users, users should also be spatially mapped. We use the following formula to continuously label user preferences:

$$P_{u,k} = \frac{\sum_{i=1}^{M} R_{u,i} * I_{i,k}}{M} \tag{16}$$

Where $P_{u,k}$ represents the user's preference for the k-th attribute of the product, $R_{u,i}$ represents the user's rating of the product i, $I_{i,k}$ represents the k-th attribute value of the product i and M represents the number of products the user has rated.

4.5 Recommendation Algorithm for Supporting RFN

Goods and users are mapped to points in n-dimensional space in Sects. 4.2 and 4.4. We apply RFN query to optimize the results recommended to users in this section.

The following formula indicates the degree of matching between user preferences and commodity attributes:

$$MC = \frac{1}{\sqrt{\sum_{i=1}^{n} (u_i - p_i)^2}} \tag{17}$$

The larger the MC, the closer the product is to the user's preference. The algorithm is described as follows:

Algorithm 2 : Based on RFN user preference recommendation algorithm BRURA

Input : Continuous spatial user set U, commodity collection I

Output : Recommended List Collection L

1.   *L←Ø*//Initialize the list set L to null
2.   **for** *i←0* **to** *U.number* **do**
3.   *UPC(u)*//Preference continuity for each user
4.   **end for**
5.   **for** *i←0* **to** *I.number* **do**
6.   *Ii.knn←Compute_Knn(Ii)*//Calculate the knn of the commodity Ii
7.   **for** *k←0* **to** *U.number* **do**
8.   **if** *Uk ∈Ii.knn*//If the user belongs to the KNN of the commodity Ii
9.   *Lk.add(Ii)*//Add item Ii to the k*th* list item of the list set
10.   **end if**
11.   **end for**
12.   **end for**
13.   **return** *L*//return the recommendation list set

---

5 Experiments

5.1 Experimental Data Set

In this experiment, the MovieLens 100k data set was used. MovieLens is a web-based research recommendation system, which is used to receive users' ratings of films and provide corresponding movie recommendation lists.

5.2 Evaluation Criteria and Method

The evaluation method used in the experiment is the leave-one-out method commonly used in machine learning. The average absolute error MAE evaluates the accuracy of the prediction by calculating the error between the product attributes recommended for the user and the actual user preferences. Assume that pi indicates the user's preference for the i-th attribute of the product and q_i indicates the degree of the i-th attribute of the product, the average absolute error MAE of the user is obtained by the following formula:

$$q_i = \frac{\sum_{j=1}^{k} q_j}{k} \tag{18}$$

$$MAE = \sum_{i=1}^{n} |p_i - q_i| \qquad (19)$$

Where q_j represents the *j*-th product recommended to the user. (18) gets the average attribute of the product recommended to the user, and then get the average absolute error MAE by formula (19).

According to the formula, users with more evaluations will have larger weights, while new users with less evaluations will have very few weights.

To balance all users, we use the mean absolute user error (MAUE) to better reflect the accuracy of system recommendation.

$$MAUE = \frac{\sum_{u=1}^{m} MAE_u}{m} \qquad (20)$$

5.3   Experimental Results and Analysis

In this experiment, we first measure the deviation of the different numbers of similar neighbors and reverse k furthest neighbors, and then measure the deviation based on the power law filtering. The weight of activity and accuracy is respectively $\alpha = 0.5$ and $\beta = 0.5$ in the expert evaluation formula. According to experiments, when the number of experts is greater than 10, the accuracy of the experiment will be significantly reduced. Therefore, the number of expert users we select in this experiment is 10.

Figure 1 analyzes the accuracy comparison based on different numbers of expert users m. It can be found that when the number of similar neighbors reaches a certain value, increasing the number m will lead to a decrease in system accuracy, m represents the number of expert users. Figure 2 analyzes the accuracy comparison based on different k farthest users. It can be found that properly increasing the number of the k furthest users of the product can increase the accuracy of the system recommendation. Figure 3, 4 is the error presentation that recommends the movie with great potential to the user after filtering by power law.

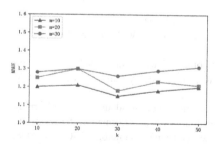
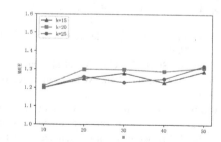

Fig. 1. Variation trend of accuracy with k value

Fig. 2. Variation trend of accuracy with m value

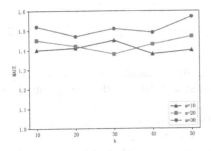

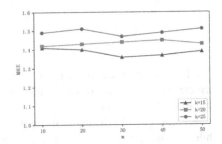

Fig. 3. Trend of accuracy change after power law filtering

Fig. 4. Trend of accuracy change after power law filtering

6   Conclusion

This paper proposes an expert-user collaborative filtering algorithm based on RFN. To fully tap the value potential of popular commodities in the niche, we use the power law to filter product information, and spatially map product attributes and user preferences at the same time, finally combine spatial query ideas to make recommendation. Experiments prove that the recommendation system of this paper has a high accuracy, which can make non-mainstream products attract more target users, and implement an algorithm for recommending popular commodities in the niche, which is of great significance for the application of the recommendation system.

References

1. Bobadilla, J., Ortega, F., Hernando, A.: Recommender systems survey. Knowl.-Based Syst. **46**(1), 109–132 (2013)
2. Xu, H.-L., Wu, X., Li, X.-D., Yan, B.-P.: Comparison study of internet recommendation system. J. Softw. **20**(2), 350–363 (2009)
3. Hussein, T.: Context-aware recommender systems. In: ACM Conference on Recommender Systems, pp. 349–350. ACM (2011)
4. Meng, X.W., Xun, H.U., Wang, L.C., et al.: Mobile recommender systems and their applications. J. Softw. **24**(1), 91–108 (2013)
5. Anderson, C.: The long tail. Wired Mag. **12**(10), 170–177 (2004)
6. Newman, M.E.J.: Power laws, Pareto distributions and Zipf's law. Contemp. Phys. **46**(5), 323–351 (2005)
7. Yang, X., Steck, H., Guo, Y., et al.: On top-k recommendation using social networks. In: Proceedings of the Sixth ACM Conference on Recommender Systems, pp. 67–74. ACM (2012)
8. Khribi, M.K., Jemni, M., Nasraoui, O.: Automatic recommendations for e-learning personalization based on web usage mining techniques and information retrieval. In: Eighth IEEE International Conference on Advanced Learning Technologies, ICALT 2008, pp. 241–245. IEEE (2008)

9. Abel, F., Gao, Q., Houben, G.-J., Tao, K.: Analyzing user modeling on Twitter for personalized news recommendations. In: Konstan, Joseph A., Conejo, R., Marzo, José L., Oliver, N. (eds.) UMAP 2011. LNCS, vol. 6787, pp. 1–12. Springer, Heidelberg (2011). https://doi.org/10.1007/978-3-642-22362-4_1

10. Yao, B., Li, F., Kumar, P.: Reverse furthest neighbors in spatial databases. In: IEEE International Conference on Data Engineering, pp. 664–675. IEEE (2009)

11. Liu, J., Chen, H., Furuse, K., Kitagawa, H.: An efficient algorithm for arbitrary reverse furthest neighbor queries. In: Sheng, Quan Z., Wang, G., Jensen, Christian S., Xu, G. (eds.) APWeb 2012. LNCS, vol. 7235, pp. 60–72. Springer, Heidelberg (2012). https://doi.org/10.1007/978-3-642-29253-8_6

12. Wang, S., Cheema, M.A., Lin, X., et al.: Efficiently computing reverse k furthest neighbors. In: ICDE 2016, pp. 1110–1121 (2016)

13. Goldberg, D., Nichols, D., Oki, B.M., et al.: Using collaborative filtering to weave an information tapestry. Commun. ACM 35(12), 61–70 (1992)

14. Sarwar, B., Karypis, G., Konstan, J., et al.: Item-based collaborative filtering recommendation algorithms. In: Proceedings of the 10th International Conference on World Wide Web, pp. 285–295. ACM (2001)

15. Zhao, Z.D., Shang, M.S.: User-based collaborative-filtering recommendation algorithms on hadoop. In: International Conference on Knowledge Discovery and Data Mining, pp. 478–481. IEEE (2010)

16. Pirasteh, P., Jung, Jason J., Hwang, D.: Item-based collaborative filtering with attribute correlation: a case study on movie recommendation. In: Nguyen, N.T., Attachoo, B., Trawiński, B., Somboonviwat, K. (eds.) ACIIDS 2014. LNCS (LNAI), vol. 8398, pp. 245–252. Springer, Cham (2014). https://doi.org/10.1007/978-3-319-05458-2_26

17. Vozalis, E.G., Konstantinos, G.M.: Recommender systems: an experimental comparison of two filtering algorithms. In: Proceedings of the 9th Panhellenic Conference in Informatics, PCI 2003 (2003)

18. Ma, H., Jia, M., Zhang, D., et al.: Combining tag correlation and user social relation for microblog recommendation. Inf. Sci. Int. J. 385(C), 325–337 (2017)

19. Guo, D., Zhao, H.: Matrix factorization recommendation algorithm fusing tag popularity and time weight. Minicomput. Syst. 37(2), 293–297 (2016)

20. Faloutsos, M., Faloutsos, P., Faloutsos, C.: On power-law relationships of the internet topology. In: ACM SIGCOMM Computer Communication Review, vol. 29, no. 4, pp. 251–262. ACM (1999)

21. Clementi, F., Gallegati, M.: Power law tails in the Italian personal income distribution. Phys. A: Stat. Mech. Appl. 350(2–4), 427–438 (2005)

22. Clauset, A., Shalizi, C.R., Newman, M.E.J.: Power-law distributions in empirical data. SIAM Rev. 51(4), 661–703 (2009)

23. Zheng, W., Li, B., Wang, Y., Qin, X.: Group recommendation algorithm model combined with preference interaction. Minicomput. Syst. 39(2), 372–378 (2018)

24. Li, B., et al.: Dynamic reverse furthest neighbor querying algorithm of moving objects. In: Li, J., Li, X., Wang, S., Li, J., Sheng, Quan Z. (eds.) ADMA 2016. LNCS (LNAI), vol. 10086, pp. 266–279. Springer, Cham (2016). https://doi.org/10.1007/978-3-319-49586-6_18

25. Zheng, W., et al.: Group recommender model based on preference interaction. In: Cong, G., Peng, W.-C., Zhang, W.E., Li, C., Sun, A. (eds.) ADMA 2017. LNCS (LNAI), vol. 10604, pp. 132–147. Springer, Cham (2017). https://doi.org/10.1007/978-3-319-69179-4_10

26. Yue, L., Chen, W., Li, X., Zuo, W., Yin, M.: A survey of sentiment analysis in social media. Knowl. Inf. Syst. 1–47 (2018)

Research of Personalized Recommendation System Based on Multi-view Deep Neural Networks

Yunfei Zi[✉], Yeli Li[✉], and Huayan Sun[✉]

School of Information Engineering, Beijing Institute of Graphic Communication,
Beijing 102600, China
527260533@qq.com, 1605872754@qq.com, 603757282@qq.com

Abstract. In recent years, deep learning has made leaps in the fields of artificial intelligence, machine learning and so on, especially in the fields of speech recognition, image recognition and self-learning. The deep neural network is similar to the biological neural network, so it has the ability of high efficiency and accurate extraction of the deep hidden features of information, and can learn multiple layers of abstract features, and can learn more about Cross-domain, multi-source and heterogeneous content information. This paper presents an extraction feature based on multi-user-project combined depth neural network, self-learning and other advantages to achieve the model of personalized information, the model through the input multi-source heterogeneous data characteristics of in-depth neural network learning, extraction fusion collaborative filtering widely personalized generation candidate sets, and then through two of models to learn to produce a sort set, Then realize accurate, real-time, personalized recommendation. The experimental results show that the model can study and extract the user's implicit feature well, and can solve the problems of sparse and new items of traditional recommendation system to some extent, and realize more accurate, real-time and personalized recommendation.

Keywords: Deep neural networks · Personalized recommendation
Collaborative filtering · Candidate set · Sort set · Multi-view

1 Introduction

In recent years, large data, artificial intelligence, cloud computing, IoT and so on made a breakthrough, along with the rapid development of these information technology revolution, the application of these technologies based on the Internet space data shows an explosion of growth. The 2013 Data group IDC report shows that global data will be over 40 ZB in 2020 (about 4 trillion GB), and it will be 22 times more than 2011, meaning global data is growing at a rate of 58% a year. The big data economy of 2017 has also officially been upgraded to the national strategic level, the massive data brings the change development to the society, at the same time also appeared the "Information overload" problem, the latent value hidden in the big data is immeasurable, but how to excavate the valuable or the user demand information in the massive data, this is one of the challenges facing the data economy, and the best way to solve this problem is to

G. Gan et al. (Eds.): ADMA 2018, LNAI 11323, pp. 514–529, 2018.
https://doi.org/10.1007/978-3-030-05090-0_44

recommend the system. Recommendation system has been for science and technology, business, medical, urban construction, government management and other fields to solve a series of difficulties, making the recommendation system by academic, industrial and highly concerned about and widely used [1–3]. Recommendation system is based on the analysis of user behavior, mining, in the vast number of data found in the user dominance, implicit demand and interest characteristics, the user may be interested in potential information or merchandise, accurate, personalized, highly efficient recommendation to the user. Compared to the search engine, the recommendation system does not require users to have a clear description of the product or demand information, it is through the user's historical memory for the "portrait" modeling, proactive for users accurate, personalized and efficient to provide users with the interests and needs of information. Personalized recommendation system has been widely used and recognized, many internet companies make use of large data to do and achieve a super related recommendation, in order to increase revenue for the company, such as electricity quotient (Amazon, Alibaba, EBay, NetEase), search engine (Google, Baidu, Yahoo, etc.), news information (Today's headlines, Grouplens, etc.), video recommendations (YouTube, Tencent video, etc.).

The traditional recommendation algorithms include collaborative filtering recommendation, combination recommendation, content-based recommendation, and so on. Among them, collaborative filtering recommendation is the most popular and widely used, it mainly through a series of user behavior of the characteristics of the model to achieve special recommendations for users, but with the Internet data is a geometric growth, the user for the recommendation of personalization, accuracy, demand for the pre-order requirements are more and more high, the complexity of data content and so on, Also exposes its own limitations, the disadvantages of the collaborative filtering recommendation algorithm are: (1) Data sparsity problem (user evaluation of items relative to the site of the total project accounted for very low, resulting in extremely sparse user Rating matrix data; (2) New items (new users and projects are not graded, resulting in very low recommendation); (3) Scalability issues (algorithms to solve the problem of rapid scale expansion of the system). At the same time, the traditional collaborative filtering recommendation algorithm is not efficient for users and deep feature extraction of the project. Based on the recommendation of the content and highly dependent on the artificial design features, thus greatly reducing the recommended efficiency. Along with the perceived change of Internet data, more and more users and project data are composed of multi-source heterogeneous data, such as video, image, label, text and so on, through combination recommendation and popular recommendation can solve some data sparse, new items problem, but because of user and project characteristic information multi-source heterogeneous, polymorphism, Complex issues such as large-scale and multiple distributions, combined recommendations and popular recommendations are also facing severe challenges.

In recent years, deep learning in artificial intelligence, machine learning has made a leap-type breakthrough, especially in the field of speech recognition and image recognition [4–8]. The deep neural network, which is similar to the biological neural network, therefore, it has the ability of efficient and accurate extraction of deep hidden features of information, and can learn multiple layers of abstract feature representation, and can learn the information of Cross-domain, multi-source and heterogeneous content, and can deal with the sparse of recommendation system, new items, Scalability and so on, which brings a new opportunity for the recommendation system to solve the inherent problems.

This paper presents a recommendation model based on depth neural network combined with multi-user-project and collaborative filtering (multi-view-collaborative filtering integrating Deep neural Network, MV-CFIDNN) [9–15], based on the theory of Deep neural network, extracting the deep hidden features of users and project information and self-learning, optimizing the extraction model, and finally, combining multi-user-project and collaborative filtering (collaborative filtering) to provide a wide range of personalized recommendations. The experimental results show that the proposed algorithm can solve some problems such as sparsity, new items and extensibility of traditional collaborative filtering recommendation algorithm, and improve the precision and personalization of the proposed model obviously.

2   Collaborative Filtering Recommendation Model

Collaborative filtering (Collaborative filtering: CF) recommendation algorithm [16–18] is through the user-project evaluation, sharing, and other data analysis to identify the user or project similarity to achieve the principle of accurate recommendation. Collaborative filtering recommendation algorithm is divided into: (1) based on the user's collaborative filtering recommendation algorithm (user-based collaborative filtering), through the user search, purchase of products like evaluation, collection and other data to measure the same behavior of the relationship between users to achieve product recommendations. (2) The collaborative filtering recommendation algorithm based on the project (item-based collaborative filtering), through different users for different products evaluation and degree of preference to measure the tightness between products, so as to achieve product recommendations. As shown in Fig. 1, the Collaborative filtering model:

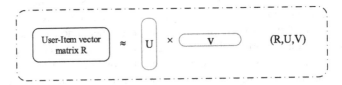

Fig. 1. Collaborative filtering recommendation framework.

2.1   Implementation of Collaborative Filtering Recommendation Algorithm

(a) Collection of user-project data.
(b) Calculating user-project similarity to identify the nearest neighbor or the highest degree of similarity to n users or projects.
(c) Recommend to the corresponding project or user the user or project with the highest degree of n similarity.

2.2 Evaluation and Calculation of Correlation Degree

In the collaborative filtration recommendation system, $D = (U, I, R)$ is used to represent the user-project data source, $U = (User_1, User_2, \ldots, User_n)$ for user set, $|U| = n$, $I = (Item_1, Item_2, \ldots, Item_m)$ for project set, $|I| = m$, The user's scoring matrix for the item is $R = m \times n$, where R_{ij} represents the user $User_i$ the project $Item_j$ score, and the Table 1 is the scoring matrix R.

The common methods of computer similarity in collaborative filtering recommendation algorithms are:

Table 1. User-project scoring matrix

USER	ITEM			
	$Item_1$	$Item_2$	...	$Item_m$
$User_1$	R_{11}	R_{12}	...	R_{1m}
$User_2$,	R_{21}	R_{22}	...	R_{2m}
...	...	...		...
$User_n$	R_{n1}	R_{n2}	...	R_{nm}

Pearson correlation coefficient (Pearson correlation coefficient): Through the user to the product evaluation data calculates two sets of data and a line's fitting degree, if the two user similarity degree high will form the diagonal. Set the user to x and y, $\bar{m}_x$ for the user x for all products of the scoring data, m_{xk} for user x to the k project score data, then the Pearson correlation coefficient formula is:

$$sim(x, y) = \frac{\sum_k (m_{xk} - \bar{m}_x)(m_{yk} - \bar{m}_y)}{\sqrt{\sum_k (m_{xk} - \bar{m}_x)^2 \sum (m_{yk} - \bar{m}_y)^2}} \tag{1}$$

After calculating the similarity, we find the nearest neighbor, Z_{um} is the nearest neighbor, the calculation of the project s prediction score.

$$m_{us} = \bar{m}_u + \frac{\overset{\circ}{\underset{\dot{a}_1 Z_{um}}{A}} sim(u, x)(m_{xs} - \bar{m}_x)}{\underset{\dot{a}_1 Z_{um}}{A} |sim(u, x)|} \tag{2}$$

The above $\bar{m}_x$ is the average score of all the scoring data for x in Z_{um}, and $\bar{m}_u$ is the mean value of the target user u based on all project scoring data.

$$\bar{m}_u = \frac{\overset{\circ}{\underset{\dot{w}_1 T_s}{A}} m_{uw}}{|T_s|} \tag{3}$$

T_s is the sum of the items, and m_{uw} indicates the value of u for product w.

3   Recommended Model of Deep Neural Networks

The recommendation system based on depth learning provides users and projects with all kinds of raw data information to input layer, in the hidden layer through the neural network learning model to learn and extract the user, the hidden characteristics of the project, finally through the implicit expression of learning to achieve user, project recommendation. The recommended structure of the depth neural network consists of three parts (Fig. 2): input layer, model layer and output layer. Input layer data information is mainly based on user's display or implicit feedback, the project information, the user produces the information, the user portrait and so on; the model layer is based on the neural network, the output layer utilizes self-learning to obtain the user, the project implicit representation, and then produces the best recommendation list by the Softmax, the inner product and other computational similarity [19–23].

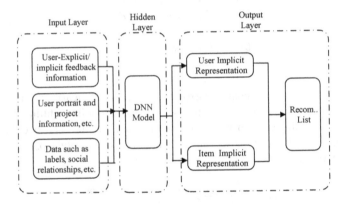

Fig. 2. Collaborative filtering recommendation framework.

Based on the depth neural network framework, two times self-learning combined with collaborative filtering CFIDNN framework (Fig. 3), the two core of CFIDNN framework are: Candidate Generation Network fusion collaborative filtering and ranking network combining collaborative filtering.

Among them, candidate set generation takes the feature of user in browsing history as input information, then retrieves a data set related to user based on multi-source heterogeneous database. This dataset is a candidate set, which is closely related to the user. This subset of candidate sets is widely personalized through collaborative filtering (CF). Then through the user, the project's many kinds of characteristic source (for example: Browsing information ID, search phrase and population geographical information) to learn computational similarity, then realize the minimum ranking set, finally based on collaborative filtering implementation recommendation.

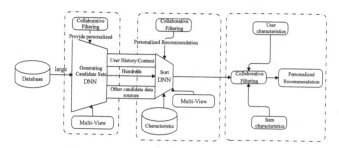

Fig. 3. Collaborative filtering recommendation framework.

3.1 Candidate Set Generation Module

Candidate Set Generation module is the problem of classifying a large multi-source heterogeneous database: A user, through its browsing record (data id), phrase searching (search tokens), user geographic information (location, browsing device), binary (login or not, gender) and continuity (age), etc. The information in the database is classified into several categories, then a number of related classification sets (the recommended probability of a project) are obtained, finally, several projects with higher probability are acquired, and personalized recommendations are achieved through collaborative filtering [24–26].

For candidate set generation, first, the history information such as user browsing and search items is mapped to vectors, then the average value is obtained by the fixed length representation. And, the input user geographic information eigenvalue optimizes personalized recommendation effect, the value of duality and the continuity eigenvalue get [0, 1] range by normalization. Secondly, all the input eigenvalues are spliced to the same vector and the spliced vectors are processed by the activation function. Finally, through the neural network training lost to Softmax classification, through the characteristics of training and the source project similarity calculation, get the highest similarity of n projects as candidate modules of the final candidate set, Fig. 4 is the candidate generation structure diagram.

Based on the generation of candidate set collaborative filtering to provide a wide range of personalization, after the group based on user-project relevance evaluation to achieve accurate, real-time, personalized recommendations.

Candidate set generation is based on the multiple-source heterogeneous database learning to choose a higher degree of user-related projects, for the prediction of user u to browse a certain information probability is:

$$P(w_t = i | U, C) = \frac{e^{v_i u}}{\sum_{j \in V} e^{v_j u}} \tag{4}$$

Among them, U is the user characteristic value, v represents multi-source heterogeneous database, vi represents the characteristic value of item I in the database, the U and vi vector have equal length, it is realized by the whole connection of the dot product in the hidden layer.

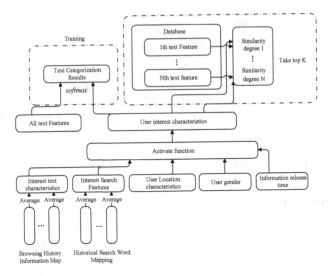

Fig. 4. Collaborative filtering recommendation framework.

Candidate Generation Fusion Collaborative filtering and sequencing generation combined with collaborative filtering model is a personalized, accurate and real-time recommendation by using the depth neural network learning feature output combined with collaborative filtering user and Project feature matrix. The frame model is shown below (Fig. 5).

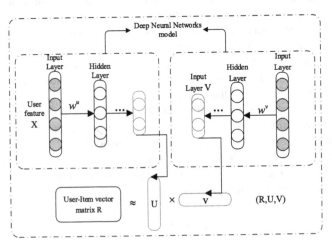

Fig. 5. Collaborative filtering recommendation framework.

3.2 Sort Build Module

The sorting generation module structure is similar to the candidate build structure, except that the sorting generation is a meticulous sorting of the candidate generation set upgrades. Similar to the traditional sorting extraction eigenvalue, the neural network sequencing is accomplished by stitching a large number of users, project-related eigenvalues (text IDs, browsing hours, etc.). The processing of eigenvalue is similar to that of candidate generation, which is based on quantization, the difference is that the ranking generation network is trained by the weighted logistic regression, and then the candidate set is graded and the higher grade K items are returned to the user or personalized recommendation by collaborative filtering [27, 28]. Figure 6 is a sort build chart.

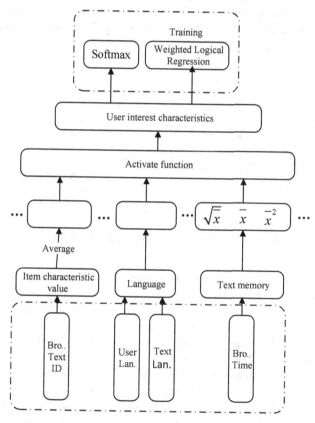

Fig. 6. Collaborative filtering recommendation framework.

The set part Softmax classification process: firstly, for the training process of the candidate generation set or the sequencing generation list, the calculation of the negative sample class is reduced to thousands of, and secondly, the recommendation stage is introduced, excluding Softmax normalization, The project score is transformed into the nearest neighbor of the point product space and the cooperative filter is calculated according to the correlation degree; Finally, select the highest user-related K as the candidate or sorted list, and then recommend the information to the user through collaborative filtering and personalized recommendation.

3.3 Multi User-Project Model

Multi-source heterogeneous features based on multi-user and multi-project are combined with two-depth neural network learning to realize personalized recommendation. The realization of the idea is; first, the quantization is mapped to the user by the original eigenvalue value, project two channels, and then use the depth of neural network model to map the user, project information vector to a hidden space, and finally, through the evaluation of similarity (such as cosine similarity method) to the user of the hidden space, the project related degree ranking, matching, So as to achieve accurate, personalized recommendations. The following figure (Fig. 7) is a multiuser-project DNN (Deep Neural network) model structure.

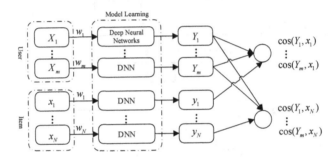

Fig. 7. Collaborative filtering recommendation framework.

The user's perspective, using its browsing history, searching (search tokens), location information, binary (login or not, gender) and continuity (age), viewing time and so on as the source eigenvalue input x_u, and then through the depth of neural network learning model to learn output implicit expression y_u. Project perspective, the use of project description, tags, types, etc. as the source eigenvalue input x_i, through the depth of neural network learning model learning output implicit y_i, where the model has a number of users, projects, respectively, m, N. The user Perspective DNN model is $f_u(x_u, w_u)$, and the DNN model is $f_i(x_i, w_i)$ for the first project perspective. If you have M sample $\{(x_{u,j}, x_{a,j})\}_{0 \leq j \leq M}$, $(x_{u,j}, x_{a,j})$, is the user you interaction with Project a, the user, the project to participate in the interaction of interactive records to learn:

$$\arg_{w_u, w_1, \ldots, w_v} \max \sum_{j=1}^{M} \frac{e^{\cos(f_u(x_{u,j}, w_u), f_a(x_{a,j}, w_a))}}{\sum_{1 \leq a \leq N} e^{\cos(f_u(x_{u,j}, w_u), f_a(x_{a,j}, w_a))}} \tag{5}$$

After the model training, learning, the user implicit representation y_u and the project implicit expression y_i, uses in the hidden space calculates the user and the project correlation degree, the rank, chooses the correlation degree high order k project and the source database collaborative filtering realizes the accurate, personalized recommendation.

3.4 Eigenvalues to Quantization

The quantification of eigenvalues is the embedding of the phrase, which maps the specially crafted text to the w dimensional space vector. First, the user, the project all correlation characteristic value merges separately, and the characteristic value quantification is the grading data then asks its average value, namely the multi-source heterogeneous raw data carries on the grading data processing and the normalization. The user specially crafted data is $(user_i, c_{i0} \oplus c_{i1} \oplus \ldots \oplus c_{in}, \frac{1}{n}\sum_{j=1}^{n} r_{ij})$: The project specially crafted data is: $(item_j, c_{0j} \oplus c_{1j} \oplus \ldots \oplus c_{mj}, \frac{1}{m}\sum_{i=1}^{m} r_{ij})$, where $\oplus$ represents a specially crafted value space connector, sorted by:

$$(user_i, d_i^{user}, \bar{r}_i^{user}) \tag{6}$$

$$(item_j, d_j^{user}, \bar{r}_j^{user}) \tag{7}$$

where d_i^{user} is the user $user_i$'s comment on item $item_1 \sim item_n$, $\bar{r}_i^{user}$ is $user_i$ the end of project $item_1 \sim item_n$ through the feature text to quantization technology, all users, the project score mean, commented data to quantify:

$$vec_i^{user} = Doc2VecC(d_i^{user}) \tag{8}$$

$$vec_j^{item} = Doc2VecC(d_j^{item}) \tag{9}$$

The $Doc2VecC$ function finally returns the w-dimensional vector, which obtains the average value by d_i^{user} the phrase in the data document, and guarantees that the last generated vector can automatically crawl the semantic feature information in the Neural network training learning. After the eigenvalues are quantified, the user i eigenvalue data is expressed as:

$$(user_i, vec_i^{user}, \bar{r}_i^{user}) \tag{10}$$

Project j Eigenvalue data is expressed as:

$$(item_j, vec_j^{item}, \bar{r}_j^{item}) \tag{11}$$

3.5 Fully Connected Layer

The full connection layer (hidden layer) enters data for the user, project source eigenvalue to quantify the value, set the hidden layer of m neurons, through the hidden layer ReLU activation function processing, obtain the vector u_i, is the user $user_i$

implicit eigenvalue, similarly, the project $item_j$ of the hidden eigenvalue vector v_j, the calculation process is as follows:

$$u_i = \text{ReLU}(w_{user_i} P_{user_i} + b_{user_i}) \tag{12}$$

$$u_i = \text{ReLU}(w_{user_i} P_{user_i} + b_{user_i}) \tag{13}$$

Which $u_i, v_j \in R^m$, w_{user_i}, w_{item_j}, respectively, the user, the project full connection layer weights, b_{user_i}, b_{item_j} respectively for the corresponding offset, for real numbers, p_t for the source eigenvalue to the quantization input after the maximum number of dimensions, $\text{ReLU}(x) = \max(0, x)$.

3.6 Matrix Decomposition

Matrix decomposition is realized by matrix multiplication, and the high-dimensional and sparse user-project ($user - item$) feature matrix is decomposed into two low dimensional eigenvalue matrices.

According to the collaborative filtering algorithm model, the user-project model is established, the User ($user$), Project ($item$) feature Matrix is learned from the historical data of the user, the project's various characteristic scores, and the prediction scoring feature matrix is obtained through the $user$, $item$ feature matrix multiplication, and finally, the real scoring matrix is fitted [29, 30]. Based on the deep neural network, this paper constructs the candidate network model combining collaborative filtering and sequencing network model, combines with collaborative filtering, uses the user browsing ID, Search tokens, user geographic information, binary value, etc. to learn, extract the user, the hidden features of the project, respectively, u_i, v_j. Then, the predictive scoring matrix $r'_{ij} \in R$ is obtained through the user and project implied feature inner product, and the formula is as follows:

$$r'_{ij} = u_i \bullet v_j^T \tag{14}$$

Finally, using the Adam Depth Learning optimization method, the prediction and the real score are fitted, for some of the projects with scoring, so that the prediction is most likely to be close to the real, this study recommended to achieve personalized recommendations for new items (not rated items to predict the real score unlimited close to the predicted value).

$$r_{ij} \approx r'_{ij} = u_i \bullet v_j^T \tag{15}$$

4   Experimental Simulation and Analysis

4.1   Causal Fuzzy Clustering

The experimental environment of the algorithm performance analysis is supported by the Windows SERVER2012 R2 operating system, the related configuration is Intel

(r) Xeon (r) Silver 4116 CPU processor, programming language python, 128G memory, dual GPU, The compiling environment is implemented in the Anaconda Jupyter notebook and matlab simulation.

4.2   Data Collection

In this paper, 2 real and real-time data sets are used to evaluate the recommended model of deep neural network fusion collaborative filtering, and the data sets are the *Amazon Movies and TV(AMT)* comment score and *Amazon Clothing(AC)* video commentary and grading respectively. Data includes user ID, item ID, and user comments, ratings. The score value is 1–5, the higher the value the more the user preference. At the same time, the experimental data are divided into the *TestSet* of training set *TrainSet* and the test set, and there is no intersection.

4.3   Evaluation Criteria

In this paper, the proposed model of depth neural network fusion collaborative filtering is used to extract the implicit features from various historical records of users and projects, and then to study and sort the characteristic values. In this paper, the RMS error (*RMSE*) is used as the index to evaluate the model, the deviation is calculated by learning the characteristic model and the real feature, and the square root is calculated with the predicted data quantity N, and the formula is as follows:

$$RMSE = \sqrt{\frac{\sum_{i,j}^{m,n} \left(r_{ij} - r'_{ij}\right)^2}{N}} \tag{16}$$

Among them, r_{ij} is the real rank value after the input characteristic value to quantify, r'_{ij} is the forecast value after studying, training, m is the user number, n is the item number, N is the evaluation component of the test concentration.

4.4   Experimental Comparison

The experiments were compared with 3 effective models, namely probabilistic Matrix factorization (PMF), LIBMF and DNNMF.

4.5   Comparative Analysis of Execution Time

Deep neural Network (DNN) recommendation algorithm and traditional collaborative filtering (CF) operating time comparison: the experimental processing data for *AMT*, *AC* real data, size of 1.88G, depth neural network input node is 1024, hidden layer 18, output node is 1024, spark cluster node is 3, Comparison method of depth neural network training and traditional collaborative filtering process data set time. The results of the experiment are shown in Fig. 8, where user indicates that the test data set is time-consuming, and the item indicates that the product test dataset is time-consuming. Obviously, DNN execution is more efficient.

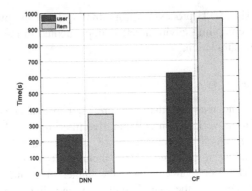

Fig. 8. Collaborative filtering recommendation framework.

4.6   Experimental Results and Analysis

In the experiment, the MV-CFIDNN model presented in this paper is evaluated under 2 real data sets, and the model is evaluated with *RMSE*, and the MV-CFIDNN is compared with PMF and LIBMF under the same experimental environment and the same data.

The parameter is set to: User, item eigenvalue weight is $\alpha = 1$, $\beta = 0.5$, MV-CFIDNN model learning rate is $lr = 0.00065$, user, project implicit characteristic is converted into $\lambda_{user} = \lambda_{item} = \lambda = 0.001$, depth neural network neuron number is 1026.

In order to compare the MV-CFIDNN model with the PMF and LIBMF model, 2 real datasets are randomly divided into 80% *TrainSet* and 20% *TestSet*, and there is no intersection between them, and the 20% data sets in *TestSet* are randomly used to validate the model parameters.

From the following figure (Fig. 9), we can know that the *RMSE* value of PMF and LIBMF is not quite the same after the test of 2 real data, but it is different from the *RMSE* value of MV-CFIDNN model, which shows that the deep neural network fusion multi-user-project and collaborative filtering model has good effect on eigenvalue

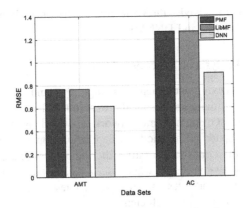

Fig. 9. Collaborative filtering recommendation framework.

extraction. The experimental results show that the *RMSE* value of the deep neural network fusion multiuser-project Collaborative filtering model (MV-CFIDNN) is reduced compared with the PMF and LIBMF models, which shows that the MV-CFIDNN model can solve the problems of sparsity and new items of the traditional algorithm model.

5 Concluding

In this paper, a MV-CFIDNN model is proposed through the fusion of the deep neural network, and the model is firstly studied by the depth neural network of the original database, and the feature value is extracted, then the candidate sets are generated, then the candidate sets are studied and extracted two times. A deep neural network learning method for generating candidate sets and sequencing set processes includes: input layer, hidden layer and output layer, in which the input layer is to quantify the input eigenvalue and the user, the project weight of the internal product transmission to the hidden layer, the hidden layer according to the value received to adjust the parameters, reset and other neural network learning, and then learn, The extracted feature values are passed to the output layer.

The predicted value is fitted with the real value by calculation. The research of recommendation system from the original artificial label set of traditional collaborative filtering recommended into the characteristics, the neural network of label Self-learning has made great progress, but there are still some questions to be studied, such as the use of multiple neural networks to combine more basic recommendation models so that the system can be intelligent and in line with human thought precision, personalized recommendation.

Acknowledgment. Fund project: National Natural Science Foundation of China (11603004), Beijing Natural Science Foundation (1173010), Beijing science and technology innovation service ability coordinated innovation project (PXM2016_014223_000025).

References

1. Peng, Y., Zhu, W., Zhao, Y., et al.: Cross-media analysis and reasoning: advances and directions. Front. Inf. Technol. Electron. Eng. **18**(1), 44–57 (2017)
2. Covington, P., Adams, J., Sargin, E.: Deep neural networks for YouTube recommendations. In: Proceedings of the 10th ACM Conference on Recommender Systems, pp. 191–198. ACM (2016)
3. Li, P., Wang, Z., Ren, Z., et al.: Neural rating regression with abstractive tips generation for recommendation (2017)
4. Song, Y., Elkahky, A.M., He, X.: Multi-rate deep learning for temporal recommendation. In: Proceedings of the 39th International ACM SIGIR Conference on Research and Development in Information Retrieval, pp. 909–912. ACM (2016)
5. Vasile, F., Smirnova, E., Conneau, A.: Meta-Prod2Vec: product embeddings using side-information for recommendation. In: ACM Conference on Recommender Systems, pp. 225–232. ACM (2016)

6. Hsieh, C.K., Yang, L., Cui, Y., et al.: Collaborative metric learning. In: Proceedings of the 26th International Conference on World Wide Web, pp. 193–201. International World Wide Web Conferences Steering Committee (2017)

7. Wang, X., He, X., Nie, L., et al.: Item silk road: recommending items from information domains to social users, pp. 185–194 (2017)

8. Roy, S., Guntuku, S.C.: Latent factor representations for cold-start video recommendation. In: Proceedings of the 10th ACM Conference on Recommender Systems, pp. 99–106. ACM (2016)

9. Zheng, L., Noroozi, V., Yu, P.S.: Joint deep modeling of users and items using reviews for recommendation. In: Proceedings of the Tenth ACM International Conference on Web Search and Data Mining, pp. 425–434. ACM (2017)

10. Wu, Y., DuBois, C., Zheng, A.X., et al.: Collaborative denoising auto-encoders for top-n recommender systems. In: Proceedings of the Ninth ACM International Conference on Web Search and Data Mining, pp. 153–162. ACM (2016)

11. Zhang, S., Yao, L., Sun, A.: Deep learning based recommender system: a survey and new perspectives (2017)

12. Liu, Q., Wu, S., Wang, L., et al.: Predicting the next location: a recurrent model with spatial and temporal contexts. In: Thirtieth AAAI Conference on Artificial Intelligence (2016)

13. Yang, C., Bai, L., Zhang, C., et al.: Bridging collaborative filtering and semi-supervised learning: a neural approach for POI recommendation. In: Proceedings of the 23rd ACM SIGKDD International Conference on Knowledge Discovery and Data Mining, pp. 1245–1254. ACM (2017)

14. Liu, Q., Wu, S., Wang, L.: Multi-behavioral sequential prediction with recurrent log-bilinear model. IEEE Trans. Knowl. Data Eng. 29(6), 1254–1267 (2017)

15. Zhuang, F., Luo, D., Yuan, N.J., et al.: Representation learning with pair-wise constraints for collaborative ranking. In: Tenth ACM International Conference on Web Search and Data Mining, pp. 567–575. ACM (2017)

16. Liu, Q., Wu, S., Wang, D., et al.: Context-aware sequential recommendation. In: 2016 IEEE 16th International Conference on Data Mining (ICDM), pp. 1053–1058. IEEE (2016)

17. Zhao, S., Zhao, T., King, I., et al.: GT-SEER: geo-temporal sequential embedding rank for point-of-interest recommendation. arXiv preprint arXiv:1606.05859 (2016)

18. Yang, C., Sun, M., Zhao, W.X., et al.: A neural network approach to joint modeling social networks and mobile trajectories. arXiv preprint arXiv:1606.08154 (2016)

19. Fang, Y., Fang, Y.: Neural citation network for context-aware citation recommendation. In: International ACM SIGIR Conference on Research and Development in Information Retrieval, pp. 1093–1096. ACM (2017)

20. Zhang, Q., Wang, J., Huang, H., Huang, X., Gong, Y.: Hashtag recommendation for multimodal microblog using co-attention network. In: IJCAI (2017)

21. Wei, J., He, J., Chen, K., et al.: Collaborative filtering and deep learning based recommendation system for cold start items. Expert Syst. Appl. 69, 29–39 (2017)

22. Wang, S., Wang, Y., Tang, J., et al.: What your images reveal: exploiting visual contents for point-of-interest recommendation. In: Proceedings of the 26th International Conference on World Wide Web, pp. 391–400. International World Wide Web Conferences Steering Committee (2017)

23. Shan, Y., Hoens, T.R., Jiao, J., et al.: Deep crossing: web-scale modeling without manually crafted combinatorial features. In: Proceedings of the 22nd ACM SIGKDD International Conference on Knowledge Discovery and Data Mining, pp. 255–262. ACM (2016)

24. Zhao, W.X., Wang, J., He, Y., et al.: Mining product adopter information from online reviews for improving product recommendation. ACM Trans. Knowl. Discov. Data (TKDD) 10(3), 29 (2016)

25. Zhu, J., Shan, Y., Mao, J.C., et al.: Deep embedding forest: forest-based serving with deep embedding features. arXiv preprint arXiv:1703.05291 (2017)
26. Guo, H., Tang, R., Ye, Y., et al.: DeepFM: a factorization-machine based neural network for CTR prediction. arXiv preprint arXiv:1703.04247 (2017)
27. Unger, M., Bar, A., Shapira, B., et al.: Towards latent context-aware recommendation systems. Knowl.-Based Syst. **104**, 165–178 (2016)
28. Xiao, J., Ye, H., He, X., et al.: Attentional factorization machines: learning the weight of feature interactions via attention networks. arXiv preprint arXiv:1708.04617 (2017)
29. Li, X., She, J.: Collaborative variational autoencoder for recommender systems. In: Proceedings of the 23rd ACM SIGKDD International Conference on Knowledge Discovery and Data Mining, pp. 305–314. ACM (2017)
30. Pan, Y., He, F., Yu, H.: Trust-aware collaborative denoising autoencoder for top-N recommendation. arXiv preprint arXiv:1703.01760 (2017)

Author Index

Printed in the United States
By Bookmasters